CHEMICAL ANALYSIS OF ECOLOGICAL MATERIALS

Chemical Analysis of Ecological Materials

EDITED BY

STEWART E. ALLEN

Formerly with
Institute of Terrestrial Ecology
Merlewood Research Station
Grange over Sands
Cumbria

SECOND EDITION
COMPLETELY REVISED

BLACKWELL SCIENTIFIC PUBLICATIONS

OXFORD LONDON EDINBURGH

BOSTON MELBOURNE

Editorial offices:
Osney Mead, Oxford OX2 0EL
(*Orders*: Tel: 0865 240201)
8 John Street, London WC1N 2ES
23 Ainslie Place, Edinburgh EH3 6AJ
3 Cambridge Center, Suite 208
Cambridge, Massachusetts 02142, USA
107 Barry Street, Carlton
Victoria 3053, Australia

First published 1974
Second edition 1989

Set by Macmillan India
Printed and bound in Great Britain by
Butler & Tanner Ltd, Frome and London

DISTRIBUTORS

USA
Publishers' Business Services
PO Box 447
Brookline Village
Massachusetts 02147
(*Orders*: Tel: (617) 524 7678)

Canada
Oxford University Press
70 Wynford Drive
Don Mills
Ontario M3C 1J9
(*Orders*: Tel: (416) 441–2941)

Australia
Blackwell Scientific Publications
(Australia) Pty Ltd
107 Barry Street
Carlton, Victoria 3053
(*Orders*: Tel: (03) 347 0300)

British Library
Cataloguing in Publication Data

Chemical analysis of ecological materials. —2nd ed., completely rev
1. Natural materials. Chemical analysis
I. Allen, Stewart E. (Stewart Ernest)
543

ISBN 0–632–01742–2

Library of Congress
Cataloging-in-Publication Data

Chemical analysis of ecological materials.
Includes index.
1. Chemistry, Analytic. 2. Soils—Analysis.
3. Plants—Analysis. 4. Water—Analysis.
I. Allen, Stewart E. (Stewart Ernest)
QD131.C4 1989 543 88–33341

ISBN 0–632–01742–2

Contents

List of Contributors, ix

Editor's Preface, xi

1 Introductory Notes, 1
S. E. ALLEN

2 Analysis of Soils, 7
H. M. GRIMSHAW

Sample collection and initial treatment, 7
  Sample replication, 8
  Sampling records, 8
  Sample collection, 9
  Time of sampling, 10
  Sample transport and storage, 10
  Drying, 11
  Reduction of sample bulk, 11
  Sieving, 12
  Reduction of particle size, 12
Preliminary and physical tests, 14
  Moisture, 14
  Loss-on-ignition, 15
  pH, 16
  Redox potential, 17
  Percolation, 19
  Density, 20
  Porosity, 21
  Saturation capacity, 21
  Particle fractionation, 21
Solution preparation, 24
  Dissolution procedures, 25
  Carbonate soils, 28
  Individual elements, 28
Soil extraction, 30
  Choice of extractants, 31
  Extraction of calcareous soils, 32
  Conditions of extraction, 32
  Common extractants, 33
  Cation exchange capacity, 34
  Total exchangeable bases, 35
  Exchangeable hydrogen, 36
  Individual elements, 36

3 Analysis of Vegetation and other Organic Materials, 46
S. E. ALLEN

Sample collection and initial treatment, 46
  Choice of component, 47
  Sampling time, 47
  Sample transport and storage, 49
  Washing, 49
  Bulk reduction, 49
  Drying, 50
  Macerating and grinding, 50
  Sources of contamination, 52
Preliminary tests, 53
  Moisture, 53
  Ash content, 54
Solution preparation, 55
  Dissolution procedures, 56
  Individual elements, 60

4 Analysis of Waters, 62
A. P. ROWLAND AND H. M. GRIMSHAW

Sample collection and storage, 62
  Collection, 63
  Preservation and storage, 65
Preliminary and general tests, 67
  Odour, turbidity and colour, 67
  Solids, 68
  pH, 69
  Alkalinity (and acidity), 69
  Conductivity, 71
  Dissolved gases, 72
Inorganic constituents, 72
  Cations, 72
  Anions, 75
  Individual elements, 76

Organic constituents, 78
Chemical oxidation demand, 78
Carbohydrates, 79
Fatty acids, 79
Chlorophyll, 79
Humic substances, 80

5 Nutrient Elements, 81
H. M. GRIMSHAW, S. E. ALLEN AND J. A. PARKINSON
Aluminium, 81
Boron, 84
Calcium, 86
Carbon, 91
Total and organic carbon, 91
Carbonate-carbon, 96
Chlorine, 97
Cobalt, 101
Copper, 104
Iron, 107
Magnesium, 111
Manganese, 114
Molybdenum, 116
Nitrogen, 118
Total organic nitrogen, 119
Inorganic nitrogen, 126
Ammonium-nitrogen, 126
Nitrate-nitrogen, 127
Nitrite-nitrogen, 132
Mineralizable nitrogen, 133
Phosphorus, 134
Phosphate-phosphorus, 135
Organic phosphorus, 139
Total phosphorus, 141
Potassium, 142
Silicon, 144
Sodium, 148
Sulphur, 150
Sulphate-sulphur, 151
Titanium, 153
Zinc, 155
Other nutrient elements, 159

6 Organic Constituents, 160
C. QUARMBY AND S. E. ALLEN
Storage and initial treatment, 161
Proximate constituents, 162
Crude fat, 163
Crude protein, 164
Soluble carbohydrates, 164
Cellulose, 166
Holocellulose, 167
α-Cellulose, 168
Hemi-cellulose, 169
Lignin, 169
Crude fibre, 171
Carbohydrates, 172
Reducing sugars, 173
Starch, 174
Chromatography of carbohydrates, 175
Nitrogenous compounds, 176
Protein, 179
Amino-acids, 179
Other nitrogenous compounds, 184
Fatty acids and lipids, 185
Total lipids, 186
Fatty acids, 187
Flavonoids and related compounds, 189
Soluble tannins, 189
Fractionation of polyphenolic compounds, 191
Other organic constituents, 194
Organic acids, 194
Chlorophyll, 195
Carotenoids, 197
Sterols, 198
Phytic acid, 198
Humus and resistant residues, 199

7 Pollutants, 201
S. E. ALLEN, J. A. PARKINSON AND A. P. ROWLAND
Heavy metals, 201
Antimony, 202
Arsenic, 203
Cadmium, 205
Chromium, 206
Lead, 207
Mercury, 209
Nickel, 213
Selenium, 214
Organic pesticides, 215
Collection and storage, 218
Extraction systems, 219
Clean-up stage, 221
Examination of extracts, 223
Substituted urea and carbamate herbicides, 227
Other pesticides, 228
Polychlorinated biphenyl compounds, 228
Other pollutants, 229
Water pollutants, 229
Cyanide, 231
Fluorine, 232
Sulphide, 234
Surfactants, 235
Polycyclic aromatic hydrocarbons, 236
Atmospheric deposition, 237

8 Instrumental Procedures, 240

J. A. PARKINSON AND A. P. ROWLAND *with* W. DAVISON (*Electroanalytical techniques*)

Colorimetry and related techniques, 241
 Turbidimetry and fluorimetry, 244
Continuous flow technique, 244
Atomic spectrometry, 249
 Flame emission, 249
 Atomic absorption, 252
 Electrothermal atomization, 256
 Emission spectrometry, 256
X-ray fluorescence spectrometry, 257
Electroanalytical techniques, 265
 Voltammetric methods, 265
 Ion-selective electrodes, 269
Chromatography, 271
 Thin-layer chromatography, 272
 Gas chromatography, 274
 High performance liquid chromatography, 278
 Ion chromatography, 280
Bomb calorimetry, 284
Automation, 289

9 Statistical Analysis and Data Processing, 292

A. P. ROWLAND AND H. M. GRIMSHAW

Basic statistical concepts, 292
 Graphs and tables, 293
 Elementary probability, 294
 Frequency distributions, 294
 Estimation, 295
 Significance, 298
 Regression and correlation, 302
 Non-parametric tests, 306
Design of experiments, 308
 Randomization, 308
 Replication, 309
 Blocking, 309
 Factorial experiments, 310
Calibration and calculation procedures, 310
 Analytical performance, 311
 Calibration, 311
 Calculation, 313
Analytical errors and quality control, 314
 Systematic errors, 314
 Random errors, 315
 Sensitivity and detection limit, 316
 Control procedures, 316
 Other methods for detection of errors, 318
Computer applications, 318
 Laboratory computing, 319
 Equipment control, 319
 Data acquisition, 320
 Data processing, 320
 Laboratory management, 320

10 Appendices, 323

I Chemical Composition of Ecological Materials, 325
II Bioenergetics, 332
III Laboratory Contamination, 335
IV Laboratory Safety, 336
V Chemical Constants, 339

References, 343

Subject Index, 361

List of Contributors

Principal contributors

S. E. ALLEN
'Fairfield', Cartmel, Grange over Sands, Cumbria LA11 6PY, UK, formerly with *Institute of Terrestrial Ecology, Merlewood Research Station, Grange over Sands, Cumbria LA11 6JU, UK*

W. DAVISON
Institute of Freshwater Ecology, The Ferry House, Ambleside, Cumbria LA22 0LP, UK

H. M. GRIMSHAW, J. A. PARKINSON, C. QUARMBY, A. P. ROWLAND
Institute of Terrestrial Ecology, Merlewood Research Station, Grange over Sands, Cumbria LA11 6JU, UK

Associated contributors

J. D. BRIDSON (Chapter 6)
Memorial University of Newfoundland, St. John's, Newfoundland, Canada A1B 3X7

M. C. FRENCH (Chapter 7)
Institute of Terrestrial Ecology, Monks Wood Experimental Station, Abbots Ripton, Huntingdon, Cambs PE17 2LS, UK

D. K. LINDLEY, J. D. ROBERTS, P. A. COWARD, VALERIE H. KENNEDY, C. WOODS
Institute of Terrestrial Ecology, Merlewood Research Station, Grange over Sands, Cumbria LA11 6JU, UK

Editor's Preface

The first edition of this handbook was produced to meet a specific need: the provision of a book on chemical analysis suitable for the ecologist. At the time there were a number of excellent analytical manuals available but none were known that dealt with ecology as an independent subject with its own requirements. The authors had themselves encountered problems through not having a specialist source of information whilst establishing an analytical chemistry service for the Nature Conservancy (now the Nature Conservancy Council). The methods which were then brought into routine use were derived partly from agricultural texts, partly with the help of colleagues in other laboratories and also after original experimental work. The resulting blend of analytical procedures and accumulated experience formed the basis of the first edition of this book.

It was clear, from the reception that the book received, that it filled a niche, subsequently confirmed by later requests for an updated edition. Unfortunately, because of competing commitments it has not been possible to comply with these requests until now, 14 years after the first edition. Not surprisingly, in the meantime there have been many changes in analytical chemistry which now have to be taken into account. In particular, instrumentation developments have been considerable. Not only have new instruments been introduced, but laboratory automation is now commonplace. Microprocessor controlled equipment was almost unknown 14 years ago and few laboratories had computerized data processing facilities.

It was a feature of the first edition that the methods had been tested and in many cases proven through routine use. As far as possible this distinction has been retained in the second edition, although the proliferation of analytical instruments has caused problems. Where experience is limited, the discussion is less detailed although references to other sources of information are generally provided.

There are a few other, mostly high cost, instrumental systems known to be effective analytical techniques which are only briefly mentioned because the authors have no experience or are only just gaining experience in their use. Examples include neutron activation and ICP spectrometry.

Despite the progress in instrument sophistication, there are still some ecologists who do not have ideal laboratory conditions and have to depend on old equipment. An attempt has been made to cater for their requirements by providing alternative, sometimes simpler procedures. In some instances the simpler test is preferable anyway and gives more meaningful results. Overall it is felt that the range of methods given is sufficient to meet most ecological requirements.

There are a number of factors that are common to most chemical methods and which have to be taken into account in order to obtain reliable results. Water purity, quality of glass and plastic are examples. Notes on these and other topics of general interest are provided at the beginning of the book in the hope that they will be applied wherever appropriate. Laboratory safety is considered in the Appendices and chemical composition and other sundry data reproduced from the first edition of the book are also included there.

The earlier division of ecological materials into three classes (soils, plant materials and waters) has been retained in this edition. Separate chapters are devoted to each of these types which enables topics such as sample collection and initial treatment, preparation of test solutions and tests specific to the material class to be considered together. Other materials (rocks, sediments, leachates, animal materials, etc.) having similarities to the main classes are mentioned where appropriate. The separate sample treatments then all lead into a single chapter (5) describing elemental procedures, mostly for nutrient elements. It is here that many of the methods, proven through routine use, are to be found.

The organic chapter (6) includes the proximate constituents which still depend on classical procedures first developed early in the century. The fractionation of organic compounds of most interest to the ecologist is also discussed in this chapter. Synthetic organic pesticides are considered in Chapter 7 together with other substances that can lead to environmental pollution. The instrumental techniques used in the methods given in this book are described in Chapter 8. The emphasis here is on practical aspects, but the theory behind the process is briefly discussed and the reader is referred to specialist works for detailed information. The final chapter discusses some of the statistical concepts that are relevant to chemical analysis and the interpretation of results obtained.

The production of most reference works depends on the support, albeit in a minor way, of many individuals and this book is no exception in this respect. Because they were so numerous it is not practicable to mention everyone by name but I would like to express my gratitude and that of all the other contributors for this support. There are, however, two individuals who deserve special thanks, these being John Beckett, for bibliographic assistance and Jennifer Chapman, for compilation help in the early stages. I would also like to say that I am indebted to my wife, Rita, for her assistance in the later stages and for her forbearance throughout.

S. E. A.

I
Introductory Notes

In these notes various conditions are considered that are common to many of the treatments and methods described in this book. They are all basic analytical requirements that are often taken for granted with little thought being paid to errors they may introduce. It is recommended therefore that these notes be quickly checked before a start is made on any of the methods. Other considerations are mentioned where relevant in the subsequent sections.

Most of the methods give precise results (with coefficient of variation of less than 5%) if the directions are followed. In certain procedures, when the element is only present in trace amounts or the method is insensitive, precision will not be as good but this is indicated where relevant. As part of the preliminary preparations, the analyst should evaluate the variation and precision limits of the method. Statistical concepts are dealt with fully in Chapter 9.

Blanks

In most determinations a correction must be made for the background concentration or 'blank' value. Two types of correction are involved in the methods in this book and should be distinguished. The first is the reagent blank, used particularly in the colorimetric methods and which functions as the zero standard. Its absorbance value forms part of the calibration curve and is therefore not subtracted from the sample results.

The other type is the treatment blank which is introduced during the preparation stage particularly by digestion acids and filter papers. This blank should be run in duplicate or triplicate in the same way as the samples and in the calculation the mean value is subtracted from the sample readings. Contamination from other sources may be more erratic and cannot always be compensated for in a blank determination. Possible sources are discussed in the text and a review of their importance in elemental analysis was prepared by Hamilton *et al.* (1972). Contamination control is also dealt with by Zief and Mitchell (1976) who discuss all aspects in great detail.

Further information about blank values is given in Chapter 9.

Reference material

It is strongly recommended that a reference material of known chemical composition be included with each batch of samples analysed. Suitable materials can sometimes be obtained from government test laboratories or commercial suppliers. If such a source is unavailable a large amount of dried, ground and well-mixed soil or plant sample should be prepared for this purpose (see Appendix I for choice of a soil or plant type). Such material is suitable for most of the analyses described in this book except for those dealing with labile constituents, for example some of the organic compounds. A reference sample is of particular value for detecting major errors and also for monitoring procedure and instrument performance. The more results that are obtained for the reference the more valuable it will become. The use of reference samples and other monitoring schemes are discussed in Chapter 9.

If possible, samples of the reference should be sent to other laboratories for comparison tests. These can often be organized at minimal cost through an exchange arrangement. Participation in inter-laboratory standard tests is always desirable (for example see Sterrett *et al.* 1987).

Reference waters are less easy to use because of chemical and microbiological changes on storage. One solution to this problem is to prepare a stock synthetic concentrate and dilute regularly. These reference solutions should be kept at just above freezing point.

Reagents

The use of high quality reagents is desirable in most methods. They are essential for preparing standard calibration solutions, and are advisable generally since contamination may be introduced through the use of inferior grade reagents. 'AnalaR' or similar analytical reagent grade is usually acceptable although for trace element analyses the high purity 'Aristar' or equivalent grade will further reduce the blank. Within the specifications stated the quality may vary from batch to batch.

Some organic solvents may be contaminated with other substances, including organic compounds with similar properties and metals picked up during manufacture or storage. Purification of these solvents is often advisable and may be done by distillation (sometimes with fractionation) or in the case of the metals it may be adequate to remove them by extraction with a strong solution of hydrochloric acid.

Ethanol is sometimes required as a reagent but because of the Excise regulations an analytical reagent grade is difficult to obtain. For most purposes industrial spirit is equally suitable, as in the preparation of indicator solutions.

Before making up standard solutions it is usually necessary to dry the reagents to drive off air equilibrium moisture. This should be followed by cooling in an efficient desiccator. If any doubt exists about reagent stability at 105 °C (the usual drying temperature), reference works such as the Handbook of Chemistry and Physics (Weast, 1987) should be consulted. Some chemical reagent manufacturers supply concentrated standard solutions in ampoules which can be diluted to give reliable working standards.

Terms such as v/v and w/v are frequently used in the text for reagent preparation and always refer to the volume or weight of reagent in the stated volume of solvent (deionized or distilled water unless otherwise stated). An expression such as $(1+3)$ means one volume of reagent is added to three volumes of solvent.

Some attention should be paid to the storage of standard and reagent solutions. Ideally it is best to keep these solutions in dark containers at a low temperature which will retard any chemical and microbial changes. Glass bottles should be made from glass that is resistant to chemical attack. Plastic containers are sometimes preferable although microbial growth occurs more readily on the surface of plastic vessels than on those made of glass. Unplasticized, high density polythene containers are recommended for most reagents.

Long term changes in the concentration of the solutions through evaporation or other causes must be taken into account. Certain ions, notably aluminium, copper, iron, molybdenum and zinc are adsorbed on to the surface of glass and some plastic containers. For these reasons both standard and reagent solutions should be frequently checked and the latter be renewed at least every month. Very dilute standard solutions should be renewed more frequently. Unstable reagents should not be kept longer than recommended in the methods given. Ammonia solutions should be kept in glass containers.

Pure water

The word 'water' as used in this text signifies that the water is of analytical quality. For some purposes normal distilled or deionized water is sufficiently pure and for physical tests it is generally possible to use tap water. However it is advisable to use the purest water available as routine. Very high purity ($>0.1\ \mu S$) is essential for some procedures and is usually prepared by a combination of treatments (e.g. reverse osmosis and deionization).

Distilled water
Water of this type will contain dissolved gases, especially CO_2 and NH_3, indeed tap water usually contains less NH_3 than distilled water. These and other undesirable gases can be removed by purging the hot water with nitrogen or distilling in the presence of either alkali (to remove CO_2) or acid (to remove NH_3). Trace amounts of metals may be leached out of the body and pipework of the still. For example iron, copper, lead and zinc may come from the metalwork and sodium, boron and zinc from the glass. Even plastic fittings may contribute organic products which can generally be removed by distilling in the presence of alkali and $KMnO_4$ and purging with nitrogen. The traditional way to produce very high quality water is to double or even triple distil from quartz stills. The simplest way to produce water of good enough quality for the tests in this manual is to first distil, then 'polish' by passing through a mixed bed resin column. This will give water of conductivity less than $0.1\ \mu S$.

Deionized water
Water that has been purified by passing through resin alone will contain trace levels of non-ionic and colloidal materials. These can be minimized by first passing the tap water through a porous ceramic or other fine filter. Silica is a characteristic impurity of deionized water. Organic matter can be leached out of the resin but it may be removed by ultraviolet irradiation in the line or otherwise by the alkali/$KMnO_4$ treatment mentioned above. It is recommended that for sensitive electroanalytical measurements, deionized water should be avoided altogether because of the possibility of ionized species coming off the resin.

Reverse osmosis water
An osmosis membrane system will purify tap water by removing a large proportion of the mineral salts and almost all the organic colloidal matter and bacteria.

Other purification methods
There are a variety of options available to add on to purification systems to remove specific fractions. A range of membrane filters can be used to remove micro-organisms and activated carbon filters will adsorb a wide range of organic compounds.

The storage of pure water may result in problems because of the absorption of gases from the atmosphere, release of substances from glass surfaces and because of bacterial and fungal growths in plastic tanks and pipes. Containers should be cleaned regularly and formaldehyde treatment may be required to control microbial contamination of resin beds and pipework.

Weighing technique

Although modern analytical balances are designed for easy and troublefree use, many analytical errors are caused by incorrect weighing techniques. Even with the rapid single pan balance, which is more stable and robust than its predecessors, some basic checks are necessary. Some notes on good weighing practice are given below.

1 Check the balance scale length frequently.
2 Return to the zero mark between weighings.
3 Keep the container to be weighed as small as possible (errors are introduced through surface adsorption of moisture on glass and porcelain).
4 In spite of 3, above, do not wipe vessels vigorously as this will introduce electrostatic charges.
5 The temperature of the object to be weighed must not differ from that in the balance case (errors may be about 0.1 mg for every 1 °C difference). Allow equilibration in the case unless dried samples are weighed.
6 Allow adequate time for samples to cool in desiccators (20 to 30 minutes minimum are required).
7 Do not overcrowd desiccators since the items to be weighed last will pick up moisture as the earlier items are removed.
8 Keep balance and its surroundings clean to avoid spillages leading to cross-contamination.
Although boxes of weights are now little used, a

high quality set is worth retaining for checking purposes.

Filter papers

Whatman cellulose filters (papers) are generally used for analytical work and charts or booklets giving information on their properties are available from the suppliers. A useful handbook is that of Meakin and Pratt (1973). The qualities of the paper used for operations described in this book are given in Table 1.1.

For many purposes glass fibre filters are preferable. Their retentivity is as good or better than that of cellulose filters and their speed is comparable to the fastest papers. Their disadvantages include fragility, higher cost and the possibility of high blank values in mineral analyses. The use of membrane filters is mentioned in the text and these are particularly useful for removing particulate material from water samples.

Glass and plastic ware

Vessels used for the preparation and measurement of standard solutions should be calibrated to NPL grade 'A' specifications. Particular care is needed over the use of stem graduated pipettes since some of these are calibrated to drain, whilst others are graduated between two fixed points. New glass vessels should always be well washed before using for the first time.

The composition and properties of glasses must be taken into account for different uses:

Soda-lime-silica

High expansion coefficient makes it susceptible to breakage by thermal shock.

Attacked by strong reagents.

Na and Ca readily leached out but content of biological trace elements (except Mo) low.

Borosilicate

Resistant to thermal shock.

Contains up to 12% B, some Zn and traces of Mo.

Greater resistance to chemical attack than soda glass.

Fused silica

Marketed under trade names

Resistant to thermal shock and more suitable for use at higher temperatures.

Least affected by chemical reagents (except HF).

Very low trace element content.

Very expensive.

Plastic vessels are frequently used in place of glass. Some of the properties of plastic materials are given in Appendix V. Particular care should be taken to check that phosphoric-ester plasticizers or fillers which have been added to the plastic (often including sodium, calcium and magnesium salts) do not cause contamination. Some plastic materials, e.g. low density polythene, are relatively porous and absorb some chemicals which may be released later. Algal and bacterial organisms may also be troublesome since there is a tendency to grow on some plastic surfaces. In relation to this there is a tendency for phosphorus in particular to be sorbed by acid-washed glass and plastic laboratory ware. For further information on the

Table 1.1. Filter paper characteristics

Paper no.	Retention (microns)	Filtering speed	Application
1	2·5	Medium	Determination of NH_3, aminoacids
41	4	Fast	Retention of gelatinous hydroxides
44	1	Slow	Most analytical operations, gives low blanks
50	1	Slow	Filtering strong acid and alkaline solutions
54	4	Fast	Coarse filtering of acid and alkaline solutions

properties of laboratory ware see Zief and Mitchell (1976).

Cleaning

The usual criterion of cleanliness for glassware is to rinse the vessel with water until an unbroken film remains. For most purposes the recommended concentrations of mild detergents are adequate to remove surface dirt. Stronger products may be needed for removing organic stains and heavy grease contamination. These substances have caustic components and often contain phosphorus so their use should be followed by thorough rinsing. Final rinsing with distilled or deionized water is always necessary after washing analytical glassware. Prolonged use of the stronger detergents, caustic solutions and hydrochloric acid as well as abrasives will result in deterioration of polished glass surfaces. Non-foaming detergents are available for use with laboratory washing machines.

Cleaning solutions can be prepared for special purposes although they are less used now since most of the proprietary products are so effective. Among the more popular solutions are:

Alkaline permanganate

Dissolve 20 g NaOH and 10 g $KMnO_4$ in 100 ml water.

Effective for some organic stains resistant to chromic acid.

Alcoholic KOH

Dissolve 25 g KOH in the minimum volume of water and make up to 1 litre with industrial alcohol.

Use for short periods only and then rinse with dilute acid.

Effective for removing grease, especially when hot.

Chromic acid

General purpose oxidizing solution.

Add powdered $K_2Cr_2O_7$ in excess to conc. HNO_3.

Acid detergent

Add 50 ml HF, 300 ml H_2SO_4 and 20 ml liquid detergent carefully to water and make up to 1 litre (HNO_3 can be used in place of H_2SO_4).

Use for short periods only and then wash well with water.

Do not use for volumetric ware.

Expression of results

In general SI units have been used in the text. However, the word litre (l) is retained in place of cubic decimetre (dm^{-3}) and millilitre (ml) is used instead of cubic centimetre (cm^{-3}).

The practice of reporting nutrient data in terms of the element rather than a compound is now very widespread and is adopted in this manual. However, some agricultural data are still occasionally reported as the oxides and so a table of conversion factors is provided in Appendix V.

Nitrogen, phosphorus and sulphur are normally determined in solution as the complex ions NH_4^+, NO_3^-, PO_4^{3-} and SO_4^{2-}, but the standards are always prepared in terms of the element. Hence in soil extracts and waters the final results are given as NH_4^+-N, NO_3^--N, PO_4^{3-}-P and SO_4^{2-}-S.

Analyses of soils and plant materials may be carried out on fresh, air-dry or oven-dry samples but the final results will normally be reported on an oven-dry basis. In this manual this is equivalent to drying to a constant weight at 105 °C. Occasionally, however, it may be desirable to express soil nutrient results on a fresh weight basis or even a fresh volume basis but this will depend on the nature of the investigation.

The expressions used for concentration may be summarized as follows:

total constituent in soil or vegetation—express as % or $\mu g\ g^{-1}$;

total constituent in water—express as $mg\ l^{-1}$ or $\mu g\ l^{-1}$;

extractable constituent in soil—express as $mg\ 100\ g^{-1}$ or $me\ 100\ g^{-1}$ (to convert $mg\ 100\ g^{-1}$ to $me\ 100\ g^{-1}$ divide the mg $100\ g^{-1}$ result by the equivalent weight of the element concerned).

The use of ppm (equivalent to $\mu g\ g^{-1}$ or $mg\ l^{-1}$ for most purposes) has been reserved for the concentrations of standards and preparation of calibration curves.

Calculation procedures and the expression of results are discussed further in Chapter 9.

Abbreviations

kg	kilograms
g	grams
mg	milligrams
μg	micrograms
l	litres
ml	millilitres
μl	microlitres
ppm	parts per million
cm	centimetres
mm	millimetres
nm	nanometres
me	milliequivalents
J	joules
kJ	kilojoules
S	siemens
μS	microsiemens
wt	weight
conc	concentrated
N	Normal
M	Molar
CEC	Cation exchange capacity
TEB	Total exchangeable bases
TDS	Total dissolved solids
TSS	Total suspended solids
TOM	Total organic matter
OFN	Oxygen-free-nitrogen

2
Analysis of Soils

The methods described in this section are intended mainly for use with mineral soils although many of the techniques are equally applicable to rock materials. The first two parts cover the collection and initial treatment of the samples together with some of the more simple physical tests. The procedures to be used in preparing solutions for elemental analyses are dealt with later in the chapter. All organic procedures including any applicable to soils are described in Chapter 6.

For ecological purposes there is more interest in those constituents available for plant uptake than in the total elemental content of soils. However, since an occasional need for total soil analyses does arise, sufficient information on how to bring soil minerals into solution using fusion and hydrofluoric acid digestion procedures is given later in this section. X-ray spectroscopy, which does not require wet preparation stages, is dealt with in Chapter 8.

The procedures used for soil extractions (p. 33) are relatively simple but the choice of extractant and the significance of the results are less easy to establish. The purpose of most soil extraction procedures is to obtain a measure of soil fertility using a chemical extractant to simulate plant uptake. Although a large amount of work has been done on their application for agricultural requirements, much less is known about the needs of natural vegetation and most of the extraction procedures are given with this reservation.

Organic soils often present a difficulty. If soils containing large amounts of organic matter are to be fused, a pre-ignition stage is desirable. Soils with very high organic contents, such as peat and plant litter are in many respects more similar to plant materials and for many purposes it may be more appropriate to use the methods given in Chapter 3.

Books dealing solely with soil analyses include Metson (1956), Jackson (1958), Page *et al.* (1982), together with Hesse (1971) who approaches the subject from the point of view of the elements. Piper (1950) and Chapman and Pratt (1961) cover both soil and plant analysis. There are many publications dealing with the analysis of various biological materials in which soil is included. Examples are Chapman and Pratt (1961), David (1978), Agricultural Development and Advisory Service (1986) and Nicholson (1984). Smith and James (1981) provide information about sampling and preparation procedures. Soils and plants form a connected system in relation to nutrient uptake by plants and some of the principles involved are discussed by Fried and Broeshart (1967).

Sample collection and initial treatment

The purpose of the investigation will determine the nature, location and number of soil samples required. In studies of plant nutrition or nutrient circulation an examination of the litter together with the soil from the top 10 or 15 cm will often be adequate. Much of the root uptake of nutrients, including that of large trees, occurs in this zone.

However, the lower horizons should also be examined to check that no significant root development has taken place. Rooting systems may occur deeper than expected in some circumstances.

Many of the processes involved in soil formation have some relevance to ecological studies. For details of these, reference should be made to Jacks (1954), Comber (1961), Bear (1964), Buckman and Brady (1969), Russell (1961), Townsend (1973), Greenland and Hayes (1981) and others. Books dealing with the examination of soils in the field include that of Clarke (1975) and the Field Handbooks and Memoirs of the Soil Survey of Great Britain deal with the examination of soils in the field. Soil profile descriptions are given by Kubiena (1953).

Sample replication

Sampling for any detailed study should allow for the inherent variability of soil materials. A high degree of spatial variability may exist over quite small areas and has been demonstrated by Ball and Williams (1968). Similar studies have been carried out by Troedsson and Tamm (1969). Often the extent of this variation is not known beforehand and hence sufficient samples must be taken to obtain a reliable measure of it. The absence of any information about site variability can make it difficult to assess the significance of any changes, such as seasonal variation, which may be implied by the data. The same difficulty would arise if, unknown to the investigator, some abnormal material was included in a composite sample.

The minimum number of field replicate samples required to obtain a result with a desired precision can readily be calculated if an estimate of the variance is available. However, the available analytical facilities may set an upper limit. In practice between five and ten replicates should be regarded as a minimum. If it is desired to estimate the site variability with the same precision as might be obtainable under the more uniform conditions of a laboratory, at least twenty samples may be required.

Two alternative approaches for sampling a particular area are in common use. One employs random numbers, often in relation to a fixed coordinate grid reference drawn up for the site being studied. The other system depends on sampling at regular intervals along a similar grid network.

The difficulty of analysing large numbers of samples often necessitates bulking, in which individual samples are combined to give a composite mixture. This method may give a satisfactory mean value for the soil but it conceals any site variability. The validity of the mean value is determined by the number of samples bulked and the procedure used. In variable soils even thirty individuals may not be sufficient (Cameron, 1971). The bulking procedure must be standardized by taking samples of equal volume from the soil depth or horizon. The presence of animal droppings can result in a localized increase in certain nutrient levels in the soil surface which might escape undetected if the site is known to be otherwise fairly uniform and only one or two composite samples are analysed. The use of group bulking to give several composite samples is preferable if this type of contamination is suspected.

All the comments given above relate to area sampling at a fixed soil depth. It is sometimes preferable to sample discrete soil horizons whose depth varies over the site but the considerations discussed above apply when deciding upon any particular sampling system. Further consideration of these and related topics is given in Chapter 9.

Sampling records

At the time of sampling, relevant information concerning both the soil and the site should be recorded. This should include sampling depth or horizon, colour, consistency and texture of the soil, aspect, drainage conditions and details of the vegetation. Any evidence of human influence should be noted, for example grazing, fertilizing, vegetation burning. A note of the national grid reference will enable later information to be obtained from the site or from maps (e.g. climate, altitude and geological features). It is normal practice to record such information on a standardized site data form.

A permanent record of a soil profile can be made by preparing a monolith. This is likely to be of more interest to the pedologist than the ecologist.

Sample collection

The method used to obtain the soil sample will depend on the nature of the investigation. Many chemical and physical tests are not invalidated if the soil structure is disturbed on sampling, but if there is a need to estimate the volume or to examine morphological or structural characteristics then a soil block or core must be obtained. Litter samples can be taken with a trowel, preferably from a quadrat to estimate weight per unit area. When examining the surface soil horizons, samples can readily be collected using a stainless steel trowel, but for deeper profile material it will be necessary to dig a pit first. The exposed profile must be cleaned to remove soil carried down from upper horizons and the sampling depth limits recorded. A narrow cylindrical trowel, preferably graduated, can be used for extracting a horizontal sample. Sampling will be easier if sequential horizon samples are obtained because the pit profile can be exposed in steps.

Soil corers are required for taking undisturbed soil samples. The simplest of these consists of a metal cylinder between 5 and 10 cm in diameter and whose length is determined by the sampling depth. The cutting edge needs to be sharp. Driving the cylinder into the ground will result in soil compression but this can be minimized by increasing the width and reducing the thickness of the metal. However, wide corers will not retain the sample on extraction and thin cylinder walls are easily damaged, so a compromise is necessary.

Quite often the corer is fitted with a horizontal 'tommy' bar to give extra leverage during insertion of the corer into the soil. Otherwise it may be necessary to drive the corer into the soil with a large mallet. This may be effective but it may then be difficult to extract the corer without digging it out. Large powered corers are sometimes used by soil scientists engaged on survey work. Even when the corer has been successfully removed from the soil it may be difficult to remove the soil core intact from the corer. This is facilitated by using a corer with a detachable cutting head whose internal diameter is slightly less than that of the main part of the corer. The soil core can be removed quite easily from these devices. A cheap expendable version can be made from extracted mild steel tubing which can be readily rolled inwards on the lathe and then sharpened. Another simple design, which will allow the corer to be removed intact, can be made by rolling thin sheet metal (about 22 gauge) and fastening the overlap to secure the cylinder. After sampling, the fastening arrangement is released which allows the metal to spring apart, giving access to the core. This model is particularly useful for peat and clay soils relatively free of stones. Some coring probes have a longitudinal slit along the side. They give some risk of inter-horizon contamination but are quite useful for exploratory work.

If gross disturbance of soil during sampling is acceptable, then the simplest design is the twist auger which can be screwed into the ground allowing the soil to be pushed in along the spiral twist. Other models include a twist or bevelled cutting head which thrusts the sample into a cylindrical collection chamber immediately behind. A shaft with a horizontal bar enables the auger to be twisted into the soil.

Contamination from sampling equipment should not be overlooked, particularly as soft materials are liable to be scratched in stony soils. Stainless steel is the most durable material for sampling devices. A simple way of reducing contamination is to coat the sampler with polyurethane varnish, but this needs renewing frequently.

Corers and augers are effective for many soil types particularly when the soil is near to field moisture capacity, but are less useful for waterlogged or dry sandy material. The sampling of peats may require precautions to minimize oxidation and in some cases it may be advisable to obtain a large block which can be sampled in the laboratory. The special corers called 'peat borers' are, in effect, long cylinders up to a metre or more in length.

Stony soils present yet another problem and in extreme cases it may be impractical to use a corer at all. If many stones are present it will be necessary to collect a large amount of soil to ensure that the sample is representative. As a rough guide the sample volume should be about 100 times the largest size of stone present. The stone content itself can be estimated by counting or weighing stones retained by a suitable mesh. Further information about the sampling of soils can be obtained from Hughes (1979) and from the Agricultural Development and Advisory Service (1986).

Time of sampling

Some consideration should be given to the most suitable time for sampling. Little evidence has been produced for any diurnal variation in chemical content, but changes with season may occur. For example, phosphorus availability may increase during the spring and summer months and there is some suggestion that inorganic nitrogen levels are higher in the spring, as shown by Saunders and Metson (1971). However, most results, especially for cations, confirm the results of Frankland *et al.* (1963) and Ball and Williams (1968) who found that spatial variability was likely to conceal any seasonal change.

Sample transport and storage

If transport and storage times are kept to a minimum, then changes in the concentration of extractable nutrients or even of some organic constituents may not be serious but certain aspects must still be taken into account.

There are some measurements such as redox potential which should be carried out in the field wherever possible. The ionic state is a soil property which may readily alter when a sample is taken, unless special precautions are taken. For example it is possible to 'fix' iron by shaking the soil with a complexing agent; or sulphide with acetate solution and dilute sulphuric acid can be used to stabilize ammonium-nitrogen. Aeration can be minimized if the sample container is sealed without leaving an air space above the soil.

When delay in transit cannot be avoided, it is desirable to keep the temperature of the sample below ambient to reduce microbial activity. Samples can be cooled with solid carbon dioxide in vacuum flasks or pre-frozen 'cold-packs' which are available from suppliers of camping equipment. Insulated containers are effective in maintaining a lowered temperature for up to two days. In general, the use of preservatives to inhibit microbial action in soils is not very practicable and may have some effect on the action of extractants and reagents used in the subsequent analysis.

Once samples are in the laboratory certain determinations particularly for pH, available nitrogen, soluble carbohydrates, amino acids and lipids should be carried out on the fresh materials as soon as possible. If this is not possible the fresh samples can often be stored for short periods at a temperature just above freezing. Bartlett and James (1980) examined methods of soil storage and recommended keeping it moist and aerobic. To prevent changes due to micro-organisms, storage below freezing may be necessary, although Allen and Grimshaw (1962) showed that low temperature storage can affect the release of some extractable constituents. In the case of inorganic nitrogen fractions, Nelson and Bremner (1972) found that storage at $-5\,^{\circ}C$ was preferable to initial air-drying although Robinson (1967) felt that no treatment was entirely satisfactory for mineral nitrogen. Gupta and Rorison (1974) also noted limitations of low temperature storage. Fresh soils are often put into polythene bags on collection and even stored in such bags. Whilst this is acceptable briefly at just above freezing point, higher temperatures can soon lead to incubation and biological changes in samples in polythene bags. For this reason cloth (calico) or similar porous materials have sometimes been recommended for storing fresh samples of biological materials.

Drying

Structural changes in soil are inevitable on drying so that the conditions can be critical. In general there is a choice between two approaches. One is rapid high temperature drying (up to 110 °C) which can lead to irreversible changes in both the lattice structure and organic fraction. The other is a longer drying at a lower temperature which may promote enzymic and biological changes in the organic fraction. The effects of drying upon fixation and release of extractable soil nutrients have been widely studied, particularly for nitrogen and phosphorus. Changes in ammonium-nitrogen have often been reported including Frye and Hutcheson (1981) who studied the mechanisms involved. Levels of extractable phosphate are sensitive even to air-drying and changes are related to the effects of drying upon the adsorption and desorption properties of the soil (Barrow and Shaw, 1980; Haynes and Swift, 1985). Molloy and Lockman (1979) also reported on phosphate together with sulphate and manganese, whilst others have studied mainly cations. Various workers have examined drying methods including Keogh and Maples (1980). Although there is no ideal procedure many workers prefer a moderately low temperature (40 °C) coupled with air-circulation over a thinly spread sample to minimize the time of drying. However, this and other recommendations should be related to the analyses required as summarized in Table 2.1.

Although storage of air-dry samples is a routine procedure in soil laboratories, some long term changes have been reported for inorganic nitrogen (Nelson and Bremner, 1972) and also for phosphate. Bartlett and James (1980) drew attention to the problems of storing dried soil.

Another aspect of handling dried soil is that rewetting may lead to a significant burst of microbial activity (Stevenson, 1956). Although this can be a problem, advantage is taken of it in certain incubation methods including that for mineralizable nitrogen.

Reduction of sample bulk

Before sieving it is often necessary to reduce the sample bulk. The traditional manual method is to

Table 2.1. Treatments of soil samples after collection

Analyses required	Recommended treatment
Hydrogen ion concentration (pH) Reduced ionic states Redox potential	Rapid examination of fresh material without delay
Organic nitrogen and phosphorus fractions Extractable nitrate, nitrite and ammonium–nitrogen Halides Amino acids, simple carbohydrates, volatile fats, humus fractions All peat extractions	Analysis of fresh material advisable (short cold storage periods usually acceptable)
Extractions of mineral soils for potassium, sodium, calcium, magnesium, iron, manganese, zinc, copper, boron Cation exchange capacity Mineralizable nitrogen Proximate organic analysis	Air drying generally suitable (40 °C)
Particle fractionation Loss-on-ignition Total concentrations of mineral constituents, phosphorus, sulphur, and generally carbon and nitrogen	Drying at 105 °C

mix thoroughly, heap into a cone and quarter on a smooth impervious base. Two opposite quarters are kept and the process continued, retaining alternate pairs of quarters each time until the required reduction has been achieved. Care needs to be taken when forming the cone to ensure that larger particles which roll down the sides of the cone are evenly distributed.

Incremental sampling is another manual method. In this the heap of mixed sample is flattened to form a 'cake' which is then divided into equal portions. Random portions are then distributed.

In general manual procedures do not select all grain sizes on an entirely random basis and mechanical methods are an improvement (Mullins and Hutchinson, 1982). One widely used device is called a riffle. In the most commonly used version the sample is poured into a large rectangular hopper which has alternating outlets allowing collection of separate sub-samples. Variants of this type of riffle contain vibrating or oscillating components to minimize bias in the distribution of particles. A spinning riffle has also been marketed. One sample reducer (Tyler Sample Splitter) systematically rejects segments of the feed material when flowing down a 45° incline.

Rotating sample dividers are probably the most reliable of the reducing devices. The sample is poured through a funnel on to an inverted cone which throws it into collection buckets revolving on a rotary turntable. The model made by Pascall is a good example of this type of divider.

The problems of sample reduction are discussed at length by Smith and James (1981), whilst Mullins and Hutchinson (1982) considered variability introduced by sub-sampling procedures. The efficiencies of different sample reduction procedures were examined statistically by Allen and Khan (1970).

Sieving

Stones, large roots and other coarse fragments are removed by passing the air-dry material through a 2 mm sieve. This size has been adopted as an international standard because the soil passing this mesh contains almost the whole of the nutritionally important fraction. All the methods given in this book for dried soils refer to < 2 mm material. Fractions of 2 mm or greater may be estimated separately as gravel (2 mm to 1 cm) and stones (> 1 cm), but otherwise they are discarded.

Before or during sieving the soil must be lightly crushed to break up aggregates but care is necessary to avoid shattering soft mineral particles such as occur in shale soils. Mechanical soil crushers are available enabling this process to be standardized. A suitable crusher is marketed by Donald Mackay Engineering (Cambridge, UK), in which steel bars rotate inside a cylinder. In the absence of such equipment gentle crushing between sheets of polythene before sieving will suffice. Brass sieves should be avoided when copper or zinc are to be estimated and stainless steel or nylon sieves are generally the most suitable.

Reduction of particle size

The term comminution is used to cover grinding, crushing, milling, pulverizing and all similar practices used to break down the sample particles. Although the < 2 mm fraction is adequate for the larger amounts of sample required, for example, in nutrient extraction procedures, it is unsuitable for sample weights in the 100 mg range usually needed for acid digestions. These and smaller weights will tend to be unrepresentative of the whole sample unless finer and more homogeneous material is prepared. Grinding to pass 150 BS mesh will meet this requirement for most soil types when, for example, determining total nitrogen and phosphorus (Rowland and Grimshaw, 1985), but the sample should pass through 250 mesh for weights under 100 mg. Kleeman (1967) and Smith and James (1981) demonstrate the need for fine grinding if sampling errors are to be avoided.

The procedure used for sample breakdown will be determined by the nature of the study. In ecological and agricultural experiments it is usually sufficient to obtain a ground sub-sample from the < 2 mm fraction of soil as indicated above. However, a geological examination of rock or soil for total mineral elements will require the entire sample to be broken down. A preliminary breakdown of rock can be achieved using impact mortars or jaw crushers. The former are suitable for smaller samples and utilize a hardened steel plug which hammers the sample in the base of the cylindrical container. Jaw crushers will be required for larger samples and a number of models are available. One widely used type crushes the sample between a fixed and a movable plate connected to an eccentric drive. Fragments up to 10 cm size can be handled in this way. Another series of machines uses heavy rollers to break up the fragments.

Once rock fragments have been reduced to say, less than 1 cm or a < 2 mm fraction of soil prepared, then further breakdown is readily achieved by a variety of grinding or milling machines. The types most frequently used in the UK include:

SWING MILL

In this type of mill (also known as a disc, vibratory or ring mill) the sample is crushed by the impact of a heavy disc and annular ring under intense vibration inside a shallow cylindrical chamber. The process is very efficient especially for mineral material. However, because of the high shearing and impact energy the machine can get quite hot. It is also relatively expensive because the grinding cylinder and contents have to be made of very hard steel alloys or of silicon carbide to minimize contamination during grinding. A well-known example of this type is the Tema mill.

HAMMER MILL

This mill is generally more suitable for vegetation, but it can also be used for soils. The sample particles are smashed by fixed or flailing hammers which revolve within a cylinder. The samples are fed through a hopper and the finely ground material passes through a sieve into a collector. These machines are less efficient than the disc mills and if a finely powdered sample is needed, the material must be repeatedly passed through progressively smaller screens. They are, however, robust and much cheaper than the disc mills.

BALL MILLS

This group includes both large rotating cylindrical mills and the generally smaller vibration mills. Both types use balls of a very hard material for breaking up the sample, but vibration mills are more effective because of their high impact energy.

Planetary ball mills are rather different in that they depend on the high rotational speed of balls in a spherical container. They are very efficient but relatively slow. Vibrating and planetary ball mills are, in addition, quite useful as sample mixers.

MORTAR AND PESTLES

Powered mortar and pestles are also referred to as end-runner mills. The grinding action of powered versions is gentle and little heat is generated so they are suitable for labile materials. However, they are rather slow in action and are not effective except for relatively small sample weights. Some models have a heavy fixed pestle and a rotating mortar and in others the reverse is the case. Pan mills are similar in action to powered mortar and pestles, in that they depend for their grinding action on heavy rollers crushing the sample.

The machines described above are only suitable for dried material. Reduction of particle size of fresh soil samples is usually accomplished by sieving rather than grinding. However, even sieving can greatly perturb natural processes in the sample including nitrification which may be significantly enhanced. Fresh litters can be reduced in a homogenizer and various models are available. However, they require a medium such as water or an organic solvent so the action is one of extraction as well as particle reduction.

Contamination during grinding is difficult to avoid although the problem is generally less serious for soils than for vegetation. Steel products may lead to contamination by iron, manganese, molybdenum and cobalt, whilst copper and zinc appear in some mill components. Ceramic balls may introduce small amounts of aluminium, boron, calcium, potassium and sodium. Tungsten carbide and agate probably result in least sample contamination although hard steel is acceptable for most purposes. Thompson and Bankston (1970) discuss the problems of contamination. Inter-sample contamination can only be minimized by adequate cleaning of the equipment during use.

Table 2.2. Characteristics of milling machines

Mill type	Jaw	Swing	Hammer	Ball	Powered mortar & pestle	Knife
Suitable for	Stony soils	Soils, chopped woody material	Organic and 'soft' soils, vegetation	Soils	Soils, chopped vegetation	Vegetation
Sample size	Large	Small	Moderate	Small	Small	Small
Final texture	Coarse	Very fine	Moderate	Fine	Fine	Moderate
Recovery	Poor	Very good (enclosed)	Low	Very good (enclosed)	Moderate to good	Good
Grinding time	Fairly rapid	Fairly rapid	Rapid	Moderate	Slow	Slow
Heat	Low	High	Moderate	Moderate	Moderate	Low
Contamination	Fairly high	Moderate	Moderate	Moderate	Low	Very low
Cleanliness	Very dusty	Clean	Dusty	Clean	Moderate	Clean
Noise	Excessive	High	High	High	Moderate	Low
Cost	Fairly high	High	Low to moderate (depends on size)	Moderate	Moderate	Moderate

Table 2.2 summarizes the characteristics of the various types of milling machine. Machines suitable for plant materials are discussed in Chapter 3.

Preliminary and physical tests

A number of preliminary analyses may be required on soil samples. Some of these are simple tests, such as moisture and loss-on-ignition, but others are more strictly physical. For convenience they are grouped together. Many physical tests carried out on soils are mainly of value to pedologists, geologists and civil engineers. However certain physical parameters are also relevant to ecological studies including, for example, soil pores which hold the soil solution containing the dissolved nutrients and gases required by plant and animal life in the soil. Information on the size and distribution of the soil pores may be obtained from measurements of percolation, bulk density and particle fractionation as described below.

Although laboratory methods are considered here, many physical tests are more appropriate to the field. Details of these and other physical tests are given by Black (1965).

Moisture

The determination of water content is important in soil studies and many methods have been developed for this purpose. Some of these are intended for use in the field, but most determinations are carried out in the laboratory.

Since moisture estimations are often used for dry weight correction of other analytical data, it is generally more convenient to express the results as percentage dry matter.

Thermal drying at between 100 and 110 °C in an air-circulation oven to constant weight is by far the most common technique in use for determining the moisture content. The main problem with oven drying is to define the dry state. Soil is a very complex material in which it is hard to differentiate between structural (lattice-bound) hydroxyl groups, water of crystallization and water in pore spaces. The retention of water by clay minerals was examined by Nutting (1943), who showed that varying drying periods were necessary for different minerals. Heating for long periods at temperatures at over 200 °C will be needed to drive off structural water from some minerals. Thus a slightly low result could be obtained by the recommended 105 °C treatment, but this may be offset by losses in weight due to volatilization, oxidation or partial breakdown of organic matter. For some purposes these factors are critical, and a true assessment of the moisture value is not possible without allowing for the nature of the material. In addition the time required for drying and particularly the temperature should always be specified with the results. Further information on moisture retention with particular reference to silicate minerals is given by Mitchell (1951). High temperature combustion methods to release water with subsequent collection in absorption trains are reviewed by Maxwell (1968).

Other methods include the Karl-Fischer titration and the immiscible solvent displacement methods, but these offer little advantage for soils in most cases. Infra-red radiation is employed in 'moisture-testers' which would appear to be applicable for fresh moisture. These and other methods were reviewed by Cope and Trickett (1965).

For a test of field moisture the porous block technique is often used. This depends on an equilibrium being achieved when water ceases to flow into a block from the surrounding soil. This condition is indicated by the change in thermal or (more usually) electrical conductivity. Although the technique is simple it is not precise and requires calibration for different soil types. It is much more reliable as a measure of soil tension. Full details of the technique are available in books on soil physics.

Fresh moisture

Weigh 10–20 g of fresh sample into a dry, weighed evaporating basin.
Place in an air-circulation oven at 105 °C and dry to constant weight (successive weighings should not differ by more than 1–2 mg).
Cool in a desiccator and weigh.
Calculate percentage fresh moisture from loss in weight.

Notes

1 The sample should not be sieved before weighing but large stones and roots should be removed.
2 This test should be well replicated.

Air-dry moisture

Weigh approximately 1 g of sieved (<2 mm) air-dried material into a weighed, dry crucible.
Place in an air-circulation oven at 105 °C and dry to constant weight (3 hours usually sufficient).
Cool in a desiccator and weigh.
Calculate percentage air-dry moisture from loss in weight.

Calculation

$$\text{Moisture (\%)} = \frac{\text{loss in wt on drying (g)} \times 100}{\text{initial sample wt (g)}}$$

Since the air-dry moisture is frequently used for correction of results to a dry weight basis it is more convenient to calculate the % dry matter as follows:

$$\text{Dry matter (\%)} = \frac{\text{oven-dry wt (g)} \times 100}{\text{initial sample wt (g)}}$$

Loss-on-ignition

The results from quantitative ashing of soils are usually expressed as 'loss-on-ignition' and as such are a rough indication of the amount of organic matter present in the soil. Loss-on-ignition is not a true measure of organic matter because at the temperature of ashing some bound water is lost from the clay minerals and is included in the overall loss. This error is relatively more serious for soils low in organic content. Multiplication of the result by a factor gives a crude estimate for the organic carbon content of the soil (p. 95), but again this is not reliable for organic-poor soils.

The choice of a suitable temperature for soil ignition is a matter of dispute. Ball (1964) and others have found that ignition at 375 °C eliminates variations due to loss of structural water from clay minerals. Losses of volatile minerals have been reported at temperatures over 500 °C. On the other hand there is a risk of incomplete combustion at too low a temperature. The present authors have found for a wide range of soils that 550 °C gives satisfactory results and this temperature is recommended here.

Procedure

Weigh approximately 1 g oven-dried material into a weighed dry crucible. (Alternatively use the oven-dry material from the moisture determination).

Place in a muffle furnace and allow the temperature to rise slowly to 550 °C.

Allow to remain at that temperature for 2 hours.

When cool transfer to a desiccator, cool to room temperature and weigh.

Calculate percentage loss-on-ignition from weight lost during combustion.

Calculation

$$\text{Loss-on-ignition (\%)} = \frac{\text{wt loss (g)} \times 100}{\text{oven-dry wt (g)}}$$

Note

A few drops of H_2O_2 can be added to promote oxidation.

pH

The pH of a soil is one of the most frequently measured parameters. Its measurement is the best known application of ion-selective electrodes, whose general principles are discussed in Chapter 8. pH gives an indication of the acidity and alkalinity which makes it valuable for soil characterization. It is conventionally defined as the reciprocal of the hydrogen ion concentration: $pH = -\log C_{H^+}$. In practice, pH electrodes record a potential which is a function of hydrogen ion activity whereby $pH = -\log a_{H^+}$. The concentration (C_{H^+}) and activity (a_{H^+}) are related by the expression $a_{H^+} = f_{H^+} C_{H^+}$ where f_{H^+} is the activity coefficient.

Various properties of the soil solution may affect the activity coefficient, including temperature, ionic size and strength and solvent density. Temperature is important and must be allowed for (see below). High ionic strengths affect the electrode response ('salt effect') but in most soils the dilutions commonly used in preparing the soil: water suspension are sufficient for the coefficient to approach unity. However, the 'salt effect' cannot always be ignored and is discussed later.

The only alternative methods for avoiding these problems are those depending on indicator solutions but they lack precision with soil suspensions.

Electrometric methods for soil pH using a glass bulb electrode are well documented, for example, by Bates (1964), Westcott (1978) and by the Standing Committee of Analysts (1978). The glass electrode potential is compared with that of a reference calomel electrode which provides a 'salt bridge' between the reference solution (saturated potassium chloride) and the sample or buffer solution. The electrode potential (E) is related to ion activity by a Nernst equation as shown in Chapter 8. In practice, glass electrodes are designed to record zero potential at pH 7.

Although modern pH meters are more reliable than earlier models, various factors still have to be taken into account to obtain reliable results. Electrode care is very important. New electrodes should be soaked for several days in a slightly acid solution (0.01M) and subsequently stored in a pH 4 buffer. Electrode performance falls gradually with time, but can sometimes be rejuvenated by dilute acid soaking.

The reference calomel electrode should be stored in weak potassium chloride solution or in a buffer where the main consideration is to avoid drying out which may allow the sinter to become blocked. A dual combination glass plus reference electrode can be similarly stored, but the variant now available as a 'pH stick' is fitted with a special cap.

Solution temperature must be allowed for. Around neutrality the error in a buffer solution is about 0.003 pH unit for every degree difference between the solution temperature and that at which the buffer was calibrated by the manufacturer. The error is greater towards the extreme ends of the pH scale. Most pH meters are temperature compensated, but this only corrects the electrode potential to the solution temperature and not to any other value. Hence for greatest accuracy the temperature of the sample and buffer solutions should be the same as that used for buffer calibration.

In ecological studies pH measurements are generally carried out on suspensions of the soil in water. Different ratios have been recommended ranging from 1:1 to 1:5 for fresh soil. Drying the soil will affect the pH regardless of the ratio. Increased dilution will minimize the activity problem, but also tends to increase the pH reading. Over a range from the paste or 'sticky point' to a ratio of 1:5 the difference in pH may exceed 0.5 (Chapman *et al.*, 1941).

As explained in Chapter 8, hydrodynamics are important. Soil suspensions should be briefly stirred initially, allowed to settle and electrodes immersed long enough for the reading to stabilize. Drifting may continue or resume after a pause particularly at high pH values where slow dissolution of particles or attack on the glass are probable causes. It is generally possible to standardize readings at a point immediately after the meter has settled down and before drifting becomes noticeable. Carbon dioxide absorption may also affect equilibria (Richards, 1954).

Soils with a high soluble salt content, as for example those subject to drying out or to heavy fertilization, may have a sufficiently high ionic strength in solution to affect the electrode response (salt effect). To minimize this effect the use of an alternative diluent such as 0.01M $CaCl_2$ (Schofield and Taylor, 1955) is often recommended, especially for agricultural soils. The recorded pH is lower than in water but more consistent over a season. Ionic strengths of soil solutions have been examined in relation to the recorded pH by Dolling and Ritchie (1985). Highly saline solutions are always troublesome and special glass electrodes should then be used. Further information about soil pH and its measurement is given by Hesse (1971), Page *et al.* (1982), Steinhardt and Mengel (1981) and Yu (1985).

The laboratory procedure described below uses a pH meter and a fresh soil:water ratio of approximately 1:2 by volume. For field measurements a portable pH meter or pH 'stick' with a robust dual electrode should be employed. In the field it may be necessary to puddle the soil–electrode contact zone using pure water to form a paste. Particular case is needed with stony soils.

Buffer solutions

The primary buffer solution is generally accepted as being a 0.05 M solution of pure dry potassium hydrogen phthalate (10.211 $g\,l^{-1}$) which has a pH of 4.002 at 20 °C and 4.008 at 25 °C (Weast, 1987). Phosphate and borate salts are also used as buffers for higher pH values. However, in practice commercial buffer tablets or solutions are widely available and should be used for soils. Solutions made from tablets should be renewed at regular intervals. Commercial solutions usually contain a preservative. Buffers of pH 4 and 7 are amongst the most widely used in soil work.

Procedure

Set up the pH meter and electrodes according to the manufacturers' instructions. Always calibrate the meter against two buffer solutions, one on each side of the expected pH range.

Half fill a 50 ml beaker with fresh soil and add water to fill.

Stir thoroughly at first and then allow to stand for a further 10 minutes.

Immerse electrode(s) into the supernatant, swirl gently and record pH value after a few seconds. Slow drifting may often occur (see above) but rapid drifting or erratic fluctuations may indicate a fault.

Rinse the electrodes with pure water between readings and touch-dry (but not wipe) with soft tissue.

Notes

1 pH measurements should be replicated where possible. Obtain the arithmetic mean via the antilogarithms of the readings.

2 Allow adequate time for the meter to stabilize and thereafter keep on standby when not in use to maintain stability.

Redox potential

Redox potential (E_h) is related to electrons as pH is to protons ($pE = -\log a_e$ where a_e represents electron activity). pE is related to reversible redox potential by the expression:

$$pE = \frac{E/2.3RT}{E_h}$$

Redox reactions are oxidation state changes which depend on the transfer of electrons and therefore the mobility of some elements will depend on the redox potential. It is generally expressed in millivolts and should be looked upon as an intensity factor since it does not give any information on capacity. The more easily oxidation or reduction occurs the greater is the negative or positive potential of the inert electrode. The potential of a reversible system containing oxidized and reduced forms is given by a Nernst derived equation:

$$E = E_0 + \frac{RT}{nF} \ln \frac{\text{(oxidized state)}}{\text{(reduced state)}}$$

where E_0 is a constant known as the standard potential. The theoretical aspects of this system are discussed in physical chemistry texts.

The relative degree of oxidation or reduction in a soil has a marked effect on the nature of that soil, its microbial and faunal populations and associated vegetation. This is illustrated by the contrast between aerated fertile brown earth and the water-logged anaerobic conditions of gleyed soils or bog peats. A parallel situation occurs in natural waters.

A crude measurement of the oxidation–reduction state of soil solutions or waters can be made using redox indicators. The usual procedure is to test with different indicators until the colour changes in the solutions match those indicated for a particular potential. For soils and waters this has limited use, not least because the masking effect of colour and turbidity in the samples. Further information about oxidation–reduction indicators can be obtained from standard texts such as Vogel (1961).

Redox potential is normally measured electrometrically using an inert metal electrode coupled with a reference electrode. Favoured electrodes are an inert platinum strip and a calomel half cell. The complete cell can be described as follows:

$^{+}H, Hg_2Cl_2, KCl/Sat KCl/Soil/Pt^{-}$

When a potentiometer is used it is necessary to incorporate a salt bridge filled with potassium chloride–agar in the assembly to provide a liquid junction between the soil and the saturated potassium chloride solution. By using a pH meter as a millivoltmeter this is not necessary since a standard calomel reference electrode used in pH determinations provides both the saturated potassium chloride solution and a bridge link. Silver–silver chloride is an alternative.

The system depends on the ability of a platinum electrode to take up or lose electrons according to whether the ions in solution are readily oxidized or reduced. The electrode must otherwise be inert to substances present in the soil, so noble metals, notably platinum, are used. A hydrogen electrode is the ultimate reference. Hydrogen gas is bubbled over a platinized platinum electrode immersed in a solution containing hydrogen ions.

Although redox measurements are relatively easy to make it is not always easy to produce meaningful results, especially for oxidizing systems. An estimation of total oxygen, particularly in waters, is usually of more value in these conditions. In addition, any assessment of redox values should always be carried out in conjunction with pH.

Notes on procedure

1 If possible carry out redox estimations in the field, taking care to exclude air and to ensure good electrode contacts.

2 Insert the electrodes directly into the water, peat, mud and other soft materials, and puddle well for denser materials.

3 If it is necessary to return the samples to the laboratory for analysis they must be relatively undisturbed and not exposed to the atmosphere. Inert gases have sometimes been used to minimize changes.

4 Use the millivolt scale of a pH meter and connect the platinum electrode in place of the glass electrode, reversing the polarity if necessary.

5 Carry out replicate determinations and vary the test conditions slightly to obtain reproducible results.

6 Clean platinum electrode surfaces by overnight immersion in 50% sulphuric or chromic acids or by heating to red heat in a spirit flame.

Ionic state
For many purposes it may be more appropriate to determine a particular ion to indicate the redox state. The ionic species present often have considerable bearing on ionic mobility for plant uptake. Apart from oxygen and hydrogen the following are the more important ionic indicators of the redox state:

$NO_3^- \leftrightarrow NO_2^- \leftrightarrow NH_4^+$
$Fe^{2+} \leftrightarrow Fe^{3+}$
$SO_4{}^{2+} \leftrightarrow S^{2-} \leftrightarrow H_2S$
$MnO_4^- \leftrightarrow Mn^{2+} \leftrightarrow Mn^{3+}$
$Cu^+ \leftrightarrow Cu^{2+}$

Redox potentials and oxygen concentrations equivalent to some of these ionic states were given by Mortimer (1942). Some organic compounds present in soils and waters, and which are often closely associated with decomposition processes, are related to the redox state.

A useful method of indicating the reducing conditions in a peat or mud profile has been described by Urquart (1966). It depends on the presence of sulphide ions and uses copper sheets electroplated with silver, which are pushed into the soil. The reducing zone is clearly marked by blackening of the silver surface.

Further details on the theory of redox systems and their measurement have been given by Ives and Janz (1961), Hesse (1971) and Bohn (1971). An example of the use of redox measurements to characterize peat profiles is given by Urquart and Gore (1973). A bibliography on soil redox potential determination has been produced by the Commonwealth Bureau of Soils (1978).

Percolation

Percolation or hydraulic conductivity is defined as the rate of movement of water through a soil. The determination is of greatest value to civil engineers, but also has some application in agriculture and in ecological studies.

The subject covers percolation into the soil following rain and includes the rate of drainage and the ease with which water is made available to roots, or lost from the soil by evapo-transpiration. The first two factors involve water movements in a horizontal or vertical direction. Amongst the factors which govern the percolation of water in a soil are pressure gradients, absorption and osmosis and the nature, distribution and size of the capillary pores between the soil particles. Permeability is another way of expressing percolation which takes into account the fluid density and viscosity. Measurement of the percolation rate is made by observation of a unidirectional flow of water through a profile, usually from the top to the bottom.

The principal sources of error with this test, especially if carried out in the laboratory, arise from the difficulty in taking undisturbed soil cores. Some compression is inevitable but this may be minimized by using cylinders of thin metal and taking the widest possible core. Channelling effects frequently occur near the cylinder edge due to disturbance of the soil. Repeating determinations until reasonably uniform results are obtained is the most reliable way of meeting this difficulty. The effectiveness of the method falls off with extreme soil types. Even a small degree of compaction makes clay soils almost impervious whilst cores of sandy soils collapse when removed from a corer.

Methods of taking soil cores and measurement of percolation and permeability are discussed more fully by Kelley *et al.* (1948), Smith and Stallman (1954) and by Reeve (1957). A simplified method is described below.

Procedure

Obtain a thin galvanized metal cylinder with a sharpened cutting edge. (A suitable material is 20 gauge galvanized steel rolled to a diameter of 20–30 cm.)

Obtain the core by driving the cylinder vertically into the soil to about half its depth (a block of wood on the top edge of the cylinder enables a heavy mallet to be used, but excessive hammering should be avoided since the soil packing may be loosened).

Dig out the cylinder plus core and trim off surplus soil level with the bottom of the cylinder.

Soak the soil by passing water through until saturated.

Set up the soil cylinder in the position shown in Fig. 2.1 and cover the surface with a coarse filter paper.
Invert a large volumetric flask above the cylinder.
Allow to settle then measure the volume of water (V ml) passing in time (t) seconds.

Calculation

$$\text{Percolation (ml s}^{-1}) = \frac{V \times D}{A \times t \times H}$$

where V = water volume passing in time t (cm^3); A = cross-sectional area of soil cylinder (cm^2); D = depth of soil core (cm); H = head of water (cm).

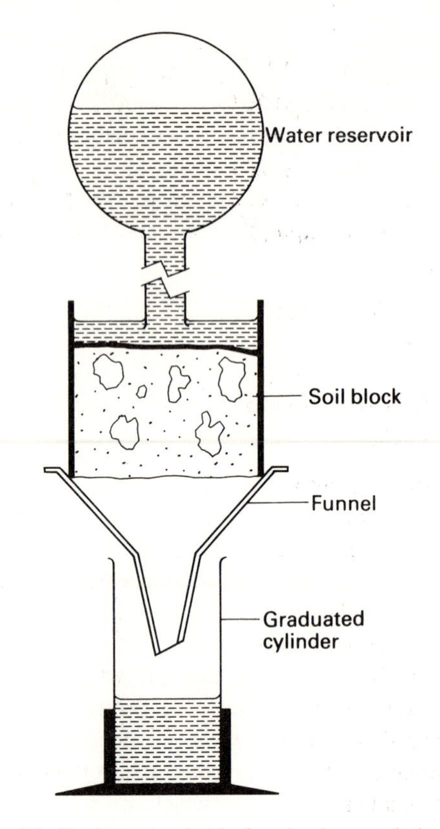

Fig. 2.1. Equipment suitable for simple percolation experiments.

Density

Bulk density

Bulk density, or apparent density, is the ratio of the mass of the soil to that of its total volume. The mass can readily be determined by drying to constant weight but the volume is less easy to measure accurately.

On sites where cores can be obtained with minimal disturbance, the use of a cylinder enables a known volume of soil to be extracted. The walls of the cylinder should be as thin as practicable to minimize any compaction. Certain light sandy and organic soils are difficult to extract in this way but can be removed after digging a pit and driving a plate horizontally to seal off the cylinder base.

Double wall cylinders are very effective for removing soil cores. In these the inner cylinder containing the sample may be withdrawn separately and the excess soil can be trimmed off cleanly at each end. Single cylinders should if possible be driven in slightly below the soil surface to allow for trimming the ends.

Cube blocks can be prepared *in situ* from some soil types, particularly amorphous peats and clays.

If stones or roots make it impracticable to use the above methods it may still be possible to obtain an irregular block. This should first be weighed and then coated with paraffin wax of known weight and specific gravity. The volume of the block is obtained by water displacement. Separate soil samples should be taken for moisture determination. A frozen soil block can also be used to give an approximate result.

Conventional methods for measuring bulk density are given in most soil texts whilst Doran and Mielke (1984) have described a relatively rapid procedure. An approximate method was given by Jeffrey (1970) who found an empirical relation between loss-on-ignition and bulk density for a number of soil types. If widely applicable this would enable a rough estimate to be obtained for shallow, stony soils. However, Harrison and Bocock (1981) produced a similar regression to that of Jeffrey (1970) but recommended that, for greatest precision, each soil type studied should have its own equation.

Particle density

This is a measure of the density of the soil particles. If the sample has been dried, gently crushed and sieved (<2 mm), a pycnometer (specific gravity bottle) may be used. The calibration of the pycnometer should be checked and about 10 g sieved soil taken. The main source of error arises from air trapped amongst the particles. This should be removed by placing the pycnometer in a vacuum desiccator and leaving overnight at reduced pressure. Some of the solvent can be placed in the desiccator for topping up the bottle.

Three other factors should be taken into account. First, the presence of a high proportion of finely divided particles results in high values due to hydration if water is used to fill the flask. The use of non-polar solvents such as benzene or toluene is recommended in such cases for accurate determinations. The specific gravity of the solvent must be known. Second, the method does not give the average particle density because the weights are governed by the relative amounts of fractions of differing densities. Third, it is assumed that the particles do not react chemically with the solvent. An alternative method for obtaining the density of coarse inert particles is to drop a known weight into a cylinder of water and measure the volume displacement.

Porosity

The amount of pore space or porosity of the soil has a marked influence on percolation. The total porosity, that is the percentage of the bulk volume not occupied by solids, may be obtained from the determination of the bulk and particle densities. The distribution of pores of different sizes may be obtained using suction techniques as described by Black (1965).

Calculation

$$\text{Porosity (\%)} = \frac{S-D}{S} \times 100$$

where S = particle density and D = bulk density.

Saturation capacitv

It is sometimes useful to know the maximum weight of water that a given mass of dry soil can hold. This is the saturation capacity and is expressed as the weight of water held by 100 g oven-dry soil. A simple method whereby a known weight of dry soil in a sintered glass crucible is allowed to become saturated with water is given below (Dewis and Freitas, 1970).

Procedure

Dry a sintered glass crucible, cool and weigh.
Immerse in water to saturate the sinter, remove and absorb excess water with a paper tissue.
Transfer about 25 g of soil into the crucible, gently tap down and level the surface.
Immerse the crucible and soil in water with the water level just above the sinter.
Leave immersed for 2 hours.
Remove, quickly dry the exposed glass surface and weigh.
Dry at 105 °C to constant weight (at least 6 hours).
Cool in a desiccator and weigh.

Calculation

$$\text{Saturation (\%)} = \frac{(A-B+C-D)}{D-A} \times 100$$

where A = wt of dry crucible; B = wt of crucible with sinter wet; C = wt of crucible with saturated soil; D = wt of dry crucible and soil.

Particle fractionation

This term is now widely used in place of the older one of mechanical analysis. It expresses the proportions of the various sizes of particles present in a soil.

The fractionation system which is described here and which is widely used is the International Scale (International Soil Science Society). The range of particle sizes on this scale is given in Table 2.3.

Grades also exist for particles retained on a 2 mm sieve and extend up to 200 mm but all are labelled 'gravel' on the International Scale. They

Table 2.3. International Scale of particle fractionation

Grade	Particle diameter (mm)
Coarse sand	0.2–2.0
Fine sand	0.02–0.2
Silt	0.002–0.02
Clay	<0.002

are of less interest in soil nutrient studies. A separate division proposed by the United States Department of Agriculture (USDA system) is one in which silt includes particles up to 0.05 mm diameter. Five grades of sand are also recognized by this system. Either system enables the relative amounts of different particles to be used for the classification of soil texture. A classification triangle based on the International system, but thought to be more appropriate for ecological purposes is shown in Fig. 2.2.

The method of fractionation generally used for the finer particles (<0.02 mm) is based on the dispersion and settlement of the particles in water. This process depends on the application of Stoke's law to sediment which may be simplified to:

$$v = Kr^2$$

where the velocity (v) of the fall of a particle through a liquid has a direct relationship to the square of the particle radius (r). The compound factor K can only be accepted as constant for a fixed or corrected temperature.

Two major assumptions have to be made if this law is applied to all soil particles. The first assumes that all particles behave as perfect spheres and, the second, that they all have the same density. It has been shown that these assumptions are acceptable in most circumstances, but soils with extreme density ranges need a different treatment. A critical review of sedimentation methods has been made by the Society for Analytical Chemistry (1968).

For the separation of particle sizes above 0.02 mm (fine and coarse sand), a successive sieving technique is generally used. A nest of sieves mounted on a vibration machine is adequate for most purposes, although the procedure must be carefully standardized. British Standard specifications for test sieves are given in BS 1796:1962. A comprehensive system for the accurate determination of size distributions from 0.1–1000 μm based on sedimentation and sieving is described by Kiff (1973). Dry and wet sieving techniques are compared by Robertson *et al.* (1984).

In the method often accepted as a standard, aliquots are taken by pipette at various times after dispersion of the soil–water suspension. These

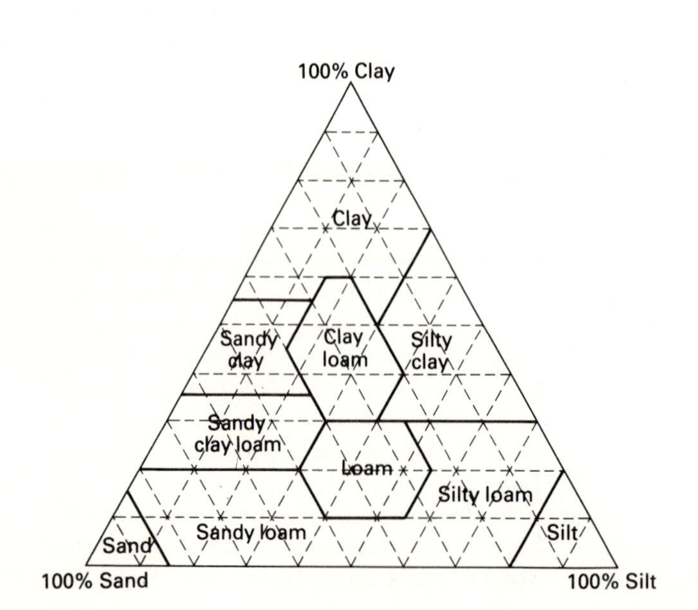

Fig. 2.2. Soil classification triangle based on the ISSS particle fractionation.

times have been chosen to correspond to the International size ranges. The dry matter content of the aliquot is then determined. This method is given by Piper (1950).

A much quicker method was proposed by Bouyoucos (1926) who inserted a hydrometer in the sample suspension to indicate the stage of settlement after a specified time. Although the method is not very precise it gives results in general agreement with those produced by the pipette method. Both methods are unsuitable for soils with a high organic or soluble salt content. Low organic soils can be tested but initial treatment is needed to break down some of the organic content. Digestion with hydrogen peroxide is generally adequate for this purpose. If carbonates are present, pre-treatment with hydrochloric acid is also necessary. High iron oxide content may be troublesome but for most purposes a separate treatment is not required. An examination of pre-treatment and dispersion methods for the hydrometer procedure has been carried out by Johnson *et al.* (1985). A modified version of the Bouyoucos procedure is given below and includes corrections for temperature.

The use of the Coulter Counter for particle size analysis has been considered by Walker and Hutka (1971) but it appears to offer no specific advantage over other methods. A bibliography covering the mechanical analysis of soil has been published by the Commonwealth Bureau of Soils (1975).

Silt and clay (hydrometer method)

Reagent

Calgon, 5% w/v.
Dissolve 50 g 'Calgon' in distilled water.
Add Na_2CO_3 to bring final pH to 9.
Dilute to 1 litre.
(Calgon is mainly sodium hexametaphosphate and is used as the dispersing agent.)

Procedure

Weigh 50 g of air-dried 2 mm sieved sample into a polythene bottle or the container of a high speed stirrer.
Add 25 ml 5% Calgon and 400 ml tap water.
Thoroughly disperse the soil, either by shaking the bottles on an end-over-end shaker for 2 hours, or by stirring for 15 min on a high speed stirrer.
Transfer to a 1 litre cylinder (tall form) and dilute to mark.
Stir for 1 minute with paddle.
Commence timing for readings with Bouyoucos soil hydrometer.
Take readings as follows (introducing hydrometer 20 sec before reading):
B 4 min 48 sec Silt and clay ($<20\ \mu$)
A 5 hours Clay ($<2\ \mu$).
Take temperature of suspension after each reading.
Add (or subtract) 0.3 units for every degree above (or below) 19.5 °C.
The moisture content of the soil should be determined at the time of weighing.

Calculation

If (50 − moisture wt) g soil are dispersed in 1 litre and B = hydrometer reading ($g\,l^{-1}$) after 4 min 48 sec, A = hydrometer reading ($g\,l^{-1}$) after 5 hours then

$$\text{Clay (\%)} = \frac{A(g\,l^{-1}) \times 100}{(50 - \text{moisture wt})g} - 1$$

(where 1 = Calgon correction)

$$\text{Silt + clay (\%)} = \frac{B(g\,l^{-1}) \times 100}{(50 - \text{moisture wt})g} - 1$$

Silt (%) = (silt + clay) (%) − clay (%)

A rough estimate of the total sand fraction up to 2 mm may be obtained as follows:

Total sand (%) = 100 − (silt + clay) (%)

If the loss-on-ignition value is known, the results may be further expressed on an organic-free basis.

Notes

1 Disperse any froth produced during mixing by adding a few drops of amyl alcohol just before inserting the hydrometer.

2 If organic matter exceeds 12%, oxidize by heating with 100 volume hydrogen peroxide. Weigh 50 g of soil into a beaker, cover with distilled water, add 20 ml hydrogen peroxide and warm until the reaction subsides. Repeat if necessary. Evaporate and weigh if loss is required, otherwise proceed as above.

3 For calcareous soils, weigh 50 g of soil into a beaker, slowly add 1 + 1 HCl, warm and stir until reaction subsides. Continue adding acid until all carbonates are neutralized, then filter and wash the soil free from excess acid. Evaporate and weigh if loss is required, otherwise proceed as above.

4 Saline soils should be well washed and then reweighed before proceeding.

5 If the samples are required for mineralogical analysis allow the sediment to settle for a further 3 hours after the 5 hour hydrometer reading has been taken. The top 10 cm of liquid is then siphoned off and evaporated to dryness for subsequent examination.

Fine and coarse sand

Procedure

After the second hydrometer reading, decant off most of the suspension, refill with water, paddle and allow to settle for 4 min 48 sec.

Decant and repeat until supernatant liquid is clear.

Transfer to a BS 72 mesh sieve supported in a funnel above a beaker.

Wash through sieve using rubber pestle and light rubbing.

Place sieve on a watch glass and dry in oven.

Tap fine sand free from underside of sieve into beaker.

Continue pestling and tapping until no further sample passes through.

Transfer coarse fractions from sieve into a previously weighed basin and dry in oven at 105 °C for 3 hours.

Cool in a desiccator and weigh.

Record weight as coarse sand and calculate.

Transfer fine sand from beaker to a previously weighed basin, evaporate to dryness on a hot plate and dry in oven at 105 °C for 3 hours.

Cool in a desiccator, and weigh.

Record weight as fine sand and calculate.

Calculation

$$\text{Coarse sand (\%)} = X \times \frac{100}{50 - \text{moisture wt}}$$

$$\text{Fine sand (\%)} = Y \times \frac{100}{50 - \text{moisture wt}}$$

where X = wt of coarse sand (g) and Y = wt of fine sand (g).

If the loss-on-ignition value is known, the results may be further expressed on an organic-free basis.

Note

If the fraction over 2 mm is required this should be separated and weighed initially. Further fractionation into gravel (2 mm to 1 cm) and stones (>1 cm) can also be carried out if desired.

Solution preparation

With the exception of the elements carbon and nitrogen (and possibly phosphorus) the ecologist does not often need a measure of the total content of elements in soils or rocks. However, since the need does arise from time to time, the appropriate procedures are discussed and presented in this manual. Although most of the methods described below are intended for use with mineral soils, they are generally applicable to ground rock materials. Large amounts of soil organic matter may give rise to reducing conditions during fusion and an initial ignition, treatment will then be necessary. The acid digestion methods are more appropriate for organic soils with low mineral content, such as peat and surface litter.

The fusion and acid digestion procedures recommended for bringing elements into solution lead into the analytical techniques described in Chapter 5. They are suitable for all the elements covered in that chapter except for carbon and nitrogen. Solution preparation procedures for the latter are given under those elements in Chapter 5.

The mineral content of rocks and soils is sometimes higher and generally much more variable than that of biological materials so analytical interferences are potentially greater. In extreme cases these problems have to be dealt with during the preparative stages.

There is an extensive literature in this field and older texts are still worth consulting including, Hillebrand *et al.* (1953), Shapiro and Brannock (1956), Jackson (1958), Riley and Williams (1959a, b, c, d), Stanton (1966) and Maxwell (1968). Page *et al.* (1982) also include methods for total elemental analyses. Hillebrand *et al.* and Page *et al.* are useful because they discuss the many interferences that can arise through the total analysis of complex materials such as soils and rocks.

Most laboratories with heavy demands for rock analyses will use X-ray fluorescence or other instrumental techniques which have the advantage of not requiring wet preparative stages. X-ray fluorescence is discussed in Chapter 7. However, it should be noted that commissioning these instruments will include comparison with the wet chemical methods.

Dissolution procedures

Both fusion and acid digestion methods have long been used to break down rock and soil materials, but no single procedure has universal applicability. The various constraints can be summarized briefly.

Emphasis is given to breaking down silicates because of their prominence in soils and many rocks, but alkaline fusions (with carbonate, hydroxide or borate) are then required rather than acidic fluxes (pyrosulphate, pyrophosphate). Fusions with sodium carbonate (high temperature) and sodium hydroxide (low temperature) are discussed below. Acid digestion must include hydrofluoric acid for silicates and a hydrofluoric–perchloric acid system is given below. Even when sample dissolution is complete, side reactions can lower the recovery of elements such as silicon and aluminium and must be allowed for (see below). Some minerals of chromium, titanium and zirconium are very resistant but may respond to an acidic fusion.

A further constraint arises when the element of interest is part of the fusion compound. Sodium and potassium are the main examples. Fusion with lithium metaborate is then an alternative (Ingemells, 1964) which has been applied to coal, ash and rock analyses (Boar and Ingram, 1970).

The crucible materials (commonly platinum, nickel or PTFE) are a further constraint. Platinum is essential for high temperature fusion but is vulnerable to attack by pre-treatment residues (hydrochloric acid liberating free chlorine), certain fusion materials (alkali hydroxides, peroxides, etc.) and sample components (transition metals depending on fusion conditions). These aspects are mentioned later where relevant. Platinum crucibles require care in both cleaning and use and texts such as Jackson (1958) should be consulted for a detailed description. Nickel is only suitable for low temperature fusion but convenient for alkaline hydroxides which cannot be used with platinum. PTFE and platinum are used for hydrofluoric acid digestions.

Care is required in the initial preparation of the sample as described earlier. Contamination must be minimal during drying, crushing and grinding. Specific pre-treatments are sometimes advised mainly to safeguard platinum. Organic soils are preferably ignited in a muffle before fusion. This helps to maintain an oxidizing flux but ignition itself can affect recoveries of non-metals. Samples high in iron oxides can be treated with aqua-regia (three volumes of hydrochloric acid to one of nitric acid) and the filtrates retained. The treatment concludes with drying and ignition (Page *et al.*, 1982).

Sodium carbonate fusion

Decomposition of silicates by anhydrous sodium carbonate is the most common fusion method. It is a high temperature fusion (900–1200 °C) requiring platinum crucibles. A large excess of sodium carbonate is required to ensure dissolution, particularly when iron and aluminium oxides are prominent and a ratio of ten to one is adopted here. An oxidizing flux must be maintained because under reducing conditions iron and manganese can fuse with platinum. It is sometimes recommended that sodium peroxide or potassium nitrate be included to ensure oxidation. Sodium peroxide is included in the method given here but only in quite small amounts because this substance too attacks platinum.

The melt is dissolved in acid, aided by warming. Hydrochloric acid should be used but transfer to sulphuric acid is required if, for example, titanium is to be determined by colorimetry (p. 154). This solution is suitable for the common soil mineral elements but is not ideal for silicon because some of it can come out of solution during the dissolution phase. Some workers, including Jackson (1958), prefer to dehydrate silica at the dissolution stage using perchloric acid and recover the silica which is later fused with sodium hydroxide. An alternative is to fuse the sample with sodium hydroxide (Shapiro and Brannock, 1956). In this method the melt is dissolved in water before transfer to hydrochloric acid and details are given later.

Reagents

1 Sodium carbonate, anhydrous.
2 Hydrochloric acid, conc.
3 Sodium peroxide.

Procedure

Weigh about 0.6 g Na_2CO_3 and 0.05 g Na_2O_2 into a clean platinum crucible.
Add 0.10 g of ground soil and mix (tapping the glass rod to remove adhered particles).
Add a further 0.30 g Na_2CO_3 on top of the mixture and cover with a lid.
Fuse over a Meker burner with a low flame initially and with the lid ajar. Increase heat until the melt is liquified, then continue for about 15 minutes (see Note **1** below).
Withdraw using tongs and swirl the molten melt around the crucible sides.
Allow to cool.
Immerse the crucible and contents in about 50 ml water in a 250 ml polypropylene beaker.
Add 3 ml HCl and heat for 30 minutes on a boiling water bath.
Wash the solution into a 100 ml volumetric flask and make up to volume when cool.
Carry out blank determinations in the same way.

Notes

1 It is important to maintain oxidizing conditions during the fusion. Only the top of the blue cone should be in contact with the crucible. The Meker burner is designed to provide a very hot oxidizing flame and is to be preferred to a muffle furnace.
2 It is sometimes suggested that the sides of the platinum crucible be rolled to crack away the fused mixture. A mould should be used later to restore the crucible shape.
3 If the melt is green, the presence of manganate is indicated. In this case reduce by adding a few drops of ethanol and mix to prevent any attack on the platinum.
4 It will be necessary to adjust the acid strength for some methods and later to make up the analytical standards containing the same strength of acid and appropriate strength of fusion reagents.

Sodium hydroxide fusion

Although sodium and potassium hydroxides fuse at a lower temperature than the carbonates, they are effective for recovering silicon. Aluminium has also been determined in this way but in general the carbonate fusion should be used for this and other elements in soil. Nickel crucibles are used because hydroxides can attack platinum (Jackson, 1958). The melt is dissolved in water before transfer to an acidic solution. Smith and Bain (1982) also discuss the sodium hydroxide fusion method.

Reagents

1 Sodium hydroxide, 25% w/v.
2 Hydrochloric acid, 1+1.

Procedure

Transfer 15 ml of NaOH solution to a clean nickel crucible (about 75 ml).
Evaporate to dryness with care.
Weigh 0.10 g of the finely ground soil into the crucible.
Cover the crucible and heat to about 600 °C for about 10 minutes, preferably over a burner.
Withdraw and swirl the melt around the crucible sides.
Allow to cool.
Add 50 ml water and allow to stand overnight to dissolve the melt.

Wash into a polypropylene beaker containing 30 ml 1 + 1 hydrochloric acid.

When cool wash the solution into a 100 ml volumetric flask and dilute to volume.

Carry out blank determinations in the same way.

Note

Nickel crucibles should be cleaned with dilute hydrochloric acid before use.

Potassium pyrosulphate fusion

This is an acidic fusion reagent (as opposed to the alkaline action of sodium hydroxide and carbonate). It is particularly effective for oxides but it is less suitable than sodium carbonate for silicates. The early stages of the fusion need to be carefully watched to avoid spattering. It has the advantage that borosilicate tubes can be used since the fusion temperature is relatively low. The tubes are weakened and can only be used a few times. Alternatively platinum crucibles may be used, although there is a slight attack on the platinum, especially at higher temperatures. Potassium bisulphate may be used instead of pyrosulphate but there tends to be more spattering during fusion.

Reagents

1 Potassium pyrosulphate, fused, powdered.
2 Hydrochloric acid, conc.

Procedure

Weigh 0.10 g of the finely ground soil into a small borosilicate tube.

Add about 0.5 g potassium pyrosulphate and mix.

Heat over a burner until melted and continue heating for about 5 minutes.

Allow to cool.

Add 5 ml of water and 1 ml HCl, then heat the tube in a boiling water bath until all the melt is in solution.

Transfer to a 50 ml or 100 ml volumetric flask and dilute to volume when cool.

Carry out blank determinations in the same way.

Note

It will be necessary to adjust the acid strength in some methods and later to make up the analytical standards using the same strength of acid and include the appropriate strength of fusion reagent.

Hydrofluoric–perchloric digestion

Hydrofluoric acid is essential for the acid dissolution of silicates although the complete breakdown of organic matter requires perchloric acid to be present. These two acids form a standard combination for treating soil and many rock materials although soils rich in organic matter should be pre-treated with a nitric–perchloric mixture (p. 59). Both platinum and PTFE crucibles can be used, preferably with a sand bath. Fume cupboard materials must be suitable for both these acids bearing in mind that hydrofluoric acid attacks glass.

The reaction with silicate forms volatile silicon tetrafluoride which is lost to the atmosphere whilst residual hydrofluoric acid is itself driven off by the perchloric acid. Many workers conclude the treatment here if dissolution is complete, although accidental drying out of perchlorates must be avoided. Others transfer to a sulphuric acid basis with fuming off of surplus perchloric acid. This enables the procedures given in Chapter 5 to be used but platinum crucibles would be needed throughout and not PTFE.

Apart from silicon this procedure is suitable for the common soil mineral elements including phosphorus although it is not recommended for aluminium because some aluminium fluoride may precipitate during the treatment. Fusion is recommended for aluminium and is essential for silicon. A small number of minerals such as zircon may resist treatment even when repeated and a special fusion is then required instead.

Reagents

1 Perchloric acid, 60%.
2 Hydrofluoric acid, 40%.
3 Nitric acid, conc.
4 Sulphuric acid, conc.

Procedure

Weigh 0.10 g of finely ground soil into a platinum or PTFE ('Teflon') crucible and moisten with a little water.

Add 1 ml 60% $HClO_4$ and 7 ml 40% HF.

Cover the crucible with the lid and digest slowly for 2 hours.
Allow to cool.
Wash the inside of the lid into the crucible with water.
Evaporate the contents until dilute fumes of perchloric acid appear.
If a large residue remains repeat the digestion or take a smaller sample weight.
Add 1 ml H_2SO_4.
Heat again to drive off $HClO_4$ leaving H_2SO_4.
Allow to cool.
Dilute and filter into a 100 ml volumetric flask and dilute to volume.
Carry out blank determinations in the same way.

Notes

1 Only platinum or PTFE vessels should be used when handling HF digests.
2 Since PTFE softens at about 300 °C, hot plates and sand baths used for the digestion should be kept at a low setting.
3 Since hydrofluoric acid etches glass, fume cupboard sash closures must first be painted over with clear varnish or covered with transparent plastic sheet.

Alternative digestion systems

Fusion and hydrofluoric methods are needed because of the chemical resistance of many minerals. An exception is phosphorus whose mineral forms in nature are all phosphates which can be attacked by strong acids. Early methods relied solely on perchloric acid but this can be criticized on recovery and safety grounds. However, the two acid oxidation methods used for vegetation (Chapter 3) can also be used for soil phosphorus and details are given on p. 59. Rowland and Grimshaw (1985) showed that the two methods were equally effective for a range of soil types but recommended the sulphuric–peroxide method which can also be used for nitrogen.

It must be stressed that these digestions are not suitable for other soil mineral elements and are not a substitute for hydrofluoric acid. The failure to recover potassium was examined by Parkinson and Grimshaw (1986) who showed that only a minor contamination of vegetation by soil could be tolerated when the sulphuric–peroxide method was used.

Most nitrogen in soil is organic and digestion methods for this element and its fractions are given in Chapter 5. Some ammonium–nitrogen is held in clay lattices as a 'non-exchangeable' fraction. This is occasionally estimated and methods including hydrofluoric acid are discussed by Page *et al.* (1982).

Carbonate soils

Carbonates are a prominent group of non-silicate minerals found in soils, particularly calcareous types, but are readily decomposed by the procedures outlined above. However, when sulphuric acid is part of the procedure, the recommended sample weights should not be exceeded if calcium carbonate is the main mineral.

The carbonate fraction itself can be estimated by hydrochloric acid treatment followed by titration as given in Chapter 5. It should be noted that the calcium and magnesium content of the filtrate is not a measure of carbonate in the soil because non-carbonate fractions of these elements will be included.

Individual elements

Some general points are applicable to all the elements listed in this section.
1 The recommended weights are only a guide and for trace element analyses large sample weights may be needed. This may mean that larger reagent quantities are also required and any increase in the residual acid concentration must be allowed for before applying the elemental methods.
2 If the soil organic matter is high it should be destroyed before fusion, preferably by ignition at 450 °C. This may give low recoveries of boron, chlorine, phosphorus and sulphur unless precautions are taken as noted. Prior treatment with a nitric–perchloric mixture is needed before the hydrofluoric–perchloric digestion.
3 The final acid strength, whether 1% (v/v) or higher (see Note **1** above) must be the same in

both standards and samples when the methods of Chapter 5 are used.

4 Blank solutions should always be prepared and are particularly important in the determination of low elemental concentrations. If possible a reference soil of known chemical content should be included with each batch of samples.

5 Because of the great variability in the mineral content of soils it is not possible to predict all the interference problems that might arise. Some are covered under individual elements but before proceeding with these, or indeed any new soil methods, it is advisable to carry out standard addition and compensation tests.

The preparation methods appropriate to individual elements are summarized here. Three elements are referred to Chapter 5 because they have specific preparation techniques.

Aluminium

Na_2CO_3 or NaOH fusion.
Take 0.1 g sample to give 100 ml solution.
Proceed as on pp. 82–84.

Boron

Na_2CO_3 fusion, and 30 ml water, then 20% w/v H_2SO_4 until melt dissolves.
Take 0.25 g sample to give 100 ml solution.
Proceed as on pp. 85–86.

Calcium

Any fusion, or HF–$HClO_4$ digestion.
Take 0.1 g (less for calcareous soils) to give 100 ml solution.
Proceed as on pp. 87–91.

Carbon

Details on pp. 91–97.

Chlorine

Na_2CO_3 fusion. Addition of a small amount of $NaNO_3$ improves the flux.
Dissolve melt in water and adjust to pH 8 with 0.1 M H_2SO_4.
A water extraction of soil is sometimes adequate.
Take 1.0 g sample to give 100 ml solution.
Proceed as on pp. 98–100.

Cobalt

Any fusion method or HF–$HClO_4$ digestion.
Take 0.5 g sample to give 100 ml solution.
Proceed as on pp. 101–104.

Copper

Na_2CO_3 or $KHSO_4$ fusion, HF–$HClO_4$ digestion.
Take 0.25 g sample to give 100 ml solution.
Proceed as on pp. 104–107.

Iron

Na_2CO_3 fusion, or HF–$HClO_4$ digestion.
Take 0.1 g sample to give 100 ml solution.
Proceed as on pp. 108–111.

Magnesium

Any fusion method, or HF–$HClO_4$ digestion.
Take 0.1 g sample to give 100 ml solution.
Proceed as on pp. 111–114.

Manganese

Any fusion method, or HF–$HClO_4$ digestion.
If solution is pink or brown, add a few drops of H_2SO_3 (5% w/v SO_2) and boil before diluting to volume.
Take 0.1 g sample to give 100 ml solution.
Proceed as on pp. 114–116.

Molybdenum

Details on pp. 116–118.

Nitrogen

Details on pp. 120–134.

Phosphorus

1 Fusion followed by boiling with H_2SO_4 to convert to *ortho*-phosphate (or HF–$HClO_4$ providing the last traces of HF are destroyed).
Take 0.1 g sample to give 100 ml solution.
2 Acid digestion with HNO_3 (2 ml), $HClO_4$ (1 ml) and H_2SO_4 (0.5 ml). Take 0.05 g sieved and ground soil to give 100 ml solution.
3 Acid digestion with H_2SO_4–H_2O_2 (see p. 59) but extend digestion time to 2 hours. Take 0.2 g soil.
Proceed as on pp. 135–142.

Potassium

Any fusion method or $HF-HClO_4$ digestion, although fusion reagents may give high blanks.
Take 0.1 g sample to give 100 ml solution.
Proceed as on pp. 142–144.

Silicon

NaOH fusion.
Take 0.1 g sample to give 100 ml solution.
Proceed as on pp. 145–148.

Sodium

$HF-HClO_4$ digestion.
Take 0.1 g to give 100 ml solution.
Proceed as on pp. 148–150.

Sulphur

Mix soil with 2.5 g Na_2CO_3, preheat at 450 °C for 30 minutes before fusion.
Addition of 0.2 g $NaNO_3$ or 0.1 g Na_2O_2 improves the flux.
Take 0.5 g sample to give 100 ml solution.
Proceed as on pp. 151–153.

Titanium

Na_2CO_3 fusion, or $HF-HClO_4$ digestion (HF not recommended in the presence of high soil P).
Evaporate solutions obtained and add 30 ml 50% v/v H_2SO_4.
Evaporate to white fume stage.
Add 10 ml 50% v/v H_2SO_4 and filter.
Take 0.5 g sample to give 100 ml solution.
Proceed as on pp. 154–155.

Zinc

Any fusion method, or $HF-HClO_4$ digestion.
Take 0.1 g sample to give 100 ml solution.
Proceed as on pp. 156–159.

Soil extraction

In assessing soil fertility the available nutrient content is considered to include the ions in true solution and a proportion of those held by the clay and organic colloids. Many attempts have been made to develop simple methods which will simulate the action of plant roots in taking up nutrient ions, but with little success. For most purposes chemical extraction techniques, although essentially empirical, are used for the measurement of available nutrients in the soil. In these methods the ions (generally cations) are displaced from adsorption sites by other replacing ions applied in great excess. The displaced ions are then estimated quantitatively.

There are serious limitations to this approach, partly because an unknown fraction of non-available nutrient is taken up by most extractants. A better understanding of the status of a soil nutrient requires a measure of both the quantity factor (amount of element in the labile pool) and the intensity factor (extent to which a soil can release the element to plants). Mild extractants of the type discussed below give an approximate measure of the intensity value which in agricultural work is represented by the crop yield. The quantity factor is equivalent to the labile pool of the nutrient element (*L*-value) and requires a growth experiment for its assessment in which the uptake of a radioactive isotope is measured. This is called the isotopically exchangeable fraction (Larson, 1967).

These concepts apply to mineral nutrient elements but most of the experimental work has been carried out for phosphorus which has a radioactive isotope of suitable half-life. Larsen (1967) and Williams (1967) describe such extraction systems for phosphorus. Labile potassium was discussed by Beckett and Tinker (1964).

The terms 'available' and 'exchangeable' are still widely used for agricultural purposes, but cannot be adequately defined in relation to conventional extractants. Availability in particular is almost impossible to quantify by these extractants and the term is not used in this text. Exchangeable cations are usually determined by displacing them from the soil exchange sites by an excess of the replacement ion. However, any free pool cations are also removed in the process, so the term 'extractable' is more appropriate in most circumstances.

The value of arbitrary extractants for agricultural purposes has been increased by repeated calibration with crop response to fertilizer application for particular soil conditions, but there are still weaknesses in existing methods as discussed by Beringer (1985). In ecological work this type of data is generally lacking for native species although an effective extractant which removes soil solutions and those held by ionic exchange or weakly bound to surface forms will give a reasonable measure of nutrient element potential.

The extractants described below are used mainly for potassium, calcium, magnesium and sometimes manganese. Sodium is occasionally included even though its nutritional significance is slight. Different extractants are normally employed for phosphorus (p. 42), although acetic acid can be used for both phosphorus and cations. Other methods are given for ammonium- and nitrate-nitrogen (p. 41). Although large numbers of different extractants have been proposed, the numbers in regular use for phosphorus and potassium in soil testing labs are now fairly small (Sabre, 1980).

The discussion below will be mainly confined to the simple extractants but it should be noted that alternative approaches such as bioassay techniques can provide helpful data in nutritional studies (Hegemann and Keenan, 1983).

Choice of extractants

Although divalent ions generally have a greater displacing power than the monovalent, small ions (particularly hydrogen) are more strongly held than the larger. These two factors may seem to balance out when choosing for extraction purposes, but in practice the monovalent are the most widely favoured by soil workers.

Ammonium ions, especially in ammonium acetate have a number of advantages as replacement cations. In excess, the ammonium ion is an effective replacement for displaced nutrient cations and is not subject to slow release. Normal ammonium acetate at pH 7 was introduced by Schollenberger (1927a) and is probably the most widely used of all soil extractants for potassium, calcium, magnesium and cation exchange capacity (CEC). Phosphorus, however, is not normally determined with this extractant. An ammonium acetate procedure is outlined below and can be applied to most soil types. Ammonium chloride (pH about 4.6) has also been employed. Tucker (1985) recommended a quaternary ammonium salt (choline chloride).

Sodium is used as a displacing ion in the well-known Morgan's reagent, which has been widely used for agricultural advisory work in Britain and elsewhere. This consists of 10% w/v sodium acetate in 3% v/v acetic acid (at pH 4.8) (Morgan, 1937). Cations and phosphorus can be determined in the extract but it is not ideal for flame work, even with standard compensation, and is less often used today than formerly. Potassium has sometimes been employed in extractants (usually as the chloride) although its use precludes determination of this particular nutrient. Lithium chloride alone or mixed with lithium acetate is also sometimes used as a soil extractant. Both sodium and potassium chloride are used for ammonium-nitrogen and a method is given later.

Divalent ions are sometimes employed for exchange extractions. Calcium chloride has been used in CEC estimations. Barium chloride was introduced by Mehlich (1945) who used it in combination with triethanolamine at a pH of 8.2 and employed it for cations and CEC. The relatively high pH makes this extractant suitable for organic, calcareous and gypsum soils.

The acid extraction methods form a different group and are not considered exchange systems in the same sense as those considered above. Acetic acid is probably the most commonly used extractant in this group. It was first introduced by Williams (1928) who adopted 0.5 M strength (2.5%) and used it for phosphorus. It has the advantage of allowing phosphorus and the cations to be determined in one extract as well as being convenient to handle analytically (Williams and Stewart, 1941). The acetic acid extraction is described below and has a similar extracting capacity to ammonium acetate on many non-calcareous soils. It also employs the same soil : extractant ratio, which is tighter than that used by Williams and hence is less suitable for phosphorus on soils having a high adsorption capacity. A further disadvantage of acetic acid is that it releases a considerable amount of calcium from calcium carbonate and should not therefore be used for estimating extractable cations in chalk and limestone soils (see also below).

Weak hydrochloric acid was formerly in common use although is less favoured now since it attacks the exchange complex. The strongly dissociated mineral acids lead to substitution of hydrogen ions in the alumino-silicate lattices which breaks down the structure and transfers aluminium to the exchange sites. Hydrochloric acid was sometimes used for extractions of the minor nutrients (notably copper and zinc) although complexing agents are now preferred for these elements. Sometimes it is useful to obtain an indication of the less available nutrient potential of the soil without carrying out a total analysis as described earlier. For this purpose an extraction with 10% nitric acid may be used.

Various studies have been carried out comparing the effectiveness of soil extraction methods including those by MacDonald *et al.* (1978) and Gillman *et al.* (1983).

Extraction of calcareous soils

Calcareous soils present serious difficulties because calcium carbonate is partially soluble even in the more neutral extractants. Bower *et al.* (1952) discussed both calcareous and saline soils in this respect and advocated a high pH sodium acetate extractant. Tobia and Milad (1956) discussed exchangeable calcium in particular and reviewed several methods. They recommend dilute hydrochloric acid brought into equilibrium with solid calcium carbonate. Raising the pH of the M ammonium acetate extractant from 7 to 9 by addition of ammonia will have a similar effect. However, the amount of extractable calcium can only serve as an approximate guide to 'available' calcium even when relatively little carbonate or calcium comes into solution. There is probably little point in separating 'true' exchangeable calcium and carbonate-calcium dissolved by mild extraction since the latter is part of the soil nutrient potential.

Magnesium is also affected in a similar way, although to a lesser extent, in the extraction of soils containing magnesium carbonate.

Conditions of extraction

Some attention must be given to the details of the extraction and the state of the sample. It is preferable to extract some soils in the fresh state especially peat (see below) or when mobile constituents such as ammonium-nitrogen (p. 41) are being examined. Nevertheless, results obtained from similar soils of differing moisture contents are not necessarily strictly comparable, even allowing for conversion of results to a uniform dry basis. Some workers prefer to handle soils previously brought to their saturation moisture level by a standardized procedure such as that outlined by Loveday (1972). In some studies it is also an advantage to express data on a fresh volume basis although this requires an estimate of the bulk densities of the soil in the field.

The extraction procedure used here assumes that the soil has been dried unless otherwise stated. Slight changes may occur on drying and sieving but in most cases these are outweighed by the advantages of standardizing conditions. There have been reports, notably by Gilliam and Richter (1985), that in some soils the release of extractable ions (especially NO_3^-, NH_4^+ and PO_4^{3-}) is increased markedly by some drying and grinding treatments. Searle and Sparling (1987) also found that air-drying (and storage) affects the amounts of phosphate extracted from soil. They found that sulphate was little affected. If in doubt a simple comparison test can be carried out on fresh soil types.

Ideally the extract should be allowed to percolate slowly down a long column containing a known amount of fresh soil until no more ions are exchanged. In practice column leaching is little used because it is so slow and inconvenient for routine work, so a shaking procedure is often used instead.

The main problem with shaking procedures is that they are essentially equilibrium processes. If the ratio of extractant volume to weight of (air-dry) soil is too tight some re-adsorption of extracted nutrients may occur during the extraction until an equilibrium is established. This is more significant with phosphate, but can occur with cations. The best extractant/soil value is one such that doubling or halving the ratio makes no difference to the final result (calculated in me or mg 100^{-1}). In this respect high ratios will be better, but if too high the nutrient content in the extract may be too low to be measured with acceptable precision. In the procedure given below for ammonium acetate and acetic acid, a ratio of 25:1 is adopted. Higher ratios may be desirable for clay soils, extending up to 40:1. The latter was used by Williams and Stewart (1941) for extracting phosphate with acetic acid. Truog's reagent for phosphorus employs 200:1 (see below).

The time of shaking should be fixed and must be sufficient to allow an equilibrium to be established. This is determined by the content in solution, the capacity of the soil to re-adsorb and the content yet remaining to be extracted. One hour is recommended here and is in excess of the time required for most soils. More nutrients may be released if the same sample is shaken a second (or even third) time with a fresh volume of extractant, but the amounts involved are generally small and for routine work a single extraction is adequate. All extractions should be carried out at room temperature except where otherwise stated.

Peat has to be treated separately from mineral soils because drying causes shrinkage and hardening and makes it difficult to re-wet. The method described here involves leaching fresh peat on a filter funnel until no more ions are displaced. This is because shaking disturbs the structure of the sample to a greater extent than with mineral soils.

In the methods outlined below the extract is separated from the soil by filtering, which is usually the most convenient approach. However, some extracts of fresh peat are preferably treated by centrifuging and sintering with a filter stick. These are discussed later.

Destruction of the extractant by evaporating to dryness and igniting, or digesting with acids, is not considered necessary for most of the analytical techniques described in Chapter 5, although it may occasionally be needed to facilitate a particular determination. For most purposes it is sufficient to compensate standards with the appropriate extractant.

Common extractants

Ammonium acetate (pH 7)

This extractant is suitable for sodium, potassium, calcium and magnesium in neutral and acidic soils and may also be used for manganese. It also forms the first stage in the determination of cation exchange capacity.

Extractant

Ammonium acetate, M, pH 7.

Measure about 200–300 ml water into a large aspirator.

Add 575 ml glacial HOAc and 600 ml 0.880 NH_3 solution and mix.

Dilute to 10 litres and mix thoroughly.

Check that the pH is 7.00 ± 0.05 and adjust with drops of acetic acid or ammonia as necessary.

Procedure

1 Mineral soils

Weigh 5 g air-dry, 2 mm sieved soil into a 250 ml container.

Add 125 ml M NH_4OAc and shake for 1 hour on a rotary shaker.

Filter through No. 44 paper into borosilicate or polythene bottles and reject the first 5–10 ml.

Run two blanks with extractant only.

2 Peat

Weigh 25 g fresh peat into a pyrex beaker.

Add about 100 ml M NH_4OAc, stir and leave about 10 minutes.

Pour supernatant liquid through No. 44 filter into dry vessel.

Do not reject any filtrate.

Wash the peat on to the filter with a further 50 ml M NH_4OAc and allow to filter.

Continue to leach with successive small additions of M NH_4OAc until 250 ml of the leachate have been obtained.

Run two blanks with extractant only.

Notes

1 The moisture content must be determined at the time of weighing in order to correct the results to a dry basis.

2 Each addition of the extractant should saturate the peat and should be followed by adequate time for filtering. The whole process will take several hours to complete.

Ammonium acetate (pH 9)

This extractant is a possible alternative for highly *calcareous* soils and may be used for the same elements as above. Relatively little calcium carbonate is dissolved at this pH but the presence of excess ammonia can be troublesome. Ventilation should be sufficient to remove ammonia fumes and avoid cross-contamination in the laboratory.

Extractant

Ammonium acetate, M, pH 9.

Prepare 10 litres as given above, but take 740 ml 0.880 NH_3 solution.

Check that the pH is 9.00 ± 0.1 and adjust if necessary.

Procedure

Extract as given above for mineral soils.

Acetic acid, 2.5% v/v

This extractant is suitable for most cations but should be reserved for non-calcareous soils only. The extraction conditions for cations are less suitable for phosphorus except on sandy soils (see later). Essentially this extraction only recovers the acid soluble ions and is not based on an exchange mechanism in the same sense as ammonium acetate, although the two systems frequently give comparable values for acid soils.

Extractant

Dilute 250 ml glacial HOAc to 10 litres and mix well.

Procedure

Extract as given above for ammonium acetate. Determine moisture content as before.

Cation exchange capacity

In this procedure the cations, particularly sodium, potassium, calcium and magnesium, are displaced from the exchange sites on the soil colloids and replaced by a cation from the initial extractant. After a washing stage the adsorbed cation is then itself displaced by a leaching solution in which it is subsequently determined. Most common procedures in use for Cation Exchange Capacity (CEC) have this sequence. In practice it is a somewhat arbitrary procedure and a meaningful interpretation will depend on the nature of the soil and the extractant system used. Bache (1985) discussed the issues involved in choosing a method.

Several cations including ammonium, sodium, potassium, calcium, barium and occasionally strontium and manganese have been used by different workers to saturate the colloids. Their main advantage seems to be that they are easily determined analytically once they have been displaced by the leaching solution. (For this reason extractants based on the hydrogen ion are not normally used.)

In practice, however, the choice for any particular soil can be narrowed quite sharply. In the first instance the cation employed should be very low in the soil concerned or a low CEC result may be obtained. Secondly, the divalent cations tend to give higher and perhaps more variable CEC results than the monovalent. Finally, none are so widely applicable that they can be used for almost any soil type without getting a misleading CEC result. In practice it seems that ammonium acetate is the most widely used and is described here. Gillman *et al.* (1983) who compared different methods for the determination of cation exchange capacity recommended the use of ammonium acetate.

The traditional CEC method using ammonium acetate involves a three-phase leaching procedure, including a washing stage in the middle whereby the excess of the initial extractant is removed. For this purpose water and alcohol are commonly used, as in the procedure given below. Nevertheless, Rich (1962) and others have criticized this operation and maintain it is incomplete in soils high in amorphous sesquioxides which tend to retain the acetate part of the extractant. Such a retention could influence the estimation of CEC, as indicated by Schollenberger and Dreibelbis (1930), Mehlich (1945) and Metson (1956). Okazaki *et al.* (1964) concluded that the interpretation of CEC data would be improved if the washing step was omitted. Bascomb (1964) extracted with a barium salt and avoided a washing step by adding magnesium sulphate to precipitate all the barium (including any adsorbed). The adsorbed barium was exchanged for magnesium and the magnesium remaining in solution estimated.

The nature of the final leaching solution is less critical, but it must have a different cation from the one used initially. In the procedure given below potassium chloride is employed.

A measurement considered by some workers to be of greater value is the effective cation exchange capacity (ECEC). This is the summation of the base cations (calcium, magnesium, potassium and sodium) which excludes aluminium and hydrogen. Rapid methods for its determination are described by Pleysier and Juo (1980) and by Edmeades and Clinton (1981). The former use silver–thiourea reagent which was shown by Chabra *et al.* (1976) to have a high affinity for soil colloids.

Reagents

1 Ammonium acetate, M, pH 7.
Prepare as earlier described.

2 Industrial alcohol, 60% v/v. This reagent should be neutral and free of ammonia.

3 Potassium chloride, 5% w/v.

Procedure

Extract 10 g air-dry sieved soil with 250 ml NH_4OAc extractant and filter as described earlier.

Remove excess extractant by repeated washings of the residue with portions of industrial alcohol solution.

Reject all washings.

Leach the residue with approximately 30 ml portions of KCl solution to displace adsorbed NH_4^+-N (dilute the first portion five-fold).

Allow time for filtering in between additions and continue until 250 ml has been collected.

Determine the NH_4^+-N content in the leachate by distillation of an aliquot with MgO followed by titration with M/140 HCl (as described for extractable ammonium-nitrogen on p. 126, or by an automated procedure, p. 125).

Calculation

If T (ml) of M/140 HCl are required for titration then:

$$\text{CEC (me 100 g}^{-1}) = \frac{T\text{(ml)} \times \text{final leachate vol (ml)}}{1.4 \times \text{aliquot (ml)} \times \text{sample wt (g)}}$$

Correct to dry weight where necessary.

Notes

1 To indicate when the alcohol washing is complete, add about 10 ml 10% NH_4Cl to the first portion of industrial alcohol and keep testing filtrate with $AgNO_3$ solution until no trace of Cl is detected.

2 Instead of leaching with KCl, the ammonium ions may be recovered by distillation after transferring the entire soil to a flask and adding MgO (see Fig. 5.4, p. 123, for an illustration of the equipment). It should be noted, however, that there is the possibility of some breakdown of soil organic nitrogen by this treatment.

Total exchangeable bases

This determination refers to the sum of the metal cations extracted by ammonium acetate or other extractant. If the individual cations are being estimated the calculated sum of sodium, potassium, calcium and magnesium may be taken as an approximate estimate of Total Exchangeable Bases (TEB). It is considered by some to be of greater value than cation exchange capacity and has been called the Effective Cation Exchange Capacity (ECEC). A direct estimate of TEB was introduced by Bray and Willhite (1929). It involves igniting the extract, dissolving the residue (mainly carbonates and oxides) in excess standard hydrochloric acid and back titrating with standard alkali. This is described below. A determination of TEB obtained in this way should agree fairly well with the TEB obtained by summation. If slightly higher it is probably due to excess aluminium hydroxide plus traces of manganese and other ions.

Reagents

1 Ammonium acetate, M, pH 7.
Prepare as described earlier.
2 Hydrochloric acid, 0.10 M.
3 Sodium carbonate, 0.05 M.

Procedure

Extract 10 g air-dry sieved soil with 250 ml M NH_4OAc and filter as described earlier.
Evaporate a suitable aliquot to a low volume in a beaker, transfer to a basin and evaporate to dryness.
Ignite in a muffle furnace at 500 °C for 2 hours. Add a known excess of 0.10 M HCl (A ml) and digest on a steam bath for 30 minutes to aid dissolution.
Titrate excess acid with 0.05 M Na_2CO_3 (B ml) using methyl orange as indicator.

Calculation

If $(A-B)$ ml acid are used then:

$$\text{TEB (me 100 g}^{-1}) = \frac{(A-B)\ \text{ml} \times \text{extractant vol (ml)} \times 10}{\text{aliquot (ml)} \times \text{sample wt (g)}}$$

Correct to dry weight where necessary.

Exchangeable hydrogen

Obtaining a valid estimate of exchangeable hydrogen presents greater problems than those encountered for CEC and the individual cations. This is partly because the metals are much easier to determine in soil extracts than hydrogen but also because the chemistry of this ion both in solution and in soils is more complex than for metal cations and a precise definition of exchangeable hydrogen is lacking. In addition the role of aluminium cannot be ignored in problems which relate to soil acidity (Coulter, 1969).

Many soil workers find it sufficient for their purposes to estimate this ion by difference, i.e. CEC − TEB (Peech *et al.*, 1947). However, this is subject to the disadvantage of all difference determinations.

The only useful direct methods rely on either a titration or pH measurement. Schollenberger and Simons (1945) titrated their ammonium acetate extracts back to the original pH (7.0) using alkali. Shaw (1951 and 1952) used calcium acetate (pH 7) as an extractant and back-titrated with barium hydroxide solution. These approaches have been fairly widely applied but lack precision because of the buffering effect of the extractant. The method can be extended by plotting the exchangeable H^+ of some contrasting soils against the pH of the filtrate obtained before titration. The graph can then be applied to other soils if the pH of the filtrate is known (Brown, 1943; Woodruff, 1948). Although such methods are relatively simple they appear to be of limited value and no direct technique is described here. The CEC and TEB estimations appear to be of more value and their difference is easily found if desired.

A method for exchangeable acidity (hydrogen + aluminium) is given by Dewis and Freitas (1970). Potassium chloride is used as extractant and the leachate is back-titrated with standard alkali. Fluoride is then added to complex the aluminium, releasing OH ions which may then be titrated with standard acid to give exchangeable aluminium. Exchangeable hydrogen can be obtained as the difference between these two results.

Individual elements

The analytical procedures described in Chapter 5 are basically applicable to soil extracts with little modification. However, the greater content of potentially interfering elements in soils necessitates treating each soil type strictly on its merits. In addition the extractant itself has some effect. Methods for individual elements are considered below and possible interferences are indicated. The extractable nutrients are normally expressed as mg 100 g^{-1} or μg 100 g^{-1} as calculated for each element in Chapter 5. For direct comparison of elements the me 100 g^{-1} value is useful and is obtained by dividing the mg 100 g^{-1} value by the equivalent weight of the element.

Aluminium

Since this element is an integral part of the structure of many primary minerals and secondary (clay) minerals such as kaolinite, illite and montmorillonite, its total estimation is of value to the mineralogist and pedologist. Extractable measurements are usually required where toxicity is suspected or if the soil profile is studied. 3% oxalic acid is suitable for the latter purpose and its use is described under iron (p. 39).

It is now considered that a range of aluminium species including basic and hydroxy-polymeric forms are brought into solution through exchange reactions (Veith, 1977). However, aluminium occurs in many different forms in soil minerals and the term 'extractable' is more appropriate. The most common extractants include barium chloride, potassium chloride and ammonium acetate at pH 4.8. Yuan and Fiskell (1959), McLean (1959), Lin and Coleman (1960) and Pratt and Bair (1961) have compared these extractants. Lee and Sharp (1985) compare extractants for podzolized, high aluminium soils. An isotopic exchange method was discussed by Koch *et al.* (1984).

An extraction method using ammonium acetate (pH 4.8) is described below.

Extractant

Ammonium acetate, M, pH 4.8.

Measure about 250 ml of water into a large aspirator.

Add 575 ml glacial HOAc and 600 ml 0.88 NH_3 solution and mix well.

Dilute to 10 litres and mix again.

Adjust to pH 4.80±0.05 using acetic acid or ammonia as necessary.

Procedure

Extract 5 g air-dried sieved soil with 125 ml NH_4OAc extractant by shaking for 30 minutes on a rotary shaker.

Filter through a No. 44 paper.

Include two blank determinations.

From this point use one of the methods described on pp. 82–84.

Notes

1 About 10 g of fresh mineral or organic soils should be taken.

2 The standards used in the methods in Chapter 5 should contain the same amount of extractant as is taken in the sample aliquots.

3 Coloured extracts are rare at pH 4.8 but where they do occur the extractant must be destroyed if a colorimetric method is subsequently used. Evaporation followed by ignition or acid digestion may be used for this. This is not required when atomic absorption spectroscopy (AAS) is used but standards must be compensated with the extractant.

Boron

Although boron is an essential minor nutrient for plant growth the requirements are small. Cases of deficiency in agriculture are reported but less is known about the uptake by natural vegetation. Hot water is commonly used for extracting boron although Cartwright *et al.* (1983) recommended a calcium chloride–mannitol solution. However, Farrar (1975) reviewed extraction techniques and found no advantage over hot water refluxing. A simple boiling water method is described here.

Procedure

Weigh 10 g air-dry sieved soil into a soda glass or polypropylene container.

Boil with 50 ml water for 5 minutes.

Filter through No. 44 paper into a polythene bottle.

Include two blank determinations.

Determine the boron in the filtrate as described in Chapter 5.

Notes

1 Pyrex and other borosilicate glass must be avoided throughout except in the last stages of the colorimetric development.

2 If the extract has a high organic content (coloured) then evaporation and low temperature (450 °C) ignition stages must be introduced.

Calcium
Soils vary greatly in their calcium content, but extraction problems are largely confined to those rich in calcium carbonate. As mentioned earlier an extraction with ammonium acetate at pH 9 is one of several alternatives. Ammonium acetate at pH 7 is suitable for most soils but the 2.5% acetic acid should be restricted to non-calcareous soils.

Chlorine
This element is not often determined in soil studies but if required may be extracted as chloride using water as described for nitrate–nitrogen on p. 41.

Cobalt
Most of the extractants mentioned for copper are also suitable for cobalt but in addition 2.5% acetic acid has been used.

Copper
Zinc and cobalt may also be conveniently grouped with copper in this discussion because all three are essential minor nutrients for plants. Plant requirements are very small compared with the major nutrients, but the total content is also relatively low in many soils and particularly so for cobalt. In a few cases it may be high enough to be toxic to some species, particularly if the pH is low. Extractants frequently used in the past included dilute mineral acids (hydrochloric for copper and zinc, nitric for cobalt) which can remove a major part of the total content and occasionally, neutral salts which remove only a smaller fraction of the total. Complexing agents are now widely used. EDTA was the first of these to be tried (Swain and Mitchell, 1960 and Kabata-Pendies, 1963) but DTPA (diethylene triaminepentaacetic acid) in association with triethanolamine (TEA) is often preferred (Soltanpur *et al.*, 1976, Lindsay and Norvell, 1978 and Edlin *et al.*, 1983). Only general recommendations are given here.

Extractants
Extractant strengths are 0.1 M hydrochloric acid, 2.5% v/v acetic acid, 0.1 M EDTA and 0.01 M DTPA in $CaCl_2$ and 0.1 M in triethanolamine (adjusted to pH 7.3).

Notes
1 Brass sieves must not be used in sample preparation.
2 The extractant/soil ratio should not be too tight or adsorption during extraction may become significant with some soil types. 10:1 is generally satisfactory for hydrochloric acid and acetic acid and 3:1 for EDTA and DTPA.
3 A time of extraction (by shaking) of 1 hour is recommended.
4 Blanks are especially important for trace element analyses and three or four should be included in every batch processed.
5 Following extraction the procedures given in Chapter 5 are generally applicable but it may occasionally be necessary to concentrate the element by an organic extraction. The standards should then be made up in the appropriate extractant. In view of the inter-elemental interference that may occur in trace analyses an extraction may be necessary to separate the element being analysed.

Iron
Iron is an essential plant nutrient but levels are so high in most soils that a value for extractable iron has little nutritional significance. However, the uptake of iron has received some attention with relevance to calcareous soils where deficiency symptoms can arise (e.g. iron chlorosis). The use of DTPA as an extractant for 'available' iron has become popular (Lindsay and Norvell, 1969). Sheldrike and McKeague (1975) compared the effectiveness of a number of methods common in the USA for removing adsorbed iron from soil.

Levels of extractable iron are probably of most value to the pedologist. This is because iron–organic complexes (involving fulvic acid or various polyphenols) are relatively mobile and can be carried down the profile to give podzol iron pans in certain soil types. Incipient pan formation is not always easy to detect and the estimation of free sesquioxides provides a convenient way of doing this (Davies *et al.*, 1960; Malcolm and McCracken, 1968).

Many methods have been proposed and most have been based on sulphite, oxalate, pyrophosphate or dithionite extractants. These are considerably stronger than the extractants described earlier for the major nutrients and some attack on the clay structure seems inevitable in most cases. Nevertheless, various workers have isolated fractions of iron including oxides that are of pedological interest. Bascomb and Thanigasalam (1978) compared acetylacetone and potassium pyrophosphate for selective extraction of organic-bound iron from soils.

Hydrosulphite reduction method

One of the best known sulphite methods is that of Deb (1950) who used sodium hydrosulphite with an acetate–tartrate extractant and considered that any attack on the clay structure was minimal. A modified version of the Deb procedure is given below. In this method iron is reduced by hydrosulphite at neutral pH with tartrate added to prevent precipitation or adsorption of iron (II). The subsequent acid treatment dissolves any iron (II) sulphide that may form.

Reagents

1 M Sodium acetate–0.2 M sodium tartrate extraction solution. Dissolve 82 g anhydrous sodium acetate and 46 g sodium tartrate in water, dilute to 1 litre and mix well.

2 Sodium hydrosulphite.

3 Hydrochloric acid, 0.05 M.

Procedure

Weigh 0.25 g dry soil into a centrifuge jar and extract with 50 ml extraction solution and 2 g $Na_2S_2O_4.2H_2O$ (sodium hydrosulphite) for 50 minutes in a water bath at 40 °C.

Centrifuge and transfer the supernatant liquid, using a filter stick, to a 100 ml volumetric flask.

Treat the residue with 20 ml 0.05 M HCl for 10 minutes with frequent swirling, then wash with water.

Transfer the supernatant acid and washings to the flask, dilute to volume and mix well.

Treat blank extractions similarly.

From this point follow the colorimetric procedure using sulphonated bathophenanthroline, as given on p. 108.

Treat standards similarly to sample extracts.

Oxalate method

McKeague and Day (1966) compared oxalate and dithionite extractants for soil classification whilst Bascomb (1968) compared pyrophosphate extractable iron and organic carbon in different soil types. Oxalic acid has long been used in studying pedological processes (Gallagher and Walsh, 1943), although it has a number of limitations (Arshad *et al.*, 1972). Ball and Beaumont (1972) have compared oxalic acid and pyrophosphate extractants for isolating hydrous iron and aluminium oxides as an aid to profile description and soil classification. They outline procedures for both types of extractant. One involving 3% oxalic acid is given here in a slightly modified form.

Extractant

Oxalic acid, 3% w/v.

Dissolve 150 g oxalic acid in water, dilute to 5 litres and mix well.

Procedure

Weigh 2 g air-dry sieved (2 mm) soil into a 250 ml screw cap polythene bottle.

Add 100 ml 3% oxalic acid. Replace cap but screw loosely.

Allow to stand undisturbed for 16 hours.

Screw tight and shake for 1 hour on an end-over-end shaker.

Filter through No. 44 paper.

Include two blank determinations.

From this point follow the bathophenanthroline procedure as given on p. 108. Only a small aliquot will be required. Treat standards similarly to sample extracts.

Notes

1 If the automated bathophenanthroline procedure is used the 3% extract should be diluted at least twenty times to avoid interference from the acid. Many soil types will require a fifty times dilution. Standards and samples must be treated alike. Atomic absorption is less sensitive than bathophenanthroline for iron, but oxalic acid extracts still need to be diluted (ten times) so that the atomizer performs effectively.

2 Because it is a sesquioxide element, there is some interest in measuring aluminium in these soil extracts. Atomic absorption, using a hot nitrous oxide flame, is suitable (see p. 109). A ten times dilution should be used and a check made for possible interference due to iron.

Magnesium

In most soils much of the magnesium is fixed in primary and secondary minerals (micas, amphibolites, illites and carbonates). Plants appear to be relatively sensitive to a slight deficiency or excess of this element, although no specific extractant appears to have met with general acceptance. Magnesium is generally grouped with potassium, calcium and sodium and the ammonium acetate and acetic acid extractions already described are frequently used.

Calcium, aluminium and iron which are also extracted may interfere with the subsequent analytical stages. Methods of suppressing these interferences are dealt with on p. 111.

Manganese

Soil manganese forms a contrast to iron and aluminium in that levels in many mineral soils are quite low and possible limiting in sandy types. In spite of this, relatively little work has been done on its extractable status in the soil although various valency states have been recognized (Page *et al.*, 1982). It has been reported that exchangeable manganese can increase on air-drying (Sherman and Harmer, 1943).

A variety of extractants have been applied to this element though none seem to be exclusively used. Perhaps the phosphate extractants such as ammonium dihydrogen phosphate have been used the most. Ammonium acetate, hydroquinone, EDTA, DTPA and others have been evaluated by many workers including Hoff and Medenski (1958), Hammes and Berger (1960), Nardishaw and Cornfield (1968), Sharpe and Parks (1982) and Sheppard and Bates (1982). These studies indicate that ammonium acetate is the most reliable.

Salcedo and Warncke (1979) considered the factors affecting manganese extractability. In the absence of more detailed studies, ammonium acetate has generally been used in the authors' laboratory partly because it is convenient to determine the common nutrient cations in the same extract. The procedure described earlier for this extractant should be followed.

Molybdenum

The comments given in the section regarding trace elements (copper, zinc and cobalt) are generally applicable to molybdenum. Molybdenum is essential for plant growth, but the requirements are very small. Very few soils contain totals of more than 2 or 3 $\mu g\,g^{-1}$. The amount that is extractable may be as little as 0.01 $\mu g\,g^{-1}$. Molybdenum differs from the other trace elements in being more available to plants in alkaline soils (Davies, 1956). This element is treated separately here mainly because the determination of 'available' molybdenum has come to be associated with one particular extractant; acid ammonium oxalate as introduced by Tamm (1922). Grigg (1953) compared this reagent with four others and found that only oxalate gave results agreeing reasonably with crop response data obtained on his particular soil type. This procedure is outlined below.

Extractant

Ammonium oxalate solution, pH 3.3.

Dissolve 249 g ammonium oxalate and 126 g oxalic acid in water and dilute to 10 litres.

Procedure

Shake 25 g of air-dried sieved soil with 250 ml ammonium oxalate solution for one or more hours on a rotary shaker (see Note **1** below).

Filter through No. 44 paper into borosilicate or polythene bottles and reject the first 20–30 ml.

Include two blank determinations.

From this point proceed as on p. 116.

Notes

1 Grigg (1953) considers an overnight shaking is necessary to attain equilibrium when extracting molybdenum. This point should be checked for the particular soils being investigated.

2 Before using the method given in Chapter 5 it may be necessary to concentrate and also, if the thiocyanate method is used, to remove iron. Proceed by evaporating to white fumes in the presence of perchloric and nitric acids, and then precipitate most of the iron by adding ammonium hydroxide to the diluted digest. If the loss of molybdenum on the precipitate is suspected then it may be separated from iron by an amyl alcohol extraction at the stannous chloride stage (Grigg, 1953). The standards should be taken through this treatment if it is adopted.

Nitrogen

Mineralization of soil organic nitrogen produces three inorganic fractions: ammonium-, nitrate-, and nitrite-nitrogen, all of which can be readily extracted from the soil. A simple salt extraction which takes up free and adsorbed ions is sufficient for ammonium-nitrogen. This will also extract nitrate-nitrogen but if nitrate is required alone then a cold water extraction is acceptable. Nitrite–nitrogen is a transient ion present in low concentrations and there is little value in a separate estimation for most purposes, although an analytical method is given in Chapter 5. Inorganic nitrogen released from organic matter over a period of time (mineralizable nitrogen) is determined by incubation as described on p. 134, although Whitehead (1981) obtained good predictive results by hot potassium chloride extraction. The direct extractions given below measure mineral nitrogen of a given point in time. This is extractable nitrogen. The phrase 'nitrogen availability' which has been used for mineralizable nitrogen is avoided here.

Ammonium-nitrogen

The extractants most favoured for ammonium-nitrogen are sodium or potassium chloride solution. A 6% solution is suitable for either salt although other concentrations in the range 1 to 10% have been used. The concentration is not too critical but the same strength should be used throughout any one investigation.

Fresh soils should be used whenever ammonium-nitrogen is required. Drying will lead to some loss of free ammonia, and even of the fixed ammonia from alkaline soils.

The extracts may pick up ammonia if this escapes into the laboratory atmosphere. Some grades of filter paper may have a measurable ammonia content so a centrifuge may be preferable for separating the soil extract. If this is not possible then No. 1 Whatman filter papers will give the acceptable blank values.

Utilization of mineral forms of nitrogen by micro-organisms is likely in any biological material. Hence it is desirable to carry out extractions soon after collection although analyses may be delayed if the samples are stored at just above 0 °C (not frozen). Alternatively, it may be convenient to extract immediately and then store the salt extracts for a short time at a low temperature. Other preservation treatments are discussed in Chapter 4.

Reagent

Potassium chloride, 6% w/v.

Procedure

Weigh out 25 g fresh or 10 g air-dried sample into an extraction bottle.

Add 250 ml 6% KCl and shake for 30 minutes.

Allow to settle for 30 minutes (protect from atmospheric contamination).

Pipette off a suitable aliquot for distillation (up to 100 ml may be required).

Run blanks with 6% KCl alone.

Determine moisture at time of weighing.

Proceed as for NH_4^+-N on p. 126.

Notes

1 Organic soils may not settle easily and a centrifuge or sinter stick should then be used to obtain a reasonably clear solution. A large aliquot may be necessary for distillation since these soils are often low in inorganic-N.

2 Even analytical grades of NaCl or KCl may have measurable ammonia contents but it is possible to obtain batches with very low levels on special request from some suppliers.

Nitrate (+nitrite)-nitrogen

Since all nitrates present in the soil are freely soluble there is little difficulty in extracting them with water. It is advisable to extract only fresh material. It has been shown in the authors' laboratory that small concentration changes take place on drying some soils although this is not so serious as for ammonia-nitrogen. The main virtue of drying is to stabilize the soil when immediate analysis is not possible. Air drying is preferable to oven drying.

The concentration of nitrates (and nitrites) in the fresh soil (or in water extracts) may soon change if analyses are not carried out shortly after collection. As with ammonium-nitrogen cold storage just above 0 °C will greatly inhibit microbial processes but storage below 0 °C should not be used.

When extractable ammonium and nitrate-nitrogen are required on the same samples the salt extraction already described for ammonium-nitrogen can be used. Nitrate may be determined in this extract directly by using the distillation method or indirectly by colorimetry following reduction to nitrite-nitrogen.

Procedure

Weigh 25 g fresh soil (about 10 g air-dried) into an extraction bottle.
Add 250 ml water.
Shake for about 10 minutes on a rotary shaker and then filter.
Include two blank extractions.
Determine NO_3-N using one of the methods described in Chapter 5 (p. 128).

TREATMENT FOR COLOURED OR TURBID SOLUTIONS

Removal of colour or turbidity may sometimes be necessary for soil extracts (or waters) before a colorimetric procedure is used. Care is needed over the choice of treatment reagents because some of these (e.g. alumina cream and activated charcoal) may also remove nitrate or nitrite ions.

For lightly coloured extracts (or waters) a simple procedure is that of Mackereth (1963), in which the sample solution is shaken with a suspension of aluminium hydroxide and then filtered. This method is less effective for strongly coloured or turbid extracts and waters. An alternative procedure was given by Metson (1956).

Phosphorus

More attention has probably been given to the study of extractable forms of phosphorus than to almost any other nutrient in the soil. Reasons for this include the vital role that it plays in metabolism coupled with the fact that levels even in fertile soils are often no more than adequate. Sources of soil phosphorus, including parent rocks and rain-water are generally low in this element. Up to half or more of the total soil content may be organic and hence this fraction is a key stage in phosphorus cycling in natural ecosystems. Harrison (1985) has stressed that plant uptake in natural systems depends on recycling via organic matter decomposition. Techniques for estimating organic phosphorus are discussed in Chapter 5.

However, phosphorus released by organic breakdown or parent mineral weathering cannot meet the needs of crops on most soils and fertilization is essential. The need to assess the fate of added phosphorus and that released naturally has directed much attention to the various inorganic fractions, the proportions held in solution, labile and solid phases and the relevance of soil processes such as adsorption–desorption, complexation and others. Adsorption and desorption of phosphate by clay mineral and hydrous oxide surfaces have been intensively studied as reviewed by White (1980).

The concepts mentioned earlier underlie many of the chemical methods. The quantity factor or the extent of the labile pool is given by the *L*-value determined by isotope exchange with ^{32}P and related to a growth study. This method is not described here but it has been widely used. Pimplaskar *et al.* (1982) applied it to hill soils and discuss the problems that arose, whilst Munyinde *et al.* (1982) considered sampling aspects of the technique.

In general the conventional extractants measure the intensity factor or extent to which a soil can release phosphate. Major considerations here include the various fractions present and the effect of adsorption–desorption processes. Extractants which release the more soluble forms include 1% citric acid (Dyer, 1894), 2.5% acetic acid (Williams and Stewart, 1941), dilute buffered sulphuric acid (Truog, 1930) acetic acid–sodium acetate buffer (Morgan, 1937) and several others, whilst sodium bicarbonate buffered at pH 8.5 has been widely used for calcareous soils (Olsen *et al.*, 1954). In addition, 0.25 M sulphuric acid has been used for calcium-bound phosphorus (Hanley, 1962). More powerful are the dilute acid–fluoride types (Bray and Kurtz, 1945) which can release some adsorbed phosphate and that bound to transition metals such as aluminium (Chang and Jackson, 1957; Hanley, 1962). Anion exchange resins have also been employed.

These procedures and others have been repeatedly compared in the literature not only with one another but with plant response and with other properties including buffer capacity as described by Holford and Cullis (1985) (also McDonald *et al.*, 1978; Flannery and Markus, 1980). In spite of their limitations many of these extractants continue to be widely used and three of them are described below.

For a survey or preliminary study one of the above extractants will be adequate. When a choice has been made the factors relating to the extraction discussed earlier (p. 32) are all relevant. In particular the use of fresh or air-dry soil can affect the amount of phosphorus extracted. A number of authorities, including Jackson (1958), Grava *et al.* (1961) and others, have pointed out that drying can affect the release of phosphorus compounds. Jackson reports that high temperature drying can double the result and even 40 °C drying can enhance the result by up to 30%. The effect varies with soil type and before a drying treatment is used some checks should be carried out. In some cases the advantages of obtaining a more representative sample might be considered to out-weigh the effects of drying itself.

The extractant/soil ratio (fixed for some, e.g. Truog's reagent), time of shaking and possibility of resorption or fixation during extraction also have a bearing on the result. The resorption factor is more significant for phosphorus than the cations and wider ratios are desirable. This feature is particularly associated with the clay minerals (aluminosilicates). It is allied to the fact that an extractant tends to release the more soluble forms of phosphorus and make it nearly impossible to decide how much of any particular form has been extracted (Cooke, 1951; Askinazi and Guinsbourg, 1957). Some workers consider that a value for extractable phosphorus has little meaning unless accompanied by a measure of the capacity of the soil to adsorb phosphorus from a standard solution (Bache and Williams, 1971).

Truog's extraction

This widely used extractant is suitable for all but the most calcareous soils.

Extractant

Truog's reagent: 0.001 M H_2SO_4 buffered at pH 3.

Dilute 200 ml 0.05 M H_2SO_4 with water to 10 litres.

Add 30 g $(NH_4)_2SO_4$ and mix well.

Procedure

Weigh 2.5 g air-dried sieved soil into a polythene bottle.

Add 500 ml Truog's reagent.

Shake for 30 minutes on a rotary shaker.

Filter through a No. 44 paper.

Include blank determinations as before.

Determine moisture at the time of weighing.

Proceed as described for phosphorus on p. 135.

Note

Since the amount of extractable phosphorus increases as a result of drying, it might be considered preferable to extract the fresh soil. In this case it is recommended that double the sample weight be taken. For wet, peaty soils an even larger weight (25 g) will be needed.

Acetic acid extraction

This is less widely used for phosphorus than some extractants but has given useful data particularly for sandy soils.

Extractant

Acetic acid, 2.5% v/v.

Dilute 250 ml glacial HOAc to 10 litres and mix well.

Procedure

Extract as described earlier for the mineral nutrients (pp. 33 and 34), but taking 5 g soil with 200 ml extractant.

Include blank determinations as before.

Determine moisture at time of weighing.

Proceed as described for phosphorus on p. 135.

Notes

1 Refer to the note under Truog's extraction (p. 43).

2 The procedure used for cations with a 25:1 ratio is less suitable for phosphorus except for sandy soils.

3 Acetic acid must not be used as an extractant for calcareous soils.

Olsen's extraction

The pH of this extractant (8.5) makes it appropriate for chalk and limestone material. Some organic matter is also extracted, but organic phosphorus will not be estimated by the recommended colorimetric procedure. Constant temperature conditions are essential for this extraction (Brar and Bishnoi, 1987).

Extractant

Olsen's reagent: 0.5 M $NaHCO_3$ buffered at pH 8.5.

Dissolve 210 g $NaHCO_3$ in water in an aspirator and add 100 ml M NaOH.

Dilute to 5 litres and mix well. Check that the pH is 8.5 ± 0.05.

Procedure

Weigh 5 g air-dry sieved soil into a polythene bottle.

Add 100 ml Olsen's reagent.

Shake for 30 minutes on a rotary shaker.

Filter through No. 44 filter paper.

Run blank extractions as before.

Determine moisture at time of weighing.

Proceed as described for phosphorus (p. 135).

Notes

1 The note under Truog's extraction applies, with the exception that Olsen's reagent is not recommended for peaty soils.

2 The extract must be neutralized with 10% v/v H_2SO_4 (0.1% nitrophenol in alcohol as indicator) before colour development in the molybdenum blue procedure.

Potassium

Although potassium is present in relatively large amounts in most soils, its soil chemistry is complex and most of it is fixed in forms not available to plants. Clay minerals are very significant in fixation and release of this element. Richards and McLean (1963) and others have shown that these effects are directly related to the extent of drying prior to extraction and for some purposes an extraction of fresh soil might be desirable. A general review of fixation was given by Agarwal (1960). A more theoretical view of soil potassium was given by Beckett and Tinker (1964) who discussed the ratios of the activities of soil cations as a measure of the labile potassium. Various extractants have been advocated (Vazhenin and Karaseva, 1959) but the usefulness of many seems to be limited. Ammonium acetate appears to be the most widely accepted for exchangeable potassium and 2.5% acetic acid gives comparable values. Both will provide the ecologist with useful information initially. The extraction procedures given earlier in this section (pp. 33–34) should be followed.

Other approaches include that of Singh *et al.* (1983), who compared a number of methods and found that a constant boiling hydrochloric acid solution provided the most satisfactory measure of the potassium likely to become available in a growing season.

Sodium

Only small amounts of this element are present in non-saline soils and it is readily soluble. This, coupled with the fact that it is not very important for plant nutrition, means that the choice of extractant is not critical. The ammonium acetate and acetic acid procedures described earlier in the section can be used.

Maritime and other saline soils form a separate group. Estimations of extractable sodium are then needed to indicate saline penetration. 2.5% acetic acid or even water is suitable for most circumstances but weak hydrochloric acid has also been used.

Sulphur

Most soil sulphur is organic in nature although it is not often determined in this form. Mineral sulphur is largely sulphate, except under reducing conditions where sulphides and sulphites may be present. Although most of the common metal sulphates in soils, apart from calcium, are largely soluble, many soils have a marked capacity to adsorb sulphate. Williams and Steinbergs (1964) considered that adsorbed sulphate was available to plants. Sulphate deficiencies in plants are relatively rare, partly on account of atmospheric depositions, and extractable values are less commonly required in soil studies.

Phosphate-based extractants are popular because they displace both the readily soluble and adsorbed sulphate which is available for plant uptake. Probert (1976) found reagents containing $Ca(H_2PO_4)_2$ to be the best extractants as he managed almost full recovery of $^{35}SO_4^{2-}$. The turbidity method (Chapter 5) can be used to determine the sulphate in the 0.01 M extractant, although either a colour blank or a clean-up stage, using washed charcoal (0.2 g), will be required to remove the background colour. Potassium dihydrogen phosphate (0.0164 M) is preferable for the ion chromatography estimation of sulphate as the divalent calcium ions tend to accumulate on the column and disrupt the equilibrium of the system.

Zinc

The methods discussed under copper are also suitable for zinc. Sedberry *et al.* (1979) evaluate chemical extraction methods which are specific to zinc.

3
Analysis of Vegetation and other Organic Materials

An examination of the chemical composition of biological materials is frequently needed in ecological investigations, especially those concerned with nutritional requirements. Analyses for diagnostic purposes prior to the application of fertilizers are generally confined to agriculture. The elemental composition varies greatly within the plant and animal kingdoms (see Appendix I). In addition considerable variation occurs according to age and component and, in the case of plant materials, the period of the year the sample was collected.

The methods described here were developed for the determination of plant nutrients and other constituents present in the concentration ranges likely to be encountered in nature. They are intended primarily for vegetation samples but are also broadly applicable to other biological materials high in organic content including litters, peat and some animal components. However, some of these materials may be more resistant to oxidation than vegetation and some of the digestion stages may need to be altered or prolonged. Material rich in oil, resinous and fatty substances, such as animal tissues, conifer products and plant seeds, in particular, will require modified digestion techniques. This problem can often be alleviated by taking smaller sample weights.

The presentation within the section basically follows that used for soils. In the first part, the collection and initial treatment of the samples is considered, leading to the descriptions of the simpler general tests. Finally the preparation of the sample solutions, for later analysis of individual elements using the procedures of Chapter 5, is described. Organic compounds are dealt with separately, in Chapter 6.

Books dealing with plant analyses include Piper (1950), Chapman and Pratt (1961), Walsh and Beaton (1973) and Nicholson (1984). Information about analytical variability as shown through inter-laboratory tests is given by Bowen (1966), Watson (1981) and by Sterrett *et al.* (1987) and the use of reference standards in plant analysis is discussed by Alvarez (1980) and Daniel (1980).

Sample collection and initial treatment

It is not always appreciated that in general field variation of biological material greatly exceeds any introduced during analysis in the laboratory. It is essential to have an estimate of the field variation if the final data are to serve the objectives of the study. This may necessitate taking more samples in the field than can readily be analysed by the available facilities. These problems were considered more fully in the corresponding section on soil sampling (p. 7) and the appropriate statistical techniques are discussed in Chapter 9.

At the time of sampling any relevant site data should be recorded. These might include vegetation density, height and form of growth, associated species, nature of soil, aspect, drainage and local topographical features. The national grid reference should always be noted so that climatic and geological features can be included later if required.

Occasionally an analysis of soil fauna, small invertebrates (including aquatic) and even birds and small mammals may be required, but the problems of sampling animal populations are not considered here. Some of the statistical principles were discussed by Elliott (1971) with particular reference to aquatic fauna.

Choice of component

The inorganic content of different parts of a plant tends to vary considerably. An example is given in Appendix I for *Quercus petraea*. In general the richest tissues are those where metabolic activity is greatest, though this does not necessarily apply to every constituent. For most plants the richest organs are the shoot apices and leaves and possibly the stems if photosynthetic. Woody tissues generally have much lower concentrations but because of their bulk may contain a large proportion of the total elemental content of the plant. Roots may accumulate heavy metals in polluted areas. Differences within similar tissues occur and most though not all nutrient concentrations are higher in younger tissues. Changes with age are discussed below under seasonal variation.

The choice of tissue for analysis is largely determined by the type of plant and purpose of the study. For small herbs and grasses it is often sufficient to sample the entire aerial growth, but larger species may require leaves or other tissues to be specified. Some stipulate that for monocotyledons, the leaf blade above the sheath junction should be examined. The sampling of grassland communities was discussed by Milner and Hughes (1968). For woody species the current years growth of leaves is often a sufficient indicator of nutrient status although data from whole leaves may not be ideal for a given study. Some workers have separated petioles and blades before analysis (Haines *et al.*, 1979; Bertoni and Morard, 1982). Others prefer to analyse green shoots, particularly when the leaves are relatively small as with *Calluna vulgaris*. Specific problems arise in sampling tree leaves, where nutrient levels have been reported to vary between sun and shade leaves and with crown position (Maclean and Robertson, 1981). In general the interpretation of data from foliar analysis of woody plants is not straightforward. This has been discussed by many workers, including Wehrmann (1963), Acquaye (1967), Lamb (1968), Van den Driessche (1974), Insley *et al.* (1981) and Bowen and Nambiar (1984). Some aspects are mentioned briefly below. Other tissues of woody plants which have been analysed include bark (Godfrey-Sam-Aggrey and Garber, 1979), xylem sap (Stark and Spitzer, 1985) and phloem (Hetherington and Owens, 1979). In the case of conifers, in general only needles are taken for analysis.

The choice of tissue for organic analyses is determined by the objectives of the study. However labile substances can be a problem and some plant (and animal) species or particular tissues decay rapidly when they are no longer living organs. In extreme cases it is worth making special arrangements to carry out tests in the field immediately after sampling. However, in most cases, this is not practicable.

Freezing or cooling in the field is a possibility. Liquid nitrogen can be taken into the field in suitable vacuum containers and the sample rapidly frozen by plunging into the nitrogen. Alternatively solid carbon dioxide may be used and is easier to handle.

In most cases the most feasible approach is to transport the sample back to the laboratory as rapidly as possible, preferably surrounded by previously frozen 'cold-packs'. In some cases it is acceptable to kill the tissue by plunging into boiling 80% ethanol which may also function as an appropriate extractant for subsequent organic analysis.

Sampling time

Fluctuation in nutrient content with time occurs in all active plant tissues and apart from their physiological significance these changes must be taken into account when sampling. The problem is best illustrated by reference to tree leaves for which most data are available.

Diurnal variation
Transference of the carbohydrate products of photosynthesis occurs daily, but concurrent changes in the elemental nutrients seem to be relatively slight. Nevertheless, Mitchell and Chandler (1939) and others considered the mid-day period to be the most suitable sampling time for trees.

Seasonal variation
Seasonal changes in concentration result mainly from a movement of nutrients into a component during growth and the reverse process when senescence approaches although individual nutrients differ in their mobilities. These changes are most evident in photosynthetic tissues such as leaves. Translocation out of leaf tissues affects nitrogen, phosphorus and potassium in particular whilst the less mobile elements such as calcium tend to be retained and even increase in apparent concentration as the leaf becomes older. Although changes of this nature vary from species to species the following trends are typical of leaves of many deciduous species in temperature climates:

1 N, P, K, S—early spring peak, summer relatively steady, autumn decline (September subsidiary peak sometimes reported);
2 Ca, Si, (B)—little initial change then rising more sharply as autumn approaches;
3 Fe, Al, Zn, (Mn)—little initial change, then increasing to an early autumn peak (usually) with a sharp fall later;
4 Mg, (Mn)—trend ill-defined but a tndency to a slight increase during the growing season.

The trends for other minor nutrients are less well-defined although most follow the pattern of **3** above. The total dry biomass of deciduous leaves increases to a steady maximum from July to September then falls in the autumn. Although it is difficult to recommend one sampling period for all circumstances, the period including the end of July and early August is an acceptable compromise. Different sampling times for different elements are seldom practical.

Seasonal changes in green stems follow those of leaves but are more muted whilst woody stems and particularly older tissues show little seasonal pattern. Less data is available for roots although it is well known that in a few species potentially toxic nutrients such as copper and zinc may accumulate in roots rather than be transported in excess to sensitive shoots. This mechanism can allow metal-tolerant ecotypes of plants such as *Agrostis spp.* to survive in contaminated soil.

In conifer needles the concentrations of many elements tend to increase during the year and for these an autumn sampling or even early winter is often preferred. Needles of differing ages, however, show differing seasonal patterns, so it is important for the sample to represent a full years growth. Some of the complexities of sampling pine foliage at different times of the year are illustrated for nitrogen by Smith *et al.* (1970).

Inter-year variation
Little data has been published on year-to-year changes but there appear to be only relatively minor differences in the concentrations of major nutrient elements. Any fluctuations that occur will be influenced by the prevailing climate and, in particular, the onset and length of the growing season.

Further information about variation in chemical composition may be obtained from Mason (1958), Bould *et al.* (1960), Guha and Mitchell (1965, 1966), and Rodin and Bazilevich (1967).

Methods of sampling
Ecologists generally use the type of equipment employed in forestry and agriculture. High-level pruners are suitable for sampling tree leaves and small branches although thicker stems will require a saw. Incremental corers enable stem samples to be obtained. Shears and secateurs are suitable for grassland and similar herbage, and are essential if the full aerial growth is to be taken and biomass recorded. It is advisable to wear polythene gloves when handling the samples especially in warm weather when perspiration is likely. The sampling of roots raises practical problems because of the need to separate from soil particles which can be tedious. Root washing machines which depend on the cleaning action of fine jets have been constructed. However, the chemical content of roots is often of significance in nutritional studies and should not be overlooked.

Sample transport and storage

Treatment of the sample during transit and storage is required to minimize any deterioration and the use of cold-packs has been mentioned. Where labile constituents are involved it is essential to maintain the samples in a condition as close as possible to that in the field.

Vegetation

The details given on p. 10 concerning methods of keeping samples cool during transit also apply to vegetation. Where this is not practicable the sample should be rapidly transported in porous containers, keeping the temperature to a minimum. Although Edwards (1965) showed that storage of fresh material for several hours in a polythene bag was acceptable, prolonged storage should be avoided, particularly at room temperature, because enzymic reactions may lead to organic breakdown. Loose packing should always be used for the same reason. Steyn (1959) examined this problem and suggests the use of porous cloth bags for plant materials. Bradfield and Bould (1963) found that storage of loosely packed fresh vegetation in paper bags from two to four days caused little change in mineral composition. However, in critical cases it may be necessary to freeze the sample in the field using solid carbon dioxide (in acetone) or liquid nitrogen and transport in vacuum containers.

If labile consituents have to be determined it is always preferable to have the material analysed without delay. Failing that, the fresh plant material should be kept at −10 °C and later freeze-dried. In some cases it may be preferable to extract or treat the fresh sample to give a more stable product which can withstand storage in cool conditions.

When the interest is in proximate organic constituents or mineral nutrients rapid air-drying is advisable. Procedures are discussed on the next page. It is generally preferable to grind the sample at this stage. Air-dried ground plant material can be stored for long periods at room temperature in well ventilated conditions. In contrast the slightest decomposition may affect biochemical analyses and it is then preferable to kill tissue in the field by plunging into boiling 80% alcohol, or preferably liquid nitrogen.

Organic soils

Organic soils will normally be treated in the same way as mineral soils. In some cases, however, such as surface litter, the sample is more akin to vegetation and may need to be treated as such, particularly for a total nutrient analysis.

Animal material

Many of the above comments also apply to animal tissues and products with the exception that this material is even less stable and rapid freezing is preferable in many cases. Freeze-drying in the laboratory enables carcass samples to be handled more easily. Sterilization additives such as formalin are sometimes used, but if the material is intended for chemical analysis there must be a check on analytical interferences or contamination from such preservatives.

Washing

Plant material which is contaminated with soil particles or dust may be cleaned by washing quickly in water or weak detergent solution (see p. 52).

Bulk reduction

Fresh and often bulky samples collected in the field frequently have to be reduced in volume before initial laboratory treatment. Their state is generally too heterogeneous for a small representative sample to be taken and the high moisture content sometimes makes them inconvenient to handle. Nevertheless the use of fresh material may be unavoidable when labile compounds have to be examined. The classical technique of heaping and quartering with retention of opposite quarters may be used but this technique has its limitations with coarse material. Woody material or dry litter may be handled by chopping to a more homogeneous state using a small chaff cutter or a guillotine which should precede any further quartering. Large mincers are sometimes used for cutting fresh material and are particularly suitable for animal tissue.

When samples are intended for mineral analysis, initial air or oven drying followed by grinding should precede bulk reduction. Even after grinding, size or density bias sometimes occurs and further sample reduction may be required. Successive mixing and quartering is then satisfactory but the mechanical sample reducers suggested for soils will minimize the human factor in this operation.

Drying

When determinations of labile compounds are not required it is normal to dry samples since they are then convenient to store and with subsequent grinding will produce a homogeneous sample suitable for analysis. No ideal time or temperature of drying can be recommended because any significant loss of moisture from a sample will almost certainly be accompanied by a slight loss of volatile constituents, and by chemical changes within the sample (Melvin and Simpson, 1963). Mayland *et al.* (1978) report dry matter losses for some vegetation. This is probably not significant for most species and mineral elements are not likely to be affected. However, elements which are readily oxidized, such as sulphur, may be more susceptible to loss (Grundon and Asher, 1981). Boron and selenium may similarly be affected. Furthermore, if proximate organic analyses are required, drying conditions must be controlled carefully and for many individual organic substances drying may have to be avoided altogether.

In view of this, the choice of drying conditions will be governed by the investigation being made. Two of the more common treatments are mentioned here. Some workers prefer to use a relatively low oven temperature (35–40 °C) and warm only until the sample is sufficiently dry to be ground. This works well if the sample is thinly spread and air circulation is good. With bulky samples, however, this process may be prolonged and enzymic or microbial changes can occur in the early stages. For this reason others prefer to dry rapidly at a higher temperature and 80 °C is often used, although other temperatures up to 105 °C are sometimes favoured. Drying at high temperatures for long periods (over 24 hours) can result in significant losses in total weight so the shortest possible drying time is always advisable.

Loss of tissue fluids and incipient charring may sometimes occur and temperatures from 60 to 80 °C have been proposed for agricultural materials to minimize such effects (Hughes, 1979). For mineral nutrients and even nitrogen there is probably little to choose between the above drying temperatures, but if proximate or other organic analyses are required the temperature should not rise above 40 °C. Samples dried at less than 40 °C may be loosely regarded as 'air-dried'. True air-drying at room temperature is not to be recommended because if the material is very damp, drying will be prolonged and the metabolic changes referred to above will be accentuated particularly in the early stages. Specialized techniques such as vacuum freeze-drying are also available and may be needed for particular studies. It should not be overlooked that any subsequent use of heat on freeze-dried material may render the sample particularly susceptible to differential losses when later ground (Grassland Research Institute, 1961).

In general the oven techniques discussed above are the most convenient for handling large numbers of vegetation samples (Greenhill, 1960).

Materials such as faeces are sometimes analysed by the methods used for plant materials but this type of sample is particularly vulnerable to changes on drying when a significant loss of labile nitrogen may occur (Sharkey, 1970).

Macerating and grinding

Only in special cases, for example for a physiological study, is it necessary to analyse individual leaves or other tissues. For most nutrient work the need to obtain a representative sample requires that a large bulk of the plant material has to be collected and, after drying, it must be finely ground. A small sub-sample can then be abstracted. An assessment of the need to grind foliage before chemical analysis, was made by Salonius *et al.* (1979) with particular reference to conifer material.

The fineness of grinding is an important consideration. It is governed by the weight of sample required for analysis and the need for this to be representative of the whole material. Opinions differ on the optimum relation between particle size and minimum sample weight, but if the methods recommended in this book are used it is generally adequate to grind to pass a mesh size of 0.5 mm (BS 35). This is the smallest mesh provided with most grinders, but in practice most of the particles are broken down to a smaller size. However, some plant materials consist of long fibrous particles which are not easily broken down to a homogeneous sample. It will be necessary to use a vibration or ball mill to shatter these materials and lengthy milling may then be needed to produce a fine powder.

The possibility of plant segregation during milling has to be taken into account. Lockman and Molloy (1977) demonstrate the importance of thoroughly mixing all the sample being ground, not neglecting the final residue left in the grinder. Particle fractionation can occur even during storage of the ground sample in containers so it is recommended that the ground material be mixed, say with a spatula, before abstracting sub-samples for further analysis. Other problems encountered during plant comminution are discussed by Samsoni (1975).

Fresh material

Machines for cutting up fresh material are variously called macerators, homogenizers or emulsifiers. A blender is an example of a macerator where rapidly rotating knives cut up the tissue into fine particles. Water or an organic solvent must be present and afterwards removed or recovered for analysis. Emulsification differs only in degree from maceration. Samples are broken down into finer particles and the resultant emulsion may be sufficiently homogeneous for samples to be taken directly from it for analysis. Alexander (1969) describes an ingenious homogenizer suitable for producing fine suspensions of fresh material.

Pathological tissue homogenizers which depend on the shearing action of a pestle in a tube are suitable for soft specimens, especially those of animal origin, and are relatively cheap to purchase. However, they are less suitable for hard and fibrous materials and are also subject to overheating. The least harmful way to homogenize fresh material is to grind after cooling under liquid nitrogen. Suitable equipment for this purpose is commercially available.

Dried material

Three types of grinder are in widespread use and their principal features are considered here. Characteristics of these and other types of mill, more suitable for soils, are summarized in Table 2.1.

Beater cross (hammer) mill

Examples of this type include the Christy and Norris Mill and the smaller Glen Creston Mill. They depend on the action of revolving beater bars to shatter the material inside a grinding chamber where the material circulates until it is fine enough to be blown through a sieve. The main advantages seem to be that large quantities of material can be handled especially in the former model where harder woody tissues may also be ground if suitably broken down or splintered first. There are certain disadvantages which should however be noted:

1 There is some risk of overheating the sample though this is not likely to be serious for mineral nutrient analysis. A differential loss of material as fine dust may occur. Bailey *et al.* (1957) reported losses of high nitrogen content particles through the pores of the bag which some models employ. It has already been mentioned that freeze-dry samples are susceptible to this type of loss. Special devices incorporating fine nylon gauze have been designed to reduce loss of dust (Kretschmer and Randolph, 1954). See also Palm and Beckwith (1956).

2 There is some risk of contamination with iron, manganese, copper, etc. depending on the metals employed in construction.

3 The larger models tend to be noisy, dusty and tiring to use over long periods. These effects can be minimized by correct siting and provision of adequate insulation and dust extraction facilities.

Cutting mill

In this type of mill a set of knife blades rotate in opposition to another set fixed round the edges of the grinding chamber. The main advantage is a high recovery of material making them suitable for small samples. There are some disadvantages to consider:

1 Large samples cannot be handled in bulk.

2 They are not suitable for harder tissue such as wood.

3 Metal contamination has been reported though this is probably less serious than in the other mills mentioned here.

Ball mill

In this type of mill the material is ground by the impact action of balls of metal, agate, tungsten carbide, porcelain, etc. rolling round a grinding chamber. The grinding action is as effective as with the other types and sample recovery is high, but contamination is likely to be more serious. Metal balls present an inherent difficulty as may be expected, but flint, porcelain or even agate balls (Hood *et al.*, 1944) can also provide their quota of elements. Ball mills do have the advantage that the entire sample can be recovered. Ring or swing mills (described on p. 13) also depend on impact shattering but they are more appropriate for harder materials such as soil.

Storage of dried, ground materials

First it may be necessary to reduce bulk by subsampling after grinding. In this case some of the recommendations already given (p. 49) may be applicable. The amount required for the different tests should be estimated and then at least as much again should be retained to allow for any repeat determinations.

The dried and ground samples should be stored in well-sealed containers to reduce any uptake of water vapour. Only a brief period of heating will then be needed before weighing for analysis. Storage in a dry warm room is desirable to avoid attack by fungal and other spores which are possible in damp conditions. Glass, metal or plastic containers can all be used for storage. For short periods it may be necessary to use polythene or lined-paper bags but the possibility of vermin attack should not be overlooked when using these materials.

The precautions needed to store fresh materials have already been mentioned (p. 49).

Sources of contamination

It is convenient to discuss this topic briefly here although other sources of contamination are mentioned at relevant places in the text. In this context the term contamination covers extraneous material which could invalidate the actual result.

Detecting possible sources of contamination in the field presents as great a problem as their elimination. Inclusion of the entangled growth of other species, the presence of animal droppings, films of insect products and soil splash are some examples of contamination. The separation of species is usually straightforward but can be tedious. Soil contamination is probably the most serious since its chemical composition is so different from plant material and it is very difficult to eliminate (Thompson and Raven, 1955). Herbaceous vegetation growing close to roads is likely to be affected by dust and in winter by salt spray. However, Guha and Mitchell (1965) using the detergent cleaning technique referred to below showed that surface contamination of tree leaves even near a main road was negligible.

As previously mentioned, for gross surface contamination washing may be necessary, but should be undertaken with care. Prolonged washing may remove large amounts of nutrient elements, notably potassium. Careful sponging with pure water is advisable. Steyn (1959) showed that the effectiveness of this procedure was enhanced by adding a mild detergent (about 0.01%) to the wash water. More recently Sonneveld and Van Dijk (1982) have looked at the effectiveness of some washing procedures for removal of contaminants. The loss of nutrient elements using this procedure is negligible. It may be necessary to check that any detergent used does not contain the elements to be analysed. Quick rinses in sonic cleaners have also been used but there is some danger of tissue disintegration by this method.

The possibility of soil contamination resulting from rain, vehicle splash or from 'animal puddling' in wet conditions is often overlooked. In extreme cases 30% or over of the sample weight of a prostrate species can be due to soil. In contrast, sampling after long periods of dry weather should be avoided because of the accumulation of wind-blown soil, and dust from urban and industrial locations.

Titanium, which is high in soil and very low in plant material is a useful contamination indicator. Cherney and Robinson (1983) and Carey *et al.* (1986) look at methods for detection and correcting for soil contamination based on the determination of titanium.

There are many other sources of field contamination. These include:

Wind blown fertilisers; even on what appears to be undisturbed land there may have been past application of lime or other chemicals.

Animal excrement; this may be suspected from unusually high nitrogen and phosphorus results and may be indicated in the field by patches of vigorous growth.

Metal contamination from nearby structures (e.g. copper from overhead power cables or zinc from wire used in experimental cages).

Corroded sampling tools and metal sampling containers.

Hot weather during the sampling period may result in perspiration affecting the plant samples and giving high sodium values. Polythene gloves should be used in these conditions.

Contamination problems are not usually so serious for animal materials, although some of the above remarks apply.

The possibility of metallic contamination during grinding has already been mentioned in the section dealing with grinding equipment and is also dealt with in the soil section. In most laboratories there are many background sources of contamination which must be taken into account. Some of these are listed in Appendix III.

Preliminary tests

A number of preliminary tests are usually carried out on samples of plant material soon after collection. It is convenient to consider these tests together.

Moisture

The method generally used for determination of moisture in plant materials is measurement of the loss in weight due to drying at a temperature just over 100 °C. The important factor in thermal drying is the maintenance of a differential between the vapour pressure of the substance to be dried and that of the atmosphere of the drying chamber, as indicated by Willets (1951) and Geary (1956). As with soils, the position is further complicated since some of the water is held in tightly bound forms in the tissue which can only be removed by increasing the temperature. This leads to other volatile components being driven off on heating. In practice a compromise is necessary and a drying time is specified which will bring the material to constant weight whilst leaving minute traces of tightly bound water. For careful work this should be found by prior investigation for a particular material and for given conditions of drying.

In spite of the difficulties the overall simplicity of gravimetric drying is a great advantage and under controlled conditions can give satisfactory results. The technique has been investigated by Reith *et al.* (1948) and Mossel (1950).

There are other methods for the determination of water. These have been described and reviewed by Paech and Tracey (1955), Von Loesecke (1957) and by Kent-Jones and Amos (1967). The most suitable of the alternatives is the Dean and Stark procedure, in which water is distilled over with a lighter immiscible solvent, and the Karl-Fischer method (Mitchell, 1951 and Kolthoff and Elving, 1961).

Biological materials examined in the laboratory are usually in one of three states of dryness:

1 Fresh (immediately after sampling).

2 Air-dry (after drying to room temperature equilibrium or in an oven at 25–40 °C).

3 Oven-dry (after drying at 105 °C until a further weight loss is minimal).

Moisture determinations are required for the first two materials. The same basic procedure is used for both the fresh and air-dry state but to ensure representative conditions it is necessary to take a larger sample weight for the fresh determination. About 1 g is adequate for air-dry samples after grinding.

Procedure

Dry a suitable container, cool in a desiccator and weigh.

Thinly spread the sample in the container and weigh rapidly.

Dry in an air-circulation oven at 105 °C to constant weight (3 hours for 1 g air-dry, ground samples, but longer periods for 5–10 g fresh materials).

Cool in a desiccator and weigh.

Calculation

$$\text{Moisture (\%)} = \frac{\text{loss in wt on drying (g)}}{\text{initial sample wt (g)}} \times 100$$

Since the air-dry moisture is frequently used for later correction of results to dry weight basis it is more convenient to calculate % dry matter as follows:

$$\text{dry matter (\%)} = \frac{\text{oven dry wt (g)}}{\text{initial sample wt (g)}} \times 100$$

Ash content

Total ash determinations for plant and similar materials are often required. The term ash refers to the residue left after combustion of the oven-dried sample and is a measure of the total mineral content. For organic soils it is generally more useful to calculate the weight lost on ignition since this is an approximate measure of the organic content (see p. 95).

The principal errors in ashing arise through losses resulting from the use of too high a temperature and from incomplete combustion when the temperature or time allowed are insufficient. 550 °C is used here as a compromise temperature.

Procedure

Pre-heat a crucible in a muffle furnace to about 550 °C.

Cool in a desiccator and weigh.

Transfer approximately 1 g of sample to the crucible and weigh.

Dry at 105 °C and weigh (the sample left from the air-dry moisture determination may conveniently be used here).

Place the crucible containing the dry sample in a cold muffle furnace and allow the temperature to rise to 550 °C.

After 2 hours at 550 °C remove, and allow to cool and then transfer to a desiccator.

Weigh when cool.

Calculation

$$\text{Ash (\%)} = \frac{\text{ash wt (g)}}{\text{oven dry wt (g)}} \times 100$$

Notes

1 The presence of blackened particles may indicate an incomplete combustion, although some care is needed here as samples high in manganese may give a dark residue on ashing.

2 It is important for the temperature to rise slowly to prevent losses if the samples suddenly catch fire.

Silica-free ash

The ash content is sometimes required on a silica-free basis. This is obtained by extracting the ash with hydrochloric acid to remove all the minerals except silica. This method is described below.

An alternative method is to warm the ash with hydrofluoric acid to volatilize the silicon as silicon tetrafluoride leaving a silica-free ash in the crucible. Further details of this method are given by Maxwell (1968).

Reagents

1 Hydrochloric acid, 10% v/v.

2 Hydrochloric acid, 25% v/v.

Procedure

Follow the ashing procedure given above.

Add 5 ml of 10% HCl to the residue left in the crucible after ignition.

Evaporate to dryness on a water bath.

Add a further 0.5 ml 10% HCl and repeat the evaporation.

Add 5 ml of 25% HCl.

Cover with a watch glass and digest for 30 minutes on a water bath.

Filter through a 9 cm No. 44 Whatman filter paper, transferring all the residue to the filter.

Wash well with warm water.

Transfer the filter to the original crucible and ash for 1 hour at 550 °C.

Cool in a desiccator and weigh.

Calculation

Silica-free ash (%) =

$$\frac{\text{loss in wt after acid extraction (g)}}{\text{oven dry wt (g)}} \times 100$$

Solution preparation

The preparation of sample solutions suitable for elemental analyses is less difficult for plant and similar materials than for mineral soil. An oxidation process is necessary for the destruction of the organic matter, involving combustion (dry-ashing) or acid oxidation (wet-ashing) before a complete elemental analysis can be carried out. Analysts often hold strong views about which method is preferable. Criticisms of wet ashing focus on the expense and safety of the oxidation reagents and the possibility of contaminants being introduced by the reagents. In the case of dry-ashing it has been claimed that the method is slow and subject to losses through volatilization, incomplete combustion or retention on crucible walls. Much contradictory data has been produced showing that low recoveries or variable results may be obtained by both processes. However, as Watling and Wardale (1977) have demonstrated both methods are capable of giving good recoveries and consistent results. Examples of both methods are described in this text and the final choice depends on local circumstances. An important consideration is the need to determine different elements in the same sample solution and wet ashing is generally more satisfactory in this respect. The destruction of organic matter prior to elemental analysis is discussed by Gorsuch (1970). Giron (1973), Watling and Wardale (1977) and Prasad and Spiers (1978) have all compared dry ashing and wet ashing techniques.

It is possible to bring many of the mineral elements into solution without complete oxidation of the plant material. The sample must be ground very finely and a long extraction time is needed. Leggett and Westerman (1973) recommend the use of trichloracetic acid under these conditions for bringing magnesium, potassium, sodium and zinc into solution and, in some cases, calcium as well. Another reagent that has been proposed for this purpose is 'Soluene', a product sold for tissue disintegration prior to liquid scintillation counting. This is a quaternary ammonium base dissolved in a hydrocarbon solvent. It is efficient for animal tissue, but its value for plant materials is more limited. Jackson *et al.* (1972) describe its use prior to atomic absorption determination of trace metals.

Extraction procedures of the type described for soils are rarely needed for plant materials, but the methods given in Chapter 2 may easily be adapted for special investigations (e.g. root exchange phenomena). Crooke (1964) describes a method for the determination of the cation exchange capacity of plant roots.

Organic soils will normally be extracted as described for soils but if a total nutrient content is required the material should be dried and ground and the solution prepared as if it were plant material.

Moisture content of sample

The air-dry sample will reach equilibrium with the moisture of the atmosphere during storage and can be weighed directly for the digestion or ashing treatments. However, the moisture content of the sample must be determined so that a dry weight correction can be applied to any subsequent analytical data (see p. 53).

Alternatively the air-dry ground material can be dried in a glass sample tube before weighing to eliminate the need for a correction. This can be carried out at 105 °C for 3 hours followed by cooling in an efficient desiccator. Keep the tubes stoppered until samples are weighed.

Dissolution procedures

Dry-ashing

This method involves complete combustion of all the organic matter followed by the dissolution of the mineral constituents using hydrochloric acid. The combustion is preferably carried out in a muffle furnace which allows greater control than is possible using gas burners.

The method has many advocates and is attractive because large samples can readily be handled and reagent blanks are generally low so that it is advantageous for trace element work. Nevertheless many workers have reported low recoveries of specific elements. Most of the basic work on dry-ashing was carried out 20 or more years ago. The earlier work was reviewed by Middleton and Stuckey (1953) and other discussions include those of Gorsuch (1959), Society for Analytical Chemistry (1960), Chapman and Pratt (1961), Doshi (1969) and Isaacs and Jones (1972).

The literature contains contradictions regarding alleged losses mainly because the factors controlling the volatilization and retention mechanisms which lead to low recoveries are not always assessed. The most important factors include the temperature and time of ignition, the composition of the sample and the structure of the crucible.

All non-metals are potentially subject to volatilization losses which, for mineral elements can be reduced by ashing at the minimum temperature required to ensure a reasonably complete combustion. Temperatures not exceeding 550 °C should be used and in the procedure given below, 550 °C for 2 hours is suggested. Many workers prefer to ash at 450 °C, but some materials may not give a clean ash even after several hours at this temperature. For estimations of non-metals generally it is further necessary to raise the base status of the sample and so inhibit the formation of volatile acidic constituents of the element concerned. In this way sulphur (by addition of magnesium nitrate or acetate), chlorine (calcium oxide) and boron (sodium hydroxide or calcium oxide) can be estimated quantitatively. Reports differ for phosphorus because base-rich samples may give a full recovery at normal ashing temperatures. However, other materials may show a significant loss and it is recommended that a magnesium salt be added whenever phosphorus is to be estimated following dry-ashing (see Note **1** below).

Certain metals are also lost by volatilization notably mercury and selenium for which dry-ashing is completely unsuitable (Gorsuch, 1959). For some metals, such as lead, volatilization losses may occur according to the composition of the sample. A high chloride content is suspect in this respect although Gorsuch (1962) showed losses differed according to the type of chloride.

Volatilization losses are aggravated by violent deflagration with local overheating which must be avoided. Samples should preferably be placed in the cold muffle and the temperature allowed to rise slowly. The use of additives such as nitric acid (to facilitate oxidation) or magnesium nitrate (to raise base status) should in general be cut to a minimum.

Procedures which avoid ignition in a furnace have been proposed in the hope that losses and contamination might be minimized. Gleit and Holland (1962) advocated low temperature ashing in a stream of oxygen activated by a high frequency electromagnetic field and instruments for this purpose are now available. Schoniger (1955, 1956) ignited a small weight of sample in an oxygen flask and dissolved the elements in a solution contained in the same flask. The latter is attractive when only small sample weights are available.

Losses may arise through combination of the element with the silica wall of the crucible and/or the siliceous residue of the sample itself. It seems trace elements, particularly copper and zinc together with iron and manganese, are most likely to be affected in this way, as suggested by Piper

(1950), Chapman and Pratt (1961) and Likens and Bormann (1970). Gorsuch (1959) noted low recoveries of copper, lead, silver and cadmium but thought that diffusion into silica following reduction by organic matter was more important for copper and lead than direct formation of silicates.

Some workers have reported unexpected increases for specific elements including iron, aluminium and boron, as noted by Williams and Vlamis (1961) and Isaacs and Jones (1972). They attribute this to contamination from the lining of the muffle furnace and so this possibility should not be overlooked.

It is therefore evident that in spite of its advantages and apparent simplicity, dry-ashing cannot be universally recommended for solution preparation and its use must be governed by the elements required and the type of material.

Reagents

1 Hydrochloric acid, 1 + 1.
2 Nitric acid, conc.

Procedure

Weigh 0.2–0.5 g of dry (or air-dry) ground sample into an acid-washed porcelain basin.
Ignite at 550 °C for 2 hours in a muffle furnace. (Refer to the ashing procedure on p. 54 for further details.)
When cool add 5 ml HCl.
Cover with a watch glass and heat on a steam bath for 15 minutes.
Add 1 ml HNO_3, evaporate to dryness and continue heating for 1 hour to dehydrate silica.
Add 1 ml 1 + 1 HCl, swirl to dissolve the residue, dilute to 10 ml with water and warm to complete dissolution.
Filter through a No. 44 filter paper into a 50 ml volumetric flask and dilute to volume.
Carry out blank determinations in the same way.

Notes

1 This solution is suitable for the determination of sodium, potassium, calcium and magnesium and often iron, manganese, aluminium, zinc and copper. If phosphorus is required the sample should first be evaporated dry with 5 ml 20% (v/w) magnesium acetate or determined separately following a mixed acid digestion (see below).
2 It is important to include nitric acid at the dissolution stage to ensure complete oxidation. The combustion of highly organic samples may have led to local reducing conditions in the crucible.

Acid oxidation

All acid digestion procedures incorporate oxidizing reagents to break down the organic matter. Advantages and disadvantages of the method in comparison with dry ashing have been mentioned. The aspect that causes most concern is the danger of explosion through using strong oxidizing agents such as perchloric acid or hydrogen peroxide. Although this danger is sometimes overrated, certain precautions must be observed. The following points, some of which are recommended by the Society for Analytical Chemistry (1959) should be observed when using perchloric acid. Provided these are followed the acid can be safely used for digestion of ecological materials.
1 Work on a small scale.
2 Never allow a digestion flask to dry out and particularly beware of charring if accompanied by loss of fluidity.
3 Do not digest materials high in fat or oily substances or anything containing hydroxy groups. If in doubt carry out a preliminary oxidation with nitric acid initially at room temperature.

The inclusion of sulphuric acid in the digestion mixture as in the procedure given below will prevent the flasks drying out. For samples rich in fats, including certain animal materials, the dehydrating properties of sulphuric acid may be a disadvantage and pre-treatment with nitric acid is then advisable. The use of perchloric acid both alone and with other acids has been reviewed a number of times, including the Society for Analytical Chemistry (1959, 1960) and Gawen (1965).

Most of the explosion incidents that have been reported have been associated with perchlorates. Metal perchlorates can detonate violently if heated to dryness particularly in the presence of organic matter.

Hydrogen peroxide is less of an explosion hazard than was formerly the case since metal peroxides are now rarely present as an impurity. Also polythene containers are used in place of glass for storage. Even so, cool storage is recommended. Hydrogen peroxide freely gives off oxygen and it should not be used near flammable substances. Eyeshields should always be worn when using any of the digestion reagents.

Another disadvantage of the acid digestion system is that it is less suitable for the determinations of trace mineral elements, since large sample weights require large volumes of acid for complete oxidation. This may give rise to a high reagent blank.

The possibility of low recoveries due to potassium precipitating as perchlorate or calcium as sulphate has been pointed out by some workers. This has not been found to occur with plant materials if the sample weights and reagent quantities are used as recommended below.

Despite these difficulties, wet oxidation mixtures are widely used to bring plant and similar materials into solution. The nitric–perchloric–sulphuric acid digestion procedure is one of the most widely used, but has the disadvantage that it is unsuitable for nitrogen. A number of procedures have been put forward to resolve this problem, for example, Bould *et al.* (1960), Faithfull (1971) and Cresser and Parsons (1979). A sulphuric–peroxide digestion which recovers nitrogen along with mineral elements is used in the authors' laboratory, together with the nitric–perchloric–sulphuric digestion, and both are described here.

The sulphuric–peroxide digestion mixture includes lithium sulphate to raise the temperature and a catalyst (selenium) to ensure quantitative conversion of organic nitrogen to ammonium sulphate. Smith (1979) reported that the mixed acid digestion resulted in higher recoveries of mineral elements than did the peroxide method, but the authors have used the two procedures regularly for many years and have found little significant difference between them for a range of material types. Haynes (1980) also reported good agreements between these two digestion procedures in comparison with an alternative sulphuric acid digestion and dry-ashing. A similar digestion mixture, but including salicylic acid has been proposed by Novazamski *et al.* (1983).

A slight modification is necessary when iron and manganese are being determined. The diluted digest must be boiled to bring the iron fully into solution (Quarmby and Grimshaw, 1967) and the manganese may require boiling digest with a little sulphurous acid. These variations are dealt with later in this section.

Some organic materials are relatively resistant to breakdown by oxidation reagents and longer digestion times will then be needed. Examples include peat, some seeds and certain other plant and animal materials. Siliceous material in litter and organic soils may cause difficulties. If soil minerals are present, potassium and aluminium may be bound up with complex silicates and will not be fully recovered by either dry-ashing or the acid digestion procedures. In these circumstances the use of hydrofluoric acid may be necessary.

The conventional Kjeldahl digestion which is normally used for soil nitrogen as described on p. 120 can also be used for vegetation (take 0.1 g ground sample).

Other reagents have been used for the oxidation of plant material. Some of these are considered by Diehl and Smith (1960). The addition of potassium permanganate to a digest produces strong oxidizing conditions for attacking resistant materials and is especially suitable for trace element analyses where an organic extraction stage is subsequently used. Swaminathan *et al.* (1981) propose the use of chromic acid as a digestion agent to give a solution suitable for nitrogen, phosphorus and the mineral elements. Dilute (0.5 M) hydrochloric acid was found by Hunt (1982) to be effective for extracting calcium, magnesium and potassium from plant material.

All acid digestions are best carried out in Kjeldahl flasks or if on a semi-micro scale, in long-necked boiling tables. Refluxing of the acid digest solution is then possible.

The digestions can be carried out over any source of heat capable of maintaining boiling under refluxing conditions. When digesting in Kjeldahl flasks, the use of silica-sheathed tubular heaters with variable heat controls facilitates the digestion. Block digestors are now widely used. In these the tubes containing the digest mixture are placed in accurately-machined holes in an aluminium alloy block. This is electrically heated in a well-insulated enclosure. Equipment of this type enables constant digestion conditions to be reproduced and the reduction in scale makes the digestions less hazardous. Prasad and Spiers (1978) compare and describe block digestors suitable for plant analysis.

Mixed acid procedure

Reagents

1 Perchloric acid, 60%.
2 Nitric acid, conc.
3 Sulphuric acid, conc.

Procedure

Weigh 0.20–0.50 g of air-dried or oven-dried ground sample into a 50 ml Kjeldahl flask.
Add 1 ml 60% $HClO_4$, 5 ml HNO_3 and 0.5 ml H_2SO_4.
Swirl gently and digest slowly at moderate heat, increasing the heat later.
Digest for 10–15 minutes after the appearance of white fumes.
Set aside to cool (the cold digest is usually colourless or occasionally pink).
Dilute to about 10 ml and boil for a few minutes if iron and manganese are required. Otherwise dilute and filter (No. 44 paper) into 50 ml volumetric flask and dilute to volume.
The residual acid is now about 1% (v/v).
Carry out blank digestions in the same way.

Notes

1 Prolonged heating at the white fume stage will lead to drying out with low recoveries.
2 The residue left on the filter paper may be recovered to give an approximate silica result (see p. 145) although it is not easy to recover silica adhering to the inside of the digestion flask.
3 The solution is suitable for the determination of Na, K, Ca, Mg, Zn, Cu, Al, P and also for Fe and Mn if the boiling stage has been included.
4 If boiling is not required, the digests can be diluted to 50 ml in the digestion flask and the sediment allowed to settle, thus avoiding the filtering stage.
5 Prior standing or pre-digestion with HNO_3 before 60% $HClO_4$ is added, is recommended for samples high in protein, fat or resinous substances.

Sulphuric acid–hydrogen peroxide procedure

Reagents

1 Sulphuric acid, conc.
2 Hydrogen peroxide, 100 volume.
3 Selenium, powder.
4 Lithium sulphate, monohydrate (purest available grade).
5 Mixed digestion reagent.
Add 350 ml H_2O_2 to 0.42 g Se and 14 g $LiSO_4$ in a litre boiling flask. Add 420 ml H_2SO_4 slowly whilst mixing and cooling. (Care is needed at this stage since considerable heat is generated.) The reagent is usable for several weeks if stored just above 0 °C.

Procedure

Weigh 0.10–0.40 g of air-dried or oven-dried ground sample into a 50 ml Kjeldahl flask.
Add 4.4 ml of the mixed digestion reagent.
Heat gently, increasing the intensity when the initial vigorous reaction has subsided.
Continue heating until the digest has cleared, allow a further 30 minutes, and leave to cool.
Dilute, transfer or filter (No. 44 paper) into a 50 ml volumetric flask, dilute to volume and mix.
(Include boiling stage for Fe and Mn as given for mixed acid digestion.)
Dilute five times prior to most analyses.
The residual acid is now about 1% (v/v).
Carry out blank digestions in the same way.

Notes

1 This solution is suitable for the same range of elements as given for the mixed acid digestion and also for N.
2 Siliceous materials may cause bumping during the digestion. To prevent frothing, Nicholson (1984) recommends adding the digestion reagent in two parts (A and B):
A—Dissolve 14 g $LiSO_4$ and 0.42 g Se in 80 ml water. Add 10 ml H_2O_2 to aid dissolution.

Slowly add 420 ml H_2SO_4 whilst mixing and cooling.

B—H_2O_2, 100 volume (30%).

Add 3.0 ml A to plant sample, mix, then add 1.5 ml B.

3 If the boiling stage (for Fe, Mn) is not required, the digests can be diluted to 50 ml in the digestion flask, mixed and sediment settled out.

4 Preparation of a digest blank to be used later with the standards will ensure that the catalyst and lithium salt are included. The undigested reagent is unsuitable for adding to standards because of the peroxide present.

Individual elements

A summary of the preparative methods for individual elements is outlined below. Details are given of those elements for which the dry-ashing and/or wet digestion procedures given above have to be modified.

If large weights have to be wet digested extra amounts of oxidizing agents may have to be added, but the amount of residual sulphuric acid should not be increased. The final residual acid after dilution is generally 1% hydrochloric or sulphuric acid and should be present in the same strength in the standards prepared later. The maximum weight for the sulphuric–peroxide digestion (0.4 g) should not be exceeded. Some trace elements (copper and zinc) can be run on the 5% digest without further dilution. Standards must be prepared accordingly. Blank solutions should always be prepared and are particularly important for the trace elements.

Aluminium

Use dry-ashing or mixed acid digestion.

Take 0.5 g sample to give 50 ml solution.

Proceed as on pp. 82–84.

Boron

Curcumin

Mix 0.5 g sample with 50 mg CaO in a porcelain or platinum crucible.

Ignite at 500 °C for 1 hour.

Add 5.0 ml conc HCl, dissolve by warming, dilute to volume (50 ml).

Methylene blue

As above but take 2.0 ml conc HCl. The sensitivity of the method is much greater if H_2SO_4 is used instead of HCl but dissolution of the calcareous ash will be difficult.

Take 0.5 g sample and dilute to about 20 ml.

For both methods proceed as on pp. 85–86.

Do not use glass at any stage.

Calcium

Use dry-ashing or either digestion method.

Take 0.25 g sample to give 50 ml solution.

Proceed as on pp. 87–91.

Carbon

Details on pp. 91–97.

Chlorine

Mix sample with 1.25 g CaO in a platinum crucible.

Add water to give paste, evaporate on water bath.

Ignite at 500 °C for 24 hours, then cool.

Extract with hot water, filter through No. 44 paper.

Return residue to crucible, ash again, dissolve in 20% HNO_3 and filter.

Combine filtrates and dilute to volume.

Take 5.0 g sample to give 50 ml solution.

Proceed as on pp. 98–100.

Cobalt

Use dry-ashing or mixed acid digestion.

Take 2.5 g sample to give 25 ml solution.

Proceed as on pp. 101–104.

Copper

Use dry-ashing or either digestion method.

Take 0.4 g sample to give 50 ml solution.

Proceed as on pp. 104–107.

Iron

Use dry-ashing or either digestion method.

After acid digestion, dilute the digest with about 20 ml of water and boil for a few minutes before making up to volume.

Take 0.25 g sample to give 50 ml solution.

Proceed as on pp. 108–111.

Magnesium

Use dry-ashing or either digestion method.

Take 0.25 g sample to give 50 ml solution.

Proceed as on pp. 111–114.

Manganese

Use dry-ashing or either digestion method.

If a pink or brown colour is present after acid digestion, boil diluted solution and add 2 or 3 drops H_2SO_3 (5% w/v SO_2) before making up to volume. Alternatively hydroxylamine hydrochloride can be added to the digest as reducing agent.

Take 0.25 g sample to give 50 ml solution.

Proceed as on pp. 114–116.

Molybdenum

Details as on pp. 116–118.

Nitrogen

Details as on p. 120.

Phosphorus

Use either wet digestion method.

Take 0.25 g sample to give 50 ml solution.

Proceed as on pp. 135–142.

Potassium

Use dry-ashing or either digestion method.

Take 0.25 g sample to give 50 ml solution.

Proceed as on pp. 142–144.

Silicon

Use NaOH fusion (p. 26).

Take 0.5 g sample to give 50 ml solution.

Proceed as on pp. 145–148.

Sodium

Use dry-ashing or either digestion method.

Take 0.25 g to give 50 ml solution.

Proceed as on pp. 148–150.

Sulphur

1 DRY-ASHING

Add 2 ml conc HNO_3 to sample in dish, stand overnight, evaporate.

Add 2 ml 10% $MgNO_3.6H_2O$.

Evaporate to dryness at 70 °C, then ash at 500 °C for 12 hours.

Add 5 ml 25% HNO_3 and filter into 50 ml volumetric flask, but do not dilute to volume at this stage.

Take 0.2 g sample to give 50 ml solution.

Proceed as on p. 151.

2 BOMB COMBUSTION

Press about 0.2 g air-dry sample into a 13 mm diameter pellet.

Weigh the pellet, place in a silica crucible in the bomb used for calorimetry (see p. 284).

Pipette 25 ml water into bomb.

Fit the ignition wire and cotton fuse.

Seal the bomb and pressurize to 25 atmospheres.

Place the bomb in the calorimeter and ignite.

Leave for 15 minutes with occasional gentle swirling.

Release pressure.

Transfer solution to a 50 ml volumetric flask, washing the walls and cap of the bomb well with small amounts of water. Dilute to volume.

Analyse for SO_4–S as described in p. 152.

Titanium

Use Na_2CO_3 fusion (p. 25) or HF–H_2SO_4 digestion (p. 27).

Evaporate solution obtained and add 15 ml 50% v/v H_2SO_4.

Evaporate to white fume stage.

Add 5 ml 50% H_2SO_4 and filter.

Take 5.0 g sample to give 50 ml solution.

Proceed as on p. 154.

Zinc

Use dry-ashing or either digestion method.

Take 0.4 g sample to give 50 ml solution.

Proceed as on pp. 156.

4
Analysis of Waters

The study of chemical and physical phenomena in freshwater habitats is primarily the concern of the limnologist. In ecological studies there is more emphasis on the whole ecosystem and its interaction with its surroundings. There are interests in common between the two sciences but the distinctions are important and may influence the analytical approach. However, apart from considering the analytical chemistry requirements of freshwater studies, there is a need to take into account the input of water chemistry in terrestrial ecological research. For example, precipitation is a significant source of nutrients to some terrestrial habitats and knowledge about surface run-off, soil solutions, leachates and drainage waters are all crucial in some studies. This section has been written with all these requirements in mind.

As far as possible the treatments and methods given in this section follow the sequence of the previous sections. Sampling and initial treatments are followed by general and physical tests, then the preparative stages are given for the elemental tests of Chapter 5. Most methods are applicable to polluted waters and individual pollutants are discussed in Chapter 6. Marine and brackish waters are not specifically included here although many of the methods can be applied after modification. Information about the chemical analysis of sea-water can be obtained from Riley and Skirrow (1965) and Strickland (1968). Most of the analytical manuals dealing with the chemical analysis of fresh waters are concerned with water quality or limnology. Two of the standard works on water quality are Department of the Environment (1972) and Standard Methods for the Examination of Water and Wastewater (1985). Other recommended books include Golterman *et al.* (1978), Minear and Keith (1982), Wetzel and Likens (1979) and Hunt and Wilson (1986). A useful small handbook for use at the bench is Máckereth *et al.* (1978).

In some ways waters are less difficult to examine than soil and plant materials. Much of the initial treatment and oxidation procedures necessary for solution preparation of solid materials can be simplified or omitted. On the other hand, problems arise which are peculiar to water samples. For example, some of the constituents may be labile and change rapidly after collection, and measures are necessary to minimize these problems. For this purpose, preservation techniques such as control of temperature or addition of a preservative may be necessary even though their use complicates the subsequent analytical procedures. Variations in microbial content and activity, chemical concentrations and pollution levels make it difficult to recommend procedures to suit every circumstance. The methods described below were developed to enable a wide variety of waters to be examined with storage conditions and preservation problems in mind.

Sample collection and storage

Since water is a relatively homogeneous medium some of the sampling difficulties encountered with soils and plant materials do not arise. However, there are problems of collection which tend to be specific for waters and these are considered below.

Site data should be recorded for all sampling locations. The information generally required includes time, date, grid reference of site, weather, temperature, method of collection and information about any local activities that might influence the results. Some of these data may only be applicable for certain classes of water samples.

Collection

Rainwater

Meterological data for rainfall are obtained using standard metal gauges which must be correctly sited with regard to the local environment. However, water collected in such containers is not always suitable for analysis and equipment of larger capacity made of inert materials may be needed for experimental work. Even here problems can arise, including those due to gross but local contamination and to the practical problem of distinguishing between wet and dry deposition as atmospheric components. In some studies, wet and dry deposition must be analysed separately and the collection of dry deposition is considered briefly in Chapter 7. Meanwhile, gross inputs and true dry depositions are considered here as contaminants of wet deposition.

1 Gross contamination from bird droppings, larger arthropods and plant debris is usually local and unpredictable and most studies will require its prevention. A simple remedy for solid matter is to include a filter plug in the funnel, but if the funnel is unattended for some time continued leaching of the trapped material might be serious. Droppings from birds perching on the rim can be eliminated by adding spikes but this will not prevent contamination from birds in flight. Use of a funnel cover is an alternative (see below). Soil splash due to heavy rain must also be excluded. Splash up to a metre has been recorded on open ground (Bilham, 1932) but a collection height of 0.5–1 metre is acceptable if the ground has a closed vegetation cover.

2 Dry deposition is hard to define but is commonly understood to include soil dust, salt particles, bacteria, algal cells, spores, minute arthropods and other fine particles. For some purposes pollutant gases such as SO_2 and NO_2 are included. The dry fraction must be included when estimating total nutrient income from the atmosphere but its input should be minimal when studying wet deposition. One precaution is to keep the funnel surface clean. Gore (1968) found that certain funnel materials, notably polythene, collected a thin greasy film which could trap dust, etc. The film can be removed by wiping with chloroform and analysed separately if desired. However, it is inevitable that an open funnel will let some dry deposition into the collector.

One solution to both problems is to provide a cover which opens only when there is precipitation. This requires control by sensors but past designs have had weaknesses. For example, a sensor activated by humidity sometimes responds too slowly whilst one sensitive to vibration may not be triggered by cloud or mist. However, a combination of these two types appears to be promising (Benham, *personal comm.*) and may be adapted to fractionate rainfall over short periods of time.

Even when these steps allow an acceptable sample to be collected, rainwater soon changes in composition if left for periods in the field. One approach is to replace the vessel by a tube of ion exchange resin selected to retain ions of interest. This is referred to again later.

Throughfall and stemflow

Methods for collecting throughfall samples are very similar to rainwater except the problems associated with solid deposition are more acute. In addition to plant debris, resinous substances are washed out of pine foliage and adhere to the surface of the containers. As before, this may be overcome by regular cleaning of the equipment.

Stemflow from trees is trapped by guttering or moulded PVC attached to the stems and sealed with an inert sealant. Large volumes of solution may be collected in periods of high rainfall, and so it may be necessary to use a stream splitter linked to a small bottle and a large bin. This will produce samples in periods of either low or high rainfall. Bark fragments and associated algae, etc. are part of the sample but tubing and gutters should be kept clean where possible.

Lake and stream waters

Standing and flowing fresh waters have many differences but are conveniently discussed together here.

Waters are a more homogeneous media than are soils and plant materials, but they still may be subject to considerable variation in quality according to spatial location, depth and time of sampling. For example, running water will change in composition according to distribution of tributaries and pollution discharge points. Large lakes may stratify according to season, giving water of different quality above and below the thermocline. In general the winter is the best time to obtain a representative sample of lake water, but heavy rain and floods alter sediment loadings significantly and should be avoided. For these and other reasons a single water sample collected at one point in time and space is rarely adequate. One possibility is to use automatic samplers programmed to collect the water at specified time intervals. Some of the problems in choosing sampling depths, times and frequency are discussed by King (1971) and Hunt and Wilson (1986). The sampling of surface waters is also considered by Kingsford *et al.* (1977). The type of equipment required to sample waters is governed partly by the nature of the analyses required. If a lake is stratified separate samples above and below the thermocline may be needed. Winter-sampling generally avoids stratification.

A number of sampling devices are available for taking water samples from small ponds and from different depths in large stratified lakes. The simplest system uses a weighted bottle which is suspended at the required depth. The stopper is then removed by a sharp pull on a separate line. However, such methods are crude and more complex equipment is available for deep sampling, running water and other circumstances. These are discussed by Rainwater and Thatcher (1960), Golterman *et al.* (1978) and in standard limnological texts.

A simple device which enables samples to be obtained with minimum disturbance was described by Mackereth (1963). A weighted tube is lowered to the required depth and a modified cycle pump is used to withdraw the water to a container above the surface.

Collecting a sample for later estimation of dissolved gases requires special care to avoid turbulence during sampling. Submersible pumps or continual peristaltic sampling are probably the most satisfactory, because a smooth flow without aeration or debubbling can be ensured.

Changes in water quality caused by the sampling equipment are often overlooked. For example, there may be reactions with the sampler fabrication material, deposition of suspended material, growth of microbial material or release of dissolved gases unless appropriate precautions are taken.

In the case of smooth flowing rivers above tidal influence it is generally considered that representative samples can be obtained from about mid-depth. Local static pools should be avoided if detached from the main flow whilst turbulent stretches are always well mixed. In any water course, the existence of industrial or urban outfalls up-stream of the sampling point should be known. These rarely have constant composition or steady rate of flow. The input of fertilizer and farm drainage is also ecologically important and may stimulate eutrophication in any standing body of fresh water.

Soil solutions

Soil solutions may be collected from various horizons in the soil profile using either lysimeters or smaller collectors. Samplers can be categorized into those collecting water freely or under tension. Samplers collecting freely include boxes, bowls, trays, funnels, guttering and larger lysimeters, or any interceptor placed in the soil which drains into a reservoir. They are usually home-made devices made from copper, stainless or galvanized steel, perspex, PVC or polypropylene where the material is chosen to minimize contamination and suit the particular application. Smaller devices are usually inserted horizontally into the soil profile from a pit or trench and collect water draining from the soil column above the collector.

Porous ceramic cups are the most widely used tensioned samplers as they are readily available and relatively cheap. The sample cups are installed vertically into a hole augered from the surface to the required depth and then the soil profile is reconstructed by backfilling. During collection, a vacuum is created within the cup to draw pore water in from the surrounding horizon.

Preservation and storage

Water samples are especially subject to alteration in chemical composition due to microbiological activity and chemical reactions. Heavily polluted waters can undergo changes in composition within an hour of collection, and most natural waters are affected to some degree. Some tests (particularly pH and dissolved gases) should, if possible, be carried out in the field.

The equilibria between inorganic nitrogen compounds and to a lesser extent between sulphur compounds are subject to changes through microbiological activity. Phosphorus is also affected by microbial uptake and it is possible for ions containing nitrogen, phosphorus and sulphur to be markedly reduced in concentration by standing for a few hours. These concentrations are sometimes so low in fresh water that changes of this type may grossly distort the original values.

Chemical changes may affect iron and manganese producing insoluble compounds in the higher valency states. Calcium carbonate can also be precipitated if the pH or alkalinity conditions are changed. This must be considered if any additives are used which upset these conditions.

Treatments are required, for example, prior to the determination of dissolved oxygen whereby bacterial activity is arrested immediately after taking the sample. However, samples taken for estimating bacterial activity itself must not be pretreated.

Glass sample containers are frequently recommended for storage since polythene vessels can be porous to gaseous constituents and have been found to absorb phosphorus. In the authors' experience these effects are negligible if smooth, high density polythene containers are used. If dissolved gases are to be estimated, glass bottles with closely fitting stoppers must be used. Borosilicate glass is generally preferable for waters requiring mineral analysis although surface exchange processes sometimes bring trace elements, especially zinc, into solution. Bowditch *et al.* (1976) also consider the suitability of containers for storage of water samples.

Thorough cleaning of containers between use is extremely important. Organisms such as algae with glutinous secretion are especially difficult to remove. Both mechanical brushes or shaking with a sand and water mixture are effective for removing surface contaminants although slight abrasion of the surface itself may be difficult to avoid. Oxidizing chemicals may be preferable for glassware.

If only selected dissolved ions are to be estimated, they can often be recovered from the sample by ion-exchange resins. This avoids long term storage problems for the samples although if elution is delayed the resin itself should be stored under water. This procedure conveniently follows removal of suspended matter (see below) and a sample column was described in Standard Methods for the Examination of Water and Wastewater (1985). Some workers have used resins to recover ions from rainwater in the field. Whilst this concentrates the ions and avoids the difficulties of transporting and storing rainwater samples, microbial growth or organic blockage on the resin itself can be a problem.

Some of the methods commonly used for the preservation of samples are given below. In some cases it may be necessary to take sub-samples and apply a separate preservation technique to each. Other methods for storage and preservation which are appropriate for specific analyses are given in Table 4.1.

In general, to minimize possible bias of results caused by any changes occurring during storage it is important to:

1 Analyse the samples as soon as possible. Ensure that solution collectors at sites are emptied regularly.

2 Fill containers to exclude air.

3 Keep sample cool, but do not freeze.

Table 4.1. Methods for collection and preservation of water samples

	Glass bottles	Plastic containers	Exclude air	Add acid[1]	Mercuric chloride[2]	Analyse as soon as possible	Cold storage (+1 °C)	Freeze storage (−10 °C)	Fine filtration[10]
Mineral elements	✓[3]	✓[4]		✓[5]			✓	✓[6]	✓
Ammonium-N	✓	✓		✓	✓	○	✓		✓
Nitrate-N	✓	✓			✓	○	✓		
Nitrite-N	✓	✓	○		✓	○	✓		×
Phosphate	✓	[7]			✓	○	✓	×	✓
Sulphide-S[8]	✓		○			○			
Other anions	✓	✓				✓	✓	✓	✓
Silica	×	✓					✓	×	✓
Organic constituents	✓				✓	✓	✓[9]	✓	
Organic pesticides	✓	×				✓	✓[9]		
Conductivity/pH	✓	✓	○				✓	×	×
Dissolved gases	○	×	○			○[11]	✓	×	×

○ Essential ✓ Recommended × Avoid.

[1] Add 1 ml conc HCl or conc H_2SO_4/litre water. [2] Add 20 mg mercuric chloride/litre water. [3] See p. 4 about glass contaminants. [4] High density polythene most suitable; avoid plastic closures containing fillers. [5] Especially for iron. [6] Not suitable for aluminium, calcium or iron—add dilute HCl. [7] Provided the polythene has been treated (see p. 67). [8] Add Cd–Zn OAc in the field. [9] Also keep in dark. [10] See below for effect of filtering. [11] See also p. 72.

4 Filter samples if they contain suspended matter. Retain filter and residue for analysis if desired.

Physical preservation methods

Fine filtration

If the interest is only in the dissolved fraction then fine filtration, which removes many of the micro-organisms, can be applied. It will also remove fine mineral matter and any traces of turbidity which could affect a later analytical stage. In general, fine suspended material is a major factor in the instability of samples and it is most important to remove it and especially for storage. Filtering should be carried out as soon as possible after collection (within 8 hours). However, pH, alkalinity and dissolved gases must be determined on unfiltered sub-samples.

Separation of 'suspended' and 'soluble' matter by filtering raises certain problems. The separation itself is arbitrary depending on the pore size of the filter. There is no 'standard' pore size comparable to the 2 mm limit used for soils although 0.5 μm has sometimes been accepted as a convenient boundary (Golterman *et al.*, 1978). Finer filters are frequently used although during ultrafine filtration especially, the effective pore size can change during use as the filter becomes partly blocked.

Both cellulose and membrane filters are available although the conventional cellulose papers are less often used in water analysis today. Glass fibre discs or cellulose nitrate membranes are widely used. Another problem arises because papers and membranes can remove or release nutrients during filtering. The effect is more significant with phosphate and ammonium ions whose levels can be very low in some waters. An initial washing is helpful but does not eliminate the effect and blanks should be obtained to check contamination. An alternative approach is centrifuging which can be used to separate various particle sizes. Salbu (1981) has compared centrifuging with membrane filtering and noted differences for certain ions whilst Schierup and Riemann (1979) consider the general effects of filtration.

Temperature reduction

Although some microbial activity appears to continue even at 0 °C the rapid cooling of samples after collection is generally to be recommended. This can be carried out in the field using solid carbon dioxide in acetone followed by transport to the laboratory in vacuum containers. This method is not always practicable (e.g. for rain-waters) but even if another preservation technique is used, subsequent low temperature storage is usually advisable. A temperature just above freezing is an acceptable compromise. Freeze storage between −10 and −15 °C will prevent microbial changes in dissolved mineral and silica concentrations. Some deterioration can occur during thawing and a repeated freezing and thawing treatment should be avoided. Nelson and Romkens (1974) and Neis (1978) both discuss the effect of freeze-storage of water samples.

Radiation

The use of radiation for sterilization of water samples is possible, but safety, bulk and cost are factors which limit its widespread use.

Chemical preservation methods

The main difficulty with chemical preservatives is the possibility of interference with the subsequent analytical techniques. This often necessitates taking separate samples for specific analysis. Other problems arise because many potential preservatives are poisons or are relatively insoluble. It is convenient to split the chemical preservatives into two groups.

Inorganic

Although heavy metals are toxic to bacteria, only mercury is extensively used, usually as mercuric chloride. This is a very effective preservative but all three objections mentioned above often apply to its use. There are also environmental objections to the extensive use of mercury compounds. Heron (1962) recommended impregnation of polythene bottles with iodine to prevent bacterial uptake of phosphorus. This is carried out by heating a few crystals of iodine in the bottle to 60 °C or washing the bottle with a solution of iodine in potassium iodide. Ryden *et al.* (1972) suggested the use of treated polycarbonate vessels. Acidification with sulphuric or hydrochloric acid is often effective for preventing microbial activity. It is suitable when ammonia and the mineral ions are to be determined, but less so for anions as noted by Bray *et al.* (1973) who found a decreased recovery of phosphate.

Organic

Chloroform and toluene are both widely used but are not very effective as bacteriostatic agents. Dichlorethane is much more effective, but unpleasant to use as a biocidal agent. These reagents appear to be more effective for the preservation of sea water. Aliphatic esters of *p*-hydroxybenzoic acids were once recommended and similar compounds have been tried for the preservation of natural waters. Some compounds formerly used cannot now be recommended on health and safety grounds.

Preliminary and general tests

For reasons given in the previous section tests on waters should be made as soon as possible after sampling and in some cases in the field. This particularly applies to labile and gaseous constituents.

Odour, turbidity and colour

Tests for odour, turbidity and colour are more usually carried out on potable waters, however, there may be occasions when information of this kind is required for natural waters.

Odour can serve as a guide to gross pollution of water. For example, characteristic odours are associated with chlorination plants, untreated sewage and chemical industry effluents. Anaerobic conditions are also indicated by the smell of hydrogen sulphide, especially near boggy areas. It is not practicable to measure odour quantitatively but it is useful as an indicator of particular conditions.

Colour in water may be a true colour due to dissolved material or an apparent colour when suspended material is present. The latter is quite common in natural waters, seen for example when algal blooms impart a greenish tinge. In the case of drainage waters from peat or humus the brown colour may be true or apparent. Similarly an orange colour may be present in waters that have passed over iron-rich sediments or mine workings and some colours are associated with various industrial effluents. Colour is measured visually against coloured glass standards using a comparator or photometrically with a spectrophotometer. Water colours are often graded arbitrarily as in the Hazen system.

Turbidity may be used as an estimate of undissolved substances in the sample. It is generally measured by visual comparison with standards or photometrically, using a nephelometer or spectrophotometer. Turbidity and colour control light penetration in lakes which in turn affects phytoplankton populations. Specific methods such as the Secchi disc are used to estimate penetration and limnological texts give details.

Further information, including quantitative methods, on tests for odour, colour and turbidity are given in Department of the Environment (1972) and Standard Methods for the Examination of Waters and Wastewaters (1985), whilst turbidity is discussed in detail in Minear and Keith (1982).

Solids

It has long been the practice in water analysis to determine solids as dissolved, suspended and organic. Unfortunately the conventional gravimetric methods are relatively insensitive and not reliable when only a few mg l^{-1} of solids are present as in some natural waters. In these cases it will probably be of more value to estimate individual ions and total organic matter using the more sensitive methods given elsewhere in this text.

Total suspended solids (TSS)
The finer suspended matter in natural waters is usually of an organic nature representing colloidal matter which has been flocculated under the influence of bacteria and protozoa. Inorganic

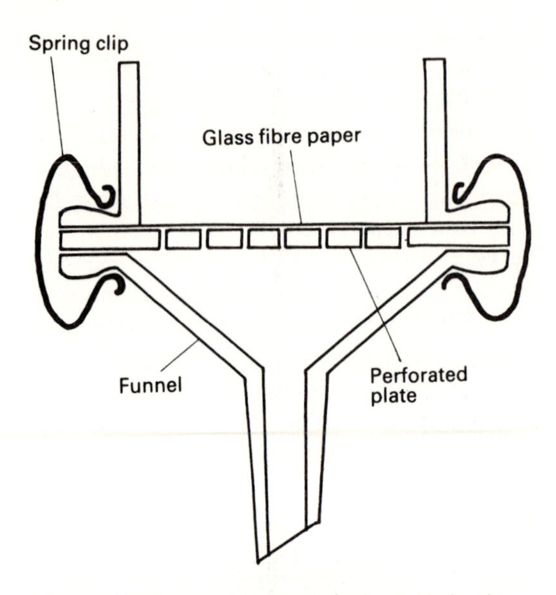

Fig. 4.1. Hartley-type Buchner funnel suitable for the determination of suspended solids in water.

suspended matter is chiefly restricted to siliceous material resulting from the erosion of mineral soils. For a given sample TSS is controlled by the porosity of the filter.

One recommended method is to filter through a weighed pad on a Gooch crucible. One useful variant uses the Hartley form of the Buchner funnel (Fig. 4.1) and glass fibre filter papers. However, for most purposes it is adequate to filter through a weighed glass fibre paper in a funnel and wash well with distilled or deionized water. The residue is dried to constant weight and then weighed.

Glass fibre papers are used since their water content is negligible and dried papers do not change in weight during weighing, in contrast to ordinary filter paper. Glass papers contain a small amount of organic matter, but this can be reduced to insignificant levels by washing in advance.

Further information about the factors affecting the determination of suspended solids can be obtained from Crane and Dewey (1980).

Total dissolved solids (TDS)
It is often convenient to determine the dissolved solids in the filtrate remaining from the TSS determination.

Procedure

Evaporate the filtrate to a small volume (taking 50 to 100 ml for streams or lakes and 500 ml for rainwater).

Transfer to a weighed basin for evaporation.

Dry at 105 °C to a constant weight.

Cool in a desiccator and weigh.

Express the result as mg l^{-1}.

Total organic matter (TOM)

Procedure

Concentrate a suitable aliquot of the water to a small bulk and transfer to a small pre-weighed evaporating basin.

Evaporate to dryness and weigh basin plus contents.

Ash in a muffle furnace, leaving at 500 °C for 1 hour. The loss in weight of the residue gives the TOM in the sample. The method is only approximate and estimates of total organic carbon are preferable when organic contents are low.

pH

The acidity (or alkalinity) of a natural water is an important property with many biological and chemical implications. Direct measurements by titration are convenient (see p. 70) but pH portrays the same property more concisely. The concept of pH is formulated solely in terms of hydrogen ions but its ecological significance is wide because the behaviour of many ions (notably aluminium) and organic compounds can be related directly to it.

pH is now almost always measured using a pH electrode which is really a specific example of a selective ion electrode. These are considered in more detail in Chapter 8. Information about the measurement of pH has already been given in Chapter 2 (p. 16), and most of the recommendations given there can be followed for waters. The main exception concerns sample mixing.

Ionic strength affects the activity coefficient and hence the pH electrode responses. High ionic strength is rarely a problem with fresh waters but low ionic strength samples such as rainwater can be difficult to analyse accurately. The difference in strength between rainwater ($<10^{-2}$ M) and the buffer gives a small but detectable error of 0.02 pH units (Tyree, 1981) but many workers consider that the electrode system is the basic problem. Reference is made later (p. 269) to the importance of constant stirring conditions and electrode geometry. Sisterson and Wurfel (1984) discuss electrode systems and note that stirring is a significant source of error. Covington *et al.* (1983) recommend the replacing of discrete samples by flowing solutions and adapt a flow cell plus electrode assembly for this purpose.

Davison (1987) outlines some recommendations which should minimize problems and highlight errors. He concludes that many of the problems associated with low ionic strength samples are related to the frits on the reference electrode, and recommends the use of reference electrodes which incorporate free flowing frits. Accuracy of measurements can be verified using dilute acid solutions (5×10^{-5} M H_2SO_4 giving a pH at 20 °C of 4.00 ± 0.02). It is most important that buffer solutions and samples are measured at the same temperature (± 2 °C).

Further information about problems encountered in the measurement of pH in waters can be obtained from the Analytical Quality Control Committee (1984) and from Hunt and Wilson (1986).

Various methods of measuring the acidity of low ionic strength samples, especially rainwaters, were described by Tyree (1981). These include titration methods which allow distinction between strong and weak acids particularly if a Gran plot is used. This was introduced (Gran, 1952) to determine equivalence points by extrapolation of a linear segment of a pH curve. This method is considered by Kramer (1982). A very different approach is that of Herczeg *et al.* (1985) who discussed electrodes and proposed that concentrations of hydrogen ions be calculated from estimates of other parameters which they consider more reliable.

Alkalinity (and acidity)

The alkalinity of water is its capacity to neutralize a strong acid, and the values obtained will depend on the pH of the titration end-point. In practice it is the bicarbonates, carbonates and hydroxides in solution that largely determine the alkalinity although there are minor contributions from silicates, phosphates and other anions. This is the conventional definition for unpolluted freshwater (Standard Methods for Examination of Water and Wastewaters, 1985; Mackereth *et al.*, 1978 and Golterman *et al.*, 1978). It is also applicable to sea water where, in addition, borates are significant (Riley and Skirrow, 1965). Rose (1983) argues for a more exact definition in terms of calcium and magnesium carbonates alone. Minear and Keith (1982) also discuss definitions.

Total alkalinity is determined by titration to the equivalence point of carbonic acid which occurs between pH 4.2 and 5.4 depending on the carbon dioxide content of the water. For routine estimation of relatively high alkalinities, titration to pH 4.5, or to a slightly lower pH with methyl orange indicator is acceptable. However, for low alkalinities the titration is preferably halted at a higher pH to allow for the effects of carbon dioxide on the equivalent point. Tables are available to select the appropriate pH if the carbon dioxide content is known (Golterman *et al.*, 1978). Standard Methods (1985) suggest pH 4.8 is suitable for an alkalinity around 160 $mg\,l^{-1}$ calcium carbonate and pH 5.1 around 30 $mg\,l^{-1}$. Appropriate indicators or a pH meter are required. The end-point for hydroxide and for conversion of carbonate to bicarbonate is reached at pH 8.3 which is easily detected with phenolphthalein indicator. The acid required to reach pH 8.3 is often recorded as 'phenolphthalein alkalinity' because total and phenolphthalein alkalinity allow the three main anions to be specified (see below).

It may be noted here that strong acid anions such as chloride, nitrate and sulphate can be collectively measured using the ion exchange method of Mackereth *et al.* (1978) which is described later. However, these anions can be estimated separately by colorimetry or ion chromatography as described elsewhere in this text.

The acidity of waters can be estimated by titrating against 0.01 M sodium hydroxide although the actual pH at the equivalence point depends on the type of acid present. Dissolved 'carbonic acid' is the main form in unpolluted waters and requires titration to pH 8.3 with phenolphthalein to complete the conversion of carbonic acid to bicarbonate. Strong acids may be present in polluted samples, supplemented by sulphuric acid from acid precipitation. The proportion of strong and weak acids can be determined if the alkali titration is combined with a Gran plot as indicated earlier.

Both alkalinity and acidity are expressed in terms of dissolved calcium carbonate, usually as $mg\,l^{-1}$ although equivalents or molarities are also of value.

Alkalinity method

Reagents

1 Hydrochloric acid.
Stock solution, approx. 0.1 M: dilute 8.9 ml conc. HCl to 1 litre with water; standardize against 0.05 M Na_2CO_3 using 0.1% bromophenol blue indicator (prepared by dissolving 0.1 g in 1.5 ml 0.1 M NaOH and diluting to 100 ml).
Working standard, 0.01 M: dilute the standardized stock solution.

2 Phenolphthalein indicator.
0.5% w/v in 50% v/v industrial methylated spirit (IMS).

3 Methyl orange indicator.
0.5% w/v in water.

4 Bromocresol green-methyl red indicator.
Dissolve 0.02 g methyl red and 0.1 g bromocresol green in 100 ml 95% v/v IMS. This indicator is preferable to methyl orange at low alkalinities.

Procedure

Measure out a suitable aliquot of the water sample (10–100 ml) into a conical flask.

If pH > 8.3 add 3–4 drops of phenolphthalein indicator and titrate against 0.01 M HCl.

To the same solution (or if pH between 4.5 and 8.3) add a few drops of methyl orange or bromocresol green-methyl red indicator.

Titrate against 0.01 M HCl until mid-colour change and note the total volume added.

Calculation

Total alkalinity (mg l^{-1} as $CaCO_3$)

$$= \frac{\text{total titre (ml)} \times 500}{\text{aliquot (ml)}} = A$$

Phenolphthalein alkalinity (mg l^{-1} as $CaCO_3$)

$$= \frac{\text{titre (ml)} \times 500}{\text{aliquot (ml)}} = B$$

Notes

1 The sample should not be diluted or concentrated and preferably not filtered before titration.
2 The various forms of alkalinity may be calculated by referring to nomographic tables such as those mentioned later in the section under carbon dioxide. In general OH^- is present if $B > A/2$, HCO_3^- present if $B < A/2$ and CO_3^{2-} present if $B > 0$ but $< A$.

Conductivity

Conductivity is a property of water governed by the total ionic content. Although it is non-specific and varies with the proportion of species present, it is often measured, because of its value in characterizing waters. It expresses the resistance of a 1 cm cube of water to the passage of a current, usually at 25 °C (specific resistance). However, the geometry of a particular cell may not meet this requirement exactly and it usually measures a fraction of the specific resistance known as the cell constant (*c*). This constant is provided by the manufacturer and must be checked and possibly adjusted before any samples are run.

The specific conductance or 'conductivity' is the reciprocal of the resistance (I/R) in Siemens (formerly mhos) or more exactly C/R for a given cell. The digital readout of modern conductivity meters is in C/R units recorded as $\mu S\ cm^{-1}$ for most freshwaters but often $mS\ cm^{-1}$ for a brackish sample.

Precautions must be taken to obtain valid results. The cell constant can be checked by reading standard solutions of potassium chloride at the temperature (usually 25 °C) for which conductivity values are published. Values at 25 °C in $\mu S\ cm^{-1}$ are as follows:

10^{-1} M	12900
10^{-2} M	1412
10^{-3} M	147
5×10^{-4} M	73.9

The calibration procedure, using a 10^{-2} M standard is given in Note **2** below.

Temperature has a noticeable effect on conductivity and must not be overlooked. Where possible, samples should be run at 25 °C but small deviations are acceptable if the meter has effective temperature compensation. The latter can be checked by running the 10^{-2} M standard at a range of temperatures, including 25 °C. However, large temperature deviations should be avoided for exact work.

Samples should not be filtered, particularly using paper filters, and should not be diluted because the correction for this step is not straightforward. Cells having small, widely separated plates (hence a large cell constant) are available and are advised for high salt concentrations.

Conductivity is discussed in various texts, including Standard Methods for the Examination of Water and Wastewater (1985), Golterman *et al.* (1978), Mackereth *et al.* (1978) and in greater detail by Minear and Keith (1982). Accuracy was evaluated by the Analytical Quality Control Committee (1984).

Procedure

Add the unfiltered sample to two tubes or bottles and bring to the required temperature (preferably 25 °C) by immersion in a water bath.
Immerse the electrodes into each tube in turn.
Record the conductivity in the second tube (having used the first as a rinse).
Check the sample temperature just after immersion of the electrode.

Notes

1 Older models were generally calibrated as reciprocal resistance ($1/R$). (Conductivity is obtained from the equation C/R where C = cell constant.)
2 The cell constant is normally given for each

instrument but may be checked against standard (0.01 M) KCl solution as follows:

Dissolve 0.7455 g dried KCl in double distilled water and dilute to 1 litre.

Nearly fill three small clean dry bottles, stopper and immerse in a bath at 25 °C. Rinse the electrode in the first two bottles and use the third to record the conductivity. If the reading is not 1412 $\mu S\ cm^{-1}$ at 25 °C adjust the cell constant as given in the manufacturer's instructions. If the new constant differs markedly from the specification the cell may be defective.

3 Soda glass tubes should not be used because ions may be taken up from the glass.

4 Ensure that the platinum electrodes are intact and not distorted. Clean as necessary with chromic acid. Keep the cell in water when not in use.

5 The solution should be perfectly still during the test.

Dissolved gases

This section is concerned with dissolved gases especially oxygen and carbon dioxide. If at all possible it is advisable to determine dissolved gases in the field. This is now more practicable, especially for oxygen, since portable instruments are available. If field analyses cannot be carried out the samples should be kept in glass bottles after all the air has been excluded. Biological uptake of oxygen is a separate problem which is discussed below.

Oxygen

A dissolved oxygen meter using a solid silver–lead electrode was developed by Mackereth (1964). This is now available from a number of manufacturers and is the most convenient technique in use. Full instructions are supplied with the instrument. Details are also given by Mackereth *et al.* (1978) and Golterman *et al.* (1978).

Until recently the standard method for dissolved oxygen in water was that originally described by Winkler (1888). In this method the oxygen first oxidizes manganous sulphate and the product liberates iodine from acidified potassium iodide. The iodine, equivalent to dissolved oxygen, is determined by titration against standard sodium thiosulphate. Details of the method together with corrections for interferences are given in Standard Methods for the Examination of Water and Wastewater (1985).

A polarograph method is also available. This depends on the reduction of oxygen at a dropping mercury electrode which produces a polarogram with two steps at about −0.05 V and −1.0 V. The first step is the more suitable for quantitative use. Potassium chloride should be added to the water sample to adjust the molarity to 0.1, with a trace of methyl red for maximum suppression.

Both of these methods have, to a large extent, been superceded by the oxygen meter.

Carbon dioxide

The simplest method for the determination of carbon dioxide is to use one of the nomographic procedures. It is necessary to know the amount of total solids, the bicarbonate alkalinity, pH and temperature of sample. Suitable nomographs are given in Standard Methods for the Examination of Water and Wastewater (1985).

If a gas chromatograph is available it provides the most convenient and sensitive method for carbon dioxide. It is particularly appropriate when other dissolved gases need to be measured (for example, oxygen, nitrogen and methane). Details about gas chromatography are given in Chapter 8. Unlike oxygen, carbon dioxide forms equilibria with the anions, HCO_3^- and CO_3^{2-}, which dominate the chemistry of many base-rich waters. Basic theory of the system is given by Golterman *et al.* (1978).

Inorganic constituents

Cations

Dissolved fraction

Provided the water is first clarified it is normally possible to use the methods described in Chapter 5 directly for sodium, potassium, calcium, magnesium and some of the other elements. It is recommended that the water be first filtered through a membrane or glass fibre filter to minimize storage problems and to prevent blockage of sampler probes, flame atomizers, narrow tubing and joints of continuous flow systems.

Blank values can be reduced by prior washing of the filter and rejection of the first few millimetres of filtrate. The blank itself can be prepared by filtering distilled or de-ionized water but it must be recognized that this solution does not fully simulate a natural water.

It is often possible to determine the major nutrient cations directly on the filtrate but for the minor elements some form of sample concentration may be necessary. Evaporation to a known volume is acceptable, but time-consuming and the residual solution may become viscous, cloudy or darker in colour. If this occurs it is preferable to use an extraction procedure (see below).

Turbidity and colour are often troublesome even in the water that has not been evaporated. Membrane filters are used to remove turbidity, but they are rather expensive. Alternative treatments such as flocculation and adsorption are less often used today. Should they be used it will be necessary to check that the ions being tested are not themselves removed in the process.

Certain waters, especially those associated with peaty soils are often strongly coloured and the type of treatment discussed above is insufficient to decolorize the water completely. Colorimetric methods are affected by a background colour but this can be tolerated if not too intense and if the adsorption peak is well separated from that of the colour developed by the chromogenic reagent. If the colours are similar it may be better to choose an alternative approach although attempts to destroy the colour itself must be avoided when only the ions in true solution are being estimated. It is sometimes possible to correct for colour by obtaining a 'colour blank'. This is the sample solution run with all reagents except the chromogenic substance. The apparent concentration of element is later predicted and subtracted from the sample value. However, this will lead to error if the pH or composition of the chromogenic reagent itself alters the intensity of the background colour.

Concentration by solvent extraction

It is possible to determine trace metal levels directly by graphite furnace atomic absorption or ICP spectrometry. However, under some circumstances it may be necessary to concentrate the sample using a solvent extraction technique. This can be recommended for iron, manganese, aluminium, cobalt, molybdenum, nickel and lead and other minor cations. The formation of metal–organic complexes enables most minor elements to be readily extracted from an aqueous media by an immiscible organic solvent. Not many solvent complex systems are suitable for extracting all trace metals simultaneously and in this respect the method compares unfavourably with ion-exchange. On the other hand this specificity can be an advantage since it enables a trace metal to be separated from other constituents likely to interfere at a later stage of the estimation. Another advantage of extractions is that complexation is usually so efficient that very few extractions are needed quantitatively to transfer the trace metals from many litres of water to a few millilitres of the organic solvent.

One of the most widely used extraction agents is dithizone (diphenylthiocarbazone) which can be used to extract cobalt, copper, iron, manganese, zinc, lead, nickel, cadmium and mercury. The extraction can be made specific or collective by control of the pH or addition of masking agents to prevent particular metals being complexed.

Other reagents which are useful for the extraction of heavy metals include diethyldithiocarbamic acid and ammonium pyrrolidinedithiocarbamate (APDC). The former is only suitable for basic or neutral solutions, but is very effective for concentration of iron, nickel, cobalt and zinc from large volumes of water. The application of APDC to extractions was first described by Malissa and Schoffman (1955). It forms complexes with more than 30 elements and can be used in acid solutions. This makes it especially suitable for atomic absorption determinations, as shown by Parker *et al.* (1967) and Watson (1968). Christian and Feldman (1970) discuss the use of APDC and describe its application in atomic absorption. A method for copper determination using APDC is given on p. 107. A complexing agent which has been used by Frei *et al.* (1973) for the extraction of some transition elements is pyridine-2-aldehyde-2-quinohydrazone (PAQH).

Concentration by ion exchange

Ion exchange resins can be used for the concentration of dilute solutions. The strong resins of the cross-linked polystyrene type incorporating sulphonic acid residues are particularly suitable for this purpose. These are usually obtained in the sodium form but with an affinity for iron and other metals so that they must be well washed with warm 10% hydrochloric acid to convert them to the hydrogen form before use. Some exchange resins incorporate chelating agents such as iminodiacetic acid groups and are of value for the concentration of trace metals such as iron and copper. Booklets describing the properties and application of ion exchange resins are available from most chemical suppliers and resin manufacturers and these should be consulted for optimum operating conditions.

The above discussion has been confined to the metal cations in solution. The anionic elements are dealt with later.

Total content

Since many ecological studies are concerned with nutrient income and loss, the measurement of the total (dissolved and suspended) content of individual elements in water is frequently required. Most of the techniques discussed above are not appropriate for this purpose and the only treatment is greatly to reduce the sample volume and digest with acid solutions (or in some cases evaporate and ignite), to bring the particulate matter into solution.

The suspended matter may be either mineral or organic in origin, but since it is finely dispersed and its total weight is generally small, the two acid digestion procedures described for plant materials are satisfactory for bringing most metals into solution. The total content of individual trace elements may be determined in this way. The method given below can be applied to most waters, although varying aliquots will be required. The final solution is also suitable for total phosphorus but other non-metals must be determined by separate procedures. A possible alternative to this is oxidation by ultra-violet radiation which Henriksen (1970) suggested for total nitrogen and phosphorus in fresh water but also found applicable to cations including iron. However, his radiation time of 4 hours is considerably longer than that required for evaporation as given below.

Procedure

Add 0.5 ml conc. H_2SO_4 to a suitable volume of unfiltered water in a 500 ml Taylor flask.

Boil down to the first signs of white fumes.

Cool, add 1 ml 60% $HClO_4$ and 5 ml conc HNO_3.

Digest until clear, but avoid drying out and cool.

Dilute, transfer or filter into a 50 ml volumetric flask, dilute to volume and mix.

Alternatively dilute to 50 ml using a dispenser and allow any silica to settle.

Prepare digestion blanks in the same way.

Notes

1 Suitable volumes for digestion range from 50 ml for lake waters to 300 ml for rainwater but not more than 2.0 mg calcium should be present.

2 As an alternative, the sulphuric acid–hydrogen peroxide digestion mixture described on p. 59 may be used. Add 0.9 ml digestion mixture to water sample, heat to reduce volume and digest to white fume stage. This will allow nitrogen to be estimated along with phosphorus and the cations.

3 Blank for the digestion acids should always be prepared although the inclusion of distilled or deionized water blank at the boiling down stage is of debatable value.

If data are required for the nutrient content of the suspended matter this must be separated from the water by filter or centrifuge as discussed earlier. The residue can then be digested by the procedures given for vegetation, or even soil if it is mainly mineral in nature.

Anions

Dissolved fraction

Anions such as sulphate, chloride, nitrate, and orthophosphate are often estimated and separate methods are given in Chapter 5. A book dealing specifically with the determination of anions is by Williams (1979). It is not often necessary to concentrate major anions but others such as phosphate may be concentrated using organic solvents and a technique for this element using n-butanol is given on p. 137. Extraction reagents such as dithizone are not applicable to anions.

Collective determinations of anions are of value. Weak acid salts (largely bicarbonate) are estimated as alkalinity, and no prior treatment of the sample is necessary or desirable. Strong acid salts (largely sulphate, chloride and nitrate) can also be estimated together and the method by Mackereth (1963) is presented here.

A sample of water is allowed to pass through a strongly acid ion exchange resin in the H^+ form. The cations are exchanged for hydrogen ions and the percolate contains a solution of the free acids corresponding to the salts present. Their concentration can then be determined by titration. The weak acid salts are undissociated at pH 4.5 and will not influence the titration.

A modification of the method can be used for chlorides. If the top half of an ion-exchange column is in the silver form, chloride will precipitate leaving nitrates and sulphates to be converted to free acids. Chlorides are then obtained by difference and a separate determination enables sulphate to be calculated. Contents of other strong anions are negligible.

Reagents

1 Potassium hydroxide, 0.01 M.
Standardize regularly.

2 Bromo-cresol green—methyl red indicator.
Dissolve 0.02 g methyl red and 0.1 g bromo-cresol green in 100 ml 95% v/v industrial methylated spirit.

3 Exchange resin, 50–100 BS mesh.
Use a strongly acidic cation exchange resin.

(a) *Acid exchanger*: prepare the hydrogen form by washing with about 6 successive volumes of 20% v/v HCl. Wash with corresponding volumes of water before use. Store under water.

(b) *Silver exchanger*: prepare by allowing 3% $AgNO_3$ to percolate through a column of the fresh resin (about 25 g). Continue leaching until the eluate turns a NaCl solution cloudy. Wash well with water until no silver nitrate is detected in the eluate. Store the resin under water in a stoppered flask.

Procedure

1 Total strong acid salts.

Prepare the exchange columns using 20 cm × 7 mm glass tubing as shown in Fig. 4.2.

Slurry the acid exchanger-resin into the column and allow it to settle free of air bubbles.

Pipette 100 ml of the water sample on to the column, supporting the pipette stem by the rubber sleeve.

Allow to percolate through the column at about 5 ml min^{-1}.

Discard the first 30 ml and collect the remainder.

Pipette a suitable aliquot of the final eluate and add a few drops of indicator solution.

Titrate against 0.01 M KOH using a 5 ml burette (= *A* ml).

2 Sulphate + nitrate.

Prepare a composite exchange column of similar size with the acid exchanger in the lower half and the silver exchanger above (see Fig. 4.2).

Pipette 100 ml of the water sample and allow to percolate as before.

Discard the first 30 ml.

Titrate a suitable aliquot of the final eluate as before (= *B* ml).

Calculation

$$\text{Total strong acid salts (me l}^{-1}\text{)} = \frac{\text{titre } A\text{(ml)} \times 10}{\text{eluate aliquot (ml)}}$$

$$\text{Sulphate + nitrate (me l}^{-1}\text{)} = \frac{\text{titre } B\text{(ml)} \times 10}{\text{eluate aliquot (ml)}}$$

Obtain chloride by difference.

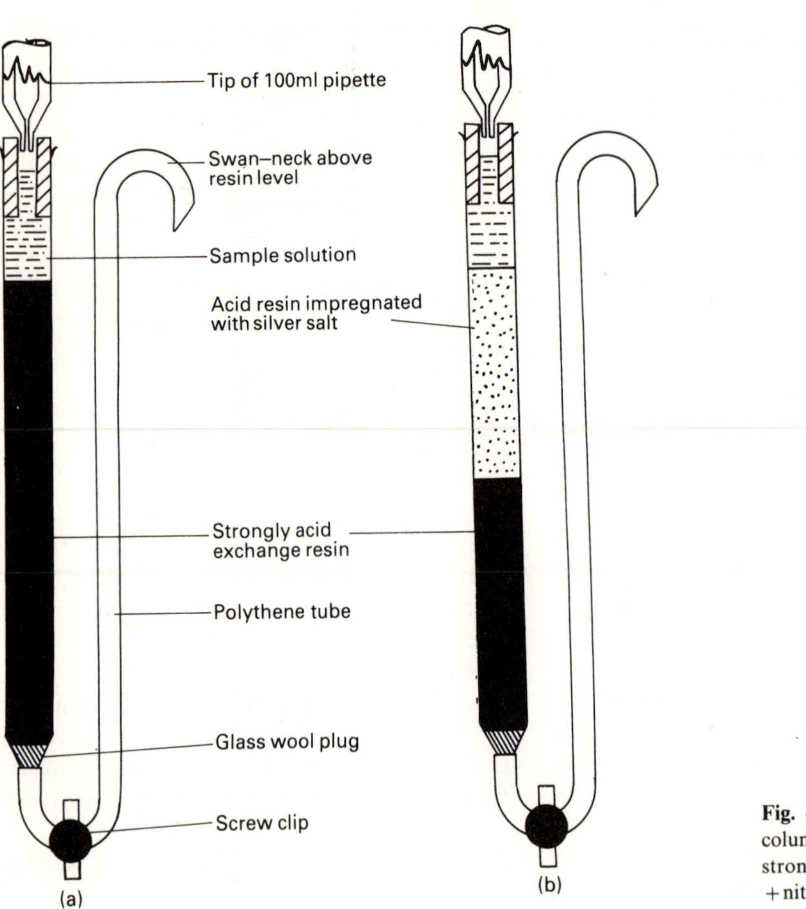

Fig. 4.2. Mackereth ion-exchange columns for estimating (a) total strong acid salts and (b) sulphate + nitrate in waters.

Note
The resins will need regenerating after about 100 determinations on samples containing around 1 me l^{-1} of ions.

Total content
Interest in total contents of non-metals in water is largely confined to nitrogen and phosphorus and occasionally other elements. Brief notes are included below under individual elements.

Individual elements

Estimates of all the elements listed for soils and plant materials may be required in one form or another in natural waters. Many of them are measured either directly as the ionic form or as a total content following digestion and the appropriate methods given in Chapter 5 should be followed in conjunction with the notes given earlier in the present section. These elements include aluminium (cation or total), boron (borate), calcium (cation or total), cobalt (total), copper (total), magnesium (cation or total), molybdenum (total), potassium (cation or total), sodium (cation or total) and zinc (total).

There are a number of points to note for other elements and these are dealt with below.

Carbon
Inorganic carbon is usually determined as carbonate and bicarbonate as described for alkalinity (p. 70) although other anions will be included.

Total carbon is estimated by combustion although total organic carbon (TOC) is of more interest to the ecologist. Older methods used a dry combustion unit to obtain total carbon. This involved evaporating a sample aliquot and transference to the combustion boat. Corrections for inorganic carbon (estimated separately) gave a value for TOC. Modern methods are instrumental using high temperature catalytic oxidation. Both organic and inorganic fractions are recorded by the same instrument. Further details about these methods are given by Johnston (1974) and Croll (1974). Dissolved organic carbon is also required. Chemical oxidation (with persulphate) or photochemical oxidation (UV) are in use. One firm has produced a continuous flow method which combines both oxidation systems.

Chlorine

Upland waters appear to contain this element only as the chloride ion. An automated colorimetric method and ion chromatography can both be used for the measurement of chloride (pp. 99 and 280). Lowland waters may also contain traces of residual chlorine derived from water treatment processes (hypochlorite ion) or from the use of pesticides (organochlorine compounds).

Iron

Various forms are present in waters but the total content is low and they are not easy to distinguish quantitatively. They are, however, important in water chemistry. The redox state of the water governs the balance between iron (II) and iron (III) which may be determined directly although care must be taken during sampling to prevent further oxidation (Bray *et al.*, 1973). Total iron may also be measured by this method using the wet digestion as given on p. 59. An ion chromatographic method is available which separates iron (II) and (III) as well as other transition metals.

Manganese

This element is similar to iron in that various forms may be present including more than one valency state and the total content is low. The methods given on p. 114 will enable a total to be measured following a wet digestion (p. 59). The proportion in solution may be determined directly but the valency states will not be distinguished.

Nitrogen

This is normally determined directly as nitrate (+nitrite), as ammonium or as organic nitrogen following digestion. Automated colorimetry and ion chromatography can be used to measure nitrate as described on pp. 130 and 280. The sulphuric–peroxide method for total nitrogen in vegetation can be adapted for waters as mentioned earlier. Henriksen (1970) utilized ultraviolet radiation to oxidize organic and ammonium nitrogen and estimated total nitrogen as nitrate + nitrite.

Phosphorus

This element is present in both inorganic and organic forms. It occurs naturally in the inorganic form as orthophosphate but synthetic phosphates may also be present following pollution. Orthophosphate may be estimated directly by the molybdenum blue method on p. 135 but the other forms will not be included. Total phosphorus can be determined following either of the wet digestion methods adopted for total cations in water (p. 59). Levels down to 5 $\mu g\ l^{-1}$ or lower can be estimated using solvent extraction of a complex formed with molybdenum blue. Methods for the determination of total phosphorus in waters containing particulate matter are compared by O'Connor and Syers (1975). Organic phosphorus can be estimated approximately as the difference between the total and dissolved values.

Silicon

This is present in soluble and colloidal forms and also in the skeletal structure of certain organisms such as diatoms. The soluble form may be measured directly as given on p. 145. The total is less often required but an approximate value may be obtained gravimetrically as suggested on p. 145. A more exact value will require an NaOH fusion (p. 26). A report on silica in waters was produced by the Standing Committee of Analysts (1981).

Sulphur

This element is found in both colloidal and dissolved forms. The inorganic fraction is almost all sulphate which can be measured by the Mackereth ion-exchange method (p. 75) or more commonly by using an ion analyser (p. 280) and at higher levels by turbidimetry (p. 151).

Sulphide ions are found under reducing conditions and since their presence may also be associated with industrial effluent and sewage discharge they are discussed further in Chapter 7. Total sulphur is not often required but it can be estimated following a specific dry-ashing procedure (p. 61). Total sulphur by wet oxidation with nitric and perchloric acids may give a low recovery in some cases.

Organic constituents

Most waters contain natural organic products although the concentration may be very low. In some cases organic pollutants may also be present. The colour of the water generally provides a rough indication of the presence of the natural organic compounds, usually humic products. Methods to estimate total or dissolved organic carbon are outlined in Chapter 5. If sufficient organic matter is present it can be estimated following evaporation and loss-on-ignition. The loss-on-ignition stage follows the procedure given on p. 15. Alternatively after evaporation the organic carbon titration method described on p. 96 can be used as a rough guide to organic content. Further information about the nature and analysis of organic constituents in waters can be obtained from Chian and DeWalle (1978). If the organic constituents are present in sufficient quantities, then it is possible to quantify the individual components using either gas or high performance liquid chromatography.

Chemical oxidation demand

This test was developed as a simpler alternative to the traditional oxygen demand tests, notably biochemical oxygen demand (BOD). This is a non-specific test which is still popular with public health analysts and water authorities. Although strictly not an organic compound test it is used as an indicator of organic pollution so is included here. It is a bio-assay method which estimates the oxygen used by micro-organisms during aerobic breakdown of organic pollutants. This involves incubation which is laborious to carry out and is not readily reproducible. It is not described here although details are available in many texts on water analysis.

Chemical Oxygen Demand (COD) is better in both respects but does not distinguish between biodegradable and more resistant fractions. For ecological purposes this is not always a disadvantage and COD is often used as an indirect measure of organic matter in water. It is for this reason that COD has been included in this section on organic constituents.

In the COD method organic matter is oxidized to carbon dioxide using acid dichromate as the oxidizing agent and its consumption is measured by titrating against a standard ferrous ammonium sulphate solution. In principle it is similar to the rapid titration methods for organic carbon given in Chapter 5. The effectiveness of oxidation can be improved by adding silver sulphate as a catalyst. Chloride interferes when above 200 $mg\,l^{-1}$ but up to 200 $mg\,l^{-1}$ can be tolerated if mercuric sulphate is present (Golterman *et al.*, 1978).

Procedures for total oxygen demand (TOD) are also available but these are instrumental methods based on high temperature catalytic combustion similar to methods sometimes used for total organic carbon (TOC) in waters. One weakness of the high temperature approach is that if oxygen-containing substances such as peroxides are present they can release oxygen and lower the TOD value. Both TOD and TOC are discussed by the Standing Committee of Analysts (1980).

Reagents

1 Potassium dichromate, M/24 (0.25 N).
Dissolve 12.26 g in water and dilute to 1 litre.

2 Sulphuric acid–silver sulphate reagent.
Dissolve 1 g Ag_2SO_4 in 100 ml conc H_2SO_4.

3 Mercuric sulphate.

4 Indicator reagent.
Dissolve 0.20 g *N*-phenylanthranilic acid in 100 ml 0.2% Na_2CO_3.
5 Ferrous ammonium sulphate standard.
Prepare stock solution by dissolving 9.8 g in 100 ml water, add 5 ml H_2SO_4 and dilute to 250 ml to give approximately 0.1 M solution. Standardize against M/24 dichromate solution.

Procedure

Pipette 5 ml water into a 100 ml round-bottomed reflux flask (see Note **1**).
Add H_2SO_4 if chloride >200 mg l^{-1} (see Note **2**). Otherwise omit.
Add 2 ml H_2SO_4–Ag_2SO_4 reagent slowly and mix.
Add 2 ml dichromate solution and again mix.
Add a further 4 ml H_2SO_4–Ag_2SO_4.
Fit condenser and reflux for 1 hour.
Titrate residual dichromate against ferrous ammonium sulphate until violet colour changes to a dark green end point. Titre $=S$.
Treat the blank (reagents only) in the same way. Titre $=B$.

Calculation

$$\text{COD (mg l}^{-1}) = \frac{B-S \text{ (ml)} \times \text{titrant molarity} \times 8 \times 10^3}{\text{aliquot (ml)}}$$

Notes

1 Sample volumes up to 50 ml can be taken provided the volumes or concentrations of the reagents are proportionately larger and if a larger reflux flask is used.
2 Add $HgSO_4$ to high chloride waters to maintain a 10:1 ratio with respect to Cl^-. This treatment is unsuitable for amounts of chloride in excess of 2000 mg l^{-1}.
3 With low organic matter samples, first evaporate a known volume to 5–10 ml.

Carbohydrates

Carbohydrates can be determined in waters by the anthrone method. Anthrone in sulphuric acid reacts with the furan derivatives of monosaccharides, polysaccharides and glycosides. Different sugars reach peak absorbance at different times, but 10–15 minutes is generally acceptable for hexoses which are the ones usually present. The full procedure is described on p. 166. Jermyn (1975) found that adding formic acid and hydrochloric acid improved the sensitivity and Conroy *et al.* (1981) described the use of this method for river waters.

Fatty acids

Volatile fatty acids are sometimes present in natural waters, generally originating from agricultural effluent. Except in special cases which can be ascribed to major agricultural or industrial pollution the amounts present are very low. They are usually determined by gas–liquid chromatography. It is necessary first to evaporate in the presence of magnesium oxide to a small volume. The solution is then acidified with sulphuric acid and the fatty acids separated by distillation.

Chlorophyll

Most of the chlorophyll in water is associated with green algae and is present as chlorophyll *a*. Other pigments (xanthophylls and carotenes) and phyaeoporphyrin, a breakdown product of chlorophyll may also be present.

In the determination of chlorophyll *a*, it is first necessary to separate algal and other suspended material from the water which contains the pigment. This can be done by filtering (membrane or glass fibre) or by concentrating using a centrifuge. The residue must then be macerated with a known volume of ice-cold acetone containing 0.1% magnesium carbonate. If filters are used they are macerated with the sample. Trial runs should be carried out initially to determine the amount of residual water left on the filter or in the centrifuge. The acetone volume can then be adjusted to give a final acetone concentration of 85%.

From this point chlorophyll *a* is measured using the method described on p. 196. This method includes chlorophyll *b* but it is unlikely that this can be detected in many water samples so that extinction at 660 nm alone can be obtained. It is not possible to separate phyaeophytin by the spectrophotometer method because it absorbs light in the same region of the spectrum as chlorophyll *a*.

Humic substances

Humic substances are the main cause of the yellow-brown colour of many natural waters. The colour intensity can be used both in the visible or ultra-violet region as a crude measure of the concentration of humic substances. If the concentration of these substances is high enough, they can be precipitated from iron (III) chloride. After recovery the humic material can be dried, weighed and estimated after deducting the iron content (Obenaus, 1963). A rapid method for the determination of humic substances in natural waters is given by Thurman (1984).

Adsorption techniques are generally used for recovering humic substances from waters. Inorganic adsorbents including activated charcoal, silica-gel and alumina were formerly used for this purpose, but retention is much stronger on organic resins and these are now more popular. Anion exchange (phenol–formaldehyde) and polyamide resins have been used but Mantoura and Riley (1975) report that styrene divinylbenzene copolymers are particularly effective for retaining humic and fulvic acids from natural waters when the pH is controlled at 2.2. Aiken *et al.* (1979) examined all the XAD resins and found that XAD-8 was the best resin for humic and fulvic acids. Sodium hydroxide solutions (between 0.1 and 0.4 M) are effective desorption agents.

It is advisable to carry out preliminary tests to determine the optimum adsorption and desorption conditions for particular waters. Further information about concentration and separation techniques suitable for humic and fulvic acids and for other organic solutes is given by Leenheer (1981). He also discussed the use of ultrafiltration and gel-filtration as a means of fractionating high and low molecular weight material. Gel-filtration using Sephadex columns is particularly effective for this purpose and is simple to use (Hine and Bursall, 1984).

If these organic solutes have been concentrated sufficiently, the sample fractionation technique described on p. 199 can be applied. By first passing through a gel-filtration column it is possible to fractionate even further by separating humic and fulvic acids from the lower molecular weight phenols and simpler carboxylic acids.

It is not usually practicable to precipitate or evaporate residual solutions to measure the organic fractions gravimetrically. A simple way of obtaining an approximate measure of organic carbon is to run the solutions through a carbon analyser and then to apply a correction factor. Alternatively the ultra-violet absorbance of 300 nm can be measured (Fuchs and Kohler, 1957). 430 nm can be used if the water is first made alkaline with sodium hydroxide.

5
Nutrient Elements

Analytical procedures for elements likely to be of nutritional interest to the ecologist are set out in this chapter, the elements being dealt with in alphabetical order. Some of the elements, e.g. aluminium, are not normally considered as nutrient elements, but they are dealt with in this section because of their marginal position. All compound ions are considered under the parent element, for example, nitrate methods are described under nitrogen. Methods for a few additional nutrient elements of marginal interest are outlined at the end of the chapter (Table 5.1). Some chemical elements more often associated with pollution are considered in Chapter 7. The presentation of the methods has been arranged so that the solutions prepared from soil, vegetation and water samples as described in Chapters 2 to 4 may generally be used without further preparation.

As indicated in the introduction, the procedures given in this chapter should give reproducible results and in general have a coefficient of variation of less than 5%. An indication will be given for the method where this may not be achieved.

Aluminium

Aluminium is a major element in many primary and secondary soil minerals although only a small fraction is mobile except in acid soils. Work on soil aluminium has long been related to pedological processes including podzolization. However, concern about acid deposition from the atmosphere has focussed attention upon long-term soil acidification. In particular mobile aluminium becomes an important regulator of pH in the more acid soils where the trivalent ion reacts with water to release hydrogen ions. Aluminium forms several complex ions whose presence in a given soil is related to the minerals present, whilst their solubility is governed by the pH (Coulter, 1969; Bache, 1985). Some mobile aluminium is organically bound and the proportion can be derived for various pH levels (Young and Bache, 1985). This fraction may be less toxic than the soluble inorganic forms.

This element is not essential for plant growth although a beneficial role has been claimed for some species and several families contain accumulator plants (Bowen, 1979). Attention has been given to the potential toxicity of mobile aluminium for plants growing on acid soils and tolerance mechanisms have been studied (Foy *et al.*, 1978). Aluminium also affects nitrogen nutrition and Rorison (1985) suggests that inorganic nitrogen present in a soil may influence the uptake of aluminium by a particular species.

Aluminium levels in most fresh waters are fairly low, but forms in solution are more important than the total content because they may affect nutrient cycling in the system and be potentially toxic to some freshwater organisms, especially fish. The toxic effects are dependent upon the species of fish, the concentration of calcium in the water and the pH. Driscoll (1984) devised a scheme to fractionate the aluminium into three well-defined fractions. Inorganic monomeric aluminium is the chemically labile form which has been correlated with toxicity to fish. The labile fraction is determined by difference between the organic monomeric aluminium (inorganic removed by a cation exchange column) and the total monomeric aluminium. Sullivan *et al.* (1986) used a catechol violet method for the determination of aluminium 'species'.

The concentration ranges generally encountered are given below:

Mineral soil	1–12%
Organic soil (peat)	0.05–0.5%
Soil extractions	10–200 mg 100 g^{-1}
Plant materials	0.01–0.1%
Animal tissues	0.001–0.02%
Rainwaters	2–100 $\mu g\,l^{-1}$
Freshwater	100–2000 $\mu g\,l^{-1}$

(dry weight basis where relevant).

Most of the colorimetric methods described for aluminium involve the formation of complexes, some being colloidal 'lakes'. Reagents used include aluminon, eriochrome-cyanin R, alizarin red S and catechol violet. Their effectiveness was compared by Dougan and Wilson (1974) and they recommended catechol violet, a reagent first suggested by Anton (1960). Cookson (1987) describes its use in a continuous flow procedure for aluminium. Another popular method, which is described by Rainwater and Thatcher (1960), uses ferron-*o*-phenanthroline as the complexing agent. It is, however less sensitive than the catechol violet procedure and is subject to iron interference.

Electroanalytical methods have been suggested but as they involve the formation of complexes of the same type as those used for colorimetry they offer no advantages.

Atomic absorption is the most suitable method for estimating aluminium in solution provided a high temperature nitrous oxide–acetylene flame is used to release the element. This method is also described later. Very low levels (mostly in waters) may require electrothermal atomization.

Catechol violet method

The blue complex has its highest absorbance at pH 6 and this is maintained with a hexamine buffer. The main interfering ions are phosphates, especially condensed inorganic phosphates, trivalent iron, fluoride and trivalent chromium. The last two ions are unlikely to be present in ecological materials in sufficient quantity to affect the result. Interference by iron is controlled by the formation of the 1,10-phenanthroline complex after reduction to the ferrous state with hydroxylamine. The presence of orthophosphate at levels in the sample of less than 5 $mg\,l^{-1}$ can be tolerated. At levels above this figure the samples should be diluted or standards compensated to the same phosphate levels as found in the sample. If the presence of polyphosphates is suspected, these can be hydrolysed to orthophosphate by boiling with acid.

The procedure given here is based on that of Dougan and Wilson (1974) and is suitable for most ecological material, having been adapted to soil extracts by Grigg and Morrison (1982) and to plant material digests by Wilson (1984).

The method is sensitive to about 0.01 μg aluminium.

Reagents

1 Aluminium standards.
Stock solution (1 ml ≡ 0.1 mg Al): dissolve 1.759 g potash alum ($Al_2(SO_4)_3,K_2SO_4.24\,H_2O$) in water, add 4 ml conc HNO_3 and dilute to 1 litre with water.
Working standard (1 ml ≡ 1 μg Al): dilute the stock solution 100 times.
Prepare fresh daily.

2 1,10-phenanthroline, 0.1% w/v.
Dissolve 50 g hydroxylammonium chloride in about 400 ml water. Add 0.5 g 1,10-phenanthroline hydrate and stir to dissolve. Transfer to a 500 ml volumetric flask and dilute to volume with water.

3 Pyrocatechol violet, 0.0375% w/v.
Store in a borosilicate glass bottle.

4 Hexamine buffer solution.
Dissolve 150 g hexamine in about 400 ml water, add 8.4 ml 0.88 NH_3 solution and dilute to 500 ml (see Note).

Procedure

Prepare sample solution as described on pp. 29, 37, 60 and dilute to volume.

Pipette 0–3 ml of working standard into 25 ml volumetric flasks to give a range of 0–3 μg Al.
Add acid or soil extractant to match the sample aliquots. Dilute to about 15 ml with water.
Pipette 15 ml sample solution into a 25 ml volumetric flask.
From here onwards treat standards and samples in the same way.
Add 0.5 ml 1,10-phenanthroline reagent, 1 ml pyrocatechol violet solution and 5 ml hexamine buffer solution and mix.
Adjust to pH 6.0–6.2 if necessary using 0.1 N HCl or dilute NH_3 solution.
Dilute to volume and stand for 10–15 minutes to allow full colour development.
Measure the absorbance at 581 nm or with an orange filter using water as a reference.
Prepare a calibration curve from the standard solutions and use it to obtain μg Al in the sample aliquots.
Carry out blank determinations in the same way and subtract where necessary.

Calculation

If $C = \mu g$ Al obtained from the graph then for:

1 Plant materials and soils,

$$\text{Al}\ (\%) = \frac{C(\mu g) \times \text{soln vol (ml)}}{10^4 \times \text{aliquot (ml)} \times \text{sample wt (g)}}$$

2 Soil extracts,

Extractable Al (mg 100 g^{-1}): as above $\times 10^3$
(for me 100 g^{-1} Al divide mg 100 g^{-1} result by 8.99).

3 Waters,

Al (mg l^{-1}) = $C(\mu g)$/aliquot (ml)

Apply factors for dilution or concentration and correct to dry weight where necessary.

Note

The volume of 0.88 NH_3 solution used may be varied to suit the sample type. The essential point is to ensure colour development at pH 6.0–6.2.

Atomic absorption method

A high temperature nitrous oxide–acetylene flame is generally used to assess aluminium by atomic absorption. The principles of this method are discussed in Chapter 8. Few elements interfere with the absorption of aluminium, iron being the only one likely to cause any difficulty in organic material digests or soil extracts. However, the relatively high levels of phosphorus and chloride may be troublesome. It is desirable to include an ionization buffer when using the nitrous oxide–acetylene flame.

Although the normal atomic absorption method can measure as little as 0.3 μg ml^{-1} aluminium, the sensitivity may be limiting for some vegetation digests and water samples. Electrothermal atomization, described in Chapter 8, can then be used or, in the case of waters, a concentration step will be necessary.

Reagents

Aluminium standards.
Stock solution (100 ppm Al): dissolve 1.759 g potash alum ($Al_2(SO_4)_3.K_2SO_4.24H_2O$) in water, add 4 ml conc HNO_3 and dilute to 1 litre with water.
Working standards: prepare a range of 0–20 ppm for plant digests or 0–100 ppm for soil extracts. Include acid or soil extractant as appropriate to match the sample solutions and 2% potassium chloride as an ionization buffer.

Procedure

Prepare sample solution as described on pp. 29, 37, 60 and dilute to volume.
Select the 309.3 nm wavelength and adjust fuel and nitrous oxide flows, slit-width and other settings as recommended for the instrument employed.
Fit the correct burner for the nitrous oxide flame.
Always allow the hollow cathode lamp adequate time to stabilize.
Prepare a calibration curve from the standard range by setting the top standard to a suitable readout value and the 0 ppm standard to zero deflection then aspirating each standard in turn.
Aspirate the sample solution under the same conditions as the standards.
Check the top, zero and one intermediate standard frequently.

Flush the spray chamber and burner frequently with water particularly if soil extracts are aspirated and check there is no build up of carbon on top of the burner.
Use the calibration curve to determine ppm Al in the sample solutions.
Carry out blank determinations in the same way and subtract where necessary.

Calculation
If C = ppm Al obtained from the graph then for:
1 Plant materials and soils,

$$\text{Al}\,(\%) = \frac{C(\text{ppm}) \times \text{soln vol (ml)}}{10^4 \times \text{sample wt (g)}}$$

2 Soil extracts,
Extractable Al (mg 100 g^{-1}): as above $\times 10^3$
(for me 100 g^{-1} Al divide mg 100 g^{-1} result by 8.99).
3 Waters,
Al (mg l^{-1}) = C (ppm)
Apply factors for dilution or concentration and correct to dry weight where necessary.

Notes
1 Particular care is necessary when using nitrous oxide–fuel mixtures and the instructions of the instrument manufacturer concerning gas pressures, lighting and extinguishing the flame should be followed carefully.
2 Greater sensitivity can be achieved by including an organic solvent such as isopropanol in the sample and standard solutions.

Electrothermal atomization
If the concentration of aluminium in the residual sample solution is below the limit of detection of flame atomic absorption then electrothermal atomization will generally be sufficiently sensitive to produce a reading. Details of this technique and operating conditions are given in Chapter 8.

Boron

Boron occurs in soil as primary silicate minerals, such as tourmaline, which weather only slowly to form borates. Only a small fraction of the total is available to plants. Maritime soils and those rich in clays tend to have higher available levels than sandy soils, although clays and other soil components can reduce the mobility of boron by adsorption as described by Evans and Sparks (1983). Borates are slightly volatile which may have resulted in widespread atmospheric circulation with the relative uniformity of boron in soils throughout the world.

Boron is an essential micro-nutrient for vascular plants being involved in cell structure and probably sugar translocation. Lewis (1980) proposed a role in lignification. Meristem tissues are sensitive to boron deficiency and the tolerance range between deficiency and toxicity appears to be narrow although some accumulator plants are known. Detergent waste and other industrial discharges may sometimes carry large amounts of boron.

Mineral soil	5–50 $\mu g\,g^{-1}$
Organic soil (peat)	2–20 $\mu g\,g^{-1}$
Soil extractions	0.2–5 $\mu g\,g^{-1}$
Plant materials	10–80 $\mu g\,g^{-1}$
Animal tissues	0.4–2 $\mu g\,g^{-1}$
Rainwaters	100 $\mu g\,l^{-1}$
Freshwater	10–500 $\mu g\,l^{-1}$

(dry weight basis where relevant).

Colorimetric methods are the most suitable for the low concentrations often found in plant and soil samples. A number of reagents based on anthroquinones, notably quinalizarin (Scharrer and Goltschell, 1935) and carmine (Hatcher and Wilcox, 1950) can be used but the colour must be developed in strong sulphuric acid and the temperature-colour relationship is critical. The use of a corrosive solvent such as concentrated sulphuric acid lessens the appeal of these methods, particularly for large numbers of routine samples. Curcumin can be used in an alternative procedure (Hayes and Metcalfe, 1962), which can be modified to exclude sulphuric acid (Dible *et al.*, 1954). Grinstead and Snider (1967) describe a curcumin procedure in which ammonium acetate is used to neutralize the concentrated acid. A procedure using curcumin is given here.

An alternative method also described below depends on the conversion of boron to BF_4^- and the subsequent formation of a coloured complex with methylene blue. This method measures boron in the range 0.2–25 μg and is suitable for boron in plant material and waters. Interference problems are more serious for soils and the curcumin procedure is then recommended.

Azomethine-H is a reagent for boron which has been used in an automated procedure. Further information about this reagent is given by Gaines and Mitchell (1979). An evaluation of different methods for boron determination has been carried out by Gestring and Soltanpur (1981).

A precaution which must be taken with all the boron methods is to avoid the use of borosilicate glassware, at least during the early stages of the procedure.

Curcumin method

When a strongly acid solution of curcumin is evaporated to dryness with boric acid, a dye (rose cyanin) is formed which gives maximum absorption at 550 nm. This is sufficiently far removed from the background curcumin absorption peak to allow good resolution. The red colour is intensified in the presence of oxalic acid. In soil extracts and waters sufficient nitrate may be present to interfere.

The method is suitable for organic materials, soils and waters and can detect about 0.2 μg boron.

Reagents

1 Boron standards.

Stock solution (1 ml ≡ 100 μg B): dissolve 0.5715 g dry H_3BO_3 in water and dilute to 1 litre.

Working standard (1 ml ≡ 1 μg B): dilute the stock solution 100 times.

2 Curcumin–oxalic acid reagent.

Dissolve 0.04 g powdered curcumin and 5.0 g oxalic acid in 80 ml 95% ethanol. Add 4.2 ml conc HCl and dilute to 100 ml with ethanol.

Prepare fresh weekly and store in a refrigerator.

3 Ethanol, 95% v/v.

Procedure

Prepare sample solution as described on pp. 29, 37 and 60 and dilute to volume.

Water samples do not usually require any initial treatments.

Pipette 0–20 ml of working standard into polypropylene basins to give a standard range of 0–20 μg B.

Pipette 1 ml aliquot of the sample solution into a polypropylene basin.

From this point treat standards and samples in the same way.

Add 4 ml curcumin–oxalic acid reagent and mix well.

Evaporate to dryness at 55 °C and then bake for 15 minutes.

Cool, add 100 ml ethanol solution and triturate to extract the colour.

Filter into a 25 ml volumetric flask (glassware can be used at this stage).

Wash with ethanol solution and dilute to volume.

Measure the absorbance within 1 hour at 540 nm or with a green filter using water as a reference.

Prepare a calibration curve for the standard solutions and use it to obtain μg B in the sample aliquots.

Carry out blank determinations in the same way and subtract where necessary.

Calculation

If $C = \mu g$ B obtained from the graph then for:

1 Plant materials, soils and soil extracts,

$$B(\mu g\ g^{-1}) = \frac{C\ (\mu g) \times \text{soln vol (ml)}}{\text{aliquot (ml)} \times \text{sample wt (g)}}$$

2 Waters,

$$B(\mu g\ l^{-1}) = \frac{C\ (\mu g) \times 10^3}{\text{aliquot (ml)}}$$

Apply factors for dilution or concentration and correct to dry weight where necessary.

BF_4–methylene blue method

The tetrafluoroborate ion complexes with methylene blue to give a colour which is soluble in dichloroethane. This colour may be measured to give a method for the determination of boron (Ducret, 1957). The procedure given below is based on that of Pasztor *et al.* (1960).

A sulphuric acid extraction will give the greatest sensitivity but hydrochloric may be required if the sample was ashed with a calcareous additive. Nitric and perchloric acids and strong hydrofluoric acid are unsuitable. To ensure the methylene blue complex is neither reduced nor oxidized, permanganate is added until just in excess and this excess is then reduced with ferrous ammonium sulphate. The method is not recommended for total boron in soils, but is suitable for plant material, water extracts of soils and waters.

The detection limit is about 0.1 μg boron.

Reagents

1 Boron standards.

Stock solution (1 ml ≡ 100 μg B): prepare as given in previous method.

Working standard (1 ml ≡ 1 μg B): dilute 5 ml stock solution to 20 ml with water. Add 5 ml 5% HF and allow to stand for 2 hours. Dilute to 500 ml with water.

2 Methylene blue, 0.001 M.

Dissolve 0.3739 g methylene blue in water and dilute to 1 litre.

3 1:2 Dichloroethane.

4 Potassium permanganate, 0.1 M.

5 Ferrous ammonium sulphate, 4% w/v.

Make up fresh before use.

6 Sulphuric acid, 5 M.

Add 272 ml conc H_2SO_4 carefully to about 700 ml of water and dilute to 1 litre.

7 Hydrofluoric acid, 5%.

Dilute 62.5 ml 40% HF to 500 ml with water and store at 5 °C in a polythene bottle.

Procedure

Prepare the sample solution as described on p. 60 and dilute to volume.

Pipette 0–10 ml of standard solution into small polythene bottles graduated at 20 ml and 50 ml to give a range of standards containing 0–10 μg B.

Add 2.5 ml 5 M H_2SO_4 and dilute to the 20 ml mark.

Take a suitable sample aliquot into similar bottles. (For filtered waters also add 2.5 ml 5 M H_2SO_4.)

Dilute to the 20 ml mark.

From this point treat standards and samples in the same way.

Add 5 ml 5% HF and stand for 2 hours.

Add $KMnO_4$ solution dropwise until a pink colour is obtained.

Add 2 ml $FeSO_4.(NH_4)_2SO_4.6H_2O$ solution.

Dilute to 50 ml with water and mix.

Add 10 ml methylene blue solution followed by 25 ml 1:2 dichloroethane.

Shake for 1 minute and allow the phases to separate.

Pipette 1 ml of the organic layer into a dry 25 ml volumetric flask and dilute to volume with dichloroethane (glassware can be used at this stage)..

Measure the absorbance at 660 nm or with a red filter using water as reference.

Prepare a calibration graph from the standards and use it to obtain μg B in the sample aliquot.

Carry out blank determinations in the same way and subtract where necessary.

Calculation

If $C = \mu g$ B obtained from the graph then for:

1 Plant materials (and soil extracts),

$$B(\mu g\ g^{-1}) = \frac{C(\mu g) \times \text{soln vol (ml)}}{\text{aliquot (ml)} \times \text{sample wt (g)}}$$

2 Waters,

$$B(\mu g\ l^{-1}) = \frac{C(\mu g) \times 10^3}{\text{aliquot (ml)}}$$

Apply factors for dilution or concentration and correct to dry weight where necessary.

Calcium

Calcium levels in soil vary greatly, from those where calcium carbonate is the dominant primary mineral and others where significant amounts of calcium sulphate or phosphate are present, to non-calcareous types where levels are sometimes quite low. In the latter soils a variety of minerals including feldspars, pyroxenes and amphibolites supply calcium. In most calcareous soils (pH > 6.5) the $CaCO_3$–HCO_3–CO_2 buffer system regulates the pH. In general Ca^{2+} is the dominant cation in exchange processes involving clay minerals, but in acid mineral soils it tends to be replaced by Al^{3+}. This can lead to a further loss of calcium by leaching and at pH < 5 the acidity is governed mainly by aluminium-based systems. Calcium also has a physical role in soil and is important in flocculating clay particles (McLean, 1975; Bache, 1984; Rorison and Robinson, 1984).

Calcium is a macronutrient for plants although levels vary greatly between species. In some plants inorganic deposits such as calcium oxalate are found in leaves. In cells it is prominent in the wall structure where it has a binding role and in membrane function and is also found in cell components such as the vacuole where temporary storage may occur. However, concentrations in the cytoplasm, particularly of free Ca^{2+} appear to be held much lower, almost at micronutrient level. The significance of this feature and other aspects of calcium nutrition have been reviewed by Hanson (1984) and Henler and Wayne (1985).

The element itself is relatively non-toxic to plants although it is well-known that disorders such as chlorosis can be induced in some species by excessive liming of soil. Calcium deficiencies, though rare, may occur in aluminium saturated acidic and magnesium serpentine soils. It is a dominant element in the skeletal structure of vertebrates and in the exoskeletons of many invertebrates. Some natural waters are also relatively rich in calcium and measures of 'hardness' (including magnesium) are still required in industry. Chemical equilibria in such lakes are dominated by the carbonate system referred to above. The concentration ranges usually encountered are given below.

Mineral soil	0.5–2% (excl. calcareous)
Organic soil (peat)	0.1–0.5%
Soil extractions	10–200 mg 100 g^{-1} (excl. calcareous)
Plant materials	0.3–2.5% (high in calcareous spp.)
Animal tissues	0.03–0.3% (excl. bones)
Rainwaters	0.1–3 mg l^{-1}
Freshwater	0.1–100 mg l^{-1}

(dry weight basis where relevant).

Routine methods for estimating calcium in solution are dominated by flame procedures, notably atomic absorption which is described below. Flame emission has occasionally been used. The only titrimetric methods widely used have been permanganate and EDTA. The latter is still used as a standard technique and is given below. A few colorimetric methods have been proposed but are not widely used for this element. A variety of methods for both calcium and magnesium have been reviewed by Boluja-Santos *et al.* (1984).

EDTA titration

This method uses the well-known complexing agent EDTA (the disodium salt of ethylenediaminetetra-acetic acid) which can form stable complexes with calcium, magnesium and other ions at specific pH values (Schwarzenbach *et al.*, 1946; Pribil, 1972). The original indicator described for this reaction was murexide (ammonium purpurate), but others have since been introduced to make the detection of the end-point easier. Calcein is a fluorescein complex which in ultra-violet light gives a yellow-green to brown colour change. Another, called calcon, was tested by Belcher *et al.* (1958). Details for using murexide, calcon and glyoxal-bis-(2-hydroxyanil) are given below. It appears there is little to choose between them although glyoxal may give a slightly more precise result (Cheeseman and Nicholson, 1968).

During the reaction, calcium forms a pink compound with the indicator in alkaline solutions, but as EDTA is added it preferentially complexes the calcium, releasing the indicator which gives the solution a blue-purple colour.

The end-point is subject to masking by some other ions or by a brown colour in the sample, but improved results can be obtained under controlled lighting conditions and adequate dilution of sample. The use of screening agents has been suggested but opinions on their effectiveness are varied. Although organic matter in waters may hinder end-point detection, it should not be destroyed unless a total calcium content is required.

Interference from heavy metals is not serious with most organic materials, but can be troublesome with soils and may cause difficulty in locating the end-point. The high dilution recommended in the method below often keeps the interference concentrations below critical levels, but is of less value if the calcium content is already low. Some interferences can be masked and reagents used for these purposes include the additions of cyanide for iron (II), copper, zinc and nickel; triethanolamine for iron (III) and aluminium and zirconyl chloride for phosphates. Phosphate interference may also be controlled by back titration. The use of thiols as masking agents for heavy metals was discussed by Halliday and Leonard (1987).

A photoelectric titrator is recommended for all EDTA titrations since it eliminates the subjective interpretation of the end-point although interferences leading to a premature end-point are not controlled.

About 10 μg calcium can be estimated.

Reagents

1 Calcium standard (1 ml $\equiv$ 0.1 mg Ca).
Dissolve 0.2497 g dry $CaCO_3$ in water containing approximately 1 ml conc HCl.
Warm to drive off CO_2, cool and make up to 1 litre.

2 EDTA solution (1 ml $\equiv$ 0.1 mg Ca).
Dissolve 0.931 g of disodium salt of ethylenediaminetetraacetic acid in about 1 litre of water and standardize by titrating against the Ca standard following the procedure described below. Store in a polythene container (glass is less suitable).

3 Sodium hydroxide, M.
Dissolve 40 g NaOH in water and dilute to 1 litre.

4 Alternative indicators.
Murexide: grind together 0.10 g murexide and 50 g NaCl in a mortar and store mixture in a dark bottle.
Calcon: dissolve 20 mg calcon in 50 ml methanol. Prepare fresh weekly.
Glyoxal: dissolve 0.20 g glyoxal-bis-(2-hydroxyanil) in 50 ml methanol.

Procedure

Prepare the sample solution as described on pp. 29, 38, 60 and dilute to volume.

Obtain a reference end-point by mixing about 5 ml M NaOH with indicator (about 0.1 g murexide or 5 drops of calcon or 5 drops of glyoxal) and dilute to about 100 ml with water.

Pipette up to 5 ml of the sample solution into another titration flask, then add about 100 ml water, 5 ml M NaOH and the indicator (about 0.1 g murexide or 5 drops of calcon).

Titrate with EDTA solution until the colour matches that of the reference end-point.

Carry out blank determinations in the same way and subtract where necessary.

Calculation

If T ml EDTA solution are required for the titration then for:

1 Plant materials and soils,

$$Ca(\%) = \frac{T\,(\text{ml}) \times \text{soln vol (ml)}}{10^2 \times \text{aliquot (ml)} \times \text{sample wt (g)}}$$

2 Soil extracts,
Extractable Ca (mg $100\,g^{-1}$): as above $\times 10^3$
(for me $100\,g^{-1}$Ca divide mg $100\,g^{-1}$ result by 20.04)

3 Waters,

$$Ca(mg\,l^{-1}) = \frac{T\,(\text{ml}) \times 10^2}{\text{aliquot (ml)}}$$

Apply factors for dilution or concentration and correct to dry weight where necessary.

Ion-selective electrode method

The calcium ion-selective electrode has been used for various ecological materials including soils and waters (see p. 269). Although there are relatively few chemical interferences, the medium must be suitable and solutions such as strong acids will require adjustment as recommended for the make of electrode. Measurements in solutions of base-poor materials may lack precision and the procedure is not advized for levels below about

0.5 mg l^{-1}. The ionic strength of poorly buffered waters (including most rainwaters) will need to be increased (Note **1**). In general, atomic absorption is preferred for non-calcareous solutions.

The electrode is the plastic membrane type and will have a life of a few months. However, most modern commercial electrodes can be readily rejuvenated by replacement of the membrane and renewal of the filling solution.

Reagents

1 Calcium standards.
Stock solution (1000 ppm Ca): prepare by dissolving 2.4973 g dry $CaCO_3$ in about 200 ml water containing 5 ml conc HCl. Heat to drive off CO_2, cool and make up to 1 litre.
Working standards: prepare a suitable range to at least 100 × detection limit (0.2 ppm) to allow an electrode check (Note **2**). Extra standards are needed at the lower end (Note **3**). Include acid or other soil extractant as appropriate to match the samples.

Procedure

Use for soil extracts and waters (see Note **1**).

Set up the instrument and electrodes according to the manufacturers instructions.

Ensure all solutions are at room temperature and then read standards and samples under the same conditions.

Prepare a calibration curve from the standard readings using semi-logarithmic paper with concentration on the log axis and instrument reading on the linear axis (See Note **2**).

Use the calibration curve to obtain ppm Ca in the sample solutions.

Carry out blank determinations in the same way and subtract where necessary.

Calculation

If C = ppm Ca obtained from the graph then for:

1 Soil extracts,
Extractable Ca ($\text{mg } 100 \text{ g}^{-1}$): as above × 10^3
(for $\text{me } 100 \text{ g}^{-1}$ Ca divide $\text{mg } 100 \text{ g}^{-1}$ result by 20.04)

2 Waters,
Ca (mg l^{-1}) = C (ppm)
Apply factors for dilution or concentration and correct to dry weight where necessary.

Notes

1 For low ionic strength samples such as waters it is recommended that 0.5 ml 4M KCl be added to 25 ml aliquots of samples and standards before reading.
2 Calcium is a divalent ion and one decade increase in concentration should therefore decrease the reading by about 29 mV.
3 The Nernst relationship (linear) does not hold down to the detection limit. The range below 2 ppm will be non-linear and need a separate calibration line.

Atomic absorption method

Atomic absorption spectroscopy is a very convenient method for the determination of calcium and the general principles are discussed in Chapter 8. The line at 422.7 nm is utilized in the procedure given below.

The biggest drawback in the application of flame methods for calcium has been interference from other elements when using an air–acetylene flame. Several elements give either a positive or negative error. The most important in biological material, particularly vegetation, are phosphorus and aluminium, both of which depress the absorption or emission because they form refractory compounds with calcium which are not readily dissociated in an air–acetylene flame. Depression by aluminium is particularly serious since even a low concentration will have an effect. Iron also depresses and should not be ignored in extracts of mineral soils, and in peats which are frequently relatively rich in this element. Sulphuric acid is a depressor, but hydrochloric acid gives some enhancement. Levels of these acids and soil extractants should be the same in both standards and samples. Some of these effects are complex and not fully understood. Those due to anions have been discussed by Pungor (1970). The significance of flame conditions in controlling these effects was discussed by Fassel and Becker (1969).

Various methods of controlling these interferences have been proposed in the past including separation of calcium as the oxalate, removal of phosphorus by ion-exchange resin and compensation of standards with levels of interfering elements appropriate for the samples. However, these have all been superseded by the use of releasing agents such as lanthanum or strontium as originally described by Yofe and Finklestein (1958) and Dinnin (1960). Both are effective, but only lanthanum can be used in sulphuric acid solutions because strontium sulphate is relatively insoluble.

Large amounts of lanthanum (1%) are widely recommended as being required, particularly for aluminium in soil extracts and plant solutions, but are less necessary for relatively pure waters. Sulphuric acid solutions have also been handled in this way, but this acid is itself a major depressant of calcium absorption and lower quantities of lanthanum are effective in its presence. In the method below, 400 mg l^{-1} lanthanum are used for 1% sulphuric acid solutions following wet digestion and 800 mg l^{-1} for soil extracts provided that the 1% sulphuric acid is present (Evans and Grimshaw, 1968). Adrian and Stevens (1977) also discuss the effects of different sample preparation methods on the determination of calcium in plant material by atomic absorption.

When analysing total digests or fusates of rocks and soils where aluminium and other metals may greatly exceed calcium, it is recommended that the hotter nitrous oxide–acetylene flame be used in preference to reliance upon releasing agents in the air–acetylene system. The hotter flame inhibits formation of refractory compounds so that lanthanum may be omitted, but an ionization buffer such as 1000 mg l^{-1} potassium should be included.

The method is sensitive to 0.1 mg l^{-1} or less of calcium.

Reagents

1 Calcium standards.

Stock solution (1000 ppm Ca): prepare by dissolving 2.4973 g dry $CaCO_3$ in about 200 ml water containing 5 ml conc HCl. Heat to drive off CO_2, cool and make up to a 1 litre.

Working standards: prepare a suitable range not greater than 0–100 ppm. Include sufficient stock $LaCl_3$ solution to bring the concentration of La to 400 ppm for plant digests and waters and 800 ppm for soil extracts. Include HCl (1% v/v) or extractant as appropriate and 1% H_2SO_4 in all cases.

2 Lanthanum chloride, 5000 ppm La.

Dissolve 6.6837 g $LaCl_3.7H_2O$ in water containing 1 ml 2 M HCl and dilute to 500 ml.

3 Sulphuric acid, 10% v/v.

Procedure

Prepare the sample solution as described on pp. 29, 38, 60 and dilute to volume.

Select the 422.7 nm wavelength and adjust air and gas flows, slit-width and other settings as recommended for the instrument employed.

Always allow the hollow cathode lamp adequate time to stabilize.

Prepare a calibration curve from the standard range by setting the top standard to a suitable readout value and the 0 ppm standard to zero deflection.

Add sufficient stock La solution to aliquots of the sample digests or extracts to obtain 400 ppm La in plant digests and waters, and 800 ppm in soil extracts.

Include HCl (to obtain 1% v/v) or extractant (to obtain original strength) as appropriate, and 10% H_2SO_4 to make final concentration 1% v/v in all cases.

Dilute to a convenient volume.

Aspirate the sample solution under the same conditions as the standards.

Check the top, zero and an intermediate standard frequently.

Flush the spray chamber and burner frequently with water, particularly if soil extracts are aspirated.

Use the calibration curve to obtain ppm Ca in the sample solutions.

Carry out blank determinations in the same way and subtract where necessary.

Calculation
If C = ppm Ca obtained from the graph then for:
1 Plant materials and soils,

$$\text{Ca (\%)} = \frac{C\text{(ppm)} \times \text{soln vol (ml)}}{10^4 \times \text{sample wt (g)}}$$

2 Soil extracts,
Extractable Ca (mg $100\,g^{-1}$): as above $\times 10^3$
(for me $100\,g^{-1}$ Ca divide mg $100\,g^{-1}$ result by 20.04)
3 Waters,
Ca (mg l^{-1}) = C(ppm)
Apply factors for dilution or concentration and correct to dry weight where necessary.

Carbon

Carbon occurs in a complexity of forms in nature. Although the only rock minerals are carbonates, those of calcium and magnesium make up dominant formations (limestones) and other metal carbonates occur locally. Carbon is chemically related to silicon but cannot substitute for it in forming silica and common silicate minerals contain only traces as carbonate. Elemental carbon occurs very locally as diamond and widely as graphite, whilst plant residues form coal and also hydrocarbons, which accumulate in certain strata as oil or gases (methane). However, inorganic carbon in soil is largely composed of carbonate particles plus carbon dioxide in the soil atmosphere whilst the upper horizons of most soils are dominated by organic carbon resulting from biological activity. Organic nitrogen also derives from biological processes but the compounds containing this element often break down more readily in soil than some resistant fractions which do not have this element. For this reason the carbon to nitrogen ratio has, in the past, been a useful aid in comparing soil types. As a rough guide, the carbon/nitrogen/sulphur ratio in most soils is approximately 100/10/1.

Carbon is the fundamental element of living tissue because of its extensive organic chemistry involving hydrogen, oxygen and certain nutrient elements including nitrogen and phosphorus. However, unlike the nutrient elements, the main source of carbon for a plant is atmospheric carbon dioxide (via photosynthesis) and not soil. In addition, many living organisms, notably algal and invertebrate groups, contain inorganic carbon (calcium carbonate) as a structural component.

Fresh waters contain dissolved carbon dioxide together with carbonate and bicarbonate ions which dominate the chemistry of calcareous waters. This is mentioned further under alkalinity (p. 69). The total organic carbon content is another valuable parameter in freshwater studies. The oceans also contain these fractions. Carbon dioxide in surface waters is in equilibrium with the atmosphere but the carbonate and bicarbonate contents of the whole ocean represent a carbon pool about fifty times that of the atmosphere. Ocean sediments also constitute a large reservoir of carbon.

The content of carbon dioxide in the atmosphere now attracts considerable attention because, although it is only a minor constituent at around 350 ppm, this level has been rising since the last century with possible climatic implications. Combustion of fossil fuels and the widespread destruction of the world's forests are considered responsible.

All these environments present aspects of the global carbon cycle which have been intensively studied with particular reference to terrestrial carbon (Houghton *et al.*, 1983; Woodwell, 1984) and atmospheric carbon dioxide (Trabalka, 1985).

Mineral soils	0.2–5%
Mineral soils (carbonates)	10–30%
Organic soils	30–55%
Plant materials	43–47%
Animal tissue	44–47% (excl. bone)
Rainwater (HCO_3–C)	1–10 mg l^{-1}
Freshwater (HCO_3–C)	5–25 mg l^{-1}

(dry weight basis where relevant).

Total and organic carbon

Both absolute and approximate methods for estimating carbon in vegetation and soil are very widely used. Absolute procedures rely on dry combustion or wet oxidation followed by estimation of the evolved carbon dioxide and each will be considered in turn below. The simplest methods involve acid oxidation followed by titration although crude estimates can be deduced from the loss-on-ignition value of soil as considered later. There is an extensive literature on estimating carbon and reviews include Charles and Simmons (1986) who focused upon soils and sediments in particular.

Combustion methods

Dry combustion is widely accepted as the primary technique for estimating total carbon although precautions are required and the two main fractions of carbon (organic and inorganic) may need to be distinguished. In this method the sample is burnt in a stream of oxygen, or inert carrier gas, and the resultant carbon dioxide carried to a collection point and measured. Other combustion products such as moisture and nitrogen oxides are generally removed first in a sequence of traps to prevent interferences and if carbon monoxide is produced, further oxidation in a 'catalyst' furnace is needed to ensure a full recovery of carbon as carbon dioxide.

Three combustion methods have been used. The oldest is the gas burner method as used in the traditional combustion train of furnace tube, traps and U-tubes. This system was developed in the semi-micro combustion unit of Ingram (1948, 1956) and described in the first edition of this book (Allen *et al.*, 1974). Only very small samples (a few mg) were acceptable but valid results could be obtained for organic carbon in finely ground vegetation. However, application to soil required that the inorganic (carbonate) content be known and a correction applied.

Gas burner methods have now been largely superseded by the 'carbon analysers' which normally operate at higher temperatures (around 1500 °C) with heat supplied by a heating element (resistance furnace) or high frequency radiation (induction coil). Most models are at least partly automated but not all were designed for biological materials. An example is the Laboratory Equipment Corporation (LECO) carbon analyser based on induction heating which was developed for the iron and steel industry. Application to vegetation and soil requires that the sample be combusted in the presence of metal coupling materials (accelerators) such as chips of iron and tin–copper (for soil see Tabatabai and Bremner, 1970). One or two models are designed for waters. Some instruments have the advantage of recording the organic and inorganic fraction separately for each sample although the usual application is for measuring total organic carbon (TOC) in water. A comparison of techniques used for TOC has been given by Small *et al.* (1986).

These methods incorporate various techniques for measuring carbon dioxide. The older combustion trains usually relied on a weighed U-tube containing an absorbent such as soda-lime whose increase in weight gave a gravimetric estimate. Sometimes the gas was collected volumetrically. Modern analysers usually rely on methods such as infra-red absorption or conductivity detection. Non-aqueous titration has also been used (Jones *et al.*, 1966). Some models for TOC go further in that carbon dioxide is converted to methane in a reduction furnace before measurement with a flame ionization detector.

These methods grade into the various 'elemental analysers' which are widely marketed. The systems described above will record hydrogen (as water) provided that the incoming gas stream is completely dry. Other models also measure nitrogen (CHN analysers) and some instruments (CHNOS analysers) allow oxygen or sulphur to be estimated as well.

Acid oxidation methods

Absolute methods

Wet oxidation offers a convenient and less costly alternative to the dry combustion analyser and is very frequently employed. Absolute methods are based on oxidation in an acid dichromate (or persulphate) solution with a series of traps for moisture and recovery of carbon dioxide as for dry combustion. One dichromate method which is described here, is based on those described by Shaw (1959) and Allison (1960). These employ a mixture of two acids (sulphuric and phosphoric) with potassium dichromate and gravimetric recovery of carbon dioxide. Although dichromate oxidation is less powerful than combustion, the methods give recoveries of total carbon which are comparable to those of dry combustion provided that conditions are suitably controlled. These include the acid to dichromate ratio (3:2 or 2:1), digestion temperature (120–160 °C and sometimes higher) and time (10–15 minutes in the Shaw–Allison method). One or two other precautions are mentioned below.

As with dry combustion the total carbon content of vegetation is equated with organic carbon but in soil analysis any carbonate-carbon present should preferably be determined separately and allowed for. Such corrections can be large for calcareous soil but any acid pre-treatment to remove carbonate requires care to avoid any attack on the organic matter. Allison (1960) used an anti-oxidant for pre-treatment of soil. His method allows estimation of carbonate-carbon if the carbon dioxide is collected (see later). For some purposes it may be sufficient to use a rapid (carbonate omitted) method for soil provided the organic carbon recovery is acceptable. This is discussed further below.

One variant of the Allison system measures trapped carbon dioxide by titrimetry instead of by weighing (Snyder and Trofymow, 1984). The apparatus used in the procedure described in this text is illustrated in Fig. 5.1.

It was found with both the Shaw and Allison designs that condensing water occasionally dropped on to the surface of the boiling acid mixture leading to a violent eruption of the acid into the adjacent absorption tubes. The modification shown in the diagram eliminates this difficulty.

An air-pump at the beginning of the train was found to give a more uniform airflow than a

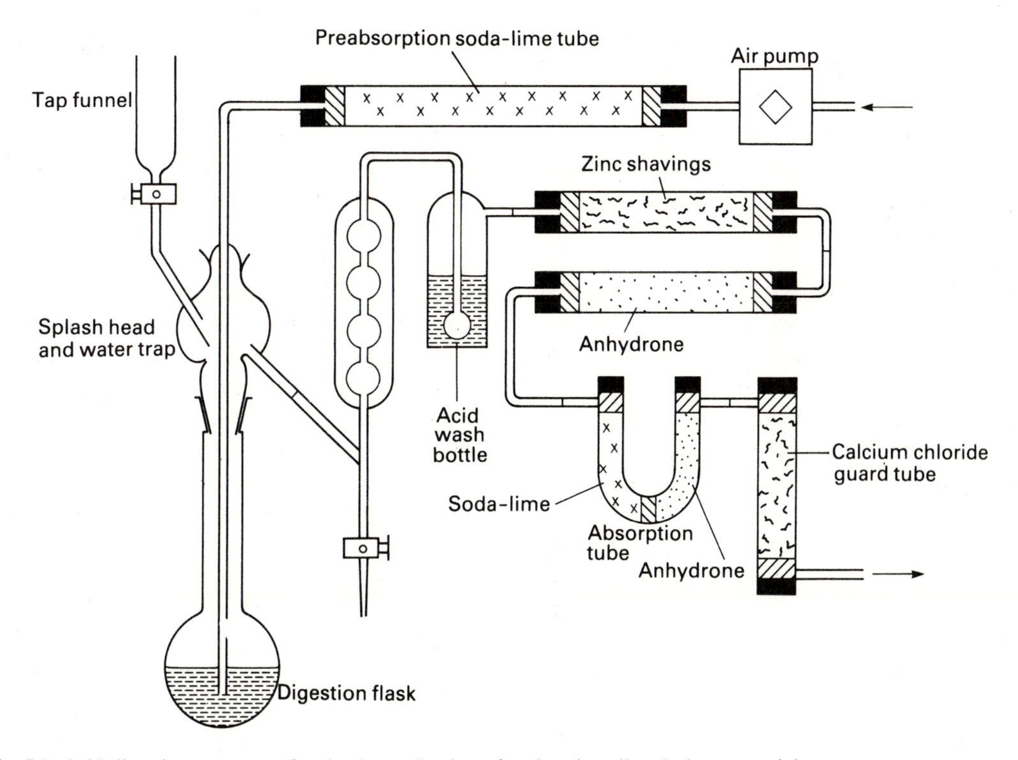

Fig. 5.1. Acid-digestion apparatus for the determination of carbon in soil and plant materials.

suction pump at the end, although more care was needed to avoid leaks at the connecting joints.

The acid digestion methods are applicable to both soils and plant materials.

Reagents

1 Soda lime.

(a) 4–10 mesh self indicating granules (for use in preabsorption tube).

(b) 10–16 mesh self indicating granules (for use in absorption tubes).

2 Magnesium perchlorate, dried ('anhydrone').

3 Potassium dichromate, powdered.

4 Zinc wire.

5 Digestion mixture.

Mix 3 parts conc H_2SO_4 with 2 parts H_3PO_4 (SG 1.75).

Procedure

Pass air through the apparatus for about 30 minutes at the start of the day.

During this time fill the U-tubes with soda lime and prepare the reagents.

Weigh the sample into the digestion flask (for plant materials take 0.10 g dry sample in a 50 ml flask, but for most soils take 0.50 g dry sample and use a 100 ml flask).

Add 3 g $K_2Cr_2O_7$ and 3 ml of water and mix.

Weigh U-tube(s).

Attach the flask and absorption tubes to the apparatus.

Open all taps and allow air to pass through at about two bubbles a second, using the H_2SO_4 bottle as an indication of the flow rate (30–35 ml min^{-1}).

Measure 25 ml of the digestion mixture into the separating funnel and allow to run slowly into the flask, closing the tap just before the last drops run in to exclude air.

Heat the flask continuously for about 15 minutes, using a small bunsen burner or spirit lamp.

Maintain the air flow at about two bubbles per second as CO_2 is evolved.

After heating, pass air through the apparatus for a further 20 minutes.

Then stop the flow, close all taps, remove the U-tube(s) and weigh.

Carry out blank determinations in the same way and subtract where necessary.

Calculation

$$C(\%) = \frac{\text{increase wt absorption tubes (g)} \times 27.29}{\text{sample wt (g)}}$$

Correct to dry weight where necessary.

Notes

1 One U-tube (with limb about 10 cm and internal diameter 1.5 cm) is sufficient to absorb the CO_2 providing the recommended flow rate is not exceeded. The other U-tube can be included in the train and checked from time to time. While in this position it is being purged ready to be brought into use.

2 Recharge the U-tubes when half the soda lime column has changed colour.

3 Particular care is needed to avoid overheating in case the acid is sucked back.

4 By setting up a duplicate apparatus adjacent to the first one it is possible to run the two concurrently and reduce the total operating time.

Rapid titration methods

Schollenberger (1927b) was the first to describe a rapid wet oxidation–titration procedure for organic carbon which he and Allison later improved. The method uses the acid dichromate system but instead of collecting carbon dioxide it relies on back titration of unused dichromate with standard ferrous ammonium sulphate and diphenylamine as an indicator. Calculation of dichromate used gave a direct estimate of organic carbon without carbonate being recorded but this advantage was greatly offset by incomplete oxidation of organic matter under the conditions generally employed.

Some of the early methods, notably that of Walkley and Black (1934), required a substantial correction to allow for the low recovery. Unlike the Schollenberger procedure no external heat was applied and recoveries differed between soil types according to the organic constituents present (Bremner and Jenkinson, 1960).

One of the more successful rapid dichromate-titration procedures was introduced by Tinsley (1950). Bremner and Jenkinson (1960) showed that, using this method, full recoveries of organic carbon were obtained for plant material although not always for other natural organic substances. Subsequently the Tinsley method was improved by Kalembasa and Jenkinson (1973) and a slightly modified version is given below. They recovered an average of 97% of the organic carbon from 22 soils and compared with other methods. The Tinsley procedure relies on oxidation under reflux which approaches the uniform enclosed conditions of the absolute methods and accounts for the high recoveries.

Reagents

1 Potassium dichromate, M/12 (0.5 N).
Dissolve 24.52 g pure $K_2Cr_2O_7$ in water and dilute to 1 litre.

2 Acid mixture (5:1 v/v).
Mix 1500 ml H_2SO_4 (SG 1.84) with 300 ml H_3PO_4 (SG 1.75).

3 Indicator reagent.
Dissolve 0.20 g *N*-phenylanthranilic acid in 100 ml 0.2% Na_2CO_3 solution.

4 Ferrous ammonium sulphate reagent 0.5 M (0.5 N).
Dissolve 196.0 g ferrous ammonium sulphate in about 300 ml water containing 20 ml H_2SO_4 (SG 1.84), dilute to 1 litre and mix well. Prepare fresh regularly. (Prepare 0.2 M if sample quantity is limiting and carbon content low.)

Procedure

1 Weigh W g air-dry sieved (2 mm) soil into a 250 ml flat-bottomed flask (W g should contain between 5 and 15 mg organic carbon.)

2 Add exactly 20 ml dichromate solution.

3 Add 30 ml acid mixture (measuring cylinder).

4 Reflux for 20 minutes with a heating mantle and condenser (BP = 165 °C).

5 Allow to cool and rinse condensor into flask with water.

6 Add 5 drops of indicator.

7 Titrate with ferrous ammonium sulphate until the violet colour changes to a dark green end-point.

8 Treat the blank (dichromate and acid only) in the same way and record the titre.

Calculation

If 1 ml 0.5 N dichromate ≡ 1.5 mg organic carbon then:

$$\text{Organic C (\%)} = \frac{(\text{blank} - \text{sample})\ \text{ml} \times 0.15}{\text{sample wt (g)}}$$

Correct to dry weight where necessary.

Carbon and organic matter from loss-on-ignition

Although loss-on-ignition is occasionally used as a guide to the carbon and organic matter content in soils, many workers consider it too unreliable, due to potential losses of volatile salts, structural water and ammonia during combustion. Ball (1964) examined the relationship between organic matter and carbon determined by the Tinsley (1950) method and loss-on-ignition at low and high temperatures. He found that an acceptable correlation could be contained for non-calcareous soils and the results were less variable than those given by the rapid titration method of Walkley and Black (1934) or similar methods. Goldin (1987), after comparison tests, considers that 600 °C loss-on-ignition is adequate for most organic matter determinations on non-calcareous soils.

Various factors are used for the conversion of organic matter (loss-on-ignition) to organic carbon, although Howard (1966) has shown that no one factor is suitable for all soils. The figure for the ratio organic matter/organic carbon which has often been accepted is 1.72 and is based on the assumption that soil organic matter contains 58% carbon. Figures suggested by Howard for different soils are 1.77 to 1.93 (mor), 1.81 to 1.83 (alluvial), 1.90 to 1.95 (peat) and 1.97 to 2.07 (mull). In cases where a large number of soils of similar nature are being examined, and independent calibration is available, the loss-on-ignition could be a convenient method for organic carbon. All values derived from loss-on-ignition results should only be considered as approximate.

Carbonate-carbon

Inorganic carbon usually occurs in soils as the carbonates of calcium and magnesium. Traces of other metal carbonates may be present locally. Dissolution of carbonate by mineral acid and determination of the carbon dioxide evolved is the basic method for estimating total carbonates in soil.

Dilute hydrochloric acid is commonly used for dissolution. Gasometric or gravimetric techniques are used for assessing carbon dioxide released although other methods are available as noted earlier. Gasometric methods include the well-known mercury manometer of Van Slyke and Folch (1940) which is still sometimes used in a modified form (Page *et al.*, 1982). A calcimeter described by Bascomb (1961) is also available which does not employ mercury. These and similar models are still available commercially but are not described further here.

Gravimetric methods collect carbon dioxide in weighed absorption tubes after removal of moisture from the gas flow as described above in the absolute methods for total carbon. One method described by Allison (1960) corrects for a weakness in acid dissolution procedures. If the reaction time is prolonged due to the presence of sparingly soluble fragments of marble or dolomite then some attack on the soil organic matter is difficult to avoid. The Allison method offsets this by including a reducing agent and is described below.

Rapid titration methods analogous to those described earlier for wet oxidation are widely used to estimate carbonates in spite of their limitations. One such method is described below.

Gravimetric method

A prolonged reaction time with mineral acid may lead to some attack upon the soil organic matter unless this is offset by the presence of a reducing agent. For example, stannous chloride could be added when hydrochloric acid is employed but the antioxidant method of Allison (1960) which uses sulphuric acid containing ferrous sulphate is given here. The apparatus required is that described earlier for the wet oxidation method.

Reagents

1 Digestion acid mixture.

Dissolve 52 g $FeSO_4.7H_2O$ in 600 ml of water. Add carefully 57 ml conc H_2SO_4 and mix. Store in a well stoppered bottle.

2 Absorbent reagents.

These reagents are the same as those described above for the absolute wet oxidation procedure for total carbon.

Procedure

The apparatus used is identical to that described for the absolute carbon method by acid oxidation (p. 92).

Weigh the soil sample, containing not more than approximately 300 mg $CaCO_3$, into a 100 ml digestion flask.

Weigh the absorption tubes and assemble as described earlier.

Add 25 ml of digestion acid mixture to the sample.

Adjust the air flow to 2 bubbles per second.

Apply heat very cautiously, using either a micro-burner or spirit lamp.

When the initial effervescence subsides, bring the solution to the boil, and continue until the total heating time is about 7 minutes.

Remove the burner, and increase the bubble rate to about 6 bubbles per second for a further 10 minutes.

Disconnect the absorption tube and weigh.

Carry out blank determinations in the same way and subtract where necessary.

Calculation

If *C* g is the increase in weight of the absorption tube, then:

$$CO_3{}^{2-}-C\,(\%) = \frac{C(g) \times 27.29}{\text{sample wt (g)}}$$

Correct to dry weight where necessary.

Rapid titration method

This method relies on dissolution with standard hydrochloric acid and back titration of unused acid with standard sodium hydroxide. An early procedure was given by Piper (1950) who noted that it gave only approximate values. One weakness is that any attack on non-carbonate minerals may consume a measurable amount of acid and give a high result. This is likely to be relatively more significant for non-calcareous soils. Another problem arises when sufficient organic matter comes into solution to make it too brown for end-point detection by indicators. In these cases the pH curve should be plotted using a pH meter and the end-point estimated from where the slope is at a maximum. This is usually practicable but slow without an automatic titrator. Interferences can distort the curve for some soils.

Literature references occasionally suggest that atomic absorption estimates of calcium and magnesium in an acid extract could be a measure of total carbonate. However, this is not recommended even though the end-point problem is avoided. Various non-carbonate minerals contain these elements and tests on different soils have shown that the resultant positive error is too great to ignore even for an approximate method.

Reagents

1 Hydrochloric acid, 0.5 M.
2 Sodium hydroxide, 0.5 M.

Procedure

Weigh 2 g air-dry sieved (2 mm) soil into a 250 ml beaker flask. (Take only 1 g of calcareous soil.)
Add 40 ml 0.5 M HCl.
Swirl and stand for at least 1 hour without heat.
Titrate excess acid with 0.5 M NaOH. Use phenolphthalein or similar indicator if solution is clear and colourless. Otherwise record change in pH (with pH meter) as alkali is run in until no further change occurs.
Plot the curve and read off the alkali volume corresponding to maximum slope.
Run blanks with 40 ml acid only.

Calculation

If 1 ml 0.5 M $HCl \equiv 25.02$ mg $CaCO_3$ then:

$$CaCO_3 (\%) = \frac{(\text{blank} - \text{sample})\text{ml} \times 2.502}{\text{sample wt (g)}}$$

Correct to dry weight where necessary.
Record in terms of CO_3-C if desired.

Chlorine

Very few rock minerals contain this element (as chloride) and it is largely absent from the common silicate minerals. Evaporate deposits of 'rock salt' near to the surface are only local in the United Kingdom and much of the chloride in soils originates from the atmosphere especially near the sea where it is the dominant anion.

Chlorine as chloride is an essential element for animal life where it contributes to ionic balance in cells and tissues. It appears to be at least a micronutrient in plants although evidence is available for only a few species. It is rarely differentiated from sodium in its effects upon plants (Jennings, 1976). Although chlorine gas is highly toxic, the ion is relatively harmless and the marine environment is rich in animal and algal life. However, only a few seed plants (e.g. halophytes) can tolerate high salinity (Flowers *et al.*, 1977; Wainwright, 1980).

Although it is a major anion in fresh waters it is less significant than bicarbonate in the chemistry of lake water. Levels in rainwater are an important indicator of marine influences in the atmosphere. Free chlorine is scarce in nature, but chlorination is a standard method for purifying water and residual chlorine is a routine estimate in the water industry (Standard Methods for the Examination of Water and Wastewater, 1985). Levels in the ocean exceed 19 $g\,l^{-1}$ although in terms of equivalents ($\simeq$0.55 $g\,eq\,l^{-1}$) it is balanced by sodium and other ions.

The concentration ranges generally encountered are given below.

Mineral soils	0.004–0.08%
Organic soils (peat)	0.03–0.2%
Soil extractions	1–40 $\mu g\,g^{-1}$ (excl. saline)
Animal tissues	0.2–0.8%
Plant materials	0.04–0.4%
Rain waters	1.0–25 $mg\,l^{-1}$
Freshwater	2.0–100 $mg\,l^{-1}$

(dry weight basis where relevant).

Although the classical gravimetric and titrimetric methods for chloride are now less used than formerly, the Volhard back-titration using thiocyanate still has its adherants. Its chief merit is its simplicity but the sensitivity is limiting. There are few colorimetric methods although the indirect thiocyanate procedure is widely used and is described below. A continuous flow method is also given.

Various electrochemical methods are available and are particularly suitable for water samples. Some of these form the basis of 'salinity' measures. For higher concentration samples, potentiometric titrations (Golterman *et al.*, 1978) and ion selective electrodes are suitable. A polarograph method is also available (Allen *et al.*, 1974). However, perhaps the most effective technique for chloride (in waters) is ion chromatography (p. 280). This procedure can also be used for plant material following oxidation in a Schoniger flask (Tabatabai and Bremner, 1970). A comparison of methods for chloride has been carried out by Steele (1978).

Thiocyanate colorimetric method

There appears to be no ideal chromogenic reagent for the determination of chloride. The thiocyanate procedure is an indirect method which involves the reaction between ferric and thiocyanate ions. Thiocyanate ions are first liberated by chloride from the largely undissociated mercuric thiocyanate and then react with excess ferric iron to produce a red colour.

Although the method has some limitations regarding the colour formation (Sandell, 1959), it is relatively rapid and convenient. It is not particularly sensitive for chloride but is applicable because this anion is one of the more abundant to be found in natural waters and soils. Several anions interfere but of these only nitrate is likely to be a problem in ecological samples and will give a small positive error. An automated variant of the method is given below.

The method is primarily intended for waters and soil but may also be used for total soil and plant analyses after fusion of the sample with sodium carbonate. The fusate must be neutralized with sulphuric acid.

The detection limit under normal conditions is of the order of 0.2 ppm Cl.

Reagents

1 Chloride standards.

Standard (1 ml ≡ 1 mg Cl): dissolve 1.6484 g dry NaCl in water and dilute to 1 litre.

2 Sodium acetate–acetic acid buffer.

Dissolve 30 g $NaOAc.3H_2O$ in 500 ml water. Add 1 ml glacial HOAc and dilute to 1 litre.

3 Ferric alum–nitric acid reagent.

Dissolve 30 g $FeNH_4(SO_4)_2.12H_2O$ in 350 ml water. Add 95 ml conc HNO_3. Bring to the boil, cool and filter. Dilute to 500 ml.

4 Mercuric thiocyanate solution.

Suspend 0.5 g $Hg(SCN)_2$ in 250 ml water. Stir for 12 hours at room temperature and filter.

Procedure

Prepare the sample solution as described on pp. 29, 38, 60, 77 and dilute to volume where appropriate. Water samples do not normally require prior treatment.

Pipette 0–10 ml standard into 50 ml volumetric flasks to give a range from 0–10 mg Cl.

Add acid or soil extractant comparable to sample aliquots.

Measure not more than 15 ml sample solution into a 50 ml volumetric flask.

From this point treat standards and samples in the same way.

Add 20 ml buffer solution and mix.

Add 10 ml acid ferric alum solution and mix.

Add 4 ml mercuric thiocyanate reagent, dilute to volume and mix.

Measure the absorbance at 460 nm or with a green filter using water as a reference.

Prepare a calibration curve from the standard solutions and use it to obtain mg Cl in the sample aliquot.

Carry out blank determinations in the same way and subtract where necessary.

Calculation
If C = mg Cl obtained from the graph then for:
1 Plant materials,

$$\text{Cl (\%)} = \frac{C\,(\text{mg}) \times \text{soln vol (ml)}}{10 \times \text{aliquot (ml)} \times \text{sample wt (g)}}$$

2 Soil extracts,
Extractable Cl (mg 100 g^{-1}): as above $\times 10^3$
3 Waters,

$$\text{Cl (mg l}^{-1}) = \frac{C\,(\text{mg}) \times 10^3}{\text{aliquot (ml)}}$$

Apply factors for dilution or concentration and correct to dry weight where necessary.

Continuous flow method

The thiocyanate method is readily adapted for continuous flow colorimetry. The flow diagram (manifold) given here in Fig. 5.2 is that needed for the AAII system described later in Chapter 8. The manifold needed for AAI operated at twice the flow rate and needed larger mixing coils. This was described in the first edition of this book (Allen *et al*., 1974).

Reagents
1 Chloride standards.
Stock solution (1000 ppm): dissolve 1.6484 g dry NaCl in water and dilute to 1 litre.
Standards: prepare a range containing from 0 to 50 ppm Cl by dilution of the stock solution. Include acid or extractant to match the sample solutions.
2 Sodium acetate–acetic acid buffer.
3 Ferric alum–nitric acid reagent.
4 Mercuric thiocyanate.
Prepare reagents **2**, **3** and **4** as given in the previous (manual) thiocyanate method.

Operating conditions
Manifold details are given in Fig. 5.2.
Wavelength for maximum absorption: 460 nm.

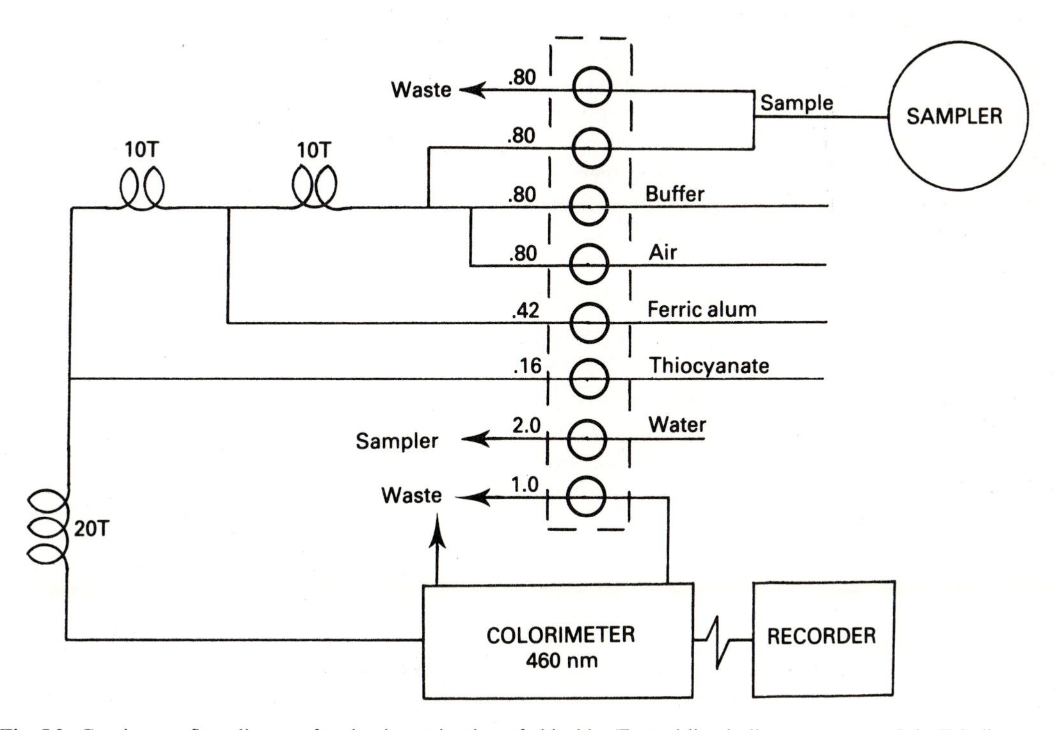

Fig. 5.2. Continuous flow diagram for the determination of chloride. (Dotted line indicates pump module. T indicates number of turns on mixing coils. All flow rates expressed as ml min^{-1}.)

Sampling rate: up to 60/hour (sample: wash ratio $=2:1$).

Calculation
If $C=$ ppm Cl obtained from the graph then for:
1 Soil extracts,

extractable Cl (mg 100 g^{-1})=

$$\frac{C\,(\text{ppm}) \times \text{sample vol (ml)}}{10 \times \text{sample wt (g)}}$$

2 Waters,

$Cl\,(mg\,l^{-1}) = C\,(ppm)$.

Apply factors for dilution or concentration and correct to dry weight where necessary.

Note
Tubing used for this method should be washed periodically with 10% HNO_3.

Ion-selective electrode method

The ISE procedure for chloride is subject to interference from cyanide, iodide, thiosulphate and bromide. However, these ions are unlikely to be present in ecological materials in sufficient quantity to cause problems. Hydroxyl ions may be troublesome and measurements should therefore be made below pH 7.

The procedure is not very sensitive, but chloride is an abundant element in nature and present in sufficient quantities in most materials to make the method a possible alternative. The sensitivity is about 0.5 ppm Cl and reproducibility should be of the order of $\pm 2.5\%$. Further information about this method can be obtained from Sekerka and Lechner (1978). See also p. 269.

Reagents
1 Chloride standards.
Stock solution (1000 ppm Cl): dissolve 1.6484 g dry NaCl in water and dilute to 1 litre.
Working standards: prepare a suitable range to at least 100 × detection limit (0.2 ppm) to allow an electrode check (Note **2**). Extra standards are needed at the lower end.

Include acid or or soil extractant as appropriate to match the samples.

Procedure
Use for soil extracts or waters.
Set up the instrument and electrodes according to the manufacturer's instructions.
Ensure all solutions are at room temperature, then read standards and samples under the same conditions.
Prepare a calibration curve from the standard readings using semi-logarithmic paper with concentration on the logarithmic axis and instrument reading on the linear axis (see Note **2**).
Use the calibration curve to obtain ppm Cl in the sample solutions.
Carry out blank determinations in the same way and subtract where necessary.

Calculation
If $C=$ ppm Cl obtained from the graph then for:

1 Soils extracts,

Extractable Cl (mg 100 g^{-1}): as above $\times 10^3$

2 Waters,

$Cl\,(mg\,l^{-1}) = C\,(ppm)$

Apply factors for dilution or concentration and correct to dry weight where necessary.

Notes
1 For low ionic strength samples such as waters it is recommended that 0.5 ml 5 M $NaNO_3$ be added to 25 ml aliquots of samples and standards before reading.
2 Chloride is a monovalent ion and a tenfold increase in concentration (one log unit) should therefore decrease the reading by about 58 mV.
3 The Nernst linear relationship does not hold down to the detection limit. The range below 5 ppm is non-linear and needs a separate calibration line.

Cobalt

Minerals containing cobalt are infrequent and it is largely absent from the common silicate minerals. Most soils contain only low levels and deficiencies occur in some parts of the world.

This element is an essential micro-nutrient for animals, being part of the structure of vitamin B_{12}, and is involved with certain enzymes. Ruminants are particularly sensitive to cobalt deficiency. Plants require only minute amounts mainly for catalytic functions and levels are often $<0.5\ \mu g\ g^{-1}$ dry weight. High levels are toxic although certain hyperaccumulator species are known for cobalt as well as for copper in the copper belt of Africa, as reviewed by Brooks *et al.* (1980). The concentration ranges normally encountered are given below.

Mineral soils	1–50 $\mu g\ g^{-1}$
Organic soils (peat)	0.2–1 $\mu g\ g^{-1}$
Soil extractions	0.05–4 $\mu g\ g^{-1}$
Plant materials	0.1–0.6 $\mu g\ g^{-1}$
Animal tissues	0.02–0.1 $\mu g\ g^{-1}$
Rainwater	0.05–0.5 $\mu g\ l^{-1}$
Freshwater	0.5–2.5 $\mu g\ l^{-1}$

(dry weight basis where relevant).

Several colorimetric reagents have been described for cobalt, but nitroso-R has usually been preferred in biological work because of its sensitivity, although other elements interfere. The procedure given below uses dithizone in chloroform to separate interferences. The polarograph is also sensitive for cobalt although the peak is very close to that of zinc and precautions are needed for ecological material as given below. Atomic absorption procedures are about as sensitive as those for copper but the much lower levels of cobalt mean that these procedures are unsuitable for ecological samples. Electrothermal atomization is then preferable and a method is suggested below. As with other micro-nutrients, contamination in any preparation stages must be avoided if valid results are to be obtained.

Preparation of the initial solution for vegetation will require increased amounts of sample and dry-ashing will be needed.

Nitroso-R-salt method

Cobalt forms a red complex with sodium 1-nitro-2-naphthol-3,6-disulphonate, better known as nitroso-R-salt. This complex is stable in boiling nitric acid which makes the reagent almost specific for the metal. Copper, iron and nickel interfere when present in very large quantities relative to cobalt. However, these may be removed by means of a preliminary extraction with dithizone in chloroform at a pH of 3.5 and then extraction of the cobalt using dithizone in carbon tetrachloride at pH 8. The nitroso-R- salt complex is formed at about pH 8 and a phosphate–borate buffer is used to maintain this pH. After colour development, boiling with dilute nitric acid destroys any heavy metal complexes which may have been formed, leaving only the cobalt colour. The method will measure about 0.1 μg cobalt in the final solution. As the levels of cobalt in most samples are very low, reproducibility is poor and $\pm 15\%$ is probably the best that can generally be achieved.

Reagents

1 Cobalt standards.
Stock solution (1 ml $\equiv$ 100 μg Co): dissolve 0.4770 g dry $CoSO_4.7H_2O$ in water and dilute to 1 litre.
Working standard (1 ml $\equiv$ 1 μg Co): dilute the stock solution 100 times.
Prepare fresh daily.

2 Citric acid, 0.2 M.
Dissolve 4.2 g citric acid in water and dilute to 100 ml.

3 Sodium hydroxide, M.
Dissolve 40 g NaOH in water and dilute to 1 litre when cool.

4 Buffer solution.
Dissolve 6.18 g H_3BO_3 and 35.62 g $Na_2HPO_4.2H_2O$ in water and add 500 ml NaOH. Dilute to 1 litre with water.

5 Dithizone, 0.5% w/v in CCl_4.

6 Dithizone, 0.20% w/v in $CHCl_3$.

7 Nitroso-R-salt, 0.2% w/v.

8 Perchloric acid, 60%.

9 Nitric acid, conc.

Procedure

Prepare the solution as described on pp. 29, 38, 60 and dilute to volume.

Pipette 0–10 ml of the standard (giving a range of standards from 0–10 μg Co) into separating funnels and add 0.2 M citric acid until the pH is 3.5.

Neutralize a sample aliquot with M NaOH and add citric acid solution until the pH is 3.5.

Transfer to a separating funnel.

From this point treat standards and samples in the same way.

Extract with 20 ml of 0.2% dithizone in $CHCl_3$.

Repeat extraction until the $CHCl_3$ layer remains green (three extractions are normally required).

Reject the $CHCl_3$ layers.

Adjust the aqueous phase to pH 8.3 by adding the buffer solution.

Extract three times with 10 ml portions of 0.05% dithizone in CCl_4.

Evaporate the combined CCl_4 extracts in a small beaker on a steam bath.

Add 3 ml 60% $HClO_4$ and 3 ml conc HNO_3 to the residue.

Cover with a watch glass and digest slowly on a hot plate until the acid solution is colourless.

Remove the watch glass and fume off the acids.

Take up the residues in 1 ml citric acid solution.

Wash the solution into a boiling tube keeping the volume about 5 ml.

Adjust the pH to 8.0 by addition of 1.2 ml buffer solution.

Add 1 ml of 0.2% nitroso-R-salt with vigorous shaking during the additions.

Boil the solution for 1 minute then add 1 ml conc HNO_3 and boil again for 1 minute.

Allow to cool and dilute to exactly 10 ml.

Measure the absorbance at 422 nm or with a blue filter using water as the reference.

Prepare a calibration curve from the standard solutions and obtain μg Co in the sample aliquots.

Carry out blank determinations in the same way and subtract where necessary.

Calculation

If $C = \mu$g Co obtained from the graph then for:

1 Plant materials, soils and soil extracts,

$$\text{Co } (\mu\text{g g}^{-1}) = \frac{C\,(\mu\text{g}) \times \text{soln vol (ml)}}{\text{aliquot (ml)} \times \text{sample wt (g)}}$$

2 Waters,

$$\text{Co } (\mu\text{g l}^{-1}) = \frac{C\,(\mu\text{g}) \times 10^3}{\text{aliquot (ml)}}$$

Apply factors for dilution or concentration and correct to dry weight where necessary.

Electrothermal atomization

If the concentration of cobalt in the residual sample solution is below the limit of detection using flame atomic absorption then electrothermal atomization will generally have sufficient sensitivity to produce a reading. Details of this technique and operating conditions suitable for most elements of interest to the ecologist are given in Chapter 8. Mitchell *et al.* (1987) propose a graphite furnace method for cobalt in acetic acid extracts of soils.

Polarographic method

Polarographic techniques are now little used for the determination of metallic elements in biological materials because atomic absorption is generally superior. The polarographic method for cobalt, as described by Nangniot (1967) is an exception because of its sensitivity, and this method is described below. As already mentioned the concentration of cobalt in most biological materials is very low and in the direct polarographic determination of cobalt the adjacent zinc step completely masks the small cobalt step. However, if dimethylglyoxime is added to an ammonium hydroxide–ammonium chloride electrolyte, a soluble cobalt complex is produced which gives a large step at $E_{\frac{1}{2}} = -1.12$ V. This is proportional to cobalt concentration and free from interferences. Nickel also reacts with dimethylglyoxime, but it is precipitated and does not interfere. The procedure described below utilizes this reaction and can either be carried out on its own or in conjunction with the determination of copper and zinc which can be determined in the same electrolyte (see Allen *et al.*, 1974).

The ultimate sensitivity depends on the relative level of zinc in the solution to be measured. Cobalt concentrations as little as 0.01–0.02 $\mu g\,ml^{-1}$ can be measured, but as levels are very low in plant materials, reproducibility better than $\pm 10\%$ is unlikely.

Reagents

1 Cobalt standards.

Stock solution (1 ml $\equiv$ 5 μg Co): dissolve 0.4770 g dry $CoSO_4.7H_2O$ in water and dilute to 1 litre. Further dilute 50 ml of this solution to 1 litre to produce the stock solution.

Working standards: pipette from the stock solution to give a range from 0–0.5 μg Co. Evaporate to dryness and take each up in 5 ml electrolyte as for the samples. (When other trace elements are to be determined in the same electrolyte, prepare a common range of standards.)

2 Dimethylglyoxime, 1% w/v.

Dissolve 1.0 g dimethylglyoxime in 100 ml ethanol.

3 Citric acid, 0.5 M.

Dissolve 26.25 g citric acid in water and dilute to 250 ml.

4 Ammonium hydroxide, 1 + 3.

5 Dithizone, 0.02% w/v in chloroform.

6 Base electrolyte.

Dissolve 50 ml 0.88 NH_3 solution and 50 g NH_4Cl in water and dilute to 500 ml.

Procedure

Prepare the sample solutions as described on pp. 29, 38, 60 and dilute to volume.

Transfer a suitable aliquot to a separating funnel.

Take a range of standard solutions and treat likewise.

Add 4 ml citric acid solution, followed by NH_4OH until the pH is between 8 and 9 (phenolphthalein is suitable as an internal indicator).

Add 10 ml dithizone reagent and shake well for 2 minutes.

Allow to settle then collect the organic phase in a 50 ml beaker.

Repeat the extraction twice using 5 ml portions of dithizone reagent.

Evaporate the combined chloroform extract to dryness on a water bath or hot plate (in a fume cupboard).

Add 3 ml HNO_3 and 3 ml 60% $HClO_4$ and cover with a watch glass.

Digest the mixture on a hot plate until colourless.

Remove the watch glass and fume off the acids (it is important to ensure that no acid is left).

Add 5 ml of the base electrolyte to the beaker and swirl to dissolve the residue.

Transfer the sample solutions and standards to the polarograph cells, subsequently following the same procedure for both.

Remove oxygen from the solution (either by passing oxygen-free N_2 or by adding a crystal of Na_2SO_3).

After reading the step or peak heights for any other elements (if required), add 0.2 ml dimethylglyoxime solution and mix by bubbling through nitrogen.

Adjust the start potential sensitivity and compensation controls on the polarograph and record the step or peak at $E_{\frac{1}{2}} = -1.12$ V.

Prepare a calibration curve from the standards and use it to obtain μg Co in the sample aliquots.

Carry out blank determinations in the same way and subtract where necessary.

Calculation

If $C = \mu g$ Co obtained from the graph then for:

1 Plant materials, soils and soil extracts,

$$\text{Co}\ (\mu g\ g^{-1}) = \frac{C\,(\mu g) \times \text{soln vol (ml)}}{\text{aliquot (ml)} \times \text{sample wt (g)}}$$

2 Waters,

$$\text{Co}\ (\mu g\ l^{-1}) = \frac{C\,(g) \times 10^3}{\text{aliquot (ml)}}$$

Apply factor for dilution or concentration and correct to dry weight where necessary.

Notes

1 Polarographic spike maxima may be suppressed by adding about 0.1 ml of 0.5% w/v gelatin solution.

2 If other ionic species interfere, the technique of standard addition is recommended. In this technique the step or peak height is measured for a known volume of sample solution. A known volume of standard solution is then added to the cell, the resultant solution is well-mixed and the new step or peak then measured. Simple calculation will then allow the concentration of cobalt in the sample to be obtained.

Copper

A wide range of minerals containing copper are known together with deposits of the native element. Several minerals are associated with ores mined in various parts of the world. The primary minerals are essentially sulphides, sometimes in combination with iron. Other copper minerals are complex oxidized products and salts although silicates rarely contain this element. However, some soil copper is held by clay minerals in an exchangeable form.

Copper is an essential micronutrient for plants and animals where it is involved with many enzyme systems (Bowen, 1979). It is a constituent of certain other proteins including the pigment haemocyanin found in crustacea and molluscs. Copper deficiencies are known in crops, but much attention has also been given to the toxic nature of the element. Certain species in contaminated (mining) areas possess tolerance mechanisms for copper and one or two plants such as the grass *Agrostis* have been intensively studied. Similar mechanisms exist for zinc and other toxic heavy metals and have been widely examined (Foy *et al.*, 1978; Woolhouse, 1983). Some species can largely exclude a toxic metal from the shoot, but other plants accumulate it there, sometimes to high levels (Baker, 1981; Brooks *et al.*, 1980). These mechanisms have a genetic basis which allows selection of varieties of potential value in reclaiming industrial waste (Bradshaw and Chadwick, 1980). Levels in natural waters can also reach toxic levels for small organisms and algae particularly in streams draining mining or other areas rich in the element. Otherwise natural water contents are low.

The concentration ranges normally encountered are:

Mineral soils	5–80 $\mu g\,g^{-1}$
Organic soils (peat)	6–40 $\mu g\,g^{-1}$
Soil extractions	0.1–3 $\mu g\,g^{-1}$
Plant materials	2.5–25 $\mu g\,g^{-1}$
Animal tissues	10–100 $\mu g\,g^{-1}$
Rainwater	0.2–2 $\mu g\,l^{-1}$
Freshwater	2–50 $\mu g\,l^{-1}$

(dry weight basis where relevant).

Copper reacts with many organic compounds, but the complex which has been most widely used is that based on sodium diethyldithiocarbamate. A method is given below and includes a dithizone extraction to separate interfering metals. The sensitivity may be limiting for the low levels of copper found in some materials.

Atomic absorption provides a simple procedure relatively free of interferences and is preferable to colorimetry, but again the sensitivity may be limiting. The direct method given below is followed by a procedure in which copper is first concentrated with APDC. An electrothermal atomizer method which is much more sensitive is also given and is recommended for low concentration samples, for example natural waters.

Polarography is relatively sensitive for copper, but generally needs a separation stage to remove interferences and it offers little advantage over the other procedures.

As with other micronutrient analyses, precautions must be taken to avoid contaminants in the laboratory and adequate blanks are essential for all methods.

Carbamate method

Copper reacts with diethyldithiocarbamates to give a brown complex. The reaction takes place over a wide pH range and the colour does not change between pH 5.7 and 9.2. The colour may be determined in aqueous solution, but the complex is not very soluble and may precipitate if the copper content is high. It is better to extract the colour into an organic solvent for colour measurement. Many of the metals which interfere are not present in biological material in sufficient quantities to affect the determination. Interference due to iron, calcium and phosphate may be prevented by the addition of citric acid.

In the procedure given below, iron and other metals are removed by extracting the copper, along with zinc and cobalt, into dithizone solution in the presence of ammonium citrate at pH 8–9. After destruction of the dithizone, the copper may be determined as the dithiocarbamate using carbon tetrachloride to extract the complex prior to colour measurement. Zinc and cobalt do not interfere.

The method is suitable for use with all ecological materials although amounts in some waters may be too low to be detected even after concentration. The method is sensitive to about 0.5 μg copper. Since copper levels are generally low, reproducibility is poor and ±10% is a realistic estimate for routine operations.

Reagents

1 Copper standards.

Stock solution (1 ml ≡ 100 μg Cu): dissolve 0.3930 g $CuSO_4.5H_2O$ in water and dilute to 1 litre.

Working standard (1 ml ≡ 10 μg Cu): dilute the stock solution 10 times.

(These standard solutions are liable to change in concentration, even during short storage periods, and must be checked or renewed frequently.)

2 Citric acid, 0.5 M.

Dissolve 52.5 g of citric acid in 500 ml water.

3 Ammonium hydroxide, 1+3.

4 Dithizone, 0.02% w/v in $CHCl_3$.

5 Dithizone, 0.02% w/v in CCl_4.

6 Sulphuric acid, 1% v/v.

7 Sodium diethyldithiocarbamate, 1% w/v.

8 Carbon tetrachloride.

9 Perchloric acid, 60%.

10 Nitric acid, conc.

Procedure

Prepare the sample solutions as described on pp. 29, 38, 60 and dilute to volume.

Pipette from 0–5 ml of the standard (giving a range of standards from 0–50 μg Cu) into separating funnels.

Transfer suitable aliquots of the sample solutions to separating funnels and from then on treat standards and samples in the same way.

Add 4 ml citric acid solution.

Adjust the pH to between 8 and 9 by adding 1+3 NH_4OH.

Add 10 ml 0.02% dithizone in $CHCl_3$ and shake the funnel for about 1 minute.

Allow the phases to separate and collect the organic (heavier) phase in a 50 ml beaker.

Repeat this extraction until the $CHCl_3$ layer remains green after shaking.

Check the pH of the aqueous phase and adjust to between 8 and 9 if necessary.

Add 10 ml 0.02% dithizone in CCl_4, shake and separate as before and add to combined $CHCl_3$ extracts.

Evaporate the extracts to dryness on a steam bath (in a fume cupboard).

Add 3 ml 60% $HClO_4$ and 3 ml conc HNO_3, cover the beaker with a watch glass and digest on a hot plate until the organic matter is destroyed.

Remove the watch glass and fume off the acid.

When cool, add 5 ml of 1% H_2SO_4 and swirl to dissolve.

Transfer to 50 ml separating funnel, add 2 ml of sodium diethyldithiocarbamate solution and mix well.

Add 10 ml CCl_4 (accurately) and mix.

Stand 10 minutes, then run off the CCl_4 phase.

Measure the absorbance at 440 nm or with a blue filter using water as a reference.

Prepare a calibration curve from the standard solutions and use it to obtain μg Cu in the sample aliquot.

Carry out blank determinations in the same way and subtract where necessary.

Calculation

If C = μg Cu obtained from the graph then for:

1 Plant materials, soils and soil extracts,

$$\text{Cu } (\mu g\ g^{-1}) = \frac{C(\mu g) \times \text{sample vol (ml)}}{\text{aliquot (ml)} \times \text{sample wt (g)}}$$

2 Waters,

$$Cu\ (\mu g\ l^{-1}) = \frac{C(\mu g) \times 10^3}{\text{aliquot (ml)}}$$

Apply factors for dilution or concentration and correct to dry weight where necessary.

Atomic absorption method

The procedure for copper by atomic absorption is virtually free from interference and is straightforward providing the sample concentration is adequate. The best sensitivity is obtained using an air–acetylene flame. For plant material it is generally necessary either to take a larger weight for the digestion (1–2 g) or to concentrate the solution. The copper content of the soils is very variable, so initial tests will be required to determine the optimum conditions. Some concentration is usually needed for water samples. A full description of the determination of copper in biological materials is given by Christian and Feldman (1970).

A concentration technique is described in the second method below. In this method ammonium pyrrolidine dithiocarbamate (APDC) is used to complex the copper (Malissa and Schoffmann, 1955). This is then extracted with 4-methyl pentan-2-one which can be aspirated directly into the flame (Parker *et al.*, 1967). This extraction can also be used for increasing the concentration of other heavy metals prior to analysis, including cadmium, cobalt, iron, mercury, lead and zinc (Brookes *et al.*, 1967; Iyengar *et al.*, 1981).

As little as 0.02 $\mu g\ ml^{-1}$ in solution can be detected using sensitive instruments. At maximum scale expansion the reproducibility can be of the order of ± 4 to 8%.

Reagents

1 Copper standard.

Stock solution (100 ppm Cu): dissolve 0.3930 g $CuSO_4.5H_2O$ in water and make up to 1 litre.

Working standards: dilute to give an intermediate standard of 10 ppm and from this dilute to produce a range from 0–1.0 ppm Cu. Include acid or soil extractant as appropriate to match the sample solutions.

2 Ammonium pyrrolidine dithiocarbamate (APDC), 1% w/v.

3 4-Methyl pentan-2-one (MIBK).

Direct method

Procedure

Prepare sample solutions as described on pp. 29, 38, 60

Select the 324.8 nm wavelength and adjust air and gas flows, slit-width and other settings as recommended for the instrument employed.

Always allow the hollow cathode lamp adequate time to stabilize.

Prepare a calibration curve from the standard range by setting the top standard to a suitable readout value and the 0 ppm standard to zero, then aspirating each standard in turn.

Aspirate the sample solutions into the flame under the same conditions as the standards.

Check the top, zero and an intermediate standard frequently.

Flush the spray chamber and burner frequently with water, particularly if soil extracts are aspirated.

Use the calibration curve to obtain ppm Cu in the sample solutions.

Carry out blank determinations in the same way and subtract where necessary.

Calculation

If C = ppm Cu obtained from the graph then for:

1 Plant materials, soils and soil extracts,

$$Cu\ (\mu g\ g^{-1}) = \frac{C\ \text{(ppm)} \times \text{soln vol (ml)}}{\text{sample wt (g)}}$$

2 Waters,

$$Cu\ (\mu g\ l^{-1}) = C\ \text{(ppm)} \times 10^3$$

Apply factors for dilution or concentration and correct to dry weight where necessary.

Note

It is desirable to use a high solids burner for soil extracts to avoid air build-up of carbon deposits.

APDC method

Procedure

(Use for low Cu concentrations.)

Prepare sample solutions as described on pp. 29, 38, 60.

Pipette suitable aliquots into narrow necked flasks.

Pipette aliquots of standard to give a range of 0–4 μg Cu in similar flasks.

Dilute standards and samples to base of neck.

Add 2 ml of 1% APDC.

Mix and leave for 2 minutes.

Add exactly 4 ml MIBK.

Shake for 1 minute and allow to separate.

Without delay aspirate the organic (lighter) phase directly into the flame.

Prepare the calibration curve as before and obtain μg Cu in the sample aliquots.

Treat the blanks in the same way and subtract where necessary.

Calculation

If $C = \mu$g Cu obtained from the graph then for:

1 Plant materials, soils and soil extracts,

$$\text{Cu } (\mu g\ g^{-1}) = \frac{C\ (\mu g) \times \text{sample vol (ml)}}{\text{aliquot (ml)} \times \text{sample wt (g)}}$$

2 Waters,

$$\text{Cu } (\mu g\ l^{-1}) = \frac{C\ (\mu g) \times 10^3}{\text{aliquot (ml)}}$$

Apply factors for dilution or concentration and correct to dry weight where necessary.

Electrothermal atomization

If the concentration of copper in the residual sample solution is below the limit of detection for flame atomic absorption then electrothermal atomization will probably give sufficient sensitivity. Details of this technique are given in Chapter 8.

Iron

Iron is a major element in various primary minerals, notably the 'ferromagnesium' group of silicates together with other compounds including oxides which form workable ores. Although less prominent in the secondary (clay) minerals it readily forms 'hydrous oxides' in soil along with aluminium. Hence most soils contain significant amounts of iron. Although ferric compounds are relatively insoluble, iron, like aluminium, can form a number of soluble hydroxy-ion species depending on the pH, and a small fraction can be mobilized by organic linkages. Ferrous ions are taken up by plants. They are most prominent under reducing conditions and are mobile unless precipitated as the sulphide.

Iron is an essential minor nutrient for plants and animals. It is involved in chlorophyll synthesis and over half the leaf iron is in the chloroplasts. Certain enzyme systems such as the oxidases depend on iron whilst in higher animals it is found in haemoglobin. In spite of its abundance sensitive plants can show iron deficiency symptoms (chlorosis) on calcareous soils or following excessive liming. The problem apparently involves several factors and has been intensively studied along with plant response mechanisms to iron stress (Decock, 1981; Cheng and Barak, 1982 and Miller *et al.*, 1984). Iron toxicities have been recorded in plants although the element is less potent than manganese and aluminium.

Although iron levels are low in natural waters it can be present there in various ionic, organic and mineral forms whose stability is related to a number of factors including redox potential and pH in water and sediments (Meltzer and Steinberg, 1983). Cycling of iron in lakes is more complex when summer stratification imposes a seasonal pattern, as studied by Davison and Tipping (1984) who also noted differences between iron and manganese.

The concentration ranges normally encountered are given below.

Mineral soils	0.5–10%
Organic soils (peat)	0.02–0.5%
Soil extractions	50–1000 $\mu g\ g^{-1}$
Plant materials	40–500 $\mu g\ g^{-1}$
Animal tissues	100–400 $\mu g\ g^{-1}$
Rainwater	5–150 $\mu g\ l^{-1}$
Freshwater	50–1000 $\mu g\ l^{-1}$

(dry weight basis where relevant).

Although many colorimetric procedures for iron have been described in the literature, only about four reagents are widely used. These are thiocyanate which reacts with iron (III) and 2,2-dipyridyl, *o*-phenanthroline and bathophenanthroline which react with iron (II). They all give red complexes which are suitable for quantitative estimates under controlled conditions. The reagents which react with iron (II) are frequently used in conjunction with a reducing agent so that total iron in solution may be determined.

Atomic absorption is also applicable for the determination of iron in biological materials and a method is given later in this Section. Both the atomic absorption and bathophenanthroline methods have been considered by the Analytical Methods Committee (1978a, b).

Contamination of samples by iron from the laboratory environment is very common unless stringent precautions are taken to prevent it (see p. 335). It is noticeable that iron results from inter-laboratory tests are generally much more variable than those of other elements.

Sulphonated bathophenanthroline method

Case (1951) introduced 4:7-diphenyl-1:10-phenanthroline (bathophenanthroline) whose iron complex is normally extracted into an organic solvent before measurement. The present method used a sulphonated form of bathophenanthroline as described by Riley and Williams (1959) and omits the complication of the organic extraction.

Riley and Williams analysed iron-rich solutions of rocks, but aliquots of plant material digests may have to be relatively large to get a measureable colour, thereby introducing significant amounts of acid. To control this and eliminate a separate neutralization stage, the strength of the buffer used by them was increased to 33% w/v sodium acetate trihydrate (giving a pH of 4.8) as described by Quarmby and Grimshaw (1967). This modification is adopted in the method described below which includes hydroxylamine hydrochloride, thereby enabling total iron to be estimated.

When analysing water the omission of the hydroxylamine reagent will enable iron (II) alone to be determined but minimum exposure of the sample to air is essential at all stages.

The method given below is applicable for most samples without modification. It is possible to measure levels down to 0.2 μg iron in the final solution. This method is readily automated and details of a continuous flow method are given later.

Reagents

1 Iron standards.

Stock solution (1 ml ≡ 0.1 mg Fe): dissolve 0.100 g clean untarnished Fe wire in about 10 ml of warm 10% H_2SO_4. When cool dilute to 1 litre with water.

Working standard (1 ml ≡ 0.001 mg Fe): dilute the stock solution 100 times. Prepare fresh at regular intervals.

2 Sulphonated bathophenanthroline reagent.

Add 4.0 ml fuming H_2SO_4 (containing 20% SO_3) to 0.4 g bathophenanthroline (4:7-diphenyl-1:10-phenanthroline). Stir until dissolved, and allow to stand 30 minutes, then pour into 400 ml water. Neutralize with NH_4OH to between pH 4 and 5 and dilute to 1 litre.

3 Sodium acetate trihydrate 33% w/v.

Dissolve 330 g $NaOAc.3H_2O$ in water and dilute to 1 litre.

4 Hydroxylamine hydrochloride, 2.5% w/v.

5 Combined reagent.

Mix 33% NaOAc solution, sulphonated bathophenanthroline reagent and 2.5% $NH_2OH.HCl$ in the ratios 4:3:1.

Procedure

Prepare the sample solution as described on pp. 29, 38, 60 and dilute to volume.

Acid digests must be diluted to about 10 ml, boiled and cooled before dilution to volume.

Pipette 0–30 ml of working standard into 50 ml volumetric flasks to give a range from 0–0.03 mg Fe.

Add acid or soil extractant comparable to sample aliquots.

Measure suitable sample aliquots of not more than 20 ml into 50 ml volumetric flasks.

From this point treat standards and samples in the same way.

Add 16 ml of combined reagent and dilute to volume.

Measure the absorbance at 536 nm or with a yellow-green filter using water as a reference.

Prepare a calibration curve from the standard solutions and use it to obtain mg Fe in the sample aliquots.

Carry out blank determinations in the same way and subtract where necessary.

Calculation

If C = mg Fe obtained from the graph then for:

1 Plant materials and soils,

$$\text{Fe (\%)} = \frac{C \text{ (mg)} \times \text{soln vol (ml)}}{10 \times \text{aliquot (ml)} \times \text{sample wt (g)}}$$

2 Soil extracts,

Extractable Fe (mg 100 g^{-1}): as above $\times 10^3$

3 Waters,

$$\text{Fe (mg l}^{-1}) = \frac{C \text{ (mg)} \times 10^3}{\text{aliquot (ml)}}$$

Apply factors for dilution or concentration and correct to dry weight where necessary.

Continuous flow method

The manual bathophenanthroline procedure is suitable for automation with only slight modification (Quarmby and Grimshaw, 1967). The combined reagent is not used. If iron (II) only is required, the hydroxylamine reagent should be replaced by water and air should be excluded from samples prior to analysis. The flow diagram given here is that needed for the AAII system described in Chapter 8. The diagram needed for AAI was described in the first edition of this book (Allen *et al.*, 1974).

Reagents

1 Iron standards.

Stock solution (100 ppm Fe): dissolve 0.100 g dry untarnished Fe wire in a few ml 10% H_2SO_4 (v/v), warming if necessary. Dilute to 1 litre.

Standard solutions: prepare a range containing 0–4 ppm. 0–1 ppm will be sufficient for water samples but will require range expansion facilities. Include 1% acid, soil extractant or water as appropriate.

2 Sulphonated bathophenanthroline

3 Sodium acetate hydrate, 33% w/v

4 Hydroxylamine hydrochloride, 0.5% w/v

Prepare **2**, **3** and **4** as instructions on p. 108.

Operating conditions

Manifold details are given in Fig. 5.3.

A 10 mm flow cell is usually suitable but a longer cell can be used.

Wavelength for maximum absorption: 536 nm.

Sampling rate: up to 60 per hour (sample: wash ratio = 2 : 1).

Calculation

If C = ppm Fe obtained from the graph then for:

1 Plant materials and soils,

$$\text{Fe (\%)} = \frac{C \text{ (ppm)} \times \text{soln vol (ml)}}{10^4 \times \text{sample wt (g)}}$$

2 Soils extracts,

Extractable Fe (mg 100 g^{-1}): as above $\times 10^3$

3 Waters,

Fe (mg l^{-1}) = C(ppm)

Apply factors for dilution and concentration and correct to dry weight where necessary.

Atomic absorption method

Atomic absorption has been found to be a satisfactory technique for the determination of iron in biological material as shown by Allan (1959, 1961) and David (1962). The most sensitive line is that at 248.3 nm. The only interference likely in the analysis of organic material is that due to sulphate (Curtis, 1969), which may be overcome by addition of sulphate to the standards, so that the level is the same as in the samples. For soil digests inter-element interferences may be more severe

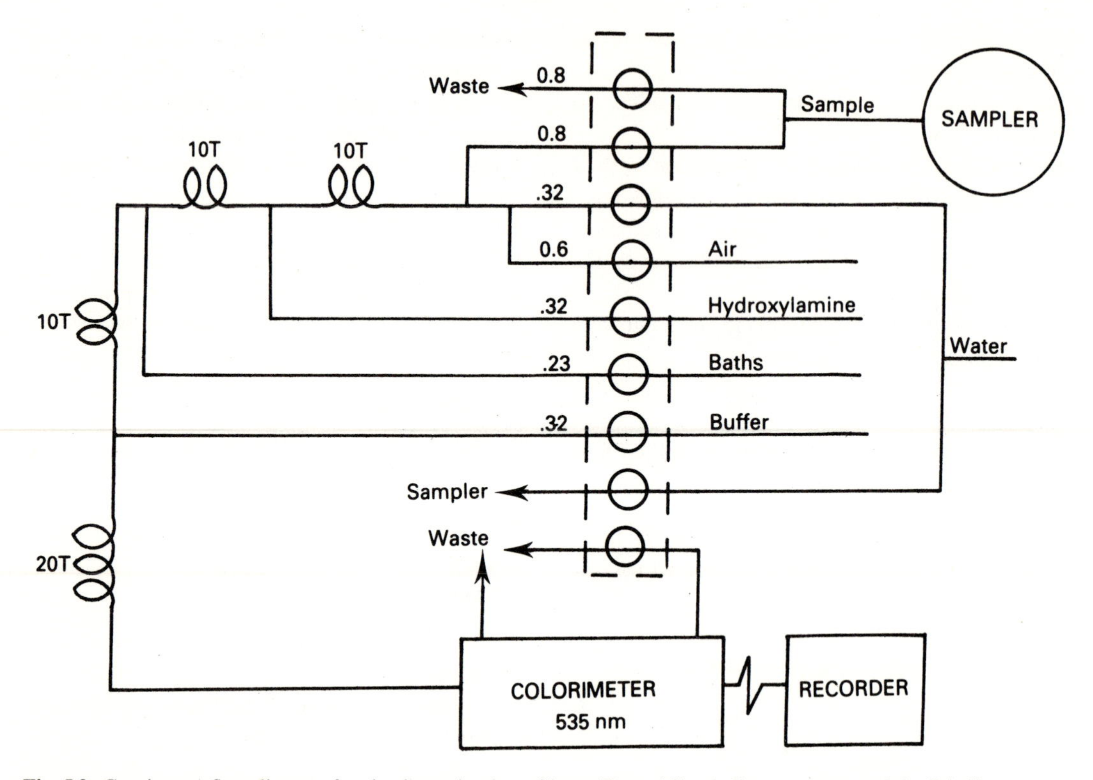

Fig. 5.3. Continuous flow diagram for the determination of iron. (Dotted line indicates pump module. T indicates, number of turns on mixing coils. All flow rates expressed as ml min^{-1}.)

and should be checked for each soil type and standards compensated accordingly. Salinas and March (1984) eliminate interfering ions by first using solvent extraction. The method is sensitive to about 0.01 $\mu g\,ml^{-1}$ iron depending on the instrument employed.

Reagents

1 Iron standards.

Stock solution (100 ppm Fe): dissolve 0.1 g clean untarnished Fe wire in about 10 ml of warm 10% H_2SO_4. When cool dilute to 1 litre.

Working standards: prepare a range of 0–20 ppm Fe. Include acid or soil extractant as appropriate to match the samples.

Procedure

Prepare the sample solution as described on pp. 29, 39, 60 and dilute to volume.

Select 248.3 nm wavelength (see Note) and adjust air and gas flows, slit-width and other settings as recommended for the instrument employed.

Always allow the hollow cathode lamp adequate time to stabilize.

Prepare a calibration curve from the standard range by setting the top standard to a suitable readout value and the 0 ppm standard to zero, then aspirating each standard in turn.

Aspirate the sample solutions under the same conditions as the standards.

Check the top, zero and an intermediate standard frequently.

Flush the spray chamber and burner frequently with water, particularly if soil extracts are aspirated.

Use the calibration curve to obtain ppm Fe in the sample solutions.

Carry out blank determinations in the same way and subtract where necessary.

Calculation
If C = ppm Fe obtained from the graph then for:
1 Plant materials and soils,

$$\text{Fe (\%)} = \frac{C\ \text{(ppm)} \times \text{soln vol (ml)}}{10^4 \times \text{sample wt (g)}}$$

2 Soil extracts,
Extractable Fe (mg $100\,g^{-1}$): as above $\times 10^3$
3 Waters,
Fe (mg l^{-1}) = C (ppm)
Apply factors for dilution or concentration and correct to dry weight where necessary.

Note
It is important to ensure that the 248.3 nm line is selected. There are two other fairly intense lines (248.8 nm and 249.1 nm) very close to this in the iron spectrum.

Magnesium

Magnesium is found in a wide range of primary silicate minerals notably the ferromagnesium group (olivine, pyroxenes and amphibolites) and in certain aluminosilicates such as the micas. Variable amounts are found in limestones and it is prominent in dolomite. It is also notable in serpentine minerals which form soils having a distinctive ecology (Proctor and Woodell, 1975). Most secondary clay minerals and chlorites contain magnesium in their structures. It has some affinity with calcium in its presence as carbonate in limestones and generally greater abundance and availability in basic soils. Both can be leached from acid soils although losses of magnesium are more readily replaced through weathering, and deficiencies in plants are not common.

The element is a macronutrient with distinctive functions in plants where it is a structural component of chlorophyll pigments. Unlike Ca^{2+}, levels of free Mg^{2+} are not unduly low in cytoplasm where it activates a much wider range of enzyme systems and is involved in ATP-dependent reactions. Magnesium is essential for animal life and low levels in herbage can induce disorders in livestock. Although it is not as common as calcium in base-rich waters, it contributes to total 'hardness' and must be included in estimates of this parameter.

The concentration ranges generally encountered are given below:

Mineral soils	0.2–2%
Organic soil (peat)	0.05–0.3%
Soil extractions	4–50 mg $100\,g^{-1}$
Plant materials	0.1–0.5%
Animal tissue	0.05–0.2%
Rainwater	0.1–2.0 mg l^{-1}
Freshwater	0.5–20 mg l^{-1}

(dry weight basis where relevant).

Although many analytical methods have been used in the past, they have almost all been superseded by atomic absorption which, in spite of some interferences, is the most reliable procedure available for magnesium. A method is described below. The EDTA titration is still in use mainly because magnesium can be estimated with calcium to give a measure of total 'hardness' in water. However, the indicator required is more prone to interferences than that used for calcium alone and magnesium has to be calculated by difference (see below). Colorimetric methods such as those based on 8-hydroxyquinoline or titan yellow are little used today. See Allen *et al.* (1974) for a description of the titan yellow method.

EDTA titration

The EDTA titration is still used in industry as a quick measure of total hardness due to calcium and magnesium salts (e.g. in boiler feed-water where high levels are unacceptable). A method for waters is described below. It is unsuitable for plant digests and soil extracts where both elements should be estimated separately by atomic absorption.

Eriochrome black T is the indicator required although precautions are required for its use. The reaction is carried out at pH 10 and a buffer solution is required. The colour change on adding EDTA is from red to blue and some magnesium should be present to obtain a satisfactory end-point. Inclusion of a small amount of the magnesium salt of EDTA in the buffer solution will meet this requirement as in the method given below, where EDTA is standardized against calcium alone. Other workers have preferred to standardize against a solution containing equivalent amounts of each element (Mackereth *et al.*, 1978).

Several ions interfere with the reaction including iron, manganese and copper, whilst phosphate tends to delay the end-point. Levels of these ions are normally low in natural waters and addition of triethanolamine will be adequate, at least for iron and manganese. Cyanide solution has been used to control iron and copper, and ferrocyanide for manganese, whilst sodium sulphide has also been recommended.

A value for magnesium alone may be obtained in two ways:

1 By difference. If calcium is estimated separately as on p. 87 then this value can be subtracted from the calcium plus magnesium result to obtain magnesium. Unfortunately Eriochrome black T will complex with other divalent ions if present and increase the apparent value for hardness. In this case magnesium by difference will be slightly over-estimated.

2 By direct titration. If calcium is removed then magnesium can be titrated alone using Eriochrome black T. Shapiro and Brannock (1956) precipitated calcium using sodium tungstate. The buffer solution and triethanolamine must be present as for total hardness.

Total hardness (calcium + magnesium)

Reagents

1 Calcium standard (1 ml ≡ 0.1 mg Ca).
Prepare as for Ca (p. 88).

2 EDTA solution (1 ml ≡ 0.1 mg Ca).
Dissolve 0.931 g di-sodium ethylenediaminetetra-acetic acid in 1 litre of water and standardize against the Ca standard in the presence of the buffer and triethanolamine as described below.

3 Indicator solution.
Dissolve 0.25 g Eriochrome black T in 50 ml industrial spirit.
Prepare fresh weekly.

4 Triethanolamine.

5 Buffer solution + magnesium.
Dissolve 67.5 g NH_4Cl in water, add 570 ml 0.88 NH_3 solution and 4 g of EDTA (magnesium salt). Dilute to 1 litre.

Procedure

Prepare the sample solution as on pp. 29, 40, 61.
Pipette up to 5 ml of the sample solution into the titration flask and dilute to about 100 ml with water.
Add 15 ml buffer solution, 10 drops of indicator and 2 ml triethanolamine.
Titrate with EDTA solution from red to a clear blue.

Calculation

If T ml EDTA solution are required for the titration then:

$$\text{Total hardness as Ca (mg l}^{-1}) = \frac{T(\text{ml}) \times 10^2}{\text{aliquot (ml)}}$$

For total hardness as Mg alone multiply by 0.6064.
Apply factors for dilution or concentration and correct to dry weight where necessary.

Notes

1 The interpretation of the end-point is easier if the titration is carried out under fluorescent lighting and further improved by using a photoelectric titrator which eliminates subjective interpretation.

2 Note that hardness values are often quoted in terms of $CaCO_3$.

Atomic absorption method

Atomic absorption is without doubt the best available method for the determination of magnesium in solution. Its advantage for magnesium lies in its greater precision, sensitivity and convenience compared with other procedures. Unlike sodium, potassium and calcium, flame emission is too insensitive for this element.

Certain elements interfere with absorption in the air–acetylene flame and present problems similar to those described for calcium. In organic materials, calcium, aluminium and phosphorus all have an effect; calcium enhances the absorption markedly whilst the effects of the other two depend on their concentration and amounts of other elements in the solution. Iron does not appear to be significant, but both sulphuric and hydrochloric acids have some effect and the same levels should be present in standards and samples. These interferences can be controlled by using lanthanum as a releasing agent. A level of 400 mg l^{-1} is employed for vegetation digests and for waters, provided 1% sulphuric acid is present as discussed for calcium. 800 mg l^{-1} of lanthanum is used for soil extracts but again 1% sulphuric acid must be present.

As recommended for calcium, total digests of rock and soil should be analysed using a nitrous oxide–acetylene flame containing an ionization buffer and omitting lanthanum.

Levels down to 0.01 mg l^{-1} or less can readily be detected for magnesium.

Reagents

1 Magnesium standards.

Stock solution (100 ppm Mg): dissolve 1.0136 g $MgSO_4 . 7H_2O$ in water containing about 1 ml H_2SO_4. Dilute to 1 litre. Avoid using old bottles of $MgSO_4 . 7H_2O$.

Working standards: prepare a range of 0–3 ppm Mg. Include sufficient stock $LaCl_3$ solution to bring the concentration of La to 400 ppm for plant digests and 800 ppm for soil extracts. Include digestion reagents or soil extractant as appropriate and 1% v/v H_2SO_4 in all cases (see below). Also include 1% HCl for dry-ashing solutions.

2 Lanthanum chloride solution, 5000 ppm La. Dissolve 6.6837 g $LaCl_3 . 7H_2O$ in water containing 1 ml 2 M HCl and dilute to 500 ml.

3 Sulphuric acid, 10% v/v.

Procedure

Prepare the sample solution as described on pp. 29, 40, 61 and dilute to volume.

Select the 285.2 nm wavelength and adjust air and gas flows, slit-width and other settings as recommended for the instrument employed.

Always allow the hollow cathode lamp adequate time to stabilize.

Prepare a calibration curve from the standard range by setting the top standard to a suitable readout value and the 0 ppm standard to zero, then aspirating each standard in turn.

Add sufficient stock La to aliquots of the sample digests or extracts to give 400 ppm La in plant digests and water samples and 800 ppm in soil extracts after dilution to a convenient volume.

Note that a 5 × dilution of a sulphuric–peroxide solution will be 1% H_2SO_4 (v/v). Mixed acid digests and other solutions (waters, soil extracts and dry-ashing solutions) all need 10% H_2SO_4 to obtain 1% (v/v).

Dilutions of soil extracts also need extractant and dry-ashing solutions require HCl to restore their original strengths.

Aspirate the sample solutions under the same conditions as the standards.

Check the top, zero and an intermediate standard frequently.

Flush the spray chamber and burner frequently with water particularly if soil extracts are aspirated.

Use the calibration curve to obtain ppm Mg in the sample solutions.

Carry out blank determinations in the same way and subtract where necessary.

Calculation

If C = ppm Mg obtained from the graph then for:

1 Plant materials and soils,

$$\text{Mg}(\%) = \frac{C(\text{ppm}) \times \text{soln vol (ml)}}{10^4 \times \text{sample wt (g)}}$$

2 Soils extracts,

Extractable Mg ($\text{mg } 100 \text{ g}^{-1}$): as above $\times 10^3$

(for me 100 g^{-1} Mg divide mg 100 g^{-1} result by 12.16)

3 Waters,

Mg (mg l^{-1}) = C (ppm)

Apply factors for dilution or concentration and correct to dry weight where necessary.

Note
It is desirable to use a high solids burner for soil extracts, particularly ammonium acetate, or undue clogging of the burner with carbon may occur.

Manganese

Although manganese is widespread it is much less abundant than iron and aluminium (the other soil sesquioxide elements) and is present in silicate minerals only as a minor element. Manganese bearing minerals, including workable ores, occur mainly in localized deposits. Three oxidation states are found in nature and the lowest (Mn^{2+}) is the main form available to plants and is more widespread in soil than ferrous iron (Fe^{2+}) because it is less easily oxidized. Soil pH also influences the form present with soluble, exchangeable and organically bound manganese being more common at lower pH values, whilst oxides (to Mn^{4+}) are more prominent under alkaline conditions. Micro-organisms have a key role in the oxidation–reduction processes.

Manganese is an essential minor nutrient for plants and animals. It is involved in enzyme activation and in some of the reactions required for photosynthesis and nitrogen metabolism. The element resembles iron in that deficiencies can occur on calcareous soil and toxicities on acid sites in species not adapted to these habitats. Concentrations of only a few mg l^{-1} in solution can affect a sensitive species.

Concentrations in natural waters are low, but various ionic, organic and mineral forms can occur depending on redox potential, pH and other factors. Although these forms resemble those of iron, Mn^{2+} is relatively more stable than Fe^{2+} and the cycling patterns of the two elements between water and sediments are not the same (Meltzer and Steinberg, 1983; Davison and Tipping, 1984).

The concentration ranges normally encountered are given below.

Mineral soils	200–2000 $\mu g\ g^{-1}$
Organic soils (peat)	50–500 $\mu g\ g^{-1}$
Soil extractions	5–500 $\mu g\ g^{-1}$
Plant materials	50–1000 $\mu g\ g^{-1}$
Animal tissues	5–50 $\mu g\ g^{-1}$
Rainwater	0.4–3 $\mu g\ l^{-1}$
Freshwater	1–80 $\mu g\ l^{-1}$

(dry weight basis where relevant).

In the past only colorimetric procedures have been widely used for this element, particularly the periodate oxidation which gives the very stable permanganate colour. Piper (1950) gives a modified wet digestion which may be linked to a periodate oxidation and can be recommended. In general, however, this method lacks sufficient sensitivity for many materials and the formaldoxime procedure is suggested as a more suitable colorimetric alternative. A modification of this is described below.

Of the other methods atomic absorption is the most convenient for routine use as well as being sensitive and relatively free from interferences. A procedure is outlined below.

Formaldoxime method

In alkaline solution manganese reacts with formaldoxime to form a brown complex which is suitable for colorimetric determination. Iron and copper can interfere by complexing with the formaldoxime, but these linkages can be broken down in warm solution. Excess phosphate may form a precipitate, but this is prevented by the addition of HEEDTA (Na-salt of *N*-(2-hydroxyethyl)ethylenediaminetriacetic acid) (Bradfield, 1957). Perchloric acid is added to prevent premature fading of the colour.

The method is applicable to most sample types although waters will probably need to be concentrated. It is possible to measure as little as 5 μg manganese in solution.

Reagents
1 Manganese standards.
Stock solution (1 ml ≡ 0.1 mg Mn): dissolve 0.4060 g $MnSO_4.4H_2O$ in water containing 1 ml conc H_2SO_4 and dilute to 1 litre.
Working standard (1 ml ≡ 0.005 mg Mn): dilute the stock solution 20 times.

2 Formaldoxime reagent.
Dissolve 20 g paraformaldehyde and 55 g hydroxylamine sulphate [$(NH_2OH)_2.H_2SO_4$)] in boiling water and dilute to 100 ml when cold.
Dilute 10 times immediately before use.
3 Sodium hydroxide, 10% w/v.
4 Perchloric acid, 10% v/v.
Dilute from 60% $HClO_4$ and prepare immediately before use.
5 HEEDTA, 10% w/v.
Prepare from the Na salt of *N*-hydroxyethylethylenediaminetriacetic acid.

Procedure

Prepare the sample solutions as described on pp. 29, 40, 61 and dilute to volume.
Pipette 0–6 ml working standard into 50 ml volumetric flasks to give a standard range of 0–0.03 mg Mn.
Add acid or extractant as appropriate to match the samples.
Pipette a suitable aliquot of the sample solution (usually 10–25 ml) into a 50 ml volumetric flask.
From this point treat standards and samples in the same way.
Add 5 ml $HClO_4$ and 5 ml HEEDTA in succession with mixing.
Adjust the pH to between 8 and 10 using NaOH.
Add 1.5 ml formaldoxime followed rapidly by a further 2 ml NaOH.
Dilute into the flask neck, mix and leave in a water bath at 65 °C for 2 hours.
Cool, dilute to volume and mix.
Measure the absorbance at 450 nm or with a blue filter using water as a reference.
Prepare a calibration curve from the standards and use it to obtain mg Mn in the sample aliquots.
Carry out blank determinations in the same way and subtract where necessary.

Calculation

If C = mg Mn obtained from the graph then for:
1 Plant materials and soils,

$$\mathrm{Mn}(\%) = \frac{C\ (\mathrm{mg}) \times \mathrm{soln\ vol\ (ml)}}{10 \times \mathrm{aliquot\ (ml)} \times \mathrm{sample\ wt\ (g)}}$$

2 Soil extracts,
Extractable Mn (mg 100 g^{-1}): as above $\times 10^3$

3 Waters,

$$\mathrm{Mn\ (mg\ l^{-1})} = \frac{C\ (\mathrm{mg}) \times 10^3}{\mathrm{aliquot\ (ml)}}$$

Apply factors for dilution or concentration and correct to dry weight where necessary.

Atomic absorption method

This procedure is recommended for manganese because of its sensitivity and convenience together with its relative freedom from interference. It is essential that the standards and samples contain manganese in the same valency state. This will be so if the mixed acid digestion is used, but not necessarily so for soil extracts. Some concentration is generally necessary for waters. The principles of atomic absorption are discussed in Chapter 8 and its application in the determination of manganese is dealt with by Christian and Feldman (1970).

Atomic absorption is suitable for all sample materials. Most instruments will determine 0.05 μg ml^{-1}.

Reagents

1 Manganese standards.
Stock solution (100 ppm Mn): dissolve 0.4060 g $MnSO_4.4H_2O$ in water containing 1 ml conc H_2SO_4 and dilute to 1 litre.
Working standards: prepare a range of standards from 0–3 ppm by dilution from the stock solution. Include acid or soil extractant as appropriate to match the sample solutions.

Procedure

Prepare sample solution as described on pp. 29, 40, 61 and dilute to volume.
Select the 279.5 nm wavelength and adjust air and gas flows, slit-width and other settings as recommended for the instrument employed.
Always allow the hollow cathode lamp adequate time to stabilize.
Prepare a calibration curve from the standard range by setting the top standard to a suitable readout value and the 0 ppm standard on zero, then aspirating each standard in turn.

Aspirate the sample solutions into the flame under the same conditions as the standards.
Check the top, zero and an intermediate standards frequently.
Flush the spray chamber and burner frequently with water, particularly if soil extracts are aspirated.
Use the calibration curve to obtain ppm Mn in the sample solutions.
Carry out blank determinations in the same way and subtract where necessary.

Calculation
If C = ppm Mn obtained from the graph then for:
1 Plant materials and soils,

$$\text{Mn } (\%) = \frac{C\,(\text{ppm}) \times \text{soln vol (ml)}}{10^4 \times \text{sample wt (g)}}$$

2 Soils extracts,
Extractable Mn (mg 100 g^{-1}): as above $\times 10^3$
3 Waters,
Mn (mg l^{-1}) = C (ppm)
Apply factors for dilution and concentration and correct to dry weight where necessary.

Note
It is desirable to use a special high solids burner for soil extracts, particularly ammonium acetate, to minimize blockage by carbon deposits.

Molybdenum

The few minerals which contain molybdenum are confined to veins in granite rocks, although weathering products are also found in sediments. Soil levels average around only 1 μg g^{-1} (Bowen, 1979) which is the lowest for elements currently accepted as essential for life. Although granite-based materials yield higher totals, the element is more available in alkaline soils and herbage contents approaching toxicity for livestock are not unknown. In general, molybdenum catalyses certain enzymic reactions and in particular it is essential for nitrogen fixation by legumes and other plants because of its role in nitrogenase activity by symbiotic organisms. It is also needed for nitrate reductase activity in plants.

The concentration ranges normally encountered are given below.

Mineral soils	0.5–10 μg g^{-1}
Organic soils (peat)	0.05–0.5 μg g^{-1}
Soil extractions	0.01–0.2 μg g^{-1}
Plant materials	0.1–0.8 μg g^{-1}
Animal tissues	0.03–0.3 μg g^{-1}
Rainwater	0.05–0.3 μg g^{-1}
Freshwater	0.1–2 μg l^{-1}

(dry weight basis where relevant).

The low levels in ecological materials pose problems in both sample preparation and analyses. The most widely used colorimetric method is that based on the thiocyanate reaction and larger samples weight of plant material are required. The method described below includes a dry-ashing procedure suitable for larger samples. An alternative colorimetric technique depends on the reaction with zinc dithiol. The product is then extracted into iso-amyl acetate before colour intensity measurement. This procedure is described by Quin and Brooks (1975).

In general polarographic methods are more sensitive and require less sample, but prior separations are required to overcome interference problems. Atomic absorption is barely sensitive enough unless the element is first concentrated by extraction with APDC (ammonium pyrrolidine dithiocarbamate) as suggested by Butler and Matthews (1966) and measured using a nitrous oxide–acetylene flame. Flameless atomic absorption can also be used for the determination of molybdenum whilst Yamada and Hattori (1987) propose the use of HPLC (High Performance Liquid Chromatography) for the determination of molybdenum in soils and plants.

Because of the low concentration of molybdenum in vegetation it is not practicable to use an acid digestion procedure to prepare the test solution and dry-ashing must be used instead.

Thiocyanate method

In acid solution and in the presence of a reducing agent, molybdenum (v) gives an amber to cherry red colour with thiocyanate. This colour forms when the thiocyanate concentration is above 0.08 M but fades at concentrations over 0.2 M.

The method given here is based on that of Ellis and Olson (1950), as modified by Jackson (1958). Acetone is used as the reducing agent. This also stabilizes the colour (Grimaldi and Wells, 1943) and limits interference by reducing iron (III) to iron (II) which does not form a coloured complex with thiocyanate. Other elements in biological materials are not likely to have a significant effect.

The method is applicable to most ecological materials, although levels will be very low in most samples. Amounts of about 0.5 μg in solution can be measured. The final precision may be no better than $\pm 15\%$ owing to the very low levels encountered.

Reagents

1 Molybdenum standards.

Stock solution (1 ml ≡ 100 μg Mo): dissolve 0.1840 g $(NH_4)_6Mo_7O_{24}.4H_2O$ (ammonium molybdate) in water and dilute to 1 litre.

Working standard (1 ml ≡ 5 μg Mo): dilute the stock solution 20 times.

2 Hydrochloric acid, 1 + 1.

3 Nitric acid, conc.

4 Ammonium hydroxide, 1 + 1.

5 Perchloric acid, 60%.

6 Potassium thiocyanate, 10% w/v.

7 Acetone.

8 Hydrochloric acid, conc.

Preparation of solution

PLANT MATERIAL

Procedure

Ash about 10 g dried ground sample as given on p. 57, dissolve and dilute to volume.

Evaporate a suitable aliquot to dryness (if necessary evaporate the whole solution).

Dissolve residue in 5 ml 1 + 1 HCl, warm and add 1 ml conc HNO_3.

Add 1 + 1 NH_4OH until pH is between 8 and 9.

Filter into a borosilicate basin and add 5 ml 1 + 1 HCl.

Evaporate to dryness to dehydrate silica.

Redissolve in 10 ml 1 + 1 HCl and filter into a 50 ml volumetric flask.

Wash with hot water to about 30 ml.

SOILS

A sodium carbonate fusion is preferable for total molybdenum in soils because it can be readily adapted to larger sample weights if necessary. See p. 25 for the basic procedure. Evaporate an aliquot of the final solution as given above.

For soil extractions refer to p. 40.

Colorimetric stage

Procedure

Pipette 0–20 ml of the working standard into 50 ml volumetric flasks, to give a range of standards from 0–100 μg Mo.

Add to each standard 5 ml conc HCl and dilute to about 30 ml.

From this point treat standards and samples in the same way.

Add 3 ml of KSCN solution and 16 ml acetone.

Mix and heat at 60–70 °C for 1–2 hours.

Cool and dilute to volume with water.

If the solution is turbid at this stage, filter.

Measure the absorbance at 470 nm or with a blue-green filter using water as a reference.

Prepare a calibration curve from the standard solutions and use it to obtain μg Mo.

Carry out blank determinations in the same way and subtract where necessary.

Calculation

If $C = \mu g$ Mo obtained from the graph then for:

1 Plant materials, soils and soil extracts,

$$\text{Mo } (\mu g\, g^{-1}) = \frac{C\ (\mu g) \times \text{soln vol (ml)}}{\text{aliquot (ml)} \times \text{sample wt (g)}}$$

2 Waters,

$$\text{Mo } (\mu g\, l^{-1}) = \frac{C\ (\mu g) \times 10^3}{\text{aliquot (ml)}}$$

Apply factors for dilution or concentration and correct to dry weight where necessary.

Polarographic method

The sensitivity of the polarographic procedure for molybdenum is fairly high, but since the amounts of this element in plant and similar materials are frequently less than 1 $\mu g\ g^{-1}$, at least 2 g of sample will be required. In the procedure given below, α-benzoinoxime is used to concentrate molybdenum and to remove interfering ions. Optimum conditions for the polarographic determination of small concentrations of molybdenum are discussed by Violanda and Cooke (1964).

Although as little as 0.01 μg molybdenum can be detected in solution, the low levels in most samples account for a poor reproducibility (±10%).

Reagents

1 Molybdenum standard.
Stock solution (1 ml ≡ 100 μg Mo): dissolve 0.1840 g $(NH_4)_6Mo_7O_{24}.4H_2O$ (ammonium molybdate) in water and dilute to 1 litre.
Working standards: dilute from the stock solution to give a standard series from 0–50 μg Mo. Evaporate these to dryness and take up in the electrolyte as for the sample solutions.
2 α-Benzoinoxime, 2% w/v.
3 Sodium hydroxide, 4 M.
Dissolve 16 g NaOH in water, cool and dilute to 100 ml.
4 Nitric acid, 2 M.
5 Ammonium nitrate, 4 M.
Dissolve 32 g NH_4NO_3 in water and dilute to 100 ml.
6 Chloroform.
7 Nitric acid, conc.
8 Perchloric acid, 60%.
9 Sulphuric acid, 10% v/v.

Procedure

Prepare the sample solutions as given on p. 117 (for waters, prior evaporation, or complexing followed by organic extraction, will be required) and dilute to volume.
Evaporate a suitable aliquot to dryness.
Take up the residue in 8 ml 10% H_2SO_4.
Filter into a separating funnel, dilute to about 50 ml.
Add 2 ml α-benzoinoxime solution and shake well for 30 seconds.
Add $CHCl_3$ and shake for 90 seconds.
Run off the $CHCl_3$ layer into a 50 ml beaker.
Repeat twice with 5 ml portions of $CHCl_3$.
Evaporate the combined $CHCl_3$ extracts.
Add 3 ml conc HNO_3 and 3 ml 60% $HClO_4$.
Cover with a watch glass and digest slowly on a hotplate until the organic matter is destroyed.
Remove the watch glass and evaporate to dryness.
To samples and standards add 0.1 ml NaOH solution and swirl to wet the residue.
Add 1 ml 2 M HNO_3 and 2 ml 4 M NH_4NO_3, mix and transfer to a polarograph cell.
Remove oxygen by passing oxygen-free nitrogen.
Adjust the start potential sensitivity and other controls of the polarograph and record the step at $E_{\frac{1}{2}} = -0.41$ V.
Prepare a calibration curve from the standard solutions and use to determine μg Mo.
Carry out blank determinations in the same way and subtract where necessary.

Calculation

If $C = \mu g$ Mo obtained from the graph then for:
1 Plant materials, soils and soil extracts,

$$\text{Mo } (\mu g\ g^{-1}) = \frac{C\ (\mu g) \times \text{soln vol (ml)}}{\text{aliquot (ml)} \times \text{sample wt (g)}}$$

2 Waters,

$$\text{Mo } (\mu g\ l^{-1}) = \frac{C\ (\mu g) \times 10^3}{\text{aliquot (ml)}}$$

Apply factors for dilution or concentration and correct to dry weight where necessary.

Nitrogen

Although nitrogen minerals are confined to nitrates in localized deposits, most rocks, and particularly silicate minerals, contain small amounts of 'fixed' ammonium ions although generally below 50 $\mu g\ g^{-1}$ (Bowen, 1979). In contrast, soil levels range from 0.1 to over 1.5% and greatly exceed those in rocks, owing to the large organic fraction built up by biological processes. Apart from fixed ammonium, a small fraction of available inorganic nitrogen is also present as ammonium, nitrite and nitrate ions. The proportion of these ions varies according to soil pH, aeration, microbial activity, plant uptake and other factors. Transformation between these forms and other processes such as mineralization of organic nitrogen, biological fixation of nitrogen and mechanisms leading to losses from soil have all been intensively studied, particularly in agriculture.

Nitrogen is essential for life and is of major significance on account of its role in amino-acids and proteins including enzymes, nucleotides, genetic material, pigments and other substances. Crop demands for this element can exceed the supplying capacity of soil and fertilization is often required to sustain yields although excess fertilizer nitrogen can have toxic effects. A limited range of higher plants can fix nitrogen symbiotically including legumes and some other species such as *Alnus*.

Levels of both ammonium- and nitrate-nitrogen are often below 1 $mg\,l^{-1}$ in fresh water and can limit plankton growth. However, in some areas concentrations of nitrate ions have risen to pollution levels due mainly to agricultural run-off. Dissolved organic nitrogen is also important in water, and unlike soil, organic contents are not high relative to the inorganic fractions. These fractions are also important in the marine environment where less is known quantitatively about the processes involved.

Whilst molecular nitrogen in the atmosphere is inert, a minute, but reactive fraction is present in both reduced (NH_3) and oxidized (NO_x) forms. The oxides resemble those of sulphur in that a proportion are derived from human activities and both are involved in the formation of acid deposition.

All these environments present aspects of the global nitrogen cycle which is the best documented of the various elemental cycles. There is an extensive literature, including a detailed summary for the United Kingdom prepared by the Royal Society Study Group (1963). Nitrogen cycling in various types of terrestrial ecosystem including forests and grasslands has long been studied.

The concentration ranges generally encountered are given below:

Mineral soils	0.1–0.5%
Organic soils (peat)	0.5–1.5%
Soil extractions NH_4^+-N	0.2–3 $mg\,100\,g^{-1}$
NO_3^--N	0.1–2 $mg\,100\,g^{-1}$
Plant materials	1–3%
Animal tissue	4–10%
Rainwaters NH_4^+-N	0.1–0.8 $mg\,l^{-1}$
NO_3^--N	0.05–0.4 $mg\,l^{-1}$
Freshwater NH_4^+-N	0.1–2 $mg\,l^{-1}$
NO_3^--N	0.05–3 $mg\,l^{-1}$

(dry weight basis where relevant).

The determination of nitrogen in ecological materials presents specific difficulties not least because its volatility restricts the way in which it can be quantitatively brought into solution. In practice, classical procedures continue to dominate the methodology, particularly the Kjeldahl digestion followed by steam distillation or colorimetry. Various adaptations of this system are presented below for estimating total, organic and inorganic forms. Specific ion electrodes are also available for ammonium and nitrate ions in waters. After preliminary treatment, the ammonia electrode can also be used to determine total nitrogen in plant material (Powers *et al.*, 1981).

Total organic nitrogen

Total organic nitrogen is probably the most frequently estimated of all nutrient elements in soils and plant materials. This is not only because of the metabolic importance of the element, but also because most nitrogen is in the organic form. This is in contrast to phosphorus and other macronutrients where a significant proportion of the soil total in each case is mineral in nature. Nevertheless, mineral nitrogen is important and its estimation is discussed later. In the case of waters, most interest centres on the ammonium- and nitrate-nitrogen fractions, but organic nitrogen is sometimes required and is discussed below.

Preparation of sample solution

Soil and plant materials

Various problems arise in preparing solutions for the estimation of total organic nitrogen and an extensive literature exists on the subject. The wet-oxidation systems containing nitric and/or perchloric acid which are normally used for breaking down organic matter are unsuitable and result in low recoveries. A better digestion system is the traditional Kjeldahl procedure and this is still probably the most widely used. A semi-micro version of this is described below and an alternative digestion procedure using a sulphuric acid–hydrogen peroxide digestion mixture is also given.

The dry-ashing method is unsuitable for nitrogen but the Dumas method is based on a modified combustion system. It is less often used than the Kjeldahl system because it is more tedious unless automated equipment is available. In the Dumas method the sample is heated strongly with copper oxide in a stream of carbon dioxide. Nitrogenous compounds in the sample form nitrogen oxides which are then reduced to nitrogen the volume of which is measured after removal of the carbon dioxide.

The Kjeldahl, Dumas and other methods have been compared by Bremner and Shaw (1958) Jacobs (1978) and Oxenham *et al.* (1983).

Semi-micro Kjeldahl digestion

The conversion of organic nitrogen to ammonia and its subsequent estimation is the basis of the well-known Kjeldahl method. Many publications have dealt with the different aspects of this procedure, including those of Bradstreet (1965), Bremner (1960) and more recently Page *et al.* (1982).

The sample is digested with sulphuric acid containing potassium or sodium sulphate to raise the reaction temperature. This digestion is now frequently carried out in tubes in heated aluminium alloy blocks which are commercially available.

A catalyst is required and copper, mercury and selenium have long been used. However, mercuric oxide is generally preferred, particularly for soils, and is used in the procedure given below. If the ammonia is subsequently estimated by distillation (see below), precautions will be necessary to prevent the formation of mercury–ammonium complexes. An objection to mercuric oxide arises from the fact that it is a more serious pollutant than the metals used in the other catalysts. The search for equally efficient alternatives has therefore continued; Williams (1973) suggested a combination of titanium dioxide and copper sulphate and Glowa (1974) proposed zirconium dioxide. A digestion procedure using peroxymonosulphuric acid as a catalyst and without added salts is proposed by Hach *et al.* (1985).

It is usually satisfactory to digest finely ground oven- or air-dry material but where much labile nitrogen is present (as in many animal and microbial tissues) it may be necessary to digest the fresh samples. It is also very important to ensure that the relatively small weights taken in semi-micro work are representative of the original material. For this reason fresh material can only be handled on a macroscale and the large amount of water present may prevent the digestion proceeding smoothly.

In general the standard Kjeldahl method recovers variable amounts of nitrate but the method described here recovers only organic- and ammonium-nitrogen. Nitrate levels in natural ecosystems can be quite low but may be significant in agricultural materials. For some studies and particularly those involving ^{15}N it is essential to recover all nitrate and nitrite-nitrogen in the procedure. The usual method is pre-treatment with a salicylic acid–sulphuric acid mixture although the recovery of nitrite has been queried (Page *et al.*, 1982). Metal reduction systems have also been used. Dervarda's alloy (aluminium, copper and zinc) is very suitable for waters (see below) but less convenient for soils because it requires an extra reflux stage. Pruden *et al.* (1985) adapted a chromium–zinc system and obtained satisfactory recoveries without a preliminary reflux.

The release of 'fixed' ammonium-nitrogen has been studied in the past, but recoveries appear to vary with the soil type and digestion procedure. Meints and Peterson (1972) obtained fairly high recoveries from upper horizons but lower from the sub-soil using a copper–selenium catalyst. Full recovery of this fraction requires a hydrofluoric acid treatment and a procedure is given by Bremner in Page *et al.* (1982).

Reagents

1 Potassium sulphate–mercuric oxide mixture.
Prepare a mixture of K_2SO_4 and HgO (NH_3-free grades) in the ratio of 20:1.
Tablets of this mixture may be obtained from Messrs. Thompson and Capper of Liverpool, UK. (In the method described below 2 g tablets are used.)
2 Sulphuric acid, conc (NH_3-free grade).

Procedure

Weigh the dried, finely ground sample into a 50 ml round bottom Kjeldahl flask. Air dried material (40 °C) is preferred.
Take: 0.100 g (plant materials, peat and litter); 0.250 g (organic rich soils); 0.500 to 1.00 g (mineral soils with low organic content).
Add 2 g K_2SO_4–HgO mixture.
Add 3 ml conc H_2SO_4 running the acid slowly down the neck while rotating the flask.
Heat the flask gently on a digestion rack until frothing subsides.
Particular care is necessary at this stage if the sample is not dry.
As the frothing subsides increase the heat until H_2SO_4 refluxes in the bottom part of the flask neck although any tendency to bump must be controlled.
After the digest becomes colourless or pale green continue heating for 30 minutes for litter and animal materials and 1 hour for peat and mineral soils.
On completion of the digestion, allow to cool until just warm and then dilute to 50 ml with water.
If distillation is to follow, insoluble residue need not be removed.
If a colorimetric method is to be used, any insoluble residue must be removed by filtration or allowed to settle overnight and an aliquot of supernatant taken.
Prepare blank digests with reagents alone.

Sulphuric acid–hydrogen peroxide digestion

This procedure was given on p. 59 as being a suitable alternative to the nitric–perchloric–sulphuric mixture for digesting plant materials prior to total nutrient analysis. Unlike the former system it is also suitable for nitrogen and is an alternative to the Kjeldahl digestion for plant materials. The sulphuric–peroxide digestion, with minor modifications is suitable for both nitrogen and phosphorus in soils (Rowland & Grimshaw, 1985). The sample weight should not exceed 0.2 g for soil and the total digestion time is extended to 2 hours. As before, excessive bumping and spitting must be avoided. The digest is diluted to 50 ml and the residue allowed to settle. A further dilution of five times is required for the automated colorimetric methods. Although this digestion is very suitable for estimating potassium and other elements in vegetation it will give low recoveries from soil and should not be used. This restriction also applies to vegetation heavily contaminated with soil (Parkinson and Grimshaw, 1986).

WATERS

Since it is not practicable to apply ion exchange, solvent extraction or precipitation steps to the determination of total organic nitrogen in water samples, an evaporation stage is unavoidable. Furthermore, since the amount of organic nitrogen present in water is so low, a large volume of sample may be required. The concentration may be as low as 0.01 $mg\,l^{-1}$ in rainwater.

The method given here uses magnesium oxide during the evaporation stage to drive off ammonium-nitrogen and Devarda's alloy is included. The alloy reduces nitrate- and nitrite-nitrogen to ammonia which is then driven off. The amounts of magnesium oxide and Devarda's alloy are kept very small to minimize the reagent blank. The optimum conditions for driving off inorganic nitrogen are therefore achieved in the later stages of the evaporation. When the sample has been reduced to a small volume, sulphuric acid is added. This forms sulphates with magnesium oxide and Devarda's alloy metals (copper, aluminium and zinc) which serve to raise the temperature making the addition of potassium sulphate unnecessary.

A small amount of mercuric oxide must be included.

Reagents
1 Devarda's alloy, powder (finely ground).
2 Magnesium oxide, powder.
3 Mercuric oxide.
4 Sulphuric acid, conc.

Procedure
Measure up to 250 ml of water sample into a 500 ml Taylor flask.
Add 0.05 g Devarda's alloy and 0.03 g MgO (see Note 1 below).
Boil down to a few ml but *do not allow to dry.*
Add 2.5 ml conc H_2SO_4 and 0.4 g HgO.
Digest to white fume stage but not to dryness.
Retain the entire digest for distillation (see below).
Prepare blanks from reagents and ammonia-free water.

Notes
1 The indophenol-blue colorimetric method (p. 124) cannot be used for solutions high in copper and distillation should follow the Devarda's alloy treatment.
2 The sulphuric acid–hydrogen peroxide procedure provides an alternative to the Kjeldahl procedure. After reducing the sample to a small volume, the digestion can be carried out with the addition of 0.9 ml of mixture. This allows estimation of the same elements as in vegetation including nitrogen recovered as organic plus ammonium-nitrogen. If MgO is added before boiling down, nitrogen will be recorded as organic nitrogen (although the excess magnesium might affect other determinations using flame equipment). Devarda's alloy is not used because removal of nitrates is unnecessary with this digestion and the temperature is controlled by lithium sulphate. Absence of the alloy also allows the indophenol-blue method to be used.

Analytical procedures

The digestion stage converts organic nitrogen to ammonium-nitrogen which, after dilution to volume will be in an approximately 5% acid solution. The classical method for estimating ammonium-nitrogen is by distillation which is described here. A colorimetric procedure is given later.

The micro-diffusion technique of Conway (1962) is suitable for estimating ammonium salts in solutions, and so is convenient for the Kjeldahl digests. Ammonia is liberated by alkali added to the sample contained in the annular compartment of a shallow dish (Conway unit). Over a period of many hours the ammonia diffuses over into standard acid in the other compartment. Micro-titration can then be used to determine the ammonium-nitrogen. Little equipment is needed for this method and large numbers of samples can be processed together. However, the method needs careful development to establish the optimum conditions. It will not be considered further in this text.

Distillation method

In this method, free ammonia is liberated from the diluted digest by steam distillation in the presence of excess alkali. If mercuric oxide has been used as a catalyst, thiosulphate must be included with the alkali to precipitate mercury and prevent the formation of mercury–ammonium complexes.

The distillate is collected in a receiver containing excess boric acid combined with an indicator solution. The ammonia is then titrated with standard hydrochloric acid to a pH of about 4.5. A versatile distillation assembly is illustrated in Fig. 5.4. Other suitable apparatus is available commercially.

Amounts as low as 10 μg of nitrogen can be measured.

Reagents
1 Nitrogen standard (1 ml $\equiv$ 0.1 mg NH_4^+-N).
Dissolve 0.1910 g of dry NH_4Cl in water and dilute to 500 ml.
1–2 ml chloroform can be added as a preservative.
2 Hydrochloric acid, M/140, (1 ml $\equiv$ 0.1 mg NH_4^+-N).
Prepare 0.1 M HCl and standardize against 0.05 M Na_2CO_3 using 0.1% bromophenol blue as indicator. Dilute the standardized acid to produce an M/140 solution.

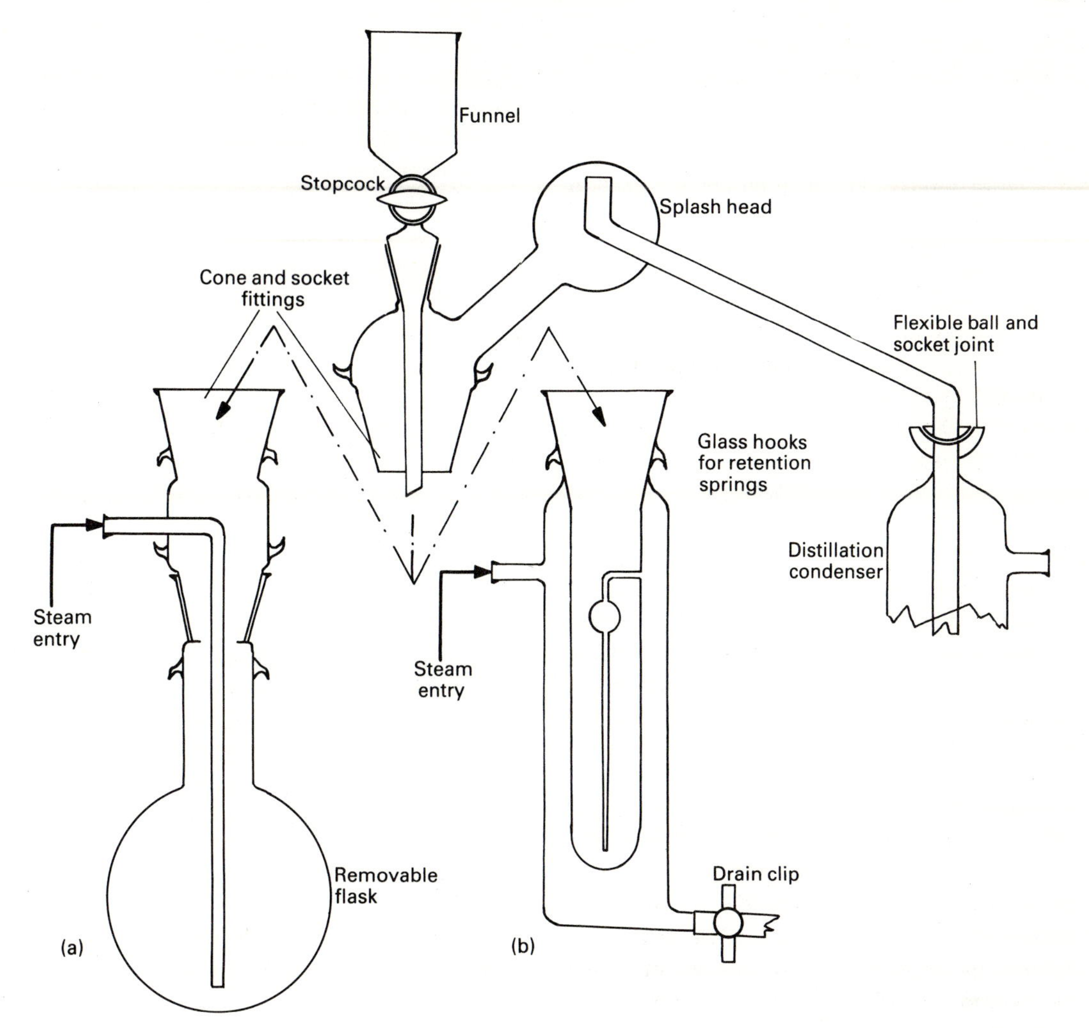

Fig. 5.4. Apparatus suitable for (a) macro and (b) semi-micro determination of nitrogen by distillation.

3 Sodium hydroxide–sodium thiosulphate mixture.

Dissolve 500 g NaOH and 25 g $Na_2S_2O_3.5H_2O$ slowly in water and dilute to 1 litre when cool.

4 Boric acid—indicator solution.

Dissolve 20 g H_3BO_3 in water, add 15 ml of BDH '4.5' indicator and dilute to 1 litre.

Distillation

A semi-micro distillation apparatus is used (as in Fig. 5.4 or similar).

Fill the steam generator with ammonia-free water and pass steam through the apparatus for approximately 30 minutes before use.

Run a steam blank to check if ready for use.

Check the distillation equipment by a recovery test with standard NH_4Cl following the procedure given below.

If satisfactory, transfer an aliquot of the sample digest to the reaction chamber through the tap funnel.

Run in 12 ml of alkali mixture and pass steam through.

Collect the distillate in a receiver containing 5 ml H_3BO_3 indicator solution.

Use a 5 or 10 ml burette and titrate against M/140 HCl from blue to a pale neutral endpoint.

Treat digestion blanks in the same way and subtract from the sample titration.

Calculation

If T ml M/140 HCl are required for the titration then for:

1 Plant materials and soils,

$$N(\%) = \frac{T\text{ (ml)} \times \text{soln vol (ml)}}{10^2 \times \text{aliquot (ml)} \times \text{sample wt (g)}}$$

2 Waters,

$$\text{Organic-N (mg l}^{-1}) = \frac{T\text{ (ml)} \times 10^2}{\text{aliquot (ml)}}$$

Note

In effect this determination is a measure of total N in plant material but soil values are better expressed as total organic nitrogen because nitrates and lattice bound ammonium-nitrogen are not included.

Indophenol-blue method

Although the indophenol-blue reaction for ammonium-nitrogen has long been known, it has been slow to supplant distillation and titration, mainly because of the difficulty of obtaining reproducible colour development. The automated version is much more satisfactory in this respect and is described later. The manual alternative will give acceptable results if the conditions are carefully controlled. In particular the time of colour development should not be exceeded because of fading. The critical factors and interferences have been discussed by many workers as shown in a review by Searle (1984).

The method presented here is based on a Technicon procedure (Rowland, 1983). In the reaction ammonium-nitrogen is oxidized by sodium hypochlorite and then coupled with a phenolic compound (sodium salicylate) to produce the indophenol-blue colour. Interference by mercury from Kjeldahl digests is controlled by the inclusion of Rochelle salt, whilst sodium nitroprusside is added as a catalyst. Excess copper may be troublesome but will be rarely noticed except in an occasional soil solution. Nevertheless, Devarda's alloy must not be used in the preparation of water samples.

Sufficient alkali is included to enable Kjeldahl digests to be run undiluted at approximately 5% acid but the sulphuric acid–hydrogen peroxide solution should be diluted five times and run at the 1% acid level.

Reagents

1 Nitrogen standards.

Stock solution (1 ml ≡ 0.1 mg NH_4^+-N): prepare as for distillation procedure.

Working standard (1 ml ≡ 0.001 mg NH_4^+-N): dilute the stock solution 100 times. Prepare fresh each day.

2 Combined reagent.

Dissolve 35 g sodium potassium tartrate (Rochelle salt), 17.5 g sodium salicylate and 0.5 g sodium nitroprusside in about 400 ml water. Add 40 ml 50% sodium hydroxide and mix to dissolve. Dilute to 1 litre, mix well and store at +2 °C.

3 Sodium hypochlorite solution, 0.15% available Cl; dilute from 14% available Cl stock (obtainable from reagent suppliers—see Note).

Procedure

Pipette 0 to 10 ml working standard into 50 ml volumetric flasks to give a range from 0 to 0.01 mg NH_4^+-N.

Add blank acid digest to match the sample aliquots.

Pipette not more than 5 ml sample solution into a 50 ml volumetric flask.

From this point treat standards and samples in the same way.

Add 40 ml combined reagent and mix.

Add 4 ml sodium hypochlorite reagent, dilute to volume and mix well.

Leave 10 minutes in a water bath at 40 °C.

Measure the absorbance at 660 nm or use an orange filter, using water as a reference.

Prepare a calibration curve from the standard values and use it to obtain mg NH_4^+-N in the sample aliquot.

Carry out blank determinations in the same way and subtract where necessary.

Calculation
If C = mg NH_4^+-N obtained from the graph then for:

1 Plant materials and soils,

$$N(\%) = \frac{C\text{ (mg)} \times \text{soln vol (ml)}}{10 \times \text{aliquot (ml)} \times \text{sample wt (g)}}$$

2 Waters,

$$\text{Organic-N (mg l}^{-1}) = \frac{C\text{ (mg)} \times 10^3}{\text{aliquot (ml)}}$$

Apply factors for dilution and correct to dry weight where necessary.

Note
Sodium hypochlorite stock solution gradually loses chlorine and should be purchased at least once a year.

Continuous flow method
As mentioned earlier the indophenol-blue procedure is more satisfactory if used in its automated form. The initial sample treatments and interferences already referred to apply to the continuous flow method. The flow diagram (manifold) shown in Fig. 5.5 is that needed for the AAII system (see Chapter 8). The manifold needed for AAI having a faster flow and larger mixing coils was described by Allen *et al.* (1974).

Reagents
1 Nitrogen standards.
Stock solution (200 ppm N): dissolve 0.7640 g dry NH_4Cl in water and dilute to 1 litre.
Standard solutions: prepare a suitable range by dilution of stock solution. Plant materials may require from 0 to 40 ppm, but waters need only 0–5 ppm. Include digestion blank, soil extractant or water as appropriate.
2 Combined reagent.
3 Sodium hypochlorite solution.
Prepare **2** and **3** as instructions on p. 124.

Operating conditions
Manifold details are given in Fig. 5.5.
Heating coil temperature 37 °C.
A 10 mm flow cell is generally suitable but longer cells can be used.
Wavelength for maximum absorption 660 nm.

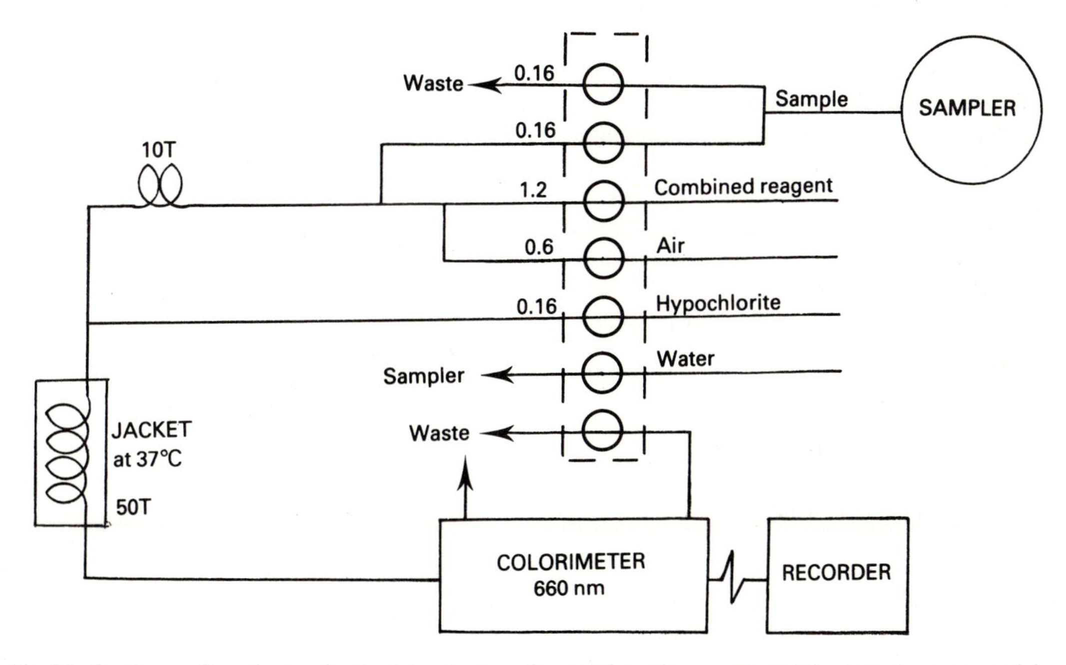

Fig. 5.5. Continuous flow diagram for the determination of ammonium-nitrogen. (Dotted line indicates pump module. T indicates number of turns on mixing coils. All flow rates expressed as ml min^{-1}.)

Sampling rate: up to 40 per hour (sample:wash ratio = 2:1).

Calculation

If C = ppm N obtained from the graph then for:

1 Plant materials and soils (organic N),

$$N(\%) = \frac{C\ (\text{ppm}) \times \text{soln vol (ml)}}{10^4 \times \text{sample wt (g)}}$$

2 Soil extracts,

Extractable NH_4^+-N (mg 100 g^{-1}): as above × 10^3

3 Waters (organic or ammonium-nitrogen),

Organic N or NH_4^+-N (mg l^{-1}) = C (ppm).

Apply factors for dilution or concentration and correct to dry weight where necessary.

Note

Wash manifold through weekly with hypochlorite (2% available Cl).

Inorganic nitrogen

It is often of value to determine inorganic-nitrogen (mainly ammonium and nitrate forms) in soils and waters. Levels are usually low compared with organic nitrogen, particularly in soils, but the inorganic fractions are important because they are the forms available for plant uptake.

Methods for the determination of ammonium and nitrate-nitrogen are given here. Total inorganic-nitrogen can be estimated directly by following the distillation procedure for nitrate (p. 128) and using the total distillate in the titration. A similar steam distillation procedure for inorganic nitrogen is recommended by Preez *et al.* (1987).

Ammonium-nitrogen

Preparation of sample solution

SOILS

Although some ammonium ions are in free solution, a further proportion are assumed to be adsorbed on to colloidal surfaces in a similar way to the metal nutrient cations. Hence a similar type of extractant is required and normally sodium or potassium chloride is employed. A method for extracting with 6% potassium chloride was given on p. 41.

WATERS

Waters require little preparation except the removal of suspended matter, particularly if colorimetric methods are to be later employed. If filter papers are used, the possibility of the paper retaining or releasing ammonium ions must be borne in mind. No. 1 grade appears to be suitable.

It is important when testing for ammonium (or nitrate and nitrite ions) in waters or other biological materials, that the sample be analysed soon after collection. Microbial activity may alter the balance between the various forms of nitrogen, or even lead to net losses. The preservation of aqueous samples is discussed in Chapter 4.

Analytical procedures

Distillation method

The distillation method for ammonium-nitrogen in soil extracts and waters uses the apparatus described for organic-nitrogen (Fig. 5.4) but with a distillation flask (a) replacing the reaction chamber (b). Magnesium oxide is used to liberate ammonia in place of sodium hydroxide. The reaction then takes place at about pH 10 which minimizes decomposition of labile organic nitrogen. Some workers have preferred to use phosphate buffers at pH 7.5–8.0 but magnesium oxide is adequate for most purposes.

Reagents

1 Standard solution (1 ml ≡ 0.1 mg NH_4^+-N).

2 Standard hydrochloric acid, M/140.

3 Boric acid–indicator solution.

4 Magnesium oxide powder.

Prepare **1**, **2** and **3** as instructions on p. 122.

Procedure

Use the distillation equipment as described earlier, or similar but with a removable flask.

Pass steam through the apparatus for about 30 minutes.

Run steam blanks and titrate with M/140 HCl. (Not more than 0.10 ml of acid should be required.)

Transfer an appropriate aliquot of the soil extract or water sample into a 500 ml distillation flask (Fig. 5.4).

Add 0.2 g MgO.

Quickly attach the flask to the distillation apparatus, close all taps and allow steam to pass through.

Collect about 50 ml of distillate in a beaker flask containing 5 ml boric acid–indicator mixture.

Titrate the distillate using M/140 HCl from blue to a pale neutral end-point.

Treat extraction blanks in the same way and subtract from the sample titre.

Calculation

If T ml M/140 HCl are required for the titration then for:

1 Soil extracts,

Extractable NH_4^+-N (mg 100 g^{-1}) =

$$\frac{T\text{ (ml)} \times \text{extractant vol (ml)} \times 10}{\text{aliquot (ml)} \times \text{sample wt (g)}}$$

2 Waters,

$$NH_4^+\text{-N(mg l}^{-1}) = \frac{T\text{ (ml)} \times 10^2}{\text{aliquot (ml)}}$$

Apply factors for dilution and correct to dry weight where necessary.

Indophenol-blue method

The manual form of this method was described earlier and may be applied directly to soil extracts and waters although the sensitivity may be limiting in some cases. If, however, the automated form is used, range expansion facilities make the method sufficiently sensitive (see p. 125).

Nitrate-nitrogen

There has been increasing interest in nitrate-nitrogen in recent years because the relatively large amounts being discharged into lakes and rivers through fertilizer drainage and industrial waste constitute a major pollution hazard. This excess nitrate is considered, with phosphorus, to be the principal factor leading to eutrophication in waters that had previously been relatively low in nutrient status. In addition, at levels above 10 mg l^{-1}, nitrate-nitrogen in drinking water becomes a potential health hazard, particularly to infants. Nitrates are reduced to nitrites in the body and these accumulate in infants because they lack the enzymes to reduce to cellular nitrogen and ammonia.

The solutions for analysis are prepared as summarized for ammonium-nitrogen except that water extracts of soils are quite adequate, although any 6% KCl extracts prepared for ammonium-nitrogen may also be used. If water extracts are prepared, a 25 : 1 water to soil ratio is used and the sample shaken for 10 minutes. Nitrate ions are, like ammonium ions, sensitive to changes on storage and precautions should be taken (pp. 66, 67). No prior treatment is needed before the determination of nitrate in water.

In the estimation of nitrate-nitrogen, indirect as well as direct methods are widely used. Two main types of indirect method include:

1 *Distillation following reduction of nitrate- and nitrite-nitrogen to ammonium-nitrogen.* The method records nitrate + nitrite-nitrogen together, and if required, total inorganic nitrogen in solution. The procedure given below uses Devarda's alloy for reduction but liberates ammonia at a lower pH ($\simeq 10$) than is employed in the normal distillation given earlier.

2 *Colorimetry through diazotization following reduction of nitrate to nitrite-nitrogen.* This is considered by many to be the best overall method for nitrate-nitrogen.

A large number of direct methods are available for nitrate-nitrogen. Many chromogenic reagents have been suggested as a basis for colorimetry but they vary in sensitivity and vulnerability to interference. One of the simplest is phenoldisulphonic acid and a method is given below, although the need for evaporation during colour formation is a barrier to automation. Other methods are similar but lack of sensitivity and interferences can be a problem. One method, using brucine, has been suggested as a possible field test (Ranney and Bartlett, 1972).

A wide range of spectroscopic and electroanalytical methods have been proposed for nitrate. Ultra-violet spectrophotometry provides a simple method for waters but organic interference must be allowed for (Standard Methods for the Examination of Waters and Waste Waters, 1985). Infra-red spectroscopy has also been applied but requires a preparative stage (Citron *et al.* 1961).

A polarographic method is available for nitrate ions but it is little used today. Details are given in Allen *et al.* (1974). Another electroanalytical technique using an ion-selective electrode is described below. Ion chromatography (IC) is becoming popular for the determination of nitrates in waters. The equipment is expensive but the technique is sensitive and very effective when several anions are required on the same sample (see p. 280). IC can be used for nitrates in soil extracts but preliminary treatment may be needed to remove organic or other potential interferences. It should be noted that all direct methods estimate nitrate ions alone and indirect approaches are needed to record nitrate plus nitrite-nitrogen as a single value.

Analytical procedures

Distillation method

The detection limit and reproducibility are similar to those given for the other distillation procedures for nitrogen compounds.

Reagents

1 Standard solution (1 ml ≡ 0.1 mg N).
2 Standard hydrochloric acid, M/140.
3 Boric acid–indicator solution.
4 Magnesium oxide, powder.
5 Devarda's alloy, finely powdered.
Prepare **1**, **2** and **3** as instructions on p. 122.

Procedure

The distillation procedure used here is basically the same as that discussed earlier but the sample chamber is replaced by a detachable distillation flask (Fig. 5.4).

Transfer an appropriate aliquot of the soil extract or water sample to the distillation flask and add 0.2 g MgO.

Suitable volumes are usually:
20–50 ml for soil extracts;
50–100 ml for stream and lake waters;
200 ml for rainwater.

Connect the flask to distillation apparatus and allow steam to pass through.

Distil over about 100 ml and reject. (If NH_4^+-N is also required, collect this distillate in 5 ml boric acid–indicator and titrate with standard acid).

Remove the flask, add 0.4 g Devarda's alloy and replace immediately.

Distil over into 5 ml boric acid–indicator, collecting about 50 ml.

Use a 5–10 ml burette and titrate against M/140 HCl from blue to a pale neutral end-point.

Carry out blank determinations and subtract where necessary.

(Total inorganic nitrogen may be obtained by adding the MgO and Devarda's alloy together and collecting the total distillate.)

Calculation

If *T* ml M/140 HCl are required for the titration then for:

1 Soil extracts,

$$\text{Extractable } NO_3^-\text{-N (mg 100 g}^{-1}) = \frac{T\text{ (ml)} \times \text{extractant vol (ml)} \times 10}{\text{aliquot (ml)} \times \text{sample wt (g)}}$$

2 Waters,

$$NO_3^-\text{-N (mg l}^{-1}) = \frac{T\text{ (ml)} \times 10^2}{\text{aliquot (ml)}}$$

Apply factors for dilution or concentration and correct to dry weight where necessary.

Note

It has been reported (Best and Cranwell, 1985) that Devarda's alloy is less effective if excessive amounts of Mg ions are extracted from the soil.

These authors recommend wide soil: solution ratios and increased levels of alloy (0.4 g).

Phenoldisulphonic acid method

The reaction between nitrate and phenol 2,4-disulphonic acid produces a nitro-derivative which on conversion to the alkaline salt gives a yellow colour. The reaction obeys the Beer–Lambert law up to 12 $mg\,l^{-1}$.

The method was first used by Harper (1924). It has the advantages of being direct and simple to use and is reasonably sensitive. Two disadvantages frequently arise with natural waters or soil extracts. If organic matter is present the sample will have a yellow-brown colour. This is similar to the phenoldisulphonic acid colour and must first be removed as described on p. 73. Alternatively the Kjeldahl reduction procedure should be used.

Chlorides tend to remove nitrate as nitrosyl chloride, giving low results and 10 $mg\,l^{-1}$ of chloride is the maximum permissible level. The chloride can be removed by precipitation after the addition of silver sulphate solution. The amount of silver sulphate added should be just enough to bring the chloride within the permitted range, since excess of the silver salt will be troublesome later. An initial titration to determine the approximate chloride content may be necessary. Only water extracts of soils are suitable for this method. Nitrates in 6% sodium chloride extracts may be estimated by distillation or by the method on p. 130.

Ammonium hydroxide is the alkali normally used for the final colour development, especially if silver salts have been added to remove chlorides. However, the presence of ammonium salts at an earlier stage has been found by Hora and Webber (1960) to affect nitrate recovery. They found that the presence of two drops of 30% potassium hydroxide prevented this. A more sensitive, but less stable method for nitrite is based on the Griess–Isolvay diazotization. A continuous flow version of this method, which can be adapted for manual operation, is given on the next few pages.

The phenoldisulphonic acid method is described below. Providing no interference is encountered the technique is sensitive to about 2 μg nitrate-nitrogen.

Reagents

1 Standard solutions.

Stock solution (1 ml ≡ 0.1 mg NO_3^--N): dissolve 0.7216 g dry KNO_3 in water and dilute to 1 litre.

Working standard (1 ml ≡ 0.01 mg NO_3^--N): dilute the stock solution 10 times.

2 Phenoldisulphonic acid.

Dissolve 25 g phenol in 150 ml conc H_2SO_4.

Add 75 ml fuming H_2SO_4 (15% free SO_3) and stir thoroughly.

Heat for 2 hours in boiling water with a glass bubble in the neck of the container.

Store in a dark vessel.

3 Ammonium hydroxide, 1 + 1.

Procedure

Pipette 0–5 ml working standard into borosilicate evaporating basins, (to give a range from 0 to 0.05 mg NO_3^--N).

Pipette an aliquot of soil extract or water sample into an evaporating basin.

From this point treat standards and samples in the same way.

Evaporate to dryness on a steam bath but do not bake.

When cool add 2 ml phenoldisulphonic acid. Swirl rapidly, to bring residue and acid into contact quickly.

Stand for 10 minutes.

Add about 20 ml water.

Add carefully 1 + 1 NH_4OH until the pH is between 10 and 11.

Filter through 9 cm Whatman No. 541 filter paper into a 50 ml volumetric flask.

Make up to volume and mix well.

Measure the absorbance at 410 nm or with a blue filter, using water as a reference.

Prepare a calibration curve from the standards and use it to determine mg NO_3^--N in the sample aliquot.

Carry out blank determinations in the same way and subtract where necessary.

Calculation
If C = mg NO_3^--N obtained from the graph then for:
1 Soil extracts,

$$\text{Extractable } NO_3^-\text{-N (mg 100 g}^{-1}) = \frac{C\text{ (mg)} \times \text{extractant vol (ml)} \times 10^2}{\text{aliquot (ml)} \times \text{sample wt (g)}}$$

2 Waters,

$$NO_3^-\text{-N (mg l}^{-1}) = \frac{C\text{ (mg)} \times 10^3}{\text{aliquot (ml)}}$$

Apply factors for dilution or concentration and correct to dry weight where necessary.

Continuous flow method
The reaction of nitrite with naphthylamine–sulphonic acid followed by the production of the azo-dye is the basis of a sensitive method for nitrates if they are first reduced to nitrites. In the manual version of the method, colour development can be difficult to control but the technique is particularly suited for automation and such a method is described here. Strictly the method gives nitrate + nitrite nitrogen and this is adequate for most ecological purposes. Nitrite, although very low in most samples can be determined if the reducing agent is replaced by water (p. 132). Hydrazine sulphate catalysed by copper is the reducing agent in the method given here but a cadmium reduction column (e.g. Willis and Gentry, 1987) is often used instead.

The flow diagram (manifold) given here is that needed for the AA II system (see Chapter 8). The manifold needed for AA I with a faster flow and larger mixing coils was described by Allen *et al.* (1974).

Reagents
1 Nitrate standards.
Stock solution (100 ppm NO_3^--N): dissolve 0.7216 g dry KNO_3 in water and dilute to 1 litre.
Standards: prepare a suitable range by dilution of the stock solution.
Include soil extractant (6% NaCl) where necessary.
2 Phenol, 0.94%.
3 Sodium hydroxide, 0.048 M.
4 Hydrazine sulphate reagent; dissolve 2 g hydrazine sulphate in 1 litre water. Separately, dissolve 0.0984 g $CuSO_4.5H_2O$ in 1 litre water. Mix equal volumes fresh each day.
5 Sulphanilic acid reagent; dissolve 2.4 g in 1.6 litres water, add 111 ml conc HCl and dilute to 2 litres. Mix equal volumes 24% acetone (v/v) with sulphanilic solution fresh daily.
6 1-Naphthylamine-7-sulphonic acid, 0.04%; warm to dissolve if necessary.
7 Sodium acetate solution; dissolve 86 g sodium acetate trihydrate in 1 litre water.

Operating conditions
Manifold details are given in Fig. 5.6.
A 10 mm flow cell is generally adequate but longer cells can be used.
Heating bath temperature: 70 °C.
Wavelength for maximum absorption: 520 nm.
Sampling rate: up to 40 per hour (sample: wash ratio = 1 : 1).

Calculation
If C = ppm (NO_3^--N + NO_2^--N) obtained from the graph then for:
1 Soil extracts,

$$\text{Extractable } NO_3^-\text{-N} + NO_2^-\text{-N (mg 100 g}^{-1}) = \frac{C\text{ (ppm)} \times \text{extractant vol (ml)}}{10 \times \text{sample wt (g)}}$$

2 Waters,

$$NO_3^-\text{-N} + NO_2^-\text{-N (mg l}^{-1}) = C\text{ (ppm)}$$

Apply dilution or concentration factors and correct to dry weight where necessary.

Note
Check the purified water before use because some deionized supplies contain nitrate.

Ion-selective electrode method
As mentioned in Chapter 8, the nitrate ion-selective electrode is suitable for soil and plant extracts and waters. A large number of ions affect the response of the electrode, but in ecological

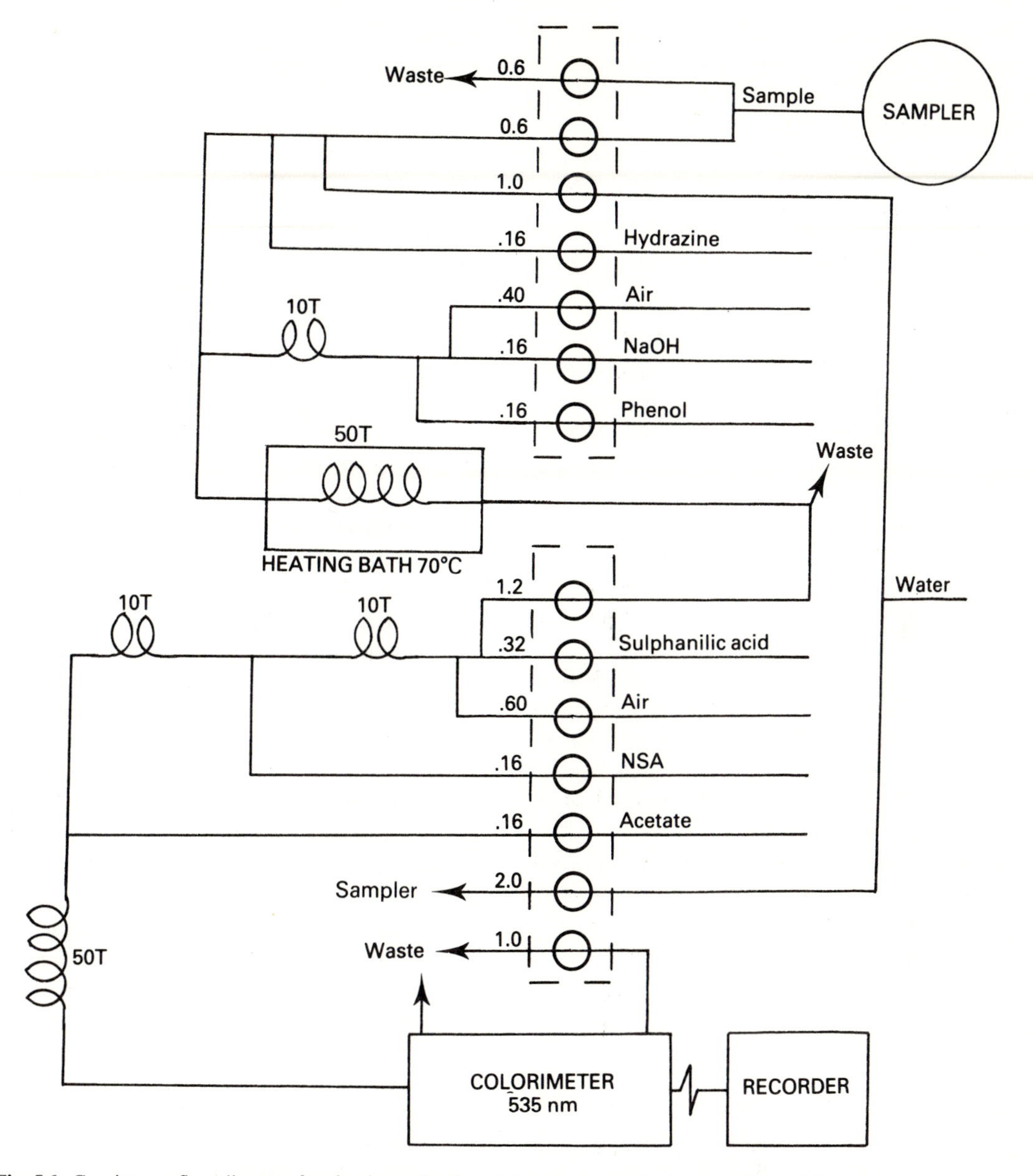

Fig. 5.6. Continuous flow diagram for the determination of (nitrate + nitrite)-nitrogen. (Dotted line indicates pump modules. T indicates number of turns on mixing coils. All flow rates expressed as ml min^{-1}.)

materials the only ones likely to be present in sufficient quantities are chloride, carbonate and bicarbonate. Chloride can be removed by the addition of sulphate whilst carbonate and bicarbonate measure acidification to pH 4.5.

The sensitivity will depend on the type of electrode used but it should be possible to measure down to 0.2 ppm nitrate. Smith (1975) and Li and Smith (1984) discuss the use of ion-selective electrodes for the determination of nitrate in soil and plant extracts.

Reagents

1 Standard solutions.

Stock solution (1000 ppm NO_3^--N): dissolve 7.216 g dry KNO_3 in water and dilute to 1 litre.

Working standards: prepare a suitable range to at least 100 × detection limit (0.2 ppm) to allow an electrode check (Note **2**). Extra standards are needed at the lower end (Note **3**).
Include acid or soil extractant as appropriate to match the samples, and adjust to pH 4.5 using 0.1 M H_2SO_4 before diluting to volume.
2 Sulphuric acid, 0.1 M.
3 Silver sulphate, 0.5% w/v.

Procedure

Prepare the sample solutions as described in Chapters 2, 3 and 4, adjust to pH 4.5 with 0.1 M H_2SO_4 and dilute to volume (see Note **1**).
Set up the instrument and electrodes according to the manufacturers' instructions.
Take 25 ml aliquots of samples and standards and add 10 ml 0.5% w/v Ag_2SO_4 to each.
Ensure all solutions are at room temperature then read standards and samples under the same conditions.
Prepare a calibration curve from the standard readings using semi-logarithmic paper with concentration on the logarithmic axis and instrument reading on the linear axis (see Note **2**).
Use the calibration curve to obtain ppm NO_3 in the sample solutions.
Carry out blank determinations in the same way and subtract where necessary.

Calculation

If C = ppm NO_3^--N obtained from the graph then for:
1 Soil extracts,

$$\text{Extractable } NO_3^-\text{-N (mg 100 g}^{-1}) = \frac{C\text{ (ppm)} \times \text{soln vol (ml)}}{10 \times \text{sample wt (g)}}$$

2 Waters,

$$NO_3^-\text{-N (mg l}^{-1}) = C\text{ (ppm)}$$

Apply factors for dilution or concentration and correct to dry weight where necessary.

Notes

1 For low ionic strength samples such as waters it is recommended that 0.5 ml 2 M $(NH_4)_2SO_4$ be added to 25 ml aliquots of the samples and standards before reading.
2 Nitrate is a monovalent ion and a tenfold increase in concentration (one log unit) should therefore decrease the reading by about 58 mV.
3 The Nernst relationship (linear) does not hold down to the detection limit. The range below 2 ppm will be non-linear and need a separate calibration line.

Nitrite-nitrogen

The levels of nitrite occurring in waters and soils are usually so low and transitory that they can often be ignored. For most purposes a combined nitrate + nitrite determination is adequate. A separate determination may, however, be required if the mechanism of nitrogen release or availability is being studied. The amount of nitrite present may be used as an indication of reducing conditions.

Since the nitrites present in soils and waters are so unstable, analysis immediately after collection is highly desirable. If this is not possible, cold storage will probably reduce the rate of conversion to nitrate. Soils must be extracted fresh with water but it will be difficult to avoid the effects of aeration.

Most commonly used methods for nitrite ions in solution involve direct colorimetry based on the Griess–Ilosvay reaction. This is a two-stage diazotization procedure whereby nitrite is diazotized to form a salt which is then coupled to form a coloured azo compound. A primary aromatic amine is needed for diazotization and sulphanilic acid is commonly used.

Naphthyl compounds are used for coupling because they provide a stable and intense colour. Two of them are α-naphthylamine and *N*-(1-naphthyl)-ethylenediamine which have been used by many workers including Mackereth *et al.* (1978) who employed the second compound in a manual procedure for natural waters. Unfortunately both of these substances are now listed as carcinogenic and their use is not recommended. An alternative reagent is 1-naphthylamine-7-sulphonic acid (Cleve's acid) which has not so far (1987) been classed as a harmful substance. The use of Cleve's acid has been described by Bunton *et al.* (1969). A method for nitrite alone is not given here but the nitrate + nitrite automated procedure described earlier can be used for the determination of nitrite by replacing the hydrazine reagent with water. The method is derived from that of Henriksen (1965) which uses Cleve's acid. Although the reaction conditions are critical, the automated procedure can be used as the basis for a manual method for nitrite. Some modifications may be needed to achieve optimum conditions especially in view of the low levels of nitrite likely to be present in most ecological materials.

The development of ion chromatography (Chapter 8) offers another direct method for nitrite although the peaks obtained for natural waters will normally be overshadowed by those of other ions present in greater amounts.

Mineralizable nitrogen

The amount of organic nitrogen mineralized and consequently available for plant growth is dependent on the activity of micro-organisms and is a key stage in the nitrogen cycle. This in turn is controlled by temperature, moisture and many other factors and shows strong seasonal trends.

The extent to which organic nitrogen can be mineralized is an important property of any soil but difficult to measure reliably. Methods for direct extractions of inorganic nitrogen using salt or water as described in Chapter 2 measure only the amounts present at a given time. Estimating amounts that could become available over a period requires specific treatment involving either incubation (biological methods) or reaction with the organic matter (chemical methods). Many procedures have been suggested (e.g. Fox and Piekielek, 1978 and Keeney in Page *et al.*, 1982). Dolmat *et al.* (1983) also review nitrogen availability tests. The terminology is not wholly agreed upon but the term 'mineralizable nitrogen' is preferred to 'nitrogen availability' used by Keeney. The two main groups of methods are discussed very briefly below and one of the older biological methods is given later.

1 *Biological methods*

In these methods the soil is incubated under conditions intended to simulate those considered to be optimal for microbial activity although in practice a wide variety of conditions have been specified. This is seen in an annotated list of nearly 30 biological methods published since 1963 and compiled by Keeney in Page *et al.* (1982). Temperatures ranged from 25 to 40 °C and incubation periods from 6 to 210 days, although two to three weeks were common. Control of the moisture content raised sufficient problems to generate two sub-groups of methods depending on whether the soil was kept at an 'optimal' moisture (aerobic) or was waterlogged (anaerobic). Secondary variations include soil pre-treatments and the use of additives. In all incubation methods the increase in mineral nitrogen during incubation (as recorded by extraction and/or distillation) is a measure of mineralizable nitrogen.

In spite of their problems biological methods seem capable of giving meaningful values if the organic nitrogen content of the soil is moderate to high. However, ecological studies can involve soils with total nitrogen contents of less than 0.1% dry basis and at these levels the incubation approach is not sufficiently sensitive.

The method given below is an aerobic procedure based on that given by Bremner in Black (1965). This has been replaced by an anaerobic procedure in the revised edition of this book (Page *et al.*, 1982).

2 *Chemical methods*

These methods are intended to simulate the mineralization processes occurring during incubation with the long time schedules of the biological methods being cut to a few hours. Many variants have been proposed and Keeney lists over 20 methods published since 1963. They show that water and a wide range of salts, acids, alkalis and oxidizing agents have been used at room temperature, 100 °C (reflux or distillation) and 121 °C (autoclave). The ammonium or total nitrogen contents of the extract or digest have both been equated with mineralizable nitrogen. Acceptable correlations are sometimes obtained between chemical and incubation data but the chemical methods remain highly arbitary and are not considered any further here.

Aerobic procedure

In this procedure the soil is prepared by drying (40 °C) and sieving (2 mm) as described in Chapter 2 for other soil tests. An additive (sand) is used and the moisture brought up to an optimal level for aerobic mineralization. The 6% KCl extraction and distillation are used to recover NH_4^+-N and $(NO_3^- + NO_2^-)$-N before and after incubation.

Reagents

1 Hydrochloric acid, M/140, 1 ml ≡ 0.1 mg N.
2 Boric acid–indicator solution.
3 Potassium chloride, 6% w/v.
4 Magnesium oxide, powder.
5 Devarda's alloy, finely powdered.
6 Sand, acid washed, 30 to 60 mesh.
Prepare **1** and **2** as instructions on p. 122.

Procedure

Weigh 5 g air-dried sieved soil into a 250 ml beaker flask (A) and similarly weigh 5 g into a soil extraction bottle (B).
Add 15 g of sand to both and swirl to mix. Then add 6 ml water to each and mix. Extract B immediately with 125 ml 6% KCl by shaking for 1 hour on a rotary shaker.
Filter through No. 44 filter paper or centrifuge.
Incubate A for 14 days at 30 °C keeping the flask sealed with a porous film.
After 14 days extract as before.
Run sand and water blanks with both A and B.
Determine total inorganic nitrogen in the extract as follows.
Pipette an aliquot (usually 20 ml) of the supernatant solution into the flask of a semi-micro distillation unit.
Add MgO and Devarda's alloy and determine the inorganic-N fraction as described earlier in this section (p. 128).

As an alternative the NH_4^+-N and $(NO_3^- + NO_2^-)$-N fractions can be estimated separately by the continuous flow methods given earlier with standards containing 6% KCl. The two results are combined to obtain total inorganic nitrogen in the extract.

Calculation

Mineralizable nitrogen (mg 100 g^{-1}) = A (mg 100 g^{-1}) − B (mg 100 g^{-1}).
Both A and B are calculated as for ammonium-nitrogen.
A = inorganic N after incubation, B = inorganic N before incubation.
Correct to dry weight where necessary.

Phosphorus

Phosphorus occurs in nature almost entirely as phosphate and resembles nitrogen in that both inorganic and organic forms are of major significance in plant–soil studies and in phosphorus cycling in natural systems (Griffiths *et al.*, 1973).

Primary mineral phosphates are widespread though rarely abundant and are dominated by the apatite group containing complex calcium phosphates. Rarer minerals such as monazite also occur in rocks and others are associated with metal ores. Although phosphate salts are soluble in acids, the apatite minerals weather only slowly and reserves in soils are low. Some of the released phosphate may react with other weathering products such as calcium, iron and aluminium to form secondary phosphates depending on soil pH and temperature. Other phosphate ions may become adsorbed ('fixed') on to mineral surfaces such as clays and hydrated iron and aluminium oxides. Adsorption–desorption processes are important regulators of the phosphate supplying capacity of a soil and have been much studied as reviewed by White (1980). The concentration of phosphate ions in the soil solution at a given time is very low but depletions due to plant uptake are replaced because of equilibria with the adsorbed or other phases. In most situations net phosphate loss from non-agricultural soils is made up by weathering income. Methods for assessing the 'labile pool' of phosphate in soil have already been discussed.

Agricultural soils usually cannot meet crop demands for phosphate, and fertilizers are essential. However, the effectiveness of phosphate fertilizers may be limited by adsorption by certain minerals and losses due to liming Haynes (1982) and by transfer to the organic form (Dalal, 1977). In turn added phosphate can affect micronutrient availability to plants (Murphy *et al.*, 1981).

In unfertilized soils, mineralization of organic phosphorus is as important a source for plants as that released from mineral phases. Enzyme (phosphatase) actively influences the rate of mineralization and has been studied in natural ecosystems (Harrison and Pearce, 1979). Other aspects of organic phosphorus in soil have also been widely examined and have been reviewed by Harrison (1986). A method for estimating this fraction is described below.

Phosphorus is an essential macronutrient for living matter and approaches nitrogen in importance even though concentrations of the latter can exceed phosphorus tenfold in plant tissues. Phosphate has both structural and metabolic functions, being a constituent of nucleic acids and certain esters and lipids whilst the inorganic ion has a key role in cell metabolism particularly in ATP–ADP cycling whereby energy is transferred in metabolic reactions (Bieleski and Ferguson, 1983). In animals, inorganic phosphate is a major structural component of bone in vertebrates.

Concentrations of phosphate ions in natural waters can be very low and form one of the limiting factors for plankton growth in spring. Inputs of effluent containing dissolved phosphate and other nutrients can lead to substantial increases in plankton populations (eutrophication). The input in rainfall is very small.

The concentration ranges normally encountered are given below.

Mineral soils	0.02–0.15%
Organic soils	0.01–0.2%
Soil extractions	0.3–8 mg 100 g^{-1}
Plant materials	0.05–0.3%
Animal tissues	0.3–4% excl. bone
Rainwater	2–50 g l^{-1}
Freshwater	5–500 g l^{-1}

(dry weight basis where relevant).

Phosphate-phosphorus

Colorimetry is almost always used for estimating phosphate-phosphorus in solution. Two basic methods are favoured and one of them (molybdenum blue) is described below. The alternative method (vanadomolybdate) was proposed by Kitson and Mellon (1944) but is less sensitive than molybdenum blue and hardly adequate for estimating trace levels in waters. It is sometimes favoured for relatively high levels of phosphate because the operating conditions are less critical than for molybdenum blue. Rann *et al.* (1987) use the method for phosphorus in plant tissue. The vanado-molybdophosphoric acid complex is yellow and formed by including ammonium vanadate along with ammonium molybdate in the reaction. The yellow colour is measured directly without chemical reduction and the absorption spectrum is broad (400–490 nm). The lower end is more sensitive, but also relatively prone to ferric iron interference so an intermediate wavelength may be preferable (Jackson, 1958). A disadvantage for soil extracts and waters is that brown colours will interfere at the wavelength used. If colour removal is not practicable the sample values may be corrected approximately by subtracting the apparent phosphate content of a 'colour blank' obtained by running the sample a second time with all reagents except the chromogenic reagent.

Molybdenum blue method

This method is based on the formation of a heteropoly-acid complex (phosphomolybdic acid) when an acid molybdate reagent is added to a solution containing orthophosphate. Reduction of this complex gives the characteristic molybdenum blue colour.

The method is very widely used because of its sensitivity but the conditions for optimum colour development are critical and have been repeatedly studied over many decades. The reduction stage is crucial and many different reducing agents have been suggested in the past. The two most widely used today are probably stannous chloride and ascorbic acid. Acidic stannous chloride appears to be the most sensitive reductant, but has a fairly critical development time. It can be used with hydrochloric or sulphuric acid, and the former is employed below. This reagent will stain glassware and tubing in time, but daily washing with dilute caustic soda followed by acid will offset this tendency. Ascorbic acid is a more stable reagent, cleaner to use and is widely employed. It is slightly less sensitive than stannous chloride although its performance can be improved.

Other important parameters are reagent concentration and the overall acidity which must be controlled (Jackson, 1958; Holman and Elliot, 1983). At high pH values the molybdate itself may give a blue colour in the absence of phosphate and alkaline samples must be neutralized with acid before reagents are added. In strong acid solutions the colour development is suppressed. Acid digests prepared as given in Chapter 3 are initially diluted to the equivalent of 1% sulphuric acid (v/v) before aliquots are taken for analysis. Standard and sample acidities must be the same. In addition only traces of oxidizing acids such as nitric and perchloric are permissible in the aliquot if the reducing stage is not to be affected. The mixed acid digestion allows for elimination of most of the perchloric acid before diluting to volume.

Certain substances, notably silicates and arsenates in solution can interfere because they react with molybdate in a similar way to phosphate. However, arsenates are rarely a problem in natural materials and the acidity of the method minimizes the effect of silicates. Iron (III) will depress the colour formation and levels are preferably kept below 10 mg l^{-1}. Up to 0.2 mg of iron can be present if an extra reduction stage, using hydrazine sulphate, is inserted.

The sensitivity of the method makes it rather vulnerable to cross-contamination from laboratory reagents in the laboratory such as phosphoric acid which is required in some colorimetric methods. Traces of phosphates in glassware can be removed by washing with dilute caustic soda followed by acid but it is preferable to keep separate glassware for this method.

In spite of these precautions the estimation of phosphate ions in true solution in natural waters is not straightforward. The molybdenum blue value is not easy to interpret because the acidity of the molybdate reagent may hydrolyse some weak bonds between phosphate and natural substances in solution and molybdate itself has come under suspicion (Tarapchak, 1983). The result is probably best regarded as 'dissolved reactive phosphorus'. A similar situation exists for the estimation of silicon in waters.

Stannous chloride reduction

Manual method

The manual method described here can be readily automated and a continuous flow version is given later. The Beer–Lambert law is obeyed beyond the range of standard values recommended below.

Reagents

1 Phosphorus standards.

Stock solution (1 ml ≡ 0.1 mg P): dissolve 0.4393 g dry KH_2PO_4 in water and dilute to 1 litre.

Working standard (1 ml ≡ 0.002 mg P): dilute the stock solution 50 times. Make up fresh at intervals.

2 Ammonium molybdate–sulphuric acid reagent.

Dissolve 25 g $(NH_4)_6Mo_7O_{24}.4H_2O$ in about 200 ml water in a beaker.

Warm slightly to dissolve.

Add carefully (with mixing and cooling) 280 ml conc H_2SO_4 to about 400 ml water.

Filter the molybdate solution into the acid mixture, mix thoroughly and make up to 1 litre when cool.

Store in the dark.

3 Stannous chloride reagent.

Dissolve 0.5 g $SnCl_2.2H_2O$ in 250 ml 2% v/v HCl. Prepare immediately before use. Note that if the reagent appears milky some stannic chloride is present in the salt and it should be discarded.

Procedure

Prepare the sample solution as described on pp. 29, 42, 61, 77 and dilute to volume.

Pipette 0–15 ml of working standard into 50 ml volumetric flasks to give a standard range from 0 to 0.3 mg P.

Include the digest acid or soil extractant comparable to that in the sample aliquot (see Note).

Pipette a suitable aliquot (normally up to 10 ml) of sample solution into a 50 ml volumetric flask.

From this point treat standards and samples in the same way.
Dilute until the flask is about two thirds full.
Add 2 ml ammonium molybdate reagent and mix.
Add 2 ml stannous chloride reagent, mix, dilute to volume and time from this stage.
Leave for 30 minutes.
Measure the absorbance at 700 nm or with a red filter using water as a reference.
Prepare a calibration curve from the standards and use it to determine mg P in the sample aliquot.
Carry out blank determinations in the same way and subtract where necessary.

Calculation

If C = mg P obtained from the graph then for:

1 Plant materials and soils,

$$\text{P}\ (\%) = \frac{C\ (\text{mg}) \times \text{soln vol (ml)}}{10 \times \text{aliquot (ml)} \times \text{sample wt (g)}}$$

2 Soil extracts,

Extractable PO_4^{3-}-P (mg 100 g^{-1}): as above $\times 10^3$.

3 Waters,

$$PO_4^{3-}\text{-P (mg l}^{-1}) = \frac{C\ (\text{mg}) \times 10^3}{\text{aliquot (ml)}}$$

Apply factors for dilution or concentration and correct to dry weight where necessary.

Note

For Olsen's soil extracts where neutralization of the solution is required (p. 44), compensate the standard solutions with neutralized extractant.

EXTRACTION TECHNIQUE FOR WATERS

Although the molybdenum blue method is sensitive, it is difficult to get precise estimates of phosphate-phosphorus at levels below 5 μg l^{-1}. Procedures for concentration with organic extractants are helpful and in the method given below the phosphomolybdate complex is extracted into n-butanol before colour development. n-Hexanol is also suitable (Mackereth, 1963). However, it should be noted that whilst marked improvements in detection limits are possible, the interpretation of the results is not always straightforward. This extraction is suitable for soils but is not usually required for plant digests. The detection limit is about 0.01 μg phosphorus as orthophosphate.

Reagents

1 Phosphorus standards
2 Ammonium molybdate–sulphuric acid reagent
3 Stannous chloride reagent
4 n-Butanol.
Prepare **1**, **2** and **3** as instructions on p. 136.

Procedure

Pipette 0–5 ml of working standard into separating funnels to give a standard range from 0 to 0.01 mg P.
Pipette a suitable aliquot (up to 100 ml) of sample water into separating funnels.
From this point treat standards and samples in the same way.
Saturate with n-butanol.
Add 2 ml acid–molybdate reagent.
Mix and allow to stand for 2 minutes.
Add 10 ml n-butanol.
Shake for 1 minute, allow to separate.
Reject aqueous layer.
Add 5 ml water, followed by 2 ml stannous chloride reagent.
Shake for 1 minute, allow to separate.
Reject aqueous layer.
Measure the absorbance at 730 nm or with a red filter using n-butanol as a reference.
Prepare a calibration curve from the standards and use it to determine mg P in the sample aliquot.
Carry out blank determinations in the same way and subtract where necessary.

Calculation

As in the previous method.

Continuous flow method

The molybdenum blue method is readily automated and is strongly recommended in this form. The vanado-molybdophosphoric acid procedure is also suitable for continuous flow analysis

(Steikel and Flannery, 1971) but the sensitivity is not adequate for estimating the very low levels found in some waters.

The method given here is adapted from the manual procedure using stannous chloride for reduction. A disadvantage of this reagent is the white deposit which it leaves on the insides of tubing, but periodic cleaning with dilute alkali and dilute acid will minimize this problem. The ascorbic acid reducing system is cleaner to use but less sensitive.

The continuous flow method is suitable for all sample solutions considered earlier except for the alkaline Olsen soil extracts. The acid of the molybdate reagent causes excess carbon dioxide to be liberated in the coils although Smith and Scott (1983) felt that a debubbler would control this.

The flow diagram (manifold) shown here is that needed for the AAII system (see Chapter 8). The time allowed for colour development is less than that accepted for the AAI system described by Allen *et al.* (1974).

Reagents

1 Phosphorus standards.

Stock solution (100 ppm P): dissolve 0.4393 g dry KH_2PO_4 in water and dilute to 1 litre.

Standards: prepare a suitable range by dilution of stock solution.

Plant materials may require 0 to 5 ppm, but 0 to 1 ppm may be sufficient for waters and soil extracts. Include 1% H_2SO_4, soil extractant or water as appropriate.

2 Ammonium molybdate–sulphuric acid reagent

3 Stannous chloride–hydrochloric acid reagent

Prepare **2** and **3** as instructions on p. 136.

Operating conditions

Manifold details are given in Fig. 5.7.

A 10 mm flow cell is normally adequate although longer cells can be used.

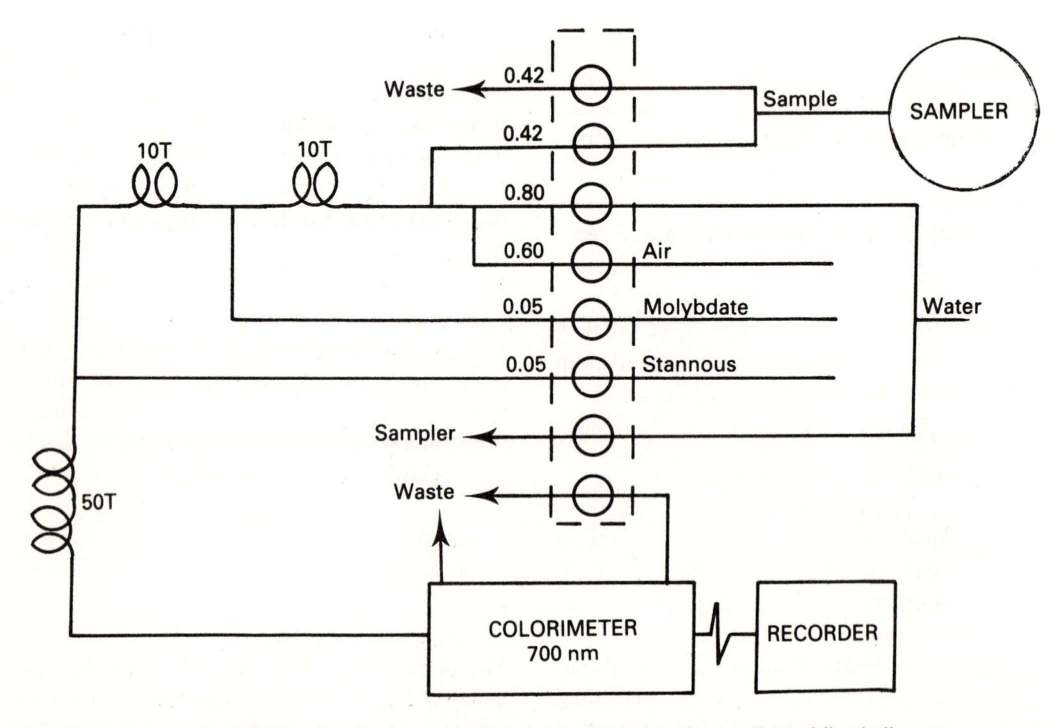

Fig. 5.7. Continuous flow diagram for the determination of phosphate-phosphorus. (Dotted line indicates pump module. T indicates number of turns on mixing coils. All flow rates expressed as ml min^{-1}.)

Wavelength for maximum absorption: 820 nm.
Sampling rate: up to 60 per hour (sample: wash ratio = 2:1).

Calculation
If C = ppm obtained from the graph then for:
1 Plant materials and soils,

$$P(\%) = \frac{C\,(\text{ppm}) \times \text{soln vol (ml)}}{10^4 \times \text{sample wt (g)}}$$

2 Soils extracts,
Extractable PO_4^{3-}-P (mg 100 g^{-1}): as above $\times 10^3$.
3 Waters,
PO_4^{3-}-P (mg l^{-1}) = C (ppm).
Apply factors for dilution or concentration and correct to dry weight where necessary.

Note
Wash daily with 5 M NaOH followed by dilute HCl to remove deposits on the tubing.

Ascorbic acid reduction
The operating conditions for ascorbic acid as normally used are not so critical as for stannous chloride and it is not subject to the opalescence which affects stannous chloride solutions. However ascorbic acid is a less sensitive reagent for the molybdenum blue colour development although its performance can be improved by various technique modifications. These include the use of a heating bath or coil at 40 °C, the addition of acetone and the use of longer path cuvettes of flow cells. The inclusion of an antimony salt has been said to improve the reduction reaction rate.

Manual methods using ascorbic acid have been given by Golterman *et al.* (1978) and Mackereth *et al.* (1978). Continuous flow procedures include those recommended by Skalar Analytical and Technicon Instruments for their instruments.

Organic phosphorus

A substantial fraction of the total phosphorus in soil is in the organic state and an estimate of this may be required. Black and Goring (1953) and Williams *et al.* (1970) have reviewed methods available. The results obtained by the various procedures are frequently different and it is difficult to establish which system gives the most reliable estimate. Saunders and Williams (1955) thought that the ignition method which is the simplest and most convenient to use gave results which were adequate for most purposes. A more recent ignition method is given by Nicholson (1984). Others have queried the use of ignition in phosphorus studies and the method seems suspect for some soil types. The risk of loss by volatilization during ignition was discussed in Chapter 3.

Various alkaline reagents have been proposed to extract organic phosphorus. Potential difficulties include incomplete extraction and hydrolytic conversion of organic phosphorus to the inorganic form. Mehta *et al.* (1954) used 0.5 M sodium hydroxide and found that more organic phosphorus was extracted if the soil was pre-extracted with concentrated hydrochloric acid. Hot alkali was used to complete the extraction but if preceded by a cold alkaline extraction the amount of organic phosphorus subjected to high temperature, and made susceptible to hydrolysis, was reduced.

Anderson (1960) adopted Mehta's extractions but found that labile organic phosphorus could be hydrolysed during the acid extraction. He proposed separating this fraction with 0.3 M sodium hydroxide prior to the acid extraction. Steward and Oades (1972) and Soltanpur *et al.* (1976) claimed that a brief period of ultrasonic vibration in sodium hydroxide was as effective as the prolonged alkaline extraction normally used. The potential value of this approach needs further confirmation. A schematic outline of the Anderson method in a slightly modified form is given in Fig. 5.8.

Extraction stage

Extractants
1 Sodium hydroxide, 0.3 M.
Dissolve 66.7 g NaOH pellets in water and dilute to 5 litres.

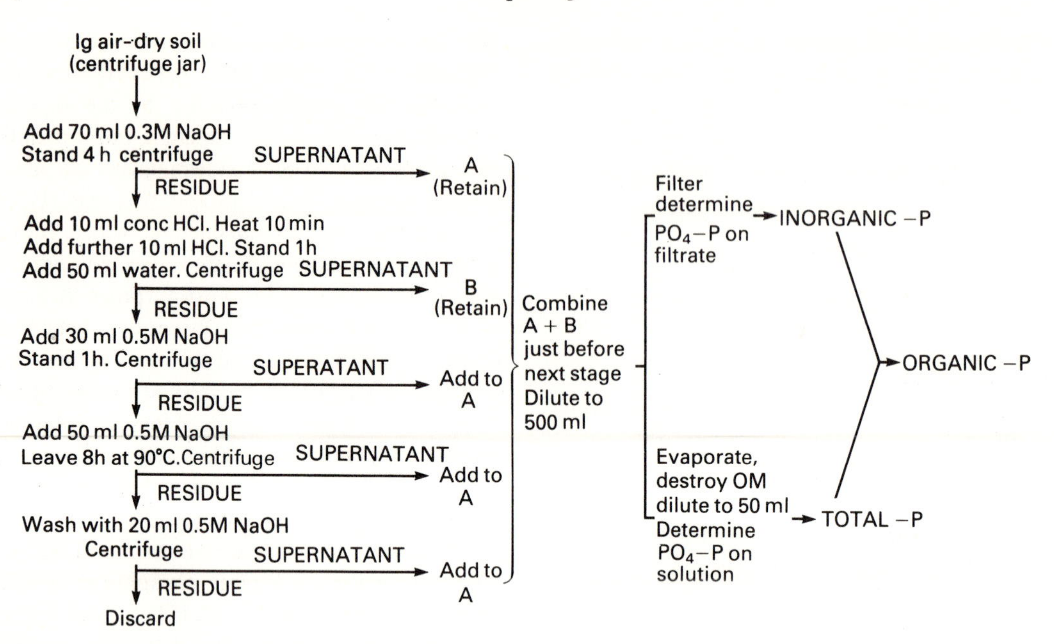

Fig. 5.8. Flow diagram for estimating organic-phosphorus in soils.

2 Sodium hydroxide, 0.5 M.

Dissolve 100 g NaOH pellets in water and dilute to 5 litres.

3 Hydrochloric acid, conc.

Procedure

Weigh 1 g air-dry, sieved soil into a centrifuge jar.

Add 70 ml 0.3 M NaOH and stand at least 4 hours with occasional swirling.

Centrifuge and decant the supernatant into a polythene bottle.

Add 10 ml conc HCl to the residue in the jar and heat 10 minutes in boiling water.

Add a further 10 ml of conc acid and stand 1 hour at room temperature.

Add about 50 ml water, centrifuge and decant into a separate polythene bottle.

Add 30 ml 0.5 M NaOH to residue and stand 1 hour at room temperature.

Centrifuge and decant into first (alkaline) bottle.

Add 50 ml 0.5 M NaOH to residue and leave for 8 hours at 90 °C in an air-circulation oven. Place glass bubble in neck to retard evaporation and top up to 50 ml with water when necessary.

Centrifuge and decant into first bottle.

Wash residue with 20 ml 0.5 M NaOH, centrifuge and decant into first bottle.

Keep acid and combined alkaline extracts separate until just prior to analysis.

Take two blanks through the same procedure.

The inorganic and total phosphorus contents of the combined acid and alkaline extracts are determined as follows and organic phosphorus found by difference.

Colorimetric stage

Inorganic phosphorus

Reagents

1 Phosphorus standards.

2 Ammonium molybdate–sulphuric acid reagent.

3 Stannous chloride reagent.

Prepare **1**, **2** and **3** as instructions on p. 136.

Procedure

Combine acid and alkaline extracts in a 500 ml volumetric flask and dilute to volume.

Mix thoroughly and immediately filter a small portion through No 44 paper (see Notes **1** and **2** below).

Pipette a suitable aliquot of filtrate into a 50 ml volumetric flask.

Pipette 0–15 ml of working standard into 50 ml volumetric flasks to give a standard range from 0 to 0.03 mg P. It is important to include an aliquot of combined acid and alkaline extractants comparable to the aliquot taken for the samples.

From this point treat standards and samples in the same way.

Develop the colour, record absorbance (700 nm) and prepare graph.

Calculation

If C = mg P obtained from the graph then:

$$\text{Inorganic P (\%)} = \frac{C\text{ (mg)} \times \text{combined vol (ml)}}{10 \times \text{aliquot (ml)} \times \text{sample wt (g)}}$$

Correct to dry weight where necessary.

Notes

1 The extracts should not be combined until just prior to the analysis. The combined extract is acidic and prompt analysis will minimize any hydrolytic conversion of organic to inorganic phosphorus.

2 The combined extract must always be well-mixed before any portion is removed to ensure that the flocculate is well dispersed.

3 The filtrates are all brown but the colour will not normally interfere at 700 nm. If it is dense however, the absorbance can be checked by diluting an aliquot to 50 ml and omitting the stannous chloride reagent.

4 The filtrates can be analysed using continuous flow colorimetry as given earlier but the tubing and coils are subject to coating with organic matter and must be cleaned regularly by passing dilute caustic soda through followed by acid.

Inorganic + organic phosphorus (combined)

Reagents

1 Phosphorus standards.

2 Ammonium molybdate–sulphuric acid reagent.

3 Stannous chloride reagent.

Prepare **1**, **2** and **3** as instructions on p. 136.

Procedure

Mix combined unfiltered extract thoroughly and immediately measure a suitable volume into a 500 ml Taylor flask.

Boil down and destroy organic matter as given for waters on p. 74.

Finally dilute to volume (50 ml).

Pipette a suitable aliquot into a 50 ml volumetric flask.

Pipette 0–15 ml of working standard into 50 ml volumetric flasks to give a standard range from 0 to 0.03 mg P. Include H_2SO_4 in the standards comparable to that in the sample aliquots.

From this point treat standards and samples in the same way.

Develop the colour and prepare graph.

Calculation

If C = mg P obtained from the graph then:

$$\text{Total P (\%)} = \frac{C\text{ (mg)} \times \text{combined vol (ml)}}{10 \times \text{aliquot (ml)} \times \text{sample wt (g)}}$$

Apply factors for dilution or concentration and correct to dry weight where necessary.

Calculation of organic phosphorus

Organic P (%) = Combined P(%) − Inorganic P (%)

Total phosphorus

Total phosphorus in ecological materials is mostly estimated as inorganic phosphate-phosphorus following a suitable dissolution procedure. More resistant forms of phosphorus in roots and soils require more drastic treatment. Both fusion and hydrofluoric–perchloric procedures are suitable for recovering phosphorus from these materials with occasional exceptions, but in practice $HF + HClO_4$ is recommended (see p. 27). Substitute methods involving perchloric acid alone have been widely used for soil phosphorus but for various reasons digestions based on nitric–perchloric–sulphuric acids (p. 59) or the sulphuric–peroxide system (p. 59) are preferred. These two digestion methods are very suitable for vegetation, but dry-ashing is less acceptable for phosphorus unless precautions are taken to minimize losses by volatilization. Modification of the two wet-digestions can also be used for total phosphorus in water.

Potassium

Potassium is prominent in primary alumino-silicates (notably certain feldspars and in micas). However, it is the two main forms, 'exchangeable' and 'non-exchangeable', held by the secondary clay minerals which are the most relevant to plant–soil relationships of this element. The non-exchangeable fraction is partly available to plants, in contrast to most primary mineral potassium and has long been studied as reviewed by Martin and Sparks (1985). This element is mobile, being readily leached from light soils low in clay minerals and also from organic matter. Many agricultural soils cannot fully meet crop demands for potassium, and fertilizers are required.

The element is essential for life and plants need it at macronutrient levels although it appears to have no structural function in animals or plants. In plant cell cytoplasm the levels are kept relatively constant as required for biochemical functions such as enzyme activation, protein syntheses and energy metabolism including phosphorylation. Concentrations can vary in the vacuole where potassium has a more biophysical role including maintenance of cell turgor. This in turn relates to stomata opening and the water economy of the plant. Uptake mechanisms have been studied and may help to explain the antagonisms sometimes reported between potassium and calcium or magnesium (Mengel and Kirkby, 1980; Leigh and Wyn Jones, 1984). Levels in freshwaters are more uniform than for sodium or calcium and largely reflect terrestrial sources. The input in rainfall is small.

The concentration ranges normally encountered are given below.

Mineral soils	0.3–2%
Organic soils (peat)	0.02–0.2%
Soil extractions	5–50 mg 100 g^{-1}
Plant materials	0.5–5%
Animal tissues	0.3–1.5%
Rainwater	0.1–1.0 mg l^{-1}

(dry weight basis where relevant).

The intense emission lines given by sodium and potassium in a flame enable low concentrations to be determined by flame photometry and the general principles are discussed later. The arc line for potassium (766 nm) is a little less intense than for sodium, but well within the capacity of simple flame photometers. Procedures for both flame emission and atomic absorption are given below. The classical gravimetric procedure using sodium cobaltnitrite has long been superseded by flame methods and no useful colorimetric reagent has been found for potassium. However, an ion-selective electrode is available for this element.

Flame emission method

Interferences are not normally serious for sodium and potassium. Residual acids (sulphuric and hydrochloric) and soil extractants have a slight effect and their levels should be the same in both standards and samples. High calcium levels give a positive error particularly with sodium and should be taken into account. Correction graphs can be prepared but dilution to reduce calcium concentrations is possibly more satisfactory. Interference by calcium is much less in 1% sulphuric acid where 10 ppm potassium can tolerate at least 200 ppm calcium.

Sodium interferes with potassium but its levels in most organic material are too low to be significant. It should however, be allowed for when analysing marine samples.

Internal standards such as lithium are usually recommended for emission methods to offset the effects of flow instability. On some instruments the internal standard is mixed automatically with the sample solution.

With sensitive instruments flame emission can be used for levels as low as 0.01 μg ml^{-1} potassium.

Reagents

1 Potassium standards.

Stock solution (1000 ppm): dissolve 1.9068 g dry KCl in water and make up to 1 litre.

Working standards: dilute to produce a suitable range between 0 and 100 ppm K. Include the appropriate amount of digest acid or soil extractant. (A much lower range is usually adequate for soil extracts and waters.)

Procedure

Prepare the sample solutions as described on pp. 30, 44, 61 and dilute to volume.

Select the 766 nm wavelength or K filter and adjust the gas pressures, slit-width and other settings as recommended for the instrument employed.

Prepare a calibration curve from the standard range by setting the top standard to a suitable readout value and the 0 ppm standard to zero, then aspirating each standard in turn.

Aspirate the sample solutions into the flame under the same conditions as the standards. Check the top, zero and an intermediate standard frequently (see Note **1**).

Flush the atomizer and burner frequently with water, particularly at the end of the run.

Use the calibration curve to determine ppm K in the sample solutions.

Carry out blank determinations in the same way and subtract where necessary.

Calculation

If C = ppm K obtained from the graph then for:

1 Plant materials and soils,

$$K\ (\%) = \frac{C\ (\text{ppm}) \times \text{soln vol (ml)}}{10^4 \times \text{sample wt (g)}}$$

2 Soil extracts,

Extractable K (mg 100 g^{-1}): as above $\times 10^3$;

(for me 100 g^{-1} K divide mg 100 g^{-1} result by 39.10)

3 Waters,

$$K\ (\text{mg l}^{-1}) = C\ (\text{ppm})$$

Apply factors for dilution or concentration and correct to dry weight where necessary.

Notes

1 Include an internal standard (e.g. 200 ppm lithium) with both standards and samples as desired.

2 Older photometers with a simple scale deflection can sometimes be run using a semi-permanent graph. However, instruments with a sampler and recorder require the calibration line to be established with each batch.

Atomic absorption method

Atomic absorption is a suitable technique for the determination of potassium. Interferences are similar to those encountered in flame emission and should be handled in the same way. Sensitivity and reproducibility are comparable to flame emission and an internal standard is not necessary. Ionization can decrease sensitivity especially in hotter flames such as nitrous oxide–acetylene. The effect is offset if a similar atom such as lithium (or sodium in marine samples) is present in excess (ionization buffer).

Reagent

1 Potassium standards.

Prepare as above.

Procedure

Prepare sample solution as described on pp. 30, 44, 61 and dilute to volume.

Select the 766.5 nm wavelength and adjust air and gas flows, slit-width and other settings as recommended for the instrument employed.

Always allow the hollow cathode lamp adequate time to stabilize.

Prepare a calibration curve from the standard range by setting the top standard to a suitable readout value and the 0 ppm standard on zero, then aspirating each standard in turn.

Aspirate the sample solutions into the flame under the same conditions as the standards (Notes **1** and **2**).

Check the top, zero and an intermediate standard frequently.

Flush the spray chamber and burner frequently with water particularly if soil extracts are aspirated.

Use the calibration curve to obtain ppm K in the sample solutions.

Carry out blank determinations in the same way and subtract where necessary.

Calculation

If C = ppm K obtained from the graph then for:

1 Plant materials and soils,

$$K(\%) = \frac{C\ (\text{ppm}) \times \text{soln vol (ml)}}{10^4 \times \text{sample wt (g)}}$$

2 Soil extracts,

Extractable K (mg $100\ g^{-1}$): as above $\times 10^3$;

(for me $100\ g^{-1}$K divide mg $100\ g^{-1}$ result by 39.10)

3 Waters,

K (mg l^{-1}) = C (ppm).

Apply factors for dilution and concentration and correct to dry weight where necessary.

Notes

1 Include an ionization buffer (e.g. 0.1% LiCl) with standards and samples, particularly when hot flames (e.g. nitrous oxide–acetylene) are used.

2 It is desirable to use a special high solids burner for soil extracts, particularly ammonium acetate to minimize blockage by carbon deposits.

Silicon

Although free silicon is not recorded in nature, silica tetrahedra (SiO_4) dominate the structures of primary and secondary silicate minerals in combination with aluminium or with iron and magnesium. These account for much of the mineral content of soils. Large numbers of silicate minerals are known although many are not common. The main primary aluminosilicates are the feldspars and micas whilst the key ferromagnesian groups are the pyroxenes and amphiboles together with olivine. All these weather to form secondary silicates or other products ('hydrous oxides') and in particular micas lead to clay silicate minerals which are of major importance in plant–soil relationships because of the nutrient elements which they hold in an exchangeable form. Silica (SiO_2) units also form important primary minerals, notably quartz, but these are inert and contain no other elements.

In contrast to soil structure this element is not of major importance in biology although it is an essential structural component (as silica) in algal groups such as the diatoms and it is present in some invertebrates, including sponges. Many grasses and sedges also accumulate silica and there are grounds for regarding it as essential for these plants (Cheng, 1982). Although silica is non-toxic in nature certain synthetic products such as asbestos are a hazard to man because the minute fibres they release to the atmosphere can induce lung diseases.

Silicon is an important nutrient in fresh and marine waters on account of algal requirements. Fresh water contains various forms, including monomeric and dimeric forms which are 'molybdate-reactive', the much less reactive polymeric forms and inert silica. Pre-treatments including filtering and the use of alkali allow partial distinction between the forms (see Standing Committee of Analysts, 1981). Three silicon values obtained directly for waters (total, molybdate-reactive and total dissolved) are considered later.

The concentration ranges (as silicon) normally encountered are given below:

Mineral soils	20–60%
Organic soils (peat)	0.2–2%
Plant materials	0.05–0.5%
Animal tissues	0.05–0.3%
Rainwater	0.01–0.1 $\mu g\, l^{-1}$
Freshwater	0.5–8 $mg\, l^{-1}$

(dry weight basis where relevant).

Total silicon in soils or plants is not often required in ecological studies but if needed it can be measured following a fusion procedure (pp. 25–27). For most purposes a rough estimate of mineral matter, as given by the ash-content following ignition, is sufficient (p. 54). However, if desired the ash can be evaporated with hydrochloric acid and a residue recovered which is an approximate measure of silica. Details of this procedure are given below. It is applicable to waters if adequate samples are taken. Sulphuric acid-based digestions also dehydrate silica effectively leaving it as a residue in the Kjeldahl flask. However, this residue tends to adhere closely to the glass and is difficult to recover quantitatively.

Colorimetric methods for silicon rely mainly on forming the 'molybdenum blue' complex similar to the procedure used for phosphorus. This procedure is most appropriate for natural waters but it can be adapted for soils and plant materials. The method given below includes control of molybdosilicic acid formation and also of interference by phosphorus. Titrimetric and atomic absorption methods have been suggested for this element, but their sensitivity is poor.

Gravimetric method

Any organic matter must first be destroyed by ignition or acid digestion. The residue is then taken up in hydrochloric acid, which decomposes silicates giving silicic acids. These are then dehydrated to silica by continued evaporation, baking and final ignition.

This method is suitable for plant (and most other organic) materials and waters but it only gives an approximate result. If the residue is volatilized with hydrofluoric acid (Standard Methods for the Examination of Waters, 1985) a better result in terms of silica is possible. However, to obtain the most accurate total silicon results, a fusion stage is necessary.

Reagent

Hydrochloric acid, conc.

Procedure

1 Plant material.

Ignite from 1 to 5 g of sample at 550 °C until the organic matter is completely destroyed.

2 Waters.

Evaporate a suitable volume (usually 200–1000 ml) to dryness on a steam bath. (If any organic matter appears to be present in the residue it should be removed by ignition.)

From this point treat all samples in the same way.

Add 5 ml conc HCl, 5 ml water, cover with a watch glass and simmer gently for a few minutes.

Remove the watch glass, evaporate to dryness and allow to bake for 15 minutes.

Repeat the addition of acid and water, simmer as before, but do not evaporate.

Filter through a small ashless filter paper.

Wash filter paper and residue well with hot water.

Ignite filter paper and residue at 550 °C.

Weigh as silica (SiO_2).

Calculation

If W (g) of residue are left after treatment then for:

1 Plant materials,

$$SiO_2\ (\%) = \frac{W(g) \times 10^2}{\text{sample wt (g)}}$$

2 Waters,

$$SiO_2\ (mg\ l^{-1}) = \frac{W(g) \times 10^6}{\text{aliquot (ml)}}$$

To express the results as Si multiply SiO_2 by 0.467.

Apply factors for concentration or correct to dry weight where necessary.

Note

If the residue is volatilized with hydrofluoric acid (Standard Methods for the Examination of Waters, 1985) or fused with alkali, a more accurate result in terms of silicon is obtained. However the best methods available for total silicon avoid recovering silica residues. For waters, the Standing Committee of Analysts (1981) recommend evaporation, fusion and redissolving before colorimetry.

Molybdenum blue method

The most convenient colorimetric procedure for silicon in solution is that based on the formation of the molybdenum blue complex. The details are similar to those followed for phosphorus. Phosphorus interference is prevented by pH control (Kahler, 1941) and the presence of oxalic or tartaric acid (Mullin and Riley, 1955).

The reaction with molybdate involves the production of two forms of molybdosilicic acid α and β depending on the acidity and other factors (Morrison and Wilson, 1963; Truesdale and

Smith, 1975). Control of their formation is desired because they reduce to the blue complexes of different absorption peaks. Morrison and Wilson showed the α-form was best reduced by an acid stannous reagent and β by acid amino-naphthol and they proposed a method based on the latter (Morrison and Wilson 1963). This procedure is given here.

The method can be used for soils and plant materials after fusion. Waters can be used directly for soluble silicon estimations but for total silicon it is necessary first to evaporate an aliquot and then to fuse with sodium carbonate (Standing Committee of Analysts, 1981), or with sodium hydroxide (see pp. 25 and 26). Levels of silicon down to 0.5 μg can be measured.

A modified procedure suitable for continuous flow equipment follows.

Reagents

(See Note **1**)

1 Silicon standards.

Stock solution (1 ml ≡ 0.1 mg Si): fuse 0.2139 g dry powdered silica (as SiO_2) with 1 g anhydrous Na_2CO_3 in a platinum crucible at 950 °C until a clear melt is obtained. After partial cooling, immerse in water in a polypropylene beaker. Warm to dissolve, cool, and dilute to 1 litre. Store in a polythene bottle.

Working standard (1 ml ≡ 0.005 mg Si): dilute the stock solution 20 times and store in a polythene bottle.

2 Ammonium molybdate–sulphuric acid reagent. Dissolve 89 g $(NH_4)_6Mo_7O_{24}.4H_2O$ in about 800 ml water. Dilute 62 ml conc H_2SO_4 to about 150 ml by careful addition to water and allow to cool. Add the acid to the molybdate solution and dilute to 1 litre.

3 Tartaric acid, 28% w/v.

4 Reducing solution.

Dissolve 2.4 g $Na_2SO_3.7H_2O$ and 0.2 g 1-amino-2-naphthol-4-sulphonic acid in about 70 ml water. Add 14 g $K_2S_2O_5$, shake until dissolved and dilute to 100 ml. Prepare fresh each week.

Procedure

Prepare the sample solutions as described on pp. 30, 61, 77 and dilute to volume.

(For soluble silicates in waters no prior treatment is needed.)

Pipette 0–6 ml working standard into clean 50 ml volumetric flasks to give a range from 0 to 0.03 mg Si.

Neutralize by adding 1+9 HCl dropwise.

Pipette a suitable aliquot of sample into a 50 ml flask.

(Solutions prepared by fusion are acidic and aliquots should be neutralized with 0.1 M NaOH using phenolphthalein as an indicator.)

From this point treat standards and samples in the same way.

Add 1.25 ml acid molybdate reagent, mix and leave for 10 minutes.

Add 1.25 ml tartaric acid solution, mix and leave for 5 minutes.

Add 1.0 ml reducing solution, mix and dilute to volume (see Note **1**).

Stand 15 minutes for a stable colour to develop.

Measure the absorbance at 810 nm or with a red filter using water as a reference.

Prepare a calibration curve from the standards and use it to determine mg Si in the sample aliquot.

Carry out blank determinations in the same way and subtract where necessary.

Calculation

If C = mg Si obtained from the graph then for:

1 Plant materials and soils,

$$\text{Si}(\%) = \frac{C\,(\text{mg}) \times \text{soln vol (ml)}}{10 \times \text{aliquot (ml)} \times \text{sample wt (g)}}$$

2 Waters,

$$\text{Si}\,(\text{mg l}^{-1}) = \frac{C\,(\text{mg}) \times 10^3}{\text{aliquot (ml)}}$$

Apply factors for dilution or concentration and correct to dry weight where necessary.

Notes

1 Deionized water generally contains small amounts of dissolved silica from the resins so distilled water is recommended for standards and reagents.

2 Glassware should be washed with a 1:1 mixture of conc HNO_3 and H_2SO_4.

3 Two values for water are readily obtained. If the water is untreated then the method records 'molybdate-reactive' silicon but the forms involved cannot be further specified. If the water is filtered, the porosity (usually 0.1–1.0 μ) provides an arbitrary definition for dissolved silicon. To measure this quantitatively, take 50 ml of the filtrate, add 5 ml M NaOH, heat for 30 minutes on a steam bath then add 15 ml of M HCl before determining silicon. These two values, with the total, allow the polymeric and insoluble fractions to be deduced, as suggested by the Standing Committee of Analysts (1981).

Continuous flow method

This method is most suited to waters but it can be used for total silicon in plant materials and soils if the solution prepared by fusion is first neutralized before putting it through the analyser. This avoids evolution of carbon dioxide in the equipment. As with the method above, care must be taken to clean glassware well and to avoid the use of deionized water.

The flow diagram (manifold) in Fig. 5.9 is that needed for the AA II system (see Chapter 8). The manifold required for the AA I with a greater flow rate and larger mixing coils was described by Allen *et al.* (1974).

Reagents

1 Silicon standards.

Stock solution (100 ppm Si): prepare by fusion as give on p. 146.

Standards: prepare a range containing 0 to 5 ppm Si.

2 Ammonium molybdate–sulphuric acid reagent.

3 Tartaric acid, 28% w/v.

4 Reducing solution.

Prepare **1**, **2** and **3** as instructions on pp. 30, 61.

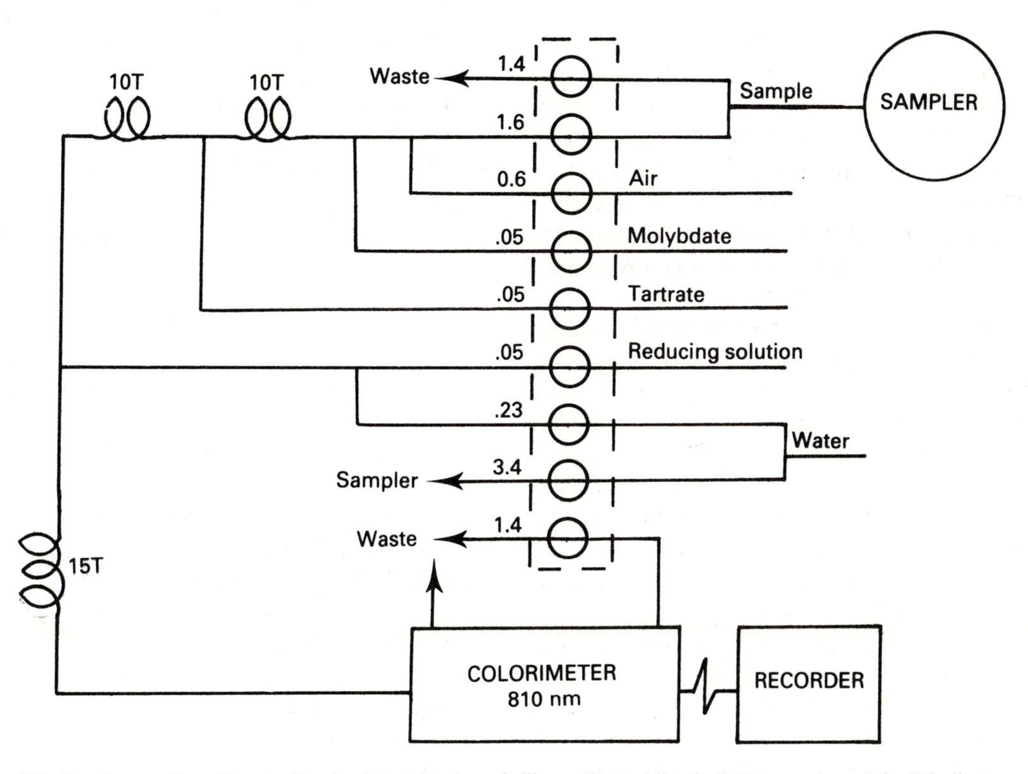

Fig. 5.9. Continuous flow diagram for the determination of silicon. (Dotted line indicates pump module. T indicates number of turns on mixing coils. All flow rates expressed as ml min^{-1}).

Operating conditions
Manifold details are given in Fig. 5.9.
A flow cell of 10 mm is generally adequate but longer cells can be used.
Wavelength for maximum absorption: 810 nm.
Sampling rate: up to 40 per hour (sample: wash ratio = 2:1)

Calculation
If C = ppm Si obtained from the graph then for:
1 Plant materials and soil,

$$\mathrm{Si}\,(\%) = \frac{C\,(\mathrm{ppm}) \times \mathrm{soln\ vol\,(ml)}}{10^4 \times \mathrm{sample\ wt\,(g)}}$$

2 Waters,
$\mathrm{Si}\,(\mathrm{mg\ l^{-1}}) = C\,(\mathrm{ppm})$
Apply factors for dilution and concentration and correct to dry weight where necessary.

Sodium

Sodium is present in certain silicate minerals including some feldspars and amphibolites but is largely absent from the minerals typical of sedimentary rocks. However, evaporation deposits of brine (rock salt) are found inland in a few local areas. It is not a structural element in clay minerals and levels in non-saline soils are relatively low. Nevertheless, sodium is usually included in any summation of total extractable cations.

The element contributes to the ionic balance in cells and tissues and is essential for animal life. The status of sodium in plants is still not entirely clear but micro-nutrient amounts may be needed by some species (Brownell, 1979). Most plants take up relatively little sodium, some inland species contain macro-nutrient levels and the ion itself is not very toxic. Nevertheless, only a few species (halophytes) can accept the salinity levels found in saltmarsh and other saline soils and the tolerance mechanisms have been intensively studied (Flowers *et al.*, 1977; Wainwright, 1980). The spread inland in the UK of some halophyte plants has been linked to the use of de-icing salt on roads. Levels of sodium in freshwaters vary according to the source, whilst the amounts in rainwater are an important indicator of marine influences in the atmosphere. Levels in the ocean ($\simeq$11 000 $\mathrm{mg\,l^{-1}}$ Na) exceed the next most abundant cation (Mg) more than eight-fold.

The concentration ranges generally encountered are (excluding saline samples) given below:

Mineral soil	0.1–2%
Organic soil (peat)	0.02–0.1%
Soil extractions	2–20 mg 100 $\mathrm{g^{-1}}$
Plant material	0.02–0.3%
Animal tissue	0.2–1.0%
Rainwater	0.5–15 $\mathrm{mg\,l^{-1}}$
Freshwater	2–100 $\mathrm{mg\,l^{-1}}$

(dry weight basis where relevant).

Some comments on the application of flame methods to the determination of sodium and potassium are given under potassium whilst the general principles are discussed in the instrumental section. Emission techniques are particularly suitable for sodium whose arc line (589 nm) is the most intense of any element at normal flame temperatures. Procedures for both flame emission and atomic absorption are given below.

The classical gravimetric technique using uranyl zinc acetate has long been displaced by flame methods and no useful colorimetric reagent has been found for sodium. However, an ion-selective electrode is available for this element.

Flame emission method

The comments on interference by acids, and soil extractants given earlier apply equally to sodium, but mutual interference from potassium is not significant with organic material. High levels of calcium may give positive errors, but these can be corrected by including corresponding levels in the standards or, preferably, by running calcareous extracts in 1% sulphuric acid to depress calcium emission. The effects of flame instability can be overcome using an internal standard (lithium) as mentioned for potassium.

The procedure will measure concentrations as low as 0.005 $\mu\mathrm{g\,ml^{-1}}$ if a sensitive instrument is used.

Reagents

1 Sodium standards.

Stock solution (100 ppm Na): dissolve 0.2542 g dry NaCl in water and make up to 1 litre.

Working standards: dilute to produce a range containing from 0 to 5 ppm Na. Include the appropriate amount of digest acid or extractant to match the samples. A greater range may be required for some waters. Make up fresh at intervals. Store all standards in borosilicate or plastic containers.

Procedure

Prepare the sample solutions as described on pp. 30, 44, 61 and dilute to volume.

Select the 589 nm wavelength or Na filter and adjust gas pressures, slit-width and other settings as recommended for the instrument employed.

Prepare a calibration curve from the standard range after setting the top standard to a suitable readout value and the 0 ppm standard to zero, then aspirating each standard in turn.

Aspirate the sample solutions into the flame under the same conditions as the standards.

Check the top, zero and an intermediate standard frequently (Note **1**).

Flush the instrument atomizer and burner frequently with water particularly at the end of the run.

Use the calibration curve to determine ppm Na in the sample solution.

Carry out blank determinations in the same way and subtract where necessary.

Calculation

If C = ppm Na obtained from the graph then for:

1 Plant materials,

$$\text{Na}\ (\%) = \frac{C\ (\text{ppm}) \times \text{soln vol (ml)}}{10^4 \times \text{sample wt (g)}}$$

2 Soil extracts,

Extractable Na (mg 100 g^{-1}): as above $\times 10^3$

(for me 100 g^{-1} Na divide mg 100 g^{-1} result by 22.99)

3 Waters,

Na (mg l^{-1}) = C (ppm)

Apply factors for dilution or concentration and correct to dry weight where necessary.

Note

Include an internal standard (e.g. 200 ppm lithium) with both standards and samples as required. Some instruments make provision automatically for this.

Atomic absorption method

Interferences are similar to the flame emission method and should be handled in the same way. Sensitivity and reproducibility are also similar to the flame emission method. As noted for potassium, ionization can reduce sensitivity when hotter flames are used but excess lithium makes a suitable ionization buffer.

Reagent

Sodium standards.

Prepare as given above.

Procedure

Prepare sample solution as described on pp. 30, 44, 61 and dilute to volume.

Select the 589.0 nm wavelength and adjust air and gas flows, slit-width and other settings as recommended for the instrument employed.

Always allow the hollow cathode lamp adequate time to stabilize.

Prepare a calibration curve from the standard range by setting the top standard to a suitable scale deflection and the 0 ppm standard on zero then aspirating each standard in turn.

Aspirate the sample solutions into the flame under the same conditions as the standards.

Check the top, zero and an intermediate standard frequently.

Flush the spray chamber and burner frequently with water particularly if soil extracts are aspirated.

Use the calibration curve to obtain ppm Na in the sample solutions.

Carry out blank determinations in the same way and subtract where necessary.

Calculation

If C = ppm Na obtained from the graph then for:

1 Plant materials and soils,

$$\text{Na (\%)} = \frac{C\text{ (ppm)} \times \text{soln vol (ml)}}{10^4 \times \text{sample wt (g)}}$$

2 Soil extracts,

Extractable Na (mg 100 g^{-1}): as above × 10^3

(for me 100 g^{-1} divide mg 100 g^{-1} result by 22.99)

3 Waters,

Na (mg l^{-1}) = C (ppm)

Apply factors for dilution and concentration and correct to dry weight where necessary.

Notes

1 Include an ionization buffer (e.g. 0.1% LiCl) with standards and samples, particularly when hot flames such as nitrous oxide–acetylene are used.

2 It is desirable to use a special high solids burner for soil extracts, particularly ammonium acetate to minimize blockage by carbon deposits.

Sulphur

Sulphur is very widely distributed in nature, where it occurs in four oxidation states. Primary silicates contain minor inclusions of sulphide and sulphate minerals and sedimentary rocks can have significant amounts of this element. Sulphide ores, deposits of elemental sulphur and certain sulphates are worked. Soil contents vary, but commonly average around 0.5% total sulphur including a significant organic fraction. Sulphate ions are available to plants and can also be held in the soil by adsorption though less strongly than phosphate. Sulphides, including H_2S may form under reducing conditions. Transformations of sulphur are important in soil (Wainwright, 1984).

This element is an essential macronutrient for plants where certain proteins, lipids and other compounds contain sulphur. Unlike phosphorus, the element links directly to carbon and hydrogen as well as oxygen in plant substances. It is essential for animals where much of it is found in protein. Plant deficiencies are known, but are rarely a problem in areas where atmospheric inputs can be significant.

Sulphate is relatively harmless to plants and can be accumulated (Rennenberg, 1984), but sulphur dioxide gas can be toxic as shown in experiments (Freer-Smith, 1985). A major proportion of atmospheric SO_2 and a significant fraction of the total acid deposition originate from fossil fuel burning (Department of the Environment, 1983). Although some of the research about the effects of pollution on vegetation has been inconclusive for both crops (Irving, 1983) and forests other work indicates that air pollution contributes to the damage seen in European forests (Innes, 1987).

The oceans contain substantial quantities of sulphate although well below the dominant anion (chloride). However, sulphate levels in fresh water can approach those of chloride depending on the relative proximity of marine and industrial sources. Atmospheric sulphuric acid of an industrial origin has been linked with falling pH values in some North European waters (Dickson, 1983), but there are conflicting views about the position in United Kingdom waters.

The concentration ranges generally encountered are given below.

Mineral soils	0.03–0.3%
Organic soils (peat)	0.03–0.4%
Soil extractions	100–500 $\mu g\ g^{-1}$
Plant material	0.08–0.5%
Animal tissue	0.2–0.8%
Rainwaters (SO_4^{2-}-S)	0.4–4 mg l^{-1}
Freshwater (SO_4^{2-}-S)	2–150 mg l^{-1}

(dry weight basis where relevant).

The classical procedure for estimating sulphate-sulphur in solution is based on the precipitation of barium sulphate which is then estimated by gravimetric, turbidimetric or titrimetric methods (Williams, 1976). The turbidimetric procedure is still a standard method in water analyses and is described below. A number of colorimetric methods have been proposed for sulphate including those based on barium chloranilate. Although these methods are more sensitive than turbidimetry, cation interferences are a problem (Gale *et al.*, 1968) and their applications have not been widespread. For samples of natural waters, ion chromatography is recommended (Busnan *et al.*, 1983).

Estimates of reduced forms such as sulphide and sulphite are sometimes required for soils and waters, but the collection and handling of the samples requires precautions because both ions readily oxidize in air. Oxidation can be minimized if air is displaced by nitrogen or carbon dioxide. Analyses should be carried out without undue delay. Methods suitable for sulphite are given by the Department of the Environment (1972) and Standard Methods for the Examination of Waters (1985). Sulphides are discussed further in Chapter 7 and a method is described there.

The quantitative recovery of sulphur from vegetation and soil is not straightforward. Acid oxidation methods have often been tried for vegetation but they are of limited value. They usually rely on combinations of nitric and perchloric acids, but the full recovery of sulphur from compounds such as methionine remains in doubt even when catalysts such as vanadium are included. The classical method is dry-ashing in the presence of a basic salt to minimize volatilization losses, and a procedure given in Chapter 3 can be followed by turbidimetry. Dry combustion methods in which the oxidation products are recovered in solution are sometimes used, including the Schoeniger flask and induction furnace combustion (e.g. LECO Sulphur Analyser) but a more promising application uses a bomb calorimeter (Parkinson, 1987). Randall and Spencer (1980) compare methods of oxidation of biological materials prior to sulphur determination.

Non-destructive procedures such as X-ray fluorescence (XRF) spectroscopy are also suitable for total sulphur provided the technique is calibrated against an absolute (chemical) method. An XRF procedure is given later.

Extractable sulphate-sulphur in soil can be estimated following the extraction method on p. 44, but reduced forms need special attention as mentioned above. The various forms of sulphur in soil can only be separated into broad groups at present (Page *et al.*, 1982). Other authors who review methods for sulphur fractions, notably organic-sulphur include Beaton *et al.* (1968), Tabatabai and Chae (1982) and Randall and Sakai (1983).

Standard methods are available for estimating atmospheric sulphur dioxide and are an aspect of air pollution analysis (see Perry and Young, 1977).

Sulphate-sulphur

The classical technique for the determination of sulphate is to precipitate barium sulphate which is then estimated gravimetrically. This is still widely used, probably because no completely satisfactory alternative has been found. A modification which is particularly suitable for large numbers of samples estimates the degree of turbidity produced by the reaction in solution.

Turbidimetric method

A number of turbidimetric methods have been described, but the one given here appears to be more precise than most and is taken from that of Butters and Chenery (1959) who modified an earlier method of Chesin and Yien (1950). Verma *et al.* (1977) also examine this technique.

The conditions for the development of turbidity need to be carefully controlled to get the best results. A fairly uniform size of barium chloride crystals is required and violent shaking at any stage is avoided.

Although barium sulphate has a low solubility product in water, sufficient comes into solution in this method to distort the lower end of the calibration line. Sufficient sulphate is added to ensure the solubility product of the barium sulphate is exceeded in every solution.

The method is applicable to soils, plant materials and waters and will detect about 0.01 mg sulphate-sulphur in 50 ml of solution.

Many attempts have been made to automate the turbidimetric procedure by continuous flow, but deposition of barium sulphate in the tubes, drift and loss of sensitivity are amongst the problems that have to be overcome. Automation using batch analysers is also difficult because of the mixing requirements. Instrument manufacturers have proposed less direct methods for continuous flow where barium is added as a complex with methylthymol blue and absorbance of uncomplexed indicator is recorded after barium sulphate has formed.

Reagents

1 Sulphur standard (1 ml ≡ 0.05 mg S).

Dissolve 0.3844 g of $MgSO_4.7H_2O$ in water and make up to 1 litre. Avoid using old bottles of this reagent.

2 Nitric acid, 25% v/v + sulphate-sulphur (1 ml ≡ 0.02 mg SO_4^{2-}-S).

Include 0.1537 g $MgSO_4.7H_2O$ in each litre of reagent.

3 Acetic acid, 50% v/v.

4 Ortho-phosphoric acid, SG 1.75.

5 Barium chloride crystals.

Grind $BaCl_2.2H_2O$ crystals to pass between 15 and 40 BS mesh.

6 Gum acacia, 0.5% w/v.

Dissolve 0.5 g gum acacia in 100 ml warm water. This reagent should be freshly prepared.

Procedure

Prepare the sample solutions as described on pp. 30, 44, 61 and dilute to volume. Water samples do not normally require prior treatment, but filter if turbid. Pipette 0 to 10 ml of the standard into 50 ml volumetric flasks to give a range of standards from 0 to 0.40 mg S.

Add acid comparable to the sample aliquots but omit magnesium nitrate (plant material).

Transfer a suitable sample aliquot into a 50 ml volumetric flask.

(Use the whole of the filtrate in the case of plant materials. For water samples include 5 ml 25% HNO_3).

From this point treat standards and samples in the same way.

Add 5 ml 50% acetic acid and 1 ml H_3PO_4 and swirl the solution to mix.

Dilute to the base of the neck and mix again.

Add 1 g $BaCl_2.2H_2O$ crystals without mixing, and leave to stand for 10 minutes. The weight is not critical but should not exceed 1 g.

Invert the flasks twice and stand for 5 minutes.

Invert twice again and leave for a further 5 minutes.

After about 10 inversions, add 1 ml gum acacia solution and make up to 50 ml.

Invert several times more and stand for $1\frac{1}{2}$ hours.

Measure the turbidity as absorbance at 470 nm or with a blue green filter using water as a reference.

Prepare a calibration curve from the standards and use it to determine mg S in the sample aliquot.

Carry out blank determinations in the same way and subtract where necessary.

Calculation

If C = mg S obtained from the graph then for:

1 Plant materials and soils,

$$S\ (\%) = \frac{C\ (\text{mg}) \times \text{soln vol (ml)}}{10 \times \text{aliquot (ml)} \times \text{sample wt (g)}}$$

2 Soil extracts,

Extractable SO_4^{2-}-S (mg l^{-1}): as above × 10^3

3 Waters,

$$SO_4^{2-}\text{-S}\ (\text{mg l}^{-1}) = \frac{C\ (\text{mg}) \times 10^3}{\text{aliquot (ml)}}$$

Apply factors for dilution or concentration and correct to dry weight where necessary.

Notes

1 Do not shake violently at any stage.

2 Strongly coloured (brown) waters should not be tested using this method, but paler colours may be corrected to some extent by a 'colour blank' (sample plus reagents excluding barium chloride) or clarified by pre-treatment with activated charcoal. Sample waters must be free of any turbidity or suspended matter.

X-ray fluorescence method

X-ray fluorescence is a rapid and convenient method for the routine determination of total sulphur in vegetation and soils, especially the former. A number of workers have developed successful procedures including Brown and Kanaris-Sotiriou (1969), McLachlan and Crawford (1970), Reed (1973), Bolton *et al.* (1973) and Bergseth and Kristiansen (1978). Vegetation requires little sample preparation beyond grinding and pressing into a suitable disc. In general, it is preferable to convert soils into a glass disc, although in some cases pressed powder discs will give acceptable results. It is recommended that the preparation treatment should be checked for each soil type encountered.

The procedure outlined below was developed using a wavelength dispersive spectrometer. However, the method can be easily modified for the purposes of energy dispersive spectrometry. The instrument is calibrated using, as standards, samples previously analysed by an established chemical procedure. A set of such standards is required. A calibration can be obtained using ratio counts against a permanent reference of an enamel such as that of Padfield and Gray (1971) containing 0.5–0.6% by weight of potassium sulphate. Calibration standards should have a similar type of matrix to the samples. The method is sensitive to 0.005% or less in vegetation.

Procedure

Treat samples and standards in the same way throughout.

Sample preparation stage

PLANT MATERIAL

Take up to 10 g of finely ground air-dry sample and press into a 40 mm diameter disc at 20 tons pressure using a suitable die and hydraulic press.

Press the disc on the same day as the measurements are to be made because vegetation pellets tend to swell on standing. Avoid touching the face of the disc with fingers.

Store in a small polythene bag until required.

SOILS

Either (i) press into a disc as for vegetation (above) or (ii) fuse with a suitable flux into a glass disc using one of the standard procedures.

Spectrometric stage

Set up the instrument according to the manufacturer's instructions using the following conditions for measurement.

X-ray tube	chromium or rhodium anode.
Tube power	60 kV 25 mA.
S line	K_α (5.37 Å, 2.307 keV).
Crystal	pentaerythritol.
Goniometer	line, 75.83°; background 77.20° and 73.00°
Collimator	coarse.
Counter	gas flow proportional.
Vacuum	on.

Evacuate the spectrometer chamber.

Count all discs at the K_α line for sufficient time to overcome counting statistics limitations.

Count all discs under the same conditions at the background positions.

Mean the background and subtract from the K_α line intensity to obtain net intensity.

Prepare a calibration curve from the standard values and use it to determine %S in the samples.

Correct to dry weight where necessary.

Titanium

Titanium minerals occur as oxides and mixed silicates and though relatively few in number they are common accessory minerals in rocks. It is a secondary element in some primary silicates. Titanium minerals are very resistant to weathering and tend to accumulate in soils where concentrations of metal in the 0.5 to 1% range are not uncommon. Titanium is relatively unavailable to plants which rarely take up more than a few $\mu g\ g^{-1}$ of the metal. For this reason it is useful as an indicator of soil contamination in plant samples and is usually estimated for this purpose. Although it is classed as a non-essential element there are occasional reports of a possible beneficial role.

The concentration ranges generally encountered are given below.

Mineral soils	0.2–1.5%
Organic soils (peat)	0.01–0.08%
Soil extractions	0.1–2 $\mu g\ g^{-1}$
Plant material	0.4–8 $\mu g\ g^{-1}$
Animal tissue	0.01–0.1 $\mu g\ g^{-1}$
Rainwater	0.03–0.2 $\mu g\ l^{-1}$
Freshwater	0.4–4 $\mu g\ l^{-1}$

(dry weight basis where relevant).

Several colorimetric methods have been proposed including the use of 'Tiron' but the most widely used is based on peroxide formation and is described below. This method is simple to use and relatively free from interferences, apart from iron which can be readily controlled. Atomic absorption can be used for titanium, but lacks sensitivity and is not suitable for vegetation. Polarography is now little used for titanium.

It is not easy to bring titanium into solution and fusion should be used even for vegetation, especially if soil particles are present (see above). The oxide minerals (TiO_2) are amongst the few which can resist a conventional hydrofluoric acid procedure. Non-destructive methods such as XRF spectroscopy are therefore preferred for this element although calibration against a destruction method is still required. The colorimetric procedure is described below.

Hydrogen peroxide method

The reaction between titanium sulphate and hydrogen peroxide was originally utilized by Weller (1894) and is still used although there remains some doubt about the structure of the yellow hydroxy-sulphate complex that forms.

Iron and molybdenum interfere with the method by enhancement of the yellow colour. However, if the iron level is high, a correction may be made by reading the sample solution at the appropriate wavelength without the addition of peroxide (Sandell, 1959). Alternatively the inclusion of phosphoric acid will control iron interference. Fluorine even in low concentrations has some bleaching effect but in most soils this is insignificant.

The yellow titanium peroxide develops best at 0.8 to 1.8 M (4–10% v/v) sulphuric acid solution. At stronger acid concentrations the colour becomes less intense and the maximum colour may shift to longer wavelengths (Sandell, 1959; Codell, 1959). The intensity of the colour is enhanced by increase in temperature so it is recommended that the reaction is carried out at a uniform temperature of 20–25 °C (Hillebrand *et al.*, 1953). The method given here is based on that of Sherman and Kanehiro, in Black (1965).

The method is applicable to plant materials, soils and waters. It can be used down to 0.05 mg.

Reagents

1 Titanium standard (1 ml ≡ 0.25 mg Ti).
Weigh 0.924 g potassium titanium oxalate into a 500 ml Taylor flask, add 2.5 g $(NH_4)_2SO_4$ and 50 ml conc H_2SO_4.
Heat gradually to boiling and boil for 5–10 minutes.
Cool and dilute to 500 ml with water.
2 Hydrogen peroxide, 100 vol.
3 Sulphuric acid, 50% v/v.
4 Phosphoric acid, SG 1.75.

Procedure

Prepare the sample solutions as described on pp. 30 and 61 and dilute to volume.
Pipette 0–10 ml of the standard into 50 ml volumetric flasks to give a range of standards from 0 to 2.5 mg Ti.
Add graded amounts of 50% H_2SO_4 to make a final acidity 6% H_2SO_4 and dilute to about 40 ml with water.
Take a 15 ml aliquot of sample solution, and dilute to about 40 ml with water.
From this point treat the standards and samples in the same way.
Add 1 ml H_3PO_4 and mix.
Add 1 ml H_2O_2 and make up to volume with water.
After shaking, measure the absorbance at 425 nm or with a blue filter using water as a reference.
Prepare a calibration curve from the standards and use it to obtain mg Ti in the sample aliquot.
Carry out blank determinations in the same way and subtract where necessary.

Calculation

If C = mg Ti obtained from the graph then:

$$\text{Ti}(\%) = \frac{C\ (\text{mg}) \times \text{soln vol (ml)}}{10 \times \text{aliquot (ml)} \times \text{sample wt (g)}}$$

Apply factors for dilution or concentration and correct to dry weight where necessary.

X-ray fluorescence method

Titanium is readily determined in vegetation and soil by X-ray fluorescence. The comments on sample preparation and calibration given under sulphur apply to titanium also. Sampes analysed by the fusion–hydrogen peroxide method (p. 154) are suitable for calibration standards. A suitable permanent reference for the ratio count method can be made by using a spot of white paint or liquid paper on an aluminium disc.

The method is sensitive to less than 10 $\mu g\,g^{-1}$.

Procedure

Set up the instrument according to the manufacturer's instructions using the following conditions for measurement.

X-ray tube	chromium anode.
Tube power	45 kV 25 mA.
Ti line	K_α(2.75 Å, 4.508 keV).
Crystal	LiF (200).
Goniometer	line 86.07 °; background 85°.
Collimator	coarse.
Counter	gas flow proportional.

Evacuate the spectrometer chamber.

Count all discs at the K_α line for sufficient time to overcome counting statistics limitations, then use the same conditions in the background position.

Subtract the background from the K_α line intensity to obtain net intensity.

Prepare a calibration curve from the standard values and use it to determine % Ti in the samples.

Correct to dry weight where necessary.

Zinc

Although the range of zinc bearing minerals is not great, the element is relatively more abundant than copper. There is only one common sulphide (ZnS), but it forms minerals which are worked in many parts of the world. Other minerals are complex oxy-products and salts and some of the less common silicates contain zinc. Clay minerals in soils can also adsorb some zinc.

This element is an essential micronutrient for plants and animals and is associated with many enzymes and with certain other proteins. Although soil contents are higher than for copper, the plant requirements are greater and deficiencies are known in crop plants. High levels are normally toxic to plants, but tolerance mechanisms have been studied in various species as discussed for copper. A few accumulator species are known. Natural waters are normally low in zinc, but industrial effluent and drainage from mining areas can be sources of pollution.

The concentration ranges normally encountered are:

Mineral soils	20–300 $\mu g\,g^{-1}$
Organic soils (peat)	10–50 $\mu g\,g^{-1}$
Soil extractions	1–40 $\mu g\,g^{-1}$
Plant materials	15–100 $\mu g\,g^{-1}$
Animal tissues	100–300 $\mu g\,g^{-1}$
Rainwater	1–15 $\mu g\,l^{-1}$
Freshwater	5–50 $\mu g\,l^{-1}$

(dry weight basis where relevant).

Relatively few colorimetric methods have been proposed for zinc and the most widely used procedure is based on dithizone although Zincon has also been used. Both are subject to interferences from other metals which must be separated from zinc by prior extractions. A procedure using dithizone is given below.

Atomic absorption is preferable to other methods because of its sensitivity and relative freedom from interferences. Pre-concentration of sample is not usually required except occasionally for waters. A procedure is outlined below.

Polarographic methods are also sensitive for zinc, but like the colorimetric procedures, they require pre-extraction to overcome interferences and offer little advantage for this element. Basically the method is the same as that described for cobalt but with the omission of the dimethylglyoxime step. Zinc gives a wave at $E_{\frac{1}{2}} = -1.33$V.

Errors due to contamination are more likely for zinc than most other elements because it is a constituent of many laboratory materials and is an impurity in some reagents. Particular care is therefore essential in the preparation stages and subsequent ashing, digestion and extraction procedures. A larger number of blanks than normal should be included.

Dithizone method

The method used is essentially that of Cowling and Miller (1941). It involves separation from interfering metals by selective extraction of zinc and measurement of the absorbance of zinc dithizonate.

Zinc and other heavy metals are first extracted with an excess of dithizone in carbon tetrachloride at about pH 10 in the presence of ammonium citrate to prevent precipitation of iron and aluminium. The dithizone extract is next shaken with 0.02 M hydrochloric acid which brings zinc and all the other metals back into the aqueous phase and leaves copper in the dithizone. An extraction with dithizone in carbon tetrachloride between pH 8.5 and 9.0 is then carried out, after addition of sodium diethyldithiocarbamate. This forms complexes with all the other metals except zinc which passes into the dithizone layer.

The method tends to give low results because some of the zinc remains in the aqueous phase during the final extraction and therefore the standard solutions should be passed through the same extraction procedures as the sample.

If the concentration of zinc is much higher than that of interfering elements, notably copper, lead, nickel and cobalt it is possible to modify the dithizone technique so that the colour intensity is measured in the carbon tetrachloride phase. Sodium diethyldithiocarbamate, and potassium cyanide may be added with the dithizone and will adequately complex the principal interfering elements at low concentrations. This procedure avoids the low recoveries referred to above but is less precise than the method given below for plant and soil materials.

The dithizone method is applicable to plant materials, soils and waters. It will measure about 0.2 μg zinc.

Reagents

1 Zinc standards.
Stock solution (1 ml ≡ 0.1 mg Zn): dissolve 0.4398 g $ZnSO_4.7H_2O$ in water and dilute to 1 litre.
Working standard (1 ml ≡ 5 μg Zn): dilute the stock solution 20 times.
2 Carbon tetrachloride.
3 Ammonium hydroxide, M.
Dilute 55 ml conc 0.88 NH_3 solution to 1 litre with water.
4 Dithizone, 0.1% w/v in CCl_4.
Filter if necessary. Store the solution in a dark bottle in a cool place.
5 Ammonium citrate, 0.5 M.
Dissolve 121.6 g of triammonium citrate in 1 litre of water.
Add NH_4OH until the pH is between 8.5 and 8.7.
6 Hydrochloric acid, approx. 0.02 M.
Dilute 1.7 ml conc HCl to 1 litre with water.
7 Sodium diethyldithiocarbamate, 0.25% w/v.
Prepare fresh, immediately before use.
8 Solution A.
Add 500 ml of 0.5 M ammonium citrate to 140 ml 0.88 NH_3 solution and dilute to 1.4 litres with water.
9 Solution B.
Mix 1 litre of 0.5 M ammonium citrate with 300 ml of M NH_4Cl and dilute with 3.2 litres of water. Immediately before use dilute 9 volumes of this solution with 1 volume of sodium diethyldithiocarbamate.

Procedure

Prepare the sample solution as described on pp. 30, 45, 61 and dilute to volume.
Pipette from 0 to 10 ml of working standard into separating funnels to give a range of standards from 0 to 50 μg Zn.
Take a suitable sample aliquot and transfer to a separating funnel.
Neutralize if necessary with M NH_4OH and 2 drops of methyl red indicator.
From this point treat standards and samples in the same way.
Dilute to between 20 and 30 ml.
Add 40 ml solution A and 10 ml of dithizone reagent.
Shake vigorously for 30 seconds and allow the phases to separate.

Run off the lower layer. (If the aqueous layer is not yellow at this stage more dithizone should be added for the next extraction).

Repeat the extraction twice and combine the CCl_4 extracts in a fresh separating funnel. Add 50 ml 0.02 M HCl and shake in the funnel for 90 seconds.

After separation discard the CCl_4 layer.

Wash the aqueous phase with 1 to 2 ml portions of CCl_4 until all traces of green dithizonate have disappeared.

Add by pipette 50 ml solution B and 10 ml dithizone reagent and shake the mixture for 1 minute.

Allow to separate and run off the lower phases into a dry clean tube.

Pipette 5 ml of this extract into a 25 ml volumetric flask and dilute to volume with CCl_4 (protect from strong light).

Measure the absorbance at 535 nm or with a yellow-green filter using water as reference.

Prepare a calibration curve from the standards and use it to obtain μg Zn in the sample aliquot.

Carry out blank determinations in the same way and subtract where necessary.

Calculation

If $C = \mu g$ Zn obtained from the graph then for:

1 Plant materials, soils and soil extracts,

$$\text{Zn } (\mu g\, g^{-1}) = \frac{C\ (\mu g) \times \text{soln vol (ml)}}{\text{aliquot (ml)} \times \text{sample wt (g)}}$$

2 Waters,

$$\text{Zn } (\mu g\, l^{-1}) = \frac{C(\mu g) \times 10^3}{\text{aliquot (ml)}}$$

Apply factors for dilution or concentration and correct to dry weight where necessary.

Atomic absorption method

Atomic absorption is a particularly suitable procedure for zinc and is to be preferred to the other techniques if an instrument is available. The principles are discussed on p. 249. Sensitivity is good and the method virtually free from interference. An air–acetylene flame is recommended. It can be applied successfully to both plant materials and soil extracts, although amounts in the latter may approach the detection limit for the method. Electrothermal atomization or some solution concentration will generally be essential for waters. A full discussion of the determination of zinc in biological materials by atomic absorption is given by Christian and Feldman (1970). Oelschlaeger and Lantzsch (1974) discuss sources of error in the determination of zinc by atomic absorption.

Most atomic absorption instruments now available will measure about 0.01 $\mu g\ ml^{-1}$.

Reagents

1 Zinc standards.

Stock solution (100 ppm Zn): dissolve 0.4398 g $ZnSO_4.7H_2O$ in water and dilute to 1 litre.

Working standards: dilute stock solution to give a range from 0 to 5 ppm Zn. Include acid or soil extractant as appropriate to match the samples.

Procedure

Prepare the sample solutions as described on pp. 30, 45, 61 and dilute to volume.

Select the 213.8 nm wavelength and adjust air and gas flows, slit-width and other settings as recommended for the instrument employed.

Always allow the hollow cathode lamp adequate time to stabilize.

Prepare a calibration curve from the standard range after setting the top standard to a suitable readout value and the 0 ppm standard to zero, then aspirating each standard in turn.

Aspirate the sample solution into the flame under the same conditions as the standards.

Check the top, zero and an intermediate standard frequently.

Flush the spray chamber and burner frequently with water, particularly if soil extracts are aspirated.

Use the calibration curve to obtain ppm Zn in the sample solution.

Table 5.1 Some minor elements of ecological interest

	Soils		Plant materials		Waters			
Element	$\mu g\ g^{-1}$	Soln prep	$\mu g\ g^{-1}$	Soln prep	$\mu g\ l^{-1}$	Treatment	Analytical techniques	References
Bi	0.2–5	Take 10 g, fuse w $K_2S_2O_7$	0.02–0.2	Take > 10 g, digest w mixed acids	0.01–2	Take > 10 l, extr w dithizone at pH > 8	C w diethyldithiocarbamate El. atomization	Stanton (1966) Brouko *et al.* (1985)
Ge	0.2 –10	Take 5 g, digest w HF followed by $KHSO_4$ fusion	< 0.1	Take > 10 g, digest w mixed acids	< 0.1	Take > 10 l, extr w CCl_4 from HCl soln	C. w phenylfluorone Reduction then At. emission	Sandell (1959) Sagar (1984) Braman & Tompkins (1978)
I	0.5 –20	Take 2 g, digest w H_2SO_4, $H_2Cr_2O_7$	0.3 –5.0	Take 10 g, mix w KOH and ash at 450 °C	0.2 –20	Take 10 l, distil in presence of ox agent	Catalytic reduction Combustion & HPLC Ion-selective Elec.	Hurst *et al.* (1983) Fukuzaki *et al.* (1979)
Li	5–100	Take 2 g, fuse w Na_2CO_3	0.01 – 0.5	Take 5 g, digest w mixed acids	0.1–2	Take 5 l, evap or ion-exch	Flame emission El. atomization	Eaton *et al.* (1982)
Sr	20 –400	Take 0.5 g, fuse w Na_2CO_3	2 –50	Take 1 g, ash 450 °C then + HCl *or* digest w mixed acids	5–100	Take 1 l, evap or ion-exch	At. absorption El. atomization	Stupar & Ajlec (1982) Salles & Curtis (1983)
Zr	40–800	Take 2 g, fuse w $KHSO_4$ or $K_2S_2O_7$	0.05–1.0	Take 20 g, ash first then fuse w $K_4P_2O_7$	0.1 –2	Take > 10 l, ppt w Al $(OH)_3$ using NH_4 (OH)	XRFS	Lieser *et al.* (1978)

C = colorimetry, El = electrothermal, At = atomic, Elec = electrode, w = with, extr = extract, ox = oxidizing, exch = exchange.
The recommended amounts for solution preparation are not taken from the references quoted.

Carry out blank determinations in the same way and subtract where necessary.

Calculation

If C = ppm Zn obtained from the graph then for:

1 Plant materials, soils and soil extracts,

$$\text{Zn } (\mu g\, g^{-1}) = \frac{C \text{ (ppm)} \times \text{soln vol (ml)}}{\text{sample wt (g)}}$$

2 Waters,

$\text{Zn } (\mu g\, l^{-1}) = C \text{ (ppm)} \times 10^3$

Apply factors for dilution or concentration and correct to dry weight where necessary.

Notes

It is desirable to use a special high solids burner for soil extracts, particularly ammonium acetate, or undue clogging with carbon will occur.

Other nutrient elements

The more important nutrient elements have been dealt with in this section. There are others which, normally, are of little interest to the ecologist. Indeed in some cases there is some doubt whether they are essential to plants and animals. Others, for example iodine, have a specific role but have, perhaps, not received so much attention because of analytical difficulties. Some of these nutrient elements are given in Table 5.1 together with summaries of sample treatments and references to analytical techniques.

There are other elements, which in small amounts are thought to be essential to wildlife yet are also considered as pollutants because they can have a harmful effect when present in certain forms or quantities. These are included in Chapter 7.

6
Organic Constituents

Ecologists are, in general, more interested in the inorganic and physical parameters than in the organic composition of biological materials. To some extent, this is because information about the former is easier to obtain and understand whereas the organic components are more complex which can lead to analytical difficulties. However some knowledge of the organic contribution is necessary because no part of the chemical ecosystem can be considered in isolation. For example, the relationship of humus products to different ecohabitats or the influence of organic decomposition products on mineral nutrient cycling are problems which may require some knowledge of the organic composition for their better understanding.

The most straightforward of the organic fractionation schemes is that for proximate constituents. These are the components which result from the primary partitioning of organic materials. This partitioning can be done in various ways geared to different requirements, for example to the needs of agriculture or to food nutrition. The methods described in this section have been found to be the most useful for ecological purposes since the components closely match those which result from the progressive degradation of biological materials. These fractionation schemes are arbitrary in nature and mostly depend on traditional laboratory procedures so are thus well-suited to the small laboratory.

In the remaining pages of the chapter the principal organic groups and compounds present in natural materials are examined and their roles discussed. Methods for separating individual compounds are considered and some of the key procedures are given in more detail. The traditional chromatographic separations were dealt with more fully in the first edition of this book. Paper and thin layer chromatography although less popular now are still used, especially for exploratory tests and in laboratories with limited instrumental facilities. Some of these procedures have therefore been reproduced here and many of the original references are also given because most of the current publications deal with instrumental procedures.

The organic composition of biomaterials is so complex and the number of individual compounds so numerous that their detailed examination becomes the job of a specialist. At this level the use of instruments such as gas chromatographs, mass spectrometers and high performance liquid chromatographs becomes essential. These instruments are now so highly developed that they can be programmed to give readouts for individual compounds. However associated experimental treatments and data interpretation are still sufficiently involved to deter any cursory investigation. Brief particulars about the applications of some instrumental procedures are given later in the section whilst technical details about the instruments are provided in Chapter 8. A general handbook of this kind cannot hope to meet the needs of the research specialist but there is now an extensive literature in this field to which the interested reader is referred. Biological and physiological problems are largely outside the scope of this book and so, with a few exceptions, methods are not given for compounds of metabolic importance.

Storage and initial treatment

Before proceeding to the analytical methods it is convenient to discuss the problems arising during the initial treatment of the samples. The nature of the treatment is of particular importance in the case of organic analysis because so many of the compounds are unstable under certain conditions. Lipids, organic acids and some carbohydrates are more labile whereas lignin, tannins and structural carbohydrates for example, are very stable. Although individual cases need special treatment there are some general precautions that should always be followed in organic analysis.

Storage

The sample should be processed as soon as possible after collection, or at least taken to an extraction or derivative stage in which it is fairly stable. If early treatment is not possible, low temperature storage is desirable. A temperature just above freezing greatly reduces microbial activity and is probably the safest for short term use. Cold storage at −10 to −20 °C is more effective in preventing all decomposition processes. However, there is a risk of physical degradation, and the denaturing of proteins below 0 °C is well-known. Successive freeze-thaw treatment should be avoided. Light and air should always be excluded during cold storage.

Apart from low temperature storage, other treatments are available to minimize microbial activity. Many of these are discussed by Sykes (1958) and include:

a The use of chemical, bacteriocidal and fungicidal agents such as phenols, certain alcohols, formaldehyde and phenyl derivatives.

b Gaseous sterilization particularly with ethylene oxide.

c Radiation treatment including gamma and ultra-violet radiation.

Although these treatments are mainly applied in microbiological studies they are of use in chemical work providing no chemical contamination problems are involved.

Drying

Degradations resulting from autolysis of the cell can be especially troublesome during the early stages of drying. Air-drying is of some help but the period taken to reach equilibrium is rather long. Drying at higher temperatures results in the breakdown of less stable molecules and the loss of volatile compounds. Drying at low temperatures under reduced pressure is an improvement, whilst freeze drying is the most satisfactory for labile materials. Some of the limitations of cold storage also apply to freeze drying. For drying small amounts of material, evacuated desiccators are convenient. Phosphorus pentoxide, and magnesium perchlorate (anhydrone) are the most effective desiccants in general use. When materials are unstable in air, drying *in vacuo* and storage under nitrogen is suggested.

Homogenization

Some physical treatment is usually necessary with biomaterials when preparing the sample for analysis. Again the nature of the compound and its source will dictate the methods used. Air-dry material is normally ground but some restriction in the range of particle size may be desirable and grinding to a fine powder should be avoided for some analyses, notably cellulose.

Many of the points made in Chapter 3 concerning the preparation of plant materials for elemental analyses are also relevant here, particularly if dried material is being processed.

The overheating which takes place in some grinding equipment may result in composition changes. This is especially so with ball mills since the sample is retained in the grinding chamber. To eliminate this effect, small ball-mills are now available which run at liquid nitrogen temperatures. Cohen (1943) reported protein changes during dry grinding and reactions between sugars and proteins were considered likely. However, these changes are probably less serious than those due to drying.

Various blenders and homogenizers are available for breaking up fresh materials. The possibility of losses should, however, be taken into account. The increased surface exposure is likely to enhance oxidation processes. If this disintegration is carried out in suspension in a suitable liquid phase, this hazard is reduced.

Physical disintegration may bring together an enzyme and its substrate which were previously separated. The enzymes involved in oxidation and hydrolysis are very active in these circumstances. Enzyme activity may however be prevented by denaturation, for example by immersion in boiling alcohol, particularly isopropanol, and by the use of selective inhibitors or by working at lower temperatures.

No treatments are suitable for all samples and constituents but the procedures below are recommended to minimize changes:

a Keep to a minimum the time between sample collection and analysis.
b Use low temperature storage in the absence of light and air.
c Use freeze or vacuum drying techniques.
d Keep the disintegration stage as short as possible.

Specific extraction and separation procedures are needed for most organic constituents. It should be borne in mind however that these also can influence the results obtained.

Proximate constituents

The fractionation of plant and similar materials into primary organic groups (the proximate components) has been carried out since the beginning of the century. The division into fats, carbohydrates and proteins has been useful for characterizing different foodstuffs since between them, they account for almost all the organic dry matter. These analyses are often linked with estimations of energy so that the three components can be assessed in terms of their contribution to the calorific value of food. Proximate analyses have also been widely used in agricultural investigations. A typical approach is that used in the study of the nutritive value of grasses and feeds for animals when it is usual to examine for:

a ether extract (equivalent to crude fat),
b crude protein (from nitrogen),
c crude fibre,
d nitrogen-free extracts (NFE) (roughly equivalent to carbohydrates),
e mineral ash.

This scheme was critically reviewed by Swift (1956) whilst Browning (1967) and McDonald *et al.* (1960) offered alternative schemes. Giger (1969) has described a semi-automated method for determining some of the fractions and Bokelman *et al.* (1983) have also used instrumental techniques and compared them with a serial extraction method.

For ecological purposes it is useful to have a proximate classification scheme that is based on natural processes such as plant and litter degradation and soil formation. Such an approach has much in common with the food and agricultural fractionation schemes but a further division of the 'carbohydrate' (NFE) component is desirable. A split into soluble carbohydrates (metabolic) and cellulose (structural) corresponds to functions in the plant. Furthermore, if one considers the mechanism of litter decomposition then a division into progressively more resistant fractions has some merit. Further partition of the cellulose, as discussed below, is sometimes of value and the most resistant component of the plant or litter is best determined as lignin. This probably has more ecological significance than the various estimates of fibre used by agricultural scientists. (However, a crude fibre method is given by way of an appendix to the lignin procedure.) Perhaps the main attraction of this extended classification is that there is evidence that it can be related to the fungal and microbial breakdown of plant litter. A similar analytical scheme suitable for the study of litter break-down or peat formation is given in Page *et al.* (1982).

A flow diagram summarizing the proximate methods described in this section is shown in Fig. 6.1. No method for mineral ash is given here since this was described in Chapter 3.

The proximate chemical composition of most plant materials fall within the ranges, given below (as percentage dry weight). Cellulose and lignin in woody materials are near the upper end of their

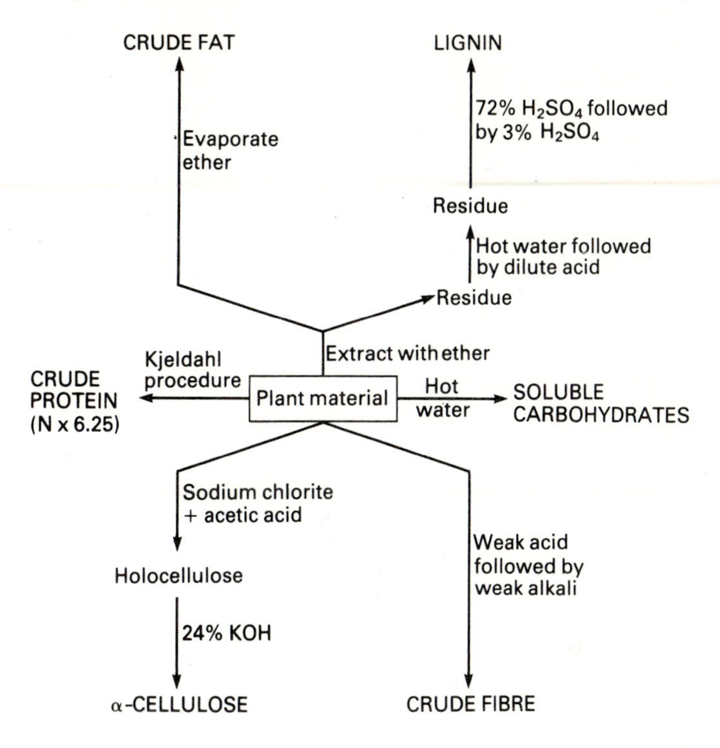

Fig. 6.1. Outline diagram showing stages in the proximate analysis of plant materials.

ranges whilst protein and soluble carbohydrates are higher in fruit and leaf tissue.

	% Dry weight
Crude fat, wax	0.5–5
Crude protein	3–15
Soluble carbohydrates	5–20
Cellulose (holo)	20–45
Cellulose (α)	5–15
Cellulose (hemi)	15–35
Lignin	10–30
Mineral ash	1–8

None of the proximate schemes guarantee that the summation of the individual results will account for exactly 100% of the original material. Nevertheless most summations seem to fall in the range 95 to 102%. A figure within this range, but preferably 97 to 100% will be sufficient to indicate that no major component has been overlooked. Failure to reach 100% may be due to several factors which are considered by Browning (1967).

Crude fat

Any detailed study of fats involves investigation of individual compounds for which specific techniques are available. This aspect is discussed on pp. 185–189. For the purpose of a proximate analysis it is sufficient to get a rough estimate of the total fat content by extraction with diethyl ether. In the method given below the ether is removed by evaporation and the residue weighed. This result is arbitrary since ether may not fully extract all fats and oils, but may include carotenoids and other ether-soluble substances. Some workers prefer to report such a result as 'ether extract' but the term 'crude fat' is adopted here. Since drying at 105 °C may result in losses of fats and oils, the use of air-dried material is preferred for extraction.

A rapid method for total lipids has been proposed by Atkinson *et al.* (1972) which in many ways appears to be superior to the traditional ether extraction. In this method the lipids are extracted in a sealed tube using a chloroform–water–methanol mixture. Although developed for animal tissues, a modification of the method has been used for sediments (Cranwell, 1978) and for litter samples (Bridson, 1985). Various solvent systems used for extraction have been critically compared (Fishwick and Wright, 1977).

Reagents

1 Diethyl ether, peroxide free.
2 Nitrogen, (oxygen free).

Procedure

Weigh 0.8 to 1.0 g of air-dried material into a 50 × 10 mm Soxhlet extraction thimble.
Transfer to 6 ml capacity Soxhlet extractor.
Weigh a clean, dry 25 ml round-bottomed flask (B14 neck) containing a glass bead.
Add about 20 ml ether to the flask, connect to the extractor and extract for 4 to 6 hours using a heating mantle.
Remove flask and place in warm water bath.
Evaporate off the ether using a stream of oxygen-free nitrogen.
Leave in vacuum oven at 40 °C for 30 minutes.
Cool in a desiccator and re-weigh.
Carry out blank determinations in the same way and subtract where necessary.

Calculation

If W = g residue in ether extract then:

$$\text{Crude fat (\%)} = \frac{W\,(\text{g}) \times 10^2}{\text{sample wt (g)}}$$

Correct to dry weight where necessary.

Note

Special care is needed when extracting with ether. Even a heating mantle can become hot enough to ignite the vapour and ether drops must not be allowed to fall on the mantle when removing the flask.

Crude protein

Detailed studies of proteins generally involve estimates of their constituent amino-acids for which chromatographic techniques are now available. These are discussed further on pp. 179–184. For a proximate analysis some estimate of the total protein content is sufficient, although it seems less easy to determine this satisfactorily than to analyse for the individual amino-acids. After extraction it is possible to estimate the total protein content using gravimetric and colorimetric procedures and these are referred to on p. 179. However, the protein solution needs to be fairly free from interfering substances, which is not always conveniently achieved with plant materials. For these reasons it has been a common practice in agricultural work to determine total organic nitrogen and multiply by a factor to obtain crude protein.

The most widely used factor appears to be 6.25. This is based on the assumption that plant proteins on average contain about 16% N. In practice a variation between 12 and 19% is possible between individual proteins and this coupled with the inclusion of non-protein nitrogen makes all factors rather arbitrary. However, the results obtained by using 6.25 seem as satisfactory as those obtained by direct chemical methods.

Procedure

Determine total organic nitrogen by the Kjeldahl digestion followed by steam distillation or the indophenol-blue colorimetric procedure (p. 119 *et seq.*).
Crude protein (%) = N (%) × 6.25.

Soluble carbohydrates

For the purpose of a proximate analysis, the carbohydrate substances in plants can be divided into two main groups; the polysaccharides whose functions are mainly structural or storage, and the simpler saccharides (sugars) which are metabolic compounds. A more detailed classification of the carbohydrates is given on p. 172.

A detailed analysis of sugars requires specific techniques which are discussed later. For proximate analyses it is sufficient to estimate the total content of simpler carbohydrates which are soluble in hot water. However, a value obtained in this way is highly arbitrary because it depends on the temperature and to a lesser extent on the time of extraction. In general, substantial amounts of polysaccharides such as starch will come into solution above 60 °C and this temperature should not be exceeded for an approximate estimate of the simpler saccharides. If 100 °C is used, an independent measure of starch will aid interpretation. It should be noted that plant tissues differ considerably in their starch contents, both within and between species. Cold water extractions have also proved useful and are still preferred by some (Thomas, 1977).

Others prefer to extract with 80% ethanol, generally for the purpose of estimating what they consider to be 'free' or 'reducing' sugars (Paech and Tracy, 1955). Ethanol has the advantage of extracting fewer interfering substances, but for proximate analyses, hot water probably gives a slightly more useful result. Acid extractants have been used (Smith *et al.*, 1964), but some hydrolysis of the polysaccharides may then occur. Enzymes such as amylase are also widely used for extraction purposes.

Several colorimetric and titrimetric methods have been applied to carbohydrates (sugars) in solution, but many of these are suitable only for specific groups. Techniques which are available for the determination of carbohydrates in solution are reviewed by Josefsson *et al.* (1972), Burney and Sieburth (1977) and Doutre *et al.* (1978). Probably the most widely used reagent is anthrone which was introduced by Dreywood (1946). A strongly acid solution of anthrone gives a blue colour with carbohydrates and provides a rapid and convenient method for routine use. Several factors influence the reaction and many variants and applications of the method have been published (e.g. Jermyn, 1975 and Conroy *et al.*, 1981).

The procedure given below is taken from Deriaz (1961). Certain precautions must be taken. Anthrone itself is not completely stable as it is susceptible to air oxidation. Roe (1955) and Zill (1956) found that the inclusion of thiourea improved stability and this was adopted by Deriaz. The sample must be adequately cooled during addition of the acid reagent or some incipient formation of the colour may occur as a result of the heat of dilution.

The time and temperature of maximum development determine the density of the final colour. It is important that the time should be checked for each type of sample analysed. Most materials need around 10 minutes, which is about the time required by glucose and other hexose sugars. Pentoses only require two or three minutes, but their content in many materials is low relative to the hexoses and may be ignored for a proximate analysis. Brown *et al.* (1957) used anthrone for the simultaneous measurement of glucose and fructose and employed two different temperatures for colour development.

The temperature of 100 °C is adopted here because the colour due to glucose is weaker at lower temperatures. It was also found that colour reproducibility was better if development occurred in the dark, so this modification too is included below. The colours are relatively stable in the cold, so rapid cooling after colour development is recommended. All absorbance values are read at 625 nm. Interferences are not serious but high levels of chloride and nitrate should be avoided. Significant amounts of nitrates give a brown colour, but rapid cooling may decrease the effect (Katz *et al.*, 1983).

An alternative reagent to anthrone is phenol–sulphuric acid which was introduced by Dubois *et al.* (1956). Doutre *et al.* (1978) compared these two reagents and preferred phenol–sulphuric acid but noted that it was not free from interference problems. Their studies showed that soil carbohydrates raised problems for both reagents and they recommended that soil extracts should be purified.

As with other organic determinations, material dried at 105 °C should not be used but air-dry samples are acceptable for proximate analyses. A hot water extraction is best carried out in a boiling flask under refluxing conditions but the conical flask procedure adopted below is an adequate substitute and is more convenient for routine use.

Reagents

1 Glucose solutions.

Stock solution (1 ml≡0.25 mg glucose): dissolve 0.250 g D-glucose (dried in a vacuum oven at 70 °C over P_2O_5) in water and dilute to 1 litre. This solution will keep for a few weeks.

Working standards: pipette a range from 0 to 20 ml stock solution into 50 ml flasks. Dilute to volume. Prepare these standards fresh daily.

2 Anthrone reagent.

Add carefully 760 ml conc H_2SO_4 to 330 ml water in a boiling flask and keep cool while mixing. Add 1 g anthrone and 1 g thiourea and dissolve using a magnetic stirrer. Transfer to a dark bottle and leave for 2 hours before use. Store at +1 °C. The reagent should be straw-coloured or pale yellow. If it becomes darker, with a green tinge, some oxidation has occurred and it should be discarded.

Procedure

1 ***Extraction (plant materials)***

Weigh 50 mg air-dry ground sample into a 100 ml conical flask.

Add about 30 ml water.

Place a glass bubble in neck and simmer *gently* on a hotplate for 2 hours.

Periodically top up to 30 ml.

Allow to cool slightly and filter through a No. 44 Whatman paper into a 50 ml volumetric flask.

Wash paper, and dilute to volume when cool.

The extract will not keep overnight, and must be prepared shortly before colour development.

Take water blank through the same stages.

2 ***Colour development***

Pipette 2 ml of each standard into a set of boiling tubes.

Pipette 2 ml of each extract or water blank into boiling tubes.

From this point treat standards and samples in the same way.

Add rapidly 10 ml anthrone reagent and mix with the tube immersed in running cold water or an ice bath.

Place tubes in a beaker of boiling water in a darkened fume cupboard and boil for 10 minutes.

Place the tubes in cold water and allow to cool, preferably in the dark.

Measure the absorbance at 625 nm or with a red filter using water as a reference.

Prepare a calibration graph from the standards and use it to obtain mg glucose in the sample aliquot.

Treat the blank determinations in the same way and subtract where necessary.

Calculation

If C=mg glucose obtained from the graph then for:

1 Plant materials,

Soluble carbohydrates (%)=

$$\frac{C\ (\text{mg}) \times \text{extract vol (ml)}}{10 \times \text{aliquot (ml)} \times \text{sample wt (g)}}$$

2 Waters,

$$\text{Soluble carbohydrates (mg l}^{-1}\text{)} = \frac{C\ (\text{mg}) \times 10^3}{\text{aliquot (ml)}}$$

Apply factors for dilution and correct to dry weight where necessary.

Cellulose

Cellulose is the most important of the polysaccharides in plants and is found in all species. It is the major structural component of cell walls in wood. Cellulose fibres have long been an important raw material in industry as well as fibres from wood (paper), seed hair (cotton), bast fibres (flax, jute) and leaf fibres (hemp). Cotton is the purest form of natural cellulose.

Cellulose is the only one of the proximate constituents that can be isolated as a relatively pure product of known chemical structure. The other proximates are all determined as mixtures, often of uncertain chemical composition.

The first stage in the isolation of cellulose is a delignification process (see below) yielding a product consisting of cellulose plus varying amounts of other polysaccharides. The latter are collectively known as hemicellulose—a term first used by Schultze (1892). The first stage product is of value because it contains the entire polysaccharide fraction of the sample. Ritter and Kurth (1934) were the first to call this fraction 'holocellulose' and to attempt to isolate it without undue loss of any polysaccharide. Their procedures for isolating holocellulose involved chlorination to break down lignin.

Chlorine dioxide and acid chlorite have also both been widely used for this purpose and an acid chlorite technique is given below.

Early procedures did not go through the holocellulose stage but attempted to isolate cellulose directly. The best known of these is the method originated by Bevan and Cross (1880) who delignified with chlorine gas and alternated this with a sodium sulphite extraction to remove the lignin products. Modifications of the method have been tried but all yield a somewhat arbitrary product which is neither holo- or α-cellulose and the procedure is little used today.

Delignification by reagents not containing chlorine has occasionally been used including nitric acid (Kurschner and Hoffer, 1929) and ethanolamine (Wise *et al.*, 1939), but they do not appear to yield products of any greater value than those obtained using chlorine.

Procedures which break down cellulose directly by wet-oxidation (acid + dichromate) followed by a back titration of excess dichromate have been advocated for fairly pure cellulose products. They cannot be used if lignin and other non-cellulose compounds are present, and hence are not generally applicable to plant materials.

The use of GLC in the determination of cellulose and hemi-cellulose has been described by Sloneker (1971).

Holocellulose

The most suitable procedure appears to be that of Wise *et al.* (1946) who delignified with sodium chlorite in very weak acetic acid at 75 °C. The details are given below. It should be noted that chlorite procedures in general have been developed for wood and may not always be ideal for leaves, etc. Juvvik (1965) however, thought litters and organic soils could be delignified in this way. All delignification procedures will attack the hemicellulose if taken to completion. To avoid this it is customary to halt the treatment when 2 to 4% lignin still remain in the holocellulose. The pH, temperature and time are all important factors in achieving this result. Some loss of polysaccharide seems unavoidable and therefore tends to limit the value of a holocellulose determination except for proximate analyses. Pre-treatment of plant materials to remove fats has been suggested but in the procedure of Wise *et al.* this is carried out only on resinous material.

The method is occasionally applied to soil organic matter, but cannot be wholly recommended because only an impure holocellulose product is obtained. In addition, the ash correction becomes more significant but is less reliably estimated by the method given above if excessive amounts of mineral soil are present.

The final dried holocellulose is corrected for ash content and nitrogen expressed as crude protein ($N \times 6.25$) even though only protein break-down products are likely to be present. Corrections for both methoxyl content (as a measure of residual lignin) and uronic acid were applied in the procedure of Wise *et al.* but these seem less important if the holocellulose result is to be supplemented by a determination of α-cellulose.

Reagents

1 Sodium chlorite.
2 Acetic acid, 10% v/v.
3 Acetone.
4 Diethyl ether.

Procedure

Weigh 0.5 g air-dry sample into a 100 ml conical flask.

Add 30 ml water, 1 ml 10% HOAc and 0.3 g sodium chlorite.

Place a glass bubble in the neck of the flask, mix, and stand in a water bath at 75 °C.

Add 10% HOAc and 0.3 g sodium chlorite at hourly intervals, swirling the flask intermittently.

Remove after 4 hours and cool immediately in ice cold water.
Filter the contents through a weighed No. 2 Pyrex sintered crucible.
Wash about 10 times with ice cold water.
Wash 3 or 4 times with acetone and finally once with ether.
Allow ether to evaporate.
Dry in oven at 105 °C for 30 minutes.
Cool in desiccator and weigh.

Corrections
Determine ash and N-content on sub-samples of holocellulose.
Multiply N result by 6.25 to give a correction for crude protein.
Subtract corrections before calculating % holocellulose.

Calculation

$$\text{Holocellulose (\%)} = \frac{\text{Corr. holocell. (g)} \times 10^2}{\text{sample wt (g)}}$$

Correct to dry weight where necessary.

Notes
1 Samples that have been previously dried at 100 °C should not be used since this may have resulted in cellulose degradation. Neither should the final holocellulose preparation be dried longer than absolutely necessary.
2 For consistent results it is preferable for the sample particle size to be within the range of 40 to 60 BS mesh.
3 For routine purposes tablets of sodium chlorite weighing 0.3 g each can be obtained commercially.

α-Cellulose

Removal of hemicellulose from holocellulose may be effected by treatment with 24% potassium hydroxide leaving a reasonably pure form of cellulose (α-cellulose) which is recovered as a white product. It is not completely pure, however, and tends to retain some mannose and xylose units which cannot be removed without attacking the cellulose structure.

The procedure given below is modified from that of Bath (1960) and appears to work best for woody tissues though it is acceptable for most leaf tissue. For decomposing tissue such as litter however, it seems less satisfactory and gives a greyish and probably degraded form of α-cellulose.

Reagents
1 Sodium chlorite.
2 Acetic acid, 10% v/v.
3 Acetic acid, 5% v/v.
4 Acetone.
5 Diethyl ether.
6 Potassium hydroxide, 24% w/v.

Procedure
Weigh 1 g air-dry sample into a conical flask.
Follow procedure for holocellulose as given above, but take 2 ml 10% HOAc and 0.6 g sodium chlorite. Omit the corrections.
Weigh the available amount of holocellulose into a 50 ml conical flask.
Add 20 ml 24% KOH and stopper with rubber bung.
Stand in water bath at 20 °C for 2 hours.
Swirl gently at intervals.
Filter through a weighed No. 2 glass sintered crucible.
Wash with water until washings are free of alkali.
Add 5 ml 5% HOAc and swirl around in crucible.
Wash again with water, acetone and finally ether.
Allow ether to evaporate.
Dry in oven at 105 °C for 30 minutes.
Cool in a desiccator and weigh as α-cellulose.

Corrections
Apply corrections as described for holocellulose.

Calculation
α-Cellulose (%) =

$$\frac{\text{Corr. } \alpha\text{-cell. (g)} \times \text{tot. uncorr. holocell. (g)} \times 10^2}{\text{holocell. sub-sample (g)} \times \text{sample wt (g)}}$$

Correct to dry weight where necessary.

Note
Prolonged drying may degrade the final product (see Note **1** of the previous method).

Hemi-cellulose

Determinations of hemi-cellulose are less satisfactory than those of cellulose. It has long been recognized that treatment of holocellulose with 5% and 24% caustic potash apparently yields two forms of hemi-cellulose (A and B). However, since all hemi-cellulose preparations are complex mixtures only a limited significance can be attached to A and B.

Hydrolyses of hemi-celluloses yield both pentoses (xylose and arabinose) and hexoses (glucose, mannose and galactose) together with uronic acid residues. It is assumed that hemi-cellulose consists of polymers of these units, such as the glucomannans, and glucuronoxylans (Browning, 1967). Doree (1947), Wise and Jahn (1952), Aspinall (1959) and Schurtz (1977) also discuss hemi-cellulose structures.

In the procedure given below (modified from Bath, 1960) the alkaline filtrate is partly neutralized and allowed to stand overnight with excess alcohol. The resultant precipitate is recovered and classed as hemi-cellulose. However, part of the hemi-cellulose remains in solution since 'hemi' plus 'α' do not equal holocellulose when the final results are compared.

For an investigation of hemi-cellulose it seems more satisfactory to treat each species separately and attempt to isolate the major sugar units by hydrolysis followed by chromatography. The procedure given below can then be regarded as a first preparative stage suitable for subsequent hydrolysis.

Reagents

1 Industrial spirit, 95% v/v.
2 Acetic acid, glacial.
3 Acetic acid, 5%.
All other reagents are as given for α-cellulose.

Procedure

Follow procedure for α-cellulose, given above, until the filtering stage is reached.
Filter through a No. 2 Pyrex sinter crucible into a reservoir containing 8 ml glacial HOAc.
Wash with water to remove all alkali.
Wash with 5 ml 5% HOAc.
Wash again with water, acetone and finally ether.
Adjust the pH of filtrate to 4.0 with 5% HOAc.
Add industrial spirit to about 3.5 times the volume of the filtrate.
Stand overnight.
Filter through a weighed No. 2 sintered crucible.
Wash with industrial spirit, acetone and ether.
Allow ether to evaporate.
Dry in oven at 105 °C for 30 minutes.
Cool in desiccator and weigh as hemi-cellulose.

Calculation

Hemi-cellulose (%) =

$$\frac{\text{wt hemi-cell. (g)} \times \text{tot. uncorr. holocell. (g)} \times 10^2}{\text{holocell. sub-sample (g)} \times \text{sample wt (g)}}$$

Correct to dry weight where necessary.

Note

Prolonged drying may degrade the final product (see Note **1** under the holocellulose method).

Lignin

For the purpose of a proximate analysis, a lignin result is frequently taken to refer to residual oraganic compounds resistant to most microbiological and chemical processes. Under these conditions it is sometimes adequate to deduct the sum of the other proximate constituents from 100. More often, however, a specific determination is preferable.

The structure of lignin is not yet completely known, but it is generally accepted as consisting mainly of phenylpropane units (C_6–C_3–C_6) built into a complex polymer closely linked to polysaccharides in the cell wall. Methods for estimating lignin usually depend on the type required. Soluble 'native' lignin is the fraction extracted by 95% ethanol (Brauns, 1939) and is about 3% dry weight of (spruce) wood. Zeichmann and Weichelt (1977) review extraction methods and favour those of Bjorkman (1956) and Pepper *et al.* (1959). Most investigators prefer to isolate the residue which is resistant to either strong acid hydrolysis or attack by alkali. Hydrolysis with 72% w/v sulphuric acid is widely used and a procedure given by Ritter *et al.* (1932) is outlined below. A few workers have used hydrochloric acid (Goss and Phillips, 1936) or alkali (Brauns and Brauns, 1960). Van Soest and Wine (1967) refer to a method for determining lignin in acid detergent fibre using potassium permanganate for the oxidation process and Van Soest (1973) reported on a collaborative study using this method. Collings *et al.* (1978) compare the permanganate method with sodium chlorite oxidation for lignin whilst the application of spectrophotometry for lignin is discussed by Van Zyl (1978).

The fraction isolated by 72% acid varies from species to species, but probably contains almost all the true lignin and is suitable for most purposes. However, some structural changes in the lignin complex seem unavoidable with a strong extractant. If a study of lignin in different species is being made, it might be useful to investigate hydrolysis with various acid concentrations.

The method given below also includes pre-treatments in which fats are removed with ether, soluble carbohydrates with water and proteins with dilute acid. Alcohol has also been used, but is perhaps best avoided as it appears to dissolve a small part of the lignin. The hydrolysis is carried out with 72% acid followed by dilution to 3% acid and gentle boiling. This breaks down the cellulose complex. It is important that the residue following pre-treatment is dried (40 °C) before the 72% acid is added. If not, the hydrolysis is affected and traces of a gelatinous substance seen at the 3% acid stage (Moon and Abou-Raya, 1952), become excessive and increase the apparent lignin retained on the sinter.

It is also necessary to correct the weight of the final dried product for ash and crude protein content. The latter is determined as N × 6.25 though this step is criticized by Brauns (1952). The preparation can be characterized for its lignin content by determination of methoxyl content but this may be omitted for a proximate analysis.

For studies involving the nutritive value of vegetation to herbivores, it may be sufficient to determine the fibre content. This should not be regarded as an alternative to lignin and is a purely arbitrary fraction composed mainly of lignin plus some cellulose. A procedure for crude fibre is given below following the lignin method.

Reagents

1 Diethyl ether.
2 Sulphuric acid, 10% v/v.
3 Sulphuric acid, 72% w/v.
Add carefully, while cooling, 720 ml conc H_2SO_4 to 540 ml water.

Procedure

1 ***Pre-extractions***

ETHER

Weigh 1 g air-dry sample on to a weighed glass fibre paper and tie into a bundle with terylene thread.
Extract with ether in a Soxhlet apparatus for about 6 hours (see Note **2**).
Allow the ether to evaporate.
Dry the bundle for 30 minutes at 105 °C.
Cool and weigh after removing thread.

WATER

Weigh about 0.80 g from the ether extracted sample into a 600 ml tall pyrex beaker.
Add about 400 ml water and boil gently for 3 hours.
Keep the volume at around 400 ml.

DILUTE ACID

Add 22 ml 10% H_2SO_4 and boil for a further hour.
Allow the contents to settle.
Remove the supernatant liquid with a No. 2 or 3 sintered filter stick and dry the residue overnight at 40 °C.
Allow the sinter stick to remain in the beaker.
Removal of the supernatant liquid should not be delayed unduly after settling is complete and an overnight delay must be avoided.

2 ***Acid hydrolysis***

Add to the dry residue 15 ml 72% H_2SO_4 at 12–15 °C, stir for 1 minute with the filter stick.
Leave the beaker in a bath for 2 hours at 18–20 °C.
Stir the contents occasionally.

Reduce the acid strength to 3% by the addition of 560 ml water.
Wash particles off sinter stick into beaker, and remove the stick.
Boil gently for 4 hours and top up when necessary.
Allow contents to settle, but do not delay filtering. Prolonged standing in acid solution may affect the quality of the final product.
Filter through a weighed No.2 sintered glass crucible.
Wash the residue free of acid with hot water.
Dry the sinter and contents for 3 hours at 105 °C.
Cool and weigh.

Corrections

Determine ash and N-content on sub-samples of the crude lignin.
Multiply N results by 6.25 to give a correction for crude protein.
Subtract corrections before calculating % lignin.

Calculation

Lignin (%) =

$$\frac{\text{corr. lignin (g)} \times \text{tot. (g)} \times 10^2}{\text{wt for water extract (g)} \times \text{sample wt (g)}}$$

where tot. is total ether extracted sample.
Correct to dry weight where necessary.

Notes

1 Air-dry material is preferable to oven-dry.
2 Several sample bundles may be extracted together in the same Soxhlet chamber.

Crude fibre

In preparing a 'crude fibre' the aim is generally to remove the more chemically reactive (and presumably more digestible) part of the samples. The residue is regarded as a measure of the roughage consumed by the animal. Ecologists often use the results in conjunction with grazing studies involving wild fauna, but there is little evidence of any digestibility relationship in these circumstances. Crude fibre contains most of the lignin and part of the cellulose fraction.

The standard crude fibre procedure used by many agricultural advisory chemists is described fully in Official Methods of Analysis of the AOAC (1984) and similar agricultural handbooks. The sample is boiled successively with 1.25% w/v sulphuric acid and 1.25% w/v sodium hydroxide. The method given below is a modified version. Moir (1971) describes an apparatus for crude fibre determination in which there is no transferring of residues. See also Selvendran *et al.* (1979).

The use of sulphuric acid alone (about 3%) has been suggested as being of greater value for digestibility studies. Such a method (Normal Acid Fibre) has been developed at the Grassland Research Institute (1961).

The crude fibre content of plant materials generally falls in the range 20 to 35%.

Reagents

1 Diethyl ether.
2 Sulphuric acid, 1.25% w/v.
3 Sodium hydroxide, 1.25% w/v.
4 Industrial spirit.

Procedure

1 *Pre-extraction*

Weigh 1 g air-dry sample on to a weighed glass fibre filter paper and tie into a bundle with terylene thread.
Extract with ether in a Soxhlet apparatus for about 6 hours (see Note **2** above).
Remove bundle, allow ether to evaporate.
Dry the bundle for 30 minutes at 105 °C, cool and weigh after removing the thread.

2 *Acid hydrolysis*

Weigh the extracted sample (approximately 0.8 g) into a tall 600 ml pyrex beaker.
Add 100 ml boiling 1.25% H_2SO_4.
Cover with a watch glass and boil gently on hotplate for 30 minutes.
Remove the acid by suction through a No. 2 sinter filter stick.
Wash three times with 50 ml boiling water and remove by sintering each time.
Leave sinter stick in beaker.

3 ***Alkali extraction***

Add 100 ml boiling 1.25% NaOH.

Cover with watch glass and boil gently on hot-plate for 30 minutes.

Remove the alkali with sinter filter stick.

Wash with 25 ml boiling 1.25% H_2SO_4.

Wash twice with 50 ml boiling water and remove by sintering each time.

Wash contents of beaker into a weighed No. 2 glass sintered crucible using about 50 ml boiling water.

Wash off sinter stick into crucible.

Wash with 30 ml industrial spirit.

Dry crucible and contents for 3 hours at 105 °C.

Cool and weigh.

Corrections

Carry out an ash determination on a large proportion of the residue.

Record as loss in weight and apply as correction to the total weight of crude fibre.

Calculation

Crude fibre (%) =

$$\frac{\text{uncorr. fibre (g)} \times \text{wt loss (g)} \times \text{tot. (g)} \times 10^2}{\text{wt ash (g)} \times \text{wt hydrol. (g)} \times \text{sample wt (g)}}$$

where tot. is total extracted material and hydrol. is weight taken for hydrolysis.

Correct to dry weight where necessary.

Carbohydrates

Carbohydrates are universally found in living tissue and are very important metabolically. In addition they provide the main structural component of plant tissue. Methods for the determination of total soluble carbohydrates and the cellulose fractions have already been given and in this part attention is given to methods of extracting the simpler carbohydrates from plant tissue and to their estimation. Details are also given for a preliminary separation by chromatography.

Carbohydrates have the general formula $C_x(H_2O)_y$ and may be classified as follows:

MONOSACCHARIDES

Pentoses, e.g. arabinose, xylose.

Hexoses, e.g. glucose, mannose, fructose.

Derivatives include:

- Ethers, e.g. glycosides;
- Esters, e.g. hydrolysable tannins, phosphate esters;
- Alcohols, e.g. glycerol;
- Acids, e.g. ascorbic acid;
- Amino sugars, e.g. glucosamine, galactosamine.

DI- AND TRI- SACCHARIDES

Reducing, e.g. maltose.

Non reducing, e.g. sucrose.

Derivatives include methyl glycosides and others.

POLYSACCHARIDES

Pentosans, e.g. xylans, arabans.

Hexosans:

- reserve (starch, inulin);
- structural (cellulose).

Complex mixed polysaccharides include:

- gums and mucilages, pectic substances, and others.

The simpler carbohydrates are particularly vulnerable to enzymic changes following sampling and therefore transit and storage periods must be kept to a minimum. If immediate analysis is not practicable brief immersion in boiling 95% ethanol will generally inhibit enzyme action. Oven drying is not entirely satisfactory because enzymic changes may be accelerated in the early stages and losses can occur at the higher temperatures. Analysis and identification of carbohydrates have been reviewed by Sturgeon (1977, 1978, 1979).

Extraction methods

All the mono-, di- and tri-saccharides are soluble in water which is normally present in the extractant. A cold aqueous extraction was described by Reifer and Melville (1947) whilst hot water may be used as in the proximate fractionation although some hydrolysis of the polysaccharides may occur. It is very difficult to avoid slight hydrolysis whenever water is present, but this will be at a minimum in aqueous ethanol which is the most favoured organic extractant for sugars. A procedure using 80% ethanol is given below. Other extraction systems have been described by Paech and Tracey (1955) and Raymakers (1974).

Procedure

Pre-extract with ether to remove fats if necessary.

Transfer about 2 g air-dry ground sample or 6 g coarsely shredded fresh material to a macerator vessel.

Add 200 ml 80% v/v ethanol.

Macerate for 10 minutes.

Filter through glass fibre paper.

Note

Materials which may need pre-extraction with ether include conifer wood and needles, larger seeds and some animal products.

It may be necessary to clarify further the solutions for sugar analysis, particularly if colorimetric procedures are to be used. The process may be combined with a stage for the removal of interfering substances. Bevenue and Washauer (1950) compared a number of treatments for these purposes including charcoal, lead acetate and Celite filter aids. Other materials used include alumina cream and cupric hydroxide (freshly precipitated). About 0.5 to 1.0 g should be added to 100 ml of the sample extract. It is advisable to run separate control tests to ensure that the clarification agents do not remove saccharides which are of interest. Ion exchange resins are useful for the removal of interfering substances. Solid phase extraction systems based on selective adsorption or elution using chemically-bonded silicas as sorbents have recently been introduced. These have attractions for the rapid clean-up of complex mixtures (van Horne, 1985). For example, in aqueous media, carbohydrates will not be retained by octadecyl (C_{18}) bonded silica whilst other, less polar compounds, will stay on the column enabling carbohydrates to be separated from many other classes of organic compounds.

Detection tests

A number of tests are available for the detection of simple carbohydrates. In the Molisch reaction a violet colour is formed by carbohydrates and glycosides when sulphuric acid is added to the sample solution containing α-naphthol. The Folin–Benedict and Fehling reactions involving the reduction of alkaline copper salts are well-known as tests for reducing sugars. A similar test depending on the reduction of ferricyanide is useful for quantitative purposes, as is the modification by Somogyi (1952) of the Folin–Benedict reaction to allow detection of slowly reacting sugars. Further information can be obtained from Official Methods of Analysis of the AOAC (1984).

Total soluble carbohydrates

The extraction and determination of total soluble carbohydrates using anthrone is described on p. 166 in the section on proximate analyses.

Reducing sugars

The reducing properties of these sugars form the basis of all the methods for this group. The Hagedorn–Jenson method (Hodge and Davis, 1952) given here is based on the quantitative oxidation by potassium ferricyanide. The ferrocyanide is precipitated as the double potassium–zinc salt whilst unused ferricyanide is estimated by adding potassium iodide to liberate iodine and then titrating with sodium thiosulphate. Many similar methods are available, some of which have been adapted for spectrophotometric or volumetric estimation.

Reagents

1 Glucose standard (1 ml ≡ 1 mg glucose).
Dry glucose in a vacuum oven at 70 °C over P_2O_5. Dissolve 1 g dried glucose in water and dilute to 1 litre.

2 Sodium thiosulphate, approx. 0.01 M.
Dissolve 2.5 g $Na_2S_2O_3.5H_2O$ in freshly boiled water and dilute to approx. 100 ml. Add 0.01 g Na_2CO_3 to stabilize the solution. Dilute 10 times before use.

3 Potassium ferricyanide–sodium carbonate reagent.
Dissolve 4.125 g $K_3Fe(CN)_6$ and 5.3 g Na_2CO_3 in water and dilute to 500 ml. Store in the dark.
4 Potassium iodide–zinc sulphate–sodium chloride reagent.
Dissolve 6.25 g KI, 12.5 g $ZnSO_4.7H_2O$, 62.5 g NaCl in water and dilute to 250 ml. Store in the dark.
5 Acetic acid, 5% v/v.
6 Starch indicator.
Add 20 ml water to 1 g starch and mix. Pour rapidly into 50 ml boiling water and continue boiling for 2 minutes. Add 20 g NaCl and make up to 100 ml with water.

Procedure
Prepare an aqueous extract of the ground plant material as described for soluble carbohydrates, taking a larger sample weight if required.
Pipette an aliquot (usually 5 ml) into a boiling tube.
Pipette 5 ml glucose standard and 5 ml water into separate boiling tubes.
From this point treat all tubes in the same way.
Add 5 ml reagent **3**, cover tubes with glass bubble and stand in boiling water for 15 minutes.
Cool for 3 minutes under running water.
Add 5 ml reagent **4** (formation of precipitate and liberation of I_2).
Add 3 ml reagent **5** and titrate with approx. 0.01 M thiosulphate.
Add a few drops of starch indicator when most of the I_2 colour has been destroyed and continue the titration until the blue colour disappears.
Treat extraction blank determinations in the same way and subtract where necessary.

Calculation
If the volumes of approx. 0.01 M thiosulphate required are:
A ml for the water blank,
B ml for the glucose standard,
and *C* ml for the sample extract;
then reducing sugars as glucose (%) =

$$\frac{A-C\ (\text{ml}) \times \text{extract vol (ml)}}{A-B\ (\text{ml}) \times \text{aliquot (ml)} \times \text{sample wt (g)} \times 2}$$

Apply factors for dilution and correct to dry weight as necessary.

Note
Glucose is employed as the standard here but other reducing sugars can be used (fructose, maltose etc.).

Starch

Starch is the major storage polysaccharide in plant materials and is widely distributed. Large and relatively constant amounts are present in some roots, seeds and fruits but the concentrations in other parts of the plant, especially the leaves are subject to diurnal variation falling to a minimum at night.

Many analytical methods are available for its determination, but most of them are not specific for starch and are not applicable to all sample types. The more precise methods involve many separation stages but for routine purposes the approximate techniques may be adequate. McCready *et al.* (1950) described a simple rapid technique in which sugars are first extracted with ethanol and the starch then hydrolysed with sulphuric acid to react with anthrone. Other procedures have been described by Ahlmwalia and Ellis (1984) [enzyme] and Lustinec *et al.* (1984) [colorimetry] whilst McRae *et al.* (1976) compared 6 methods for starch in leaf tissue.

One of the more precise methods for starch determination was originally reported by Pucher *et al.* (1948), which is also given by Official Methods of Analysis of the AOAC (1984). Water followed by perchloric acid is used to extract the starch, which after purification is hydrolysed to glucose and then determined by thiosulphate titration following the Somogyi reaction.

The rapid method given below uses the water and perchloric extraction but some separation stages are omitted and the starch is measured colorimetrically after forming the blue complex with iodine. The accuracy of the result will depend on the nature of the starch extracted.

Reagents

1 Standard solution (1 ml ≡ 1 mg starch).

Weigh 0.05 g pure starch. Add 5 ml water and heat in a water bath at 100 °C for 15 minutes. Then add 5 ml 60% $HClO_4$ rapidly while stirring. Allow to stand for 20 minutes then transfer to 50 ml volumetric flask and dilute to volume.

2 Perchloric acid, 60%.

3 Phenol red indicator, 0.1% solution in industrial spirit.

4 Sodium hydroxide, M.

5 Potassium iodide, 10% w/v.

6 Potassium iodate, 0.0125 M.

7 Acetic acid, 10% v/v.

Procedure

Weigh 250 mg dry ground plant material into a suitable test tube.

Add 200 mg fine sand, 5 ml water and mix well with a stirring rod.

Heat the tube in a boiling water bath for 15 minutes to gel the starch.

Cool to 20 to 35 °C and rapidly add 5 ml 60% $HClO_4$ whilst mixing.

Grind tissue against side of tube intermittently with rod for 20 minutes, transfer to a 100 ml volumetric flask and dilute to volume.

Mix well and allow to settle.

Transfer an appropriate aliquot (usually 5 to 15 ml) to a 50 ml volumetric flask.

Prepare a range of standards containing from 0.1 to 2.5 mg starch in 50 ml volumetric flasks. Compensate with extra 60% $HClO_4$ as necessary to ensure a uniform concentration.

From this point treat standards and samples in the same way.

Add a few drops of indicator solution and then NaOH until the solution turns red.

Add HOAc to destroy the colour and then add a further 2.5 ml.

Add 0.5 ml KI solution, 5.0 ml KIO_3 solution, shake well and dilute to volume.

Measure the absorbance at 680 nm or with a red filter using water as a reference.

Prepare a calibration curve from the standards and use it to obtain mg starch in the sample aliquot.

Carry out blank determinations in the same way.

Calculation

If C = mg starch obtained from the graph then:

$$\text{Starch (\%)} = \frac{C\text{ (mg)} \times \text{soln vol (ml)}}{10 \times \text{aliquot (ml)} \times \text{sample wt (g)}}$$

Apply factor for dilution and correct to dry weight where necessary.

Note

Ideally the starch used for the standard should have been prepared from species similar to those being tested, but the commercial product is an adequate substitute for most purposes.

Chromatography of carbohydrates

Paper and thin layer chromatography

Separation of monosaccharides and some oligosaccharides into their individual compounds can be carried out by paper chromatography. Details of the procedure are given by Jarvis and Duncan (1974). Thin layer chromatography can also be used (Gunther and Schweiger, 1968). In both cases it is necessary to remove any salts from the sample solution by using an ion exchange or electrolytic desalting process. Commercial instruments are available for this purpose.

Application of spots

It is essential for quantitative work that the spots of samples and standards are circular and are all the same size. 20 to 50 μg of each component is a suitable loading in most cases. The concentration should be such that 2 to 5 μl are applied.

Solvent systems and location reagents

Details of these are given in Tables 6.1 and 6.2, respectively.

High performance liquid chromatography

Paper chromatography can be time-consuming but the advent of HPLC and the development of suitable support materials resulted in much shorter analytical times. Sharper separation of individual carbohydrates using cation exchange resins

Table 6.1. Solvent systems for the separation of carbohydrates on chromatograms

Solvent system	Comments
Ethyl acetate: HOAc: water (9:2:2)	Flexible ratio for separation
Butyl acetate: pyridine: ethanol: water (8:2:2:1)	Suitable for simple carbohydrates
Ethyl acetate: pyridine: water (2:1:1)	
Propanol: ethyl acetate: water (7:1:2)	Good separation, temperature sensitive
iso-Propanol: n-butanol: water (7:1:2)	Good travel and separation
n-Butanol: pyridine: water (6:4:3)	Particularly suitable for oligosaccharides

and silica micro-particulate packing materials was also achieved. However, the unsuitability of early detection systems initially limited widespread application of HPLC to carbohydrates until recently. The preparation of ultra-violet absorbing derivatives was developed first, but refractive index is also popular. McBee and Maness (1983) describe a method for the determination of sugars in plant tissues by HPLC depending on the use of a cation exchange column and a refractive index detector, but Binder (1980) reported ultra-violet detection to be more sensitive than refractive index. Niesner *et al.* (1978) also examine carbohydrates by HPLC. Sugars form strong borate complexes and this has formed the basis of an anion exchange chromatography method but the analytical times are rather long. Further information on the application of HPLC in the analysis of carbohydrates can be obtained from specialist publications (see Chapter 8). Brobst and Scobell (1982) compared thin layer chromatography (TLC), gas chromatography (GC) and HPLC and considered the choice should depend on the nature of the sample.

Gas chromatography

The use of gas–liquid chromatography for carbohydrates depends on the formation of stable volatile derivatives since the carbohydrates themselves are not sufficiently volatile. Various derivatives which have been used include *O*-alkyl ethers, acetal and ketal derivatives, *O*-acetyl esters and *O*-trimethyl silyl ethers. The latter have been found to be the most successful. The applications of gas chromatography for the determination of carbohydrates have been considered by Berry (1966) and Laker (1980).

Nitrogenous compounds

Amino-acids and proteins are among the most important constituents in living tissue and account for a large proportion of the total nitrogen content. Laidlaw and Smith (1965) found amino-acids and protein contained 70% of the total nitrogen in Scots pine levels. Soils contain a lower proportion, generally between 20 and 40% according to Bremner (1949a).

Proteins are of the greatest importance metabolically and dominate the structure of enzymes. There are many different proteins each containing peptide chains of amino-acid units, which are bound into a complex molecule. Non-amino components are often associated with this structure. The complexity of proteins is such that in nutritional studies it is often more useful to examine the amino-acid content of the tissue. Therefore discussion is confined to their estimation although some comments on total protein are included.

A large number of amino-acids have been reported in living tissue. About twenty are very widely distributed and are required in protein structures. The basic formula of these is $R.CH(NH_2).COOH$ where the NH_2 group is normally attached to the α-carbon atom. This forms the assymetric centre for all but glycine. All occur naturally in the stereo-chemical L form. A list of the main acids is given in Table 6.3 and includes a note on the composition of the R

Table 6.2. Location reagents for carbohydrates

Reagent	Preparation	Use	Result
Silver nitrate	Dilute saturated aq.$AgNO_3$ 200 times with acetone Add water to redissolve	Dip chromatogram, blow dry Spray with 0.5 M NaOH, and blow dry	Most sugars give brown spots but other reducing substances may react
Phloroglucinol	0.7% phloroglucinol in acetone (A) 40% trichloroacetic in water (B) Mix 9 of A with 1 of B prior to use	Dip chromatogram, heat to 105 °C for a few minutes	Ketoses—green and yellow spots
Aniline phthalate	Dissolve 930 mg aniline and 1.6 g phthalic acid in 100 ml of n-butanol saturated with water	Spray chromatogram, dry at 105 °C for 5 minutes	Ketoses do not react, other sugars give brown and red spots. Uronic acids give fluorescent spots under UV light
Benedict's reagent	Dissolve 15 g sodium citrate, 13 g Na_2CO_3, 1 g $NaHCO_3$ in 60 ml water (A) Dissolve 1.6 g $CuSO_4.5H_2O$ in about 15 ml water (B) Mix (A) and (B) and dilute to 100 ml	Dip or spray chromatogram, dry at 105 °C for 20 minutes	Reducing sugars—orange spots
p-Anisidine hydrochloride	Dissolve 0.5 g *p*-anisidine HCl in 20 ml 60% aq. ethanol Add 90 ml n-butanol and mix	Spray chromatogram, blow dry with hot air	Aldohexoses—green and brown spots Ketohexoses—yellow spots Uronic acids—cherry red spots
Sulphanilic acid	Dissolve 2 g sulphanilic acid in 100 ml water	Spray chromatogram, heat at 100 °C for 10 minutes	Hexoses and disaccharides—yellows spots Pentoses—rose spots

Table 6.3. Common amino-acids

Amino-acid	Nature of R	Structure of R	Comments
R-CH(NH_2)COOH form			
Glycine		–H	
Alanine	Hydrocarbon	$-CH_3$	
Valine		$-CH(CH_3)_2$	
Leucine		$-CH_2CH(CH_3)_2$	
iso-Leucine		$-CH(CH_3)CH_2(CH_3)$	
Serine	Hydroxy	$-CH_2OH$	
Threonine		$-CH(OH)CH_3$	
Aspartic acid	Acidic	$-CH_2COOH$	
Glumatic acid		$-CH_2CH_2COOH$	High levels in seed proteins
Lysine	Basic	$-CH_2CH_2CH_2CH_2NH_2$	Low in seed proteins compared with leaf proteins
Ornithine		$-CH_2CH_2CH_2NH_2$	Not structural but important metabolically
Citrulline		$-CH_2CH_2CH_2NHCONH_2$	Not structural but important metabolically
Arginine		$-CH_2CH_2CH_2NHC(=NH)NH_2$	Low in seed proteins compared with leaf proteins
Cysteine	S-containing	$-CH_2SH$	
Cystine		$-CH_2$-S-S-CH_2	
Methionine		$-CH_2CH_2SCH_3$	
Phenyl alanine	Aromatic	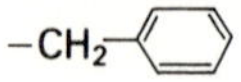 $-CH_2-$ (benzene ring)	
Tyrosine		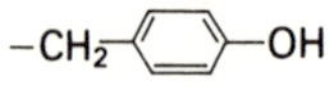 $-CH_2-$ (benzene ring) $-OH$	
Histidine	Heterocyclic	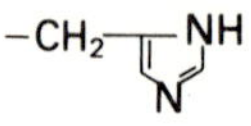 $-CH_2-$ (imidazole ring: NH, N)	Low in seed proteins compared with leaf proteins
Tryptophan		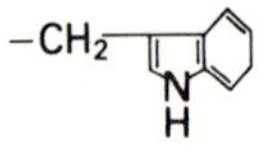 $-CH_2-$ (indole ring: N, H)	
Other amino acids			
Proline		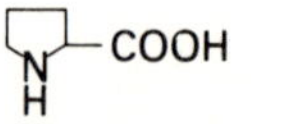 (pyrrolidine ring: N, H) –COOH	α-amino-acids which contain secondary amino-groups
Hydroxyproline		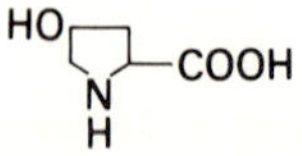 HO– (pyrrolidine ring: N, H) –COOH	
β-alanine		$NH_2CH_2CH_2COOH$	Analysis often required although they are not α-amino acids
γ-aminobutyric acid		$NH_2CH_2CH_2CH_2COOH$	

group. Individual amino-acids are readily determined by chromatography for which details are given later.

Most of the amino-acids included here are concerned in protein structure, but others have been isolated which do not have this function. However, these seem to be restricted to a limited number of species and no analytical methods are given here.

There are a number of other nitrogenous compounds found in living tissues. These include urea and related substances and various heterocyclic compounds such as purines, pyrimidines, alkaloids, porphyrins (e.g. chlorophyll) and others. Derivatives of the more simple compounds are found in all tissues, but alkaloids are restricted mainly to dicotyledons. No analytical details are given here for any of these although a method for chlorophyll is described later in the chapter.

Protein

The crude protein method given on p. 164 in which the total nitrogen is multiplied by a factor (generally 6.25) is adequate for most ecological purposes. Another approximate method measures protein-nitrogen as that fraction of the total nitrogen that remains after a Soxhlet extraction of the sample with 75% ethanol for 16 hours. However methods are available which will give a more reliable total protein figure. These involve the extraction of the protein from the sample, followed by its determination in the extract.

Extraction methods

As proteins often occur bound to other molecules, pre-extractions are usually necessary to break the bond and free the protein. Ether or petroleum ether is used to remove lipids which are the most troublesome group in this respect. In some cases the material may have to be acidified before extraction.

Protein may be extracted by repeated treatment with water or dilute alkali. Pomeranz (1965) used 3M urea solution. Other extractants include solutions of guanidine, guanidine nitrate, pyridine, phenol, thiocyanates and strong mineral acids.

Estimation of total protein

The usual procedure is to precipitate the protein from the extract followed by filtration and weighing of the precipitate. Typical precipitating agents include trichloroacetic acid, perchloric acid, and alcohol. Protein may also be precipitated by boiling. Colorimetric methods include use of the Folin phenol reagent (Lowry *et al.*, 1951, reviewed by Peterson, 1979) and also the biuret reaction between alkaline solutions of proteins and copper salts (Paech and Tracey, 1955, discussed by Rietz and Scheidegger, 1980). Robinson (1979) gives a protein method involving the use of Brilliant Blue dye.

Separation and estimation of individual proteins

Individual proteins may be purified according to molecular size by gel-filtration. Columns of cross-linked dextran gels are used where the degree of cross-linking determines the range of molecular weights which can be separated. Crystallization is used in the final stages of purification.

Electrophoresis is a useful technique and is discussed by Smith (1969). Buffers are added in which the proteins exist as amines and migrate towards the anode. Paper chromatography is less widely used for the separation of proteins than paper electrophoresis, although useful systems are given by Zweig and Whitaker (1971). HPLC methods are now available.

Amino-acids

Protein hydrolysis

With the advent of automated amino-acid analysers, the sequencing of proteins has become far easier (for a review of sequence analysis see the text edited by Neurath and Hill, 1977). When relative abundances of amino-acids are all that are required, a simple total hydrolysis will release the constituent amino-acids although low recoveries are frequently obtained. Acid hydrolysis is preferable to alkali in this respect. Losses may be reduced if the content of non-protein material, particularly carbohydrates is kept to a minimum.

The hydrolysis may be carried out either under reflux or in sealed tubes, although the former is more satisfactory when crude proteins are to be analysed. The sealed tube method is more adaptable for large numbers of analyses.

Reagents

1 Hydrochloric acid, 1 + 1.
2 Iso-propanol, 10% v/v.

Procedure

Weigh air-dried or fresh sample containing 10 to 20 mg N (plants and animals) or 30 mg (soils).

Carry out any pre-extractions that may be necessary.
Add 30 ml of 1 + 1 HCl.
Reflux for 18–24 hours.
Remove excess HCl by taking to dryness in a vacuum rotary evaporator.
Add 5–10 ml water.
Evaporate to dryness as before.
Take up residue in iso-propanol solution.
Transfer to a 25 ml volumetric flask, filtering if necessary, and make up to volume with iso-propanol solution.
Use for estimation of amino-acids.

Notes

1 Excess HCl should be removed as soon as possible after the hydrolysis is completed.
2 The time of hydrolysis will vary according to the material being analysed. It may be necessary initially to hydrolyse replicate samples for various times to determine the period necessary to give maximum amino-acid yield.
3 If not used immediately the hydrolysate should be stored in a refrigerator.

Extraction

The extraction of free amino-acids is best carried out with 70 or 80% aqueous ethanol on fresh material.

Reagents

1 Ethanol, 80% v/v.
2 Diethyl ether or petroleum ether (40 to 60°C).
3 Iso-propanol, 10% v/v.

Procedure

Weigh 5 g fresh material.
Homogenize for 3 minutes with 100 ml of ethanol solution.
Seal flask and stand overnight in the cold.
Filter or centrifuge and wash residue with a little ethanol solution.
Re-extract residue with a further 100 ml of ethanol solution.
Filter or centrifuge and bulk the filtrates.
Evaporate to dryness in a vacuum rotary evaporator.
Take up the residue in iso-propanol solution.
Transfer the solution to a 5 ml volumetric flask, filtering if necessary and dilute to volume with iso-propanol solution.
Use for estimation of amino-acids.

For samples rich in resinous substances, e.g. pine needles, it is necessary to extract with ether before taking the ethanol extract completely to dryness. The procedure given above is followed until the filtrates are bulked then:
Evaporate to 20–30 ml in a vacuum rotary evaporator.
Add 10 ml ether or petroleum ether.
Swirl and transfer to separating funnel.
Allow phases to separate then discard ether layer.
Extract once or twice more with 10 ml ether or petroleum ether discarding the solvent layers each time.
Replace aqueous layer in rotary evaporator and take to dryness.
Take up residue in iso-propanol solution.
Transfer to 5 ml volumetric flask, filtering if necessary and dilute to volume with iso-propanol solution.

Notes

1 If not used immediately the hydrolysate should be stored in a refrigerator.
2 The solutions of amino-acids obtained from the hydrolysate or the 80% ethanolic extract may be analysed in total or individually.

Total α-amino acid nitrogen

Several alternative procedures are available for the determination of total α-amino acid-N. In a 1929 publication, Van Slyke described the reaction of aliphatic-NH_2 groups with sodium nitrite and the measurement of the evolved nitrogen manometrically. Although simple, the method is not specific for α-amino acids since all free aliphatic-NH_2 groups react.

The reaction of amino-acids with ninhydrin is used in the Moore and Stein (1951) colorimetric procedure. A purple colour is formed in the presence of a reducing agent at a pH of about 5.0. Colowick and Kaplan (1957) give working details.

This reaction has the disadvantage that other nitrogeneous compounds, principally ammonia interfere. This technique is of most value after column chromatographic separation of individual amino-acids.

The method, which is described in detail here, measures the ammonia evolved during deamination by ninhydrin. The ammonia is estimated by a distillation and titration procedure similar to that described and illustrated for total nitrogen (p. 122). The method is also described by Bremner (1960) and Stevenson (1982). Simon and Jones (1983) couple this reaction with gas chromatography.

Reagents

1 Sodium hydroxide, 5 M.

2 Sodium hydroxide, 0.5 M.

3 Citric acid.

Grind to a fine powder with a pestle and mortar.

4 Ninhydrin (1,2,3-triketohydrindene monohydrate).

Grind to a fine powder with a pestle and mortar.

5 Phosphate–borate buffer, pH 11.2.

Dissolve 100 g $Na_3PO_4.12H_2O$ and 25 g $Na_2B_4O_7.10H_2O$ in water and dilute to 1 litre. Store in a tightly stoppered bottle.

6 Hydrochloric acid, M/140.

7 Boric acid–mixed indicators.

Procedure

Pipette an aliquot of amino-acid solution containing about 0.5 mg N into a 50 ml distillation flask.

Add 1 ml 0.5 M NaOH.

Place the flask in a boiling water bath for approx. 20 minutes.

Remove the flask and cool.

Add 500 mg citric acid and 100 mg ninhydrin.

Replace flask in the boiling water bath with bulb completely immersed.

Swirl after 1 minute without removing from bath.

Leave in bath for a further 9 minutes.

Remove and cool.

Transfer the flask contents to the distillation apparatus shown on p. 123.

Add 10 ml of phosphate–borate buffer and 1 ml of 5 M NaOH.

Distil, collect and titrate ammonia as described on p. 122.

Calculation

If T = ml M/140 HCl required for titration then:

α-Amino acid N(%) =

$$\frac{T(\text{ml}) \times \text{soln vol}(\text{ml})}{10^2 \times \text{aliquot}(\text{ml}) \times \text{sample wt}(\text{g})}$$

Correct to dry weight where necessary.

Fractionation of amino-acids

Chemical methods for the separation and estimation of individual amino-acids were given by Block and Bolling (1951). These have almost entirely been replaced by chromatographic separations which are considered here.

Purification

In order that chromatographic separations may be sharp, purification of the amino-acid solution is normally necessary. This includes desalting and removal of organic compounds which may interfere with the chromatogram. Desalting and electrodialysis techniques are described by Smith (1969). A common method of purification is to use ion-exchange resins and the appropriate conditions are given below. Non-polar molecules such as carbohydrates are not adsorbed on the resin. Cations are adsorbed but these are not eluted with ammonia.

Reagents

1 Ion-exchange resin.

Use a strongly acidic cation-exchange resin.

2 Ammonium hydroxide, approx. 2 M.

3 Iso-propanol, 10% v/v.

Procedure

Set up a column of the ion-exchange resin in the H^+ form containing 15 ml of settled resin.

Pass an 80% ethanolic solution of the amino-acids through the column at the rate of 3 ml minute^{-1}.

Discard the percolate.

Wash the column well using water at a similar flow rate.

Elute the amino-acids with 60–70 ml of 2 M NH_4OH at 3–6 ml minute^{-1}, collecting all the eluate.

Evaporate to dryness in a vacuum rotary evaporator.

Take up the residue in a suitable volume of 10% iso-propanol.

Use this solution for the separation of amino-acids by paper or thin layer chromatography.

Paper chromatography

As a quantitative method, paper chromatography leaves much to be desired, though with care, approximate results can be obtained. Shellard (1968) discusses the quantitative aspects of both paper and thin-layer chromatography. Owing to the large number of amino-acids often present in the hydrolysates, two-dimensional chromatography is generally used.

In the system described here, amino-acids are applied to the chromatogram as the free acid and a ninhydrin spray is used. An alternative method is available in which dinitrophenyl (DNP) derivatives of amino-acids are first prepared. These are coloured and therefore allow the progress of the separation to be followed visually. Details of the preparation of DNP derivatives and their application in chromatography are described by Smith (1969) and Zweig and Whitaker (1971).

Application of spots

It is essential for quantitative work that spots of samples and standards should be circular and all the same size. The solutions to be spotted should contain 1 to 2 mg ml^{-1} of each amino-acid when 5 μl of solutions are used.

Suitable solvent mixtures for running the chromatograms include n-butanol: acetic acid: water (12:3:5) for compact spots and as a first solvent in two-dimensional work, and phenol: water (4:1 w/v) for a good spread of R_f values and as a second solvent.

Location reagents

Ninhydrin—this is the most widely used location reagent for amino-acids and two treatments are given below.

1 0.25% w/v ninhydrin in acetone containing 5% v/v pyridine, lutidine or collidine. Spray or dip and develop colour in dark at 35°C for 1 hour.

2 0.3% ninhydrin in industrial spirit. Spray or dip and develop colour in dark at room temperature for 18 hours.

The colours produced tend to be transient, but may be stabilized by dipping or spraying with a solution of copper nitrate. (Add 1 ml saturated $Cu(NO_3)_2.3H_2O$ and 0.2 ml 10% HNO_3v/v to 100 ml ethanol. The purple spots change to a red colour which is stable for some months. Special sprays for individual amino-acids are given by Alexander and Block (1960) and Smith (1969).

Thin-layer chromatography

The application to amino-acid and peptide chemistry is well covered by Pataki (1969). Stahl (1970) also gives an account of the separation of amino-acids and their derivatives on thin layers. Solvent and loading conditions differ from paper chromatography, but other conditions are similar for both systems. Heathcote and Haworth (1969) describe the quantitative determination of amino-acids and de los Angeles (1982) gives TLC as an alternative to gas chromatography.

Layers

Silica gel G layers 250 μm thick are generally effective and microcrystalline cellulose is suitable in some circumstances . If not ready-spread the layers should not be dried in the oven, but should be allowed to come to equilibrium at room temperature overnight.

Loading

The precautions taken for the loading of paper chromatograms apply also to thin layer work. The amount of solution required is less than in paper chromatography and approximately 2 μg of each amino-acid should be applied per spot.

Solvent systems

The solvents for paper chromatograms mentioned above are suitable for cellulose thin layer work. Other solvents are given in Table 6.4. The two-dimensional solvents (2-propanol: butanone: M hydrochloric acid and 2-methylpropanol: bu-

Table 6.4. Solvent systems for the separation of amino-acids by TLC

Solvent system	Comments
Silica gel G. layers	
96% Ethanol:water (7:3)	Development time fairly short
n-Butanol:acetic acid:water (4:1:1)	Good separation efficiency; particularly suitable for two-dimensional work
Phenol:water (3:1)	20 mg NaCN added as anti-oxidant to 100 g mixture. Suitable for two-dimensional work
Methyl ethyl ketone:pyridine:water:acetic acid (70:15:15:2)	Will separate leucine and iso-leucine
Chloroform:methanol:17% NH_4OH (2:2:1)	Good separation efficiency; useful for two-dimensional work
Cellulose layers	
iso-Propanol:formic acid:water (20:1:5) Methyl ethyl ketone:t-Butanol:0.88 ammonia:water (3:5:1:1)	Good separation when used as 1st and 2nd dimensions respectively, in two-dimensional work
n-Butanol:acetone:diethylamine:water (10:10:2:5) sec-Butanol:methyl ethyl ketone:dicyclohexylamine: water (10:10:2:5) Phenol:water (3:1)	Useful when used in conjunction with each other in different chromatograms of the same sample mixture
2-Propanol:butanone:M hydrochloric acid (12:3:5) 2-Methylpropanol:butanone:propanone:methanol: water:0.88 NH_3 soln (40:20:20:1:14:5)	Improved separation in two-dimensional work

tanone: propanone: methanol: water: ammonia) are described by Haworth and Heathcote (1969) as being especially effective for the quantitative separation of amino-acids on cellulose layers. The solvents should be placed in the tanks immediately before the plates, so that the tank atmosphere is not saturated with solvent vapour.

High-performance liquid chromatography (HPLC)

Moore and Stein (1951) first described a procedure for the separation of mixtures of amino-acids by elution with buffers from ion-exchange resins, and Spackman *et al.* (1958) demonstrated that the technique was suitable for automation. Since then, many papers have appeared on the subject and the technique has been constantly improved. This was a natural progression to HPLC, the use of which for amino-acids has been reviewed by Pfeifer and Hill (1983). Vestal (1984) reviewed the application of HPLC–MS systems as demonstrated with amino-acids.

Ion-exchange chromatography can be used for primary and secondary amino-acids using colorimetric or fluorescent derivatives. The derivatives can be formed either before, or after, column separation. The ninhydrin colorimetric reaction was the original post-column derivative, but some now prefer fluorescent detection, for example Voelter and Zech (1975). Another example is the use of the phenylthiohydantoin reaction (Frank and Strubert, 1973). For greater sensitivity reverse phase is recommended but preparation of suitable derivatives is then a problem. Maximum resolution in ion-exchange chromatography depends on the use of a citrate–borate buffer system to give the appropriate pH gradient. The purity of the elution buffer can sometimes be a limiting factor.

Raltenburg (1981) discusses most amino-acid analytical procedures, whilst details about the use of HPLC techniques in particular can readily be obtained from specialist journals and books. This technique is also considered in Chapter 8.

Gas chromatography

Volatile derivatives of the amino-acids have to be prepared for GLC. Suitable procedures include

N-acyl esters of trifluoroacetyl derivatives. This method is described by Darbre and Islam (1968), whilst Weinstein (1966) and McKenzie (1981) review the use of gas chromatography for amino-acids.

Other nitrogenous compounds

In some ecological studies information is needed about the nutritional requirements and digestive processes of various mammals and birds. Two of the more important nitrogen compounds in excretory products are uric acid and creatinine. Methods for their determination are given here.

Uric acid

Reagents

1 Uric acid standard.
Stock solution (1 ml ≡ 1.5 mg uric acid): dissolve 150 mg uric acid in 100 ml of 0.5% Li_2CO_3. Make up fresh as required.
Working standard (1 ml ≡ 0.15 mg uric acid): dilute the stock solution ten times with 0.5% Li_2CO_3 solution.

2 Lithium carbonate, 0.5% w/v.
Filter if necessary.

3 Silver lactate solution.
Dissolve 3 g silver lactate in water containing 1 ml lactic acid and dilute to 100 ml with water.

4 Magnesia mixture.
Dissolve 8.75 g $MgSO_4.7H_2O$ and 17.5 g NH_4Cl in water. Add 30 ml 0.88 NH_3 solution and dilute to 100 ml with water.

5 Ammoniacal silver–magnesium reagent.
Mix together 70 ml silver lactate solution, 30 ml magnesia mixture and 100 ml 0.88 NH_3 solution. Filter and store in dark glass bottle.

6 Acid lithium chloride solution.
Dissolve 3.5 g dry LiCl in 0.1 M HCl and dilute to 100 ml with the acid.

Procedure

Weigh 1 g air-dried sample or 4–5 g fresh sample into a 250 ml calibrated flask or bottle.
Add 50 ml Li_2CO_3 solution.
Stopper tightly and shake on horizontal shaker for $1\frac{1}{2}$ hours.
Dilute to the mark with water and mix well.
Filter through a dry No. 541 paper rejecting the first few ml of filtrate.
Collect the clear runnings in a dry vessel.
Pipette 0–2.5 ml working standard into 15 ml conical centrifuge tubes to give a range of standards from 0 to 0.375 mg uric acid.
Dilute each to 5 ml with Li_2CO_3 solution.
Pipette 5 ml of filtered sample extract into a similar tube.
From this point treat standards and samples in the same way.
Add 2 ml ammoniacal silver–magnesium reagent.
Mix and allow to stand for 30 minutes.
Centrifuge at 2 500 r.p.m. for 5 minutes.
Carefully pour off supernatant liquid and discard.
Invert tube on filter paper and allow to drain for 10 minutes.
Wipe any liquid from lip of the centrifuge tube with filter paper.
Pipette into tube 10 ml of 3.5% acid LiCl solution.
Stir or shake for 2–3 minutes to dissolve uric acid.
Centrifuge as before.
Dilute supernatant solution 5 times with water.
Measure the absorbance at 282 nm with an ultraviolet lamp using acid LiCl solution diluted 5 times with water in the reference cell.
Prepare a calibration curve from the standards and use it to obtain mg uric acid in the sample aliquot.
Carry out blank determinations in the same way and subtract where necessary.

Calculation

If C = mg uric acid obtained from the graph then:

$$\text{Uric acid (\%)} = \frac{C\,(\text{mg}) \times \text{extractant vol (ml)}}{10 \times \text{aliquot (ml)} \times \text{sample wt (g)}}$$

Correct to dry weight if necessary.

Creatine and creatinine

As with uric acid, creatine and creatinine determinations may be required on excretion products of animals and to a lesser extent on other materials. The usual practice is to determine total (creatine + creatinine)-nitrogen. The colorimetric procedure given here is based on that of Owen *et al.* (1954). Some samples, especially bird droppings, will be rich in polyphenolic material which interferes with this procedure and has to be removed. This may be achieved by shaking with polyvinylpyrrolidone, (Polyclar AT). Colorimetric and enzyme methods for non-protein compounds, including creatine and creatinine, have been reviewed by Wolfschoon-Pumbo *et al.* (1982) whilst an HPLC method for creatinine is given by Soldin and Hill (1978).

Reagents

1 Creatinine standard.

Stock solution (1 ml ≡ 100 g creatinine-N): dissolve 0.0673 g creatinine in water and dilute to 250 ml with 0.1 M HCl.

Working standard (1 ml ≡ 2 μg creatinine-N): dilute the stock solution 50 times with water. Prepare fresh as required.

2 Hydrochloric acid, approx. 0.1 M.

Dilute 10 ml conc HCl to 1 litre with water.

3 Picric acid, saturated aqueous solution.

4 Sodium hydroxide, 2.5 M.

5 Alkaline picrate reagent.

Mix 27.5 ml saturated aqueous picric acid with 5.5 ml of 2.5 M NaOH and dilute to 100 ml with water. Prepare fresh as required.

6 Oxalic acid, saturated aqueous solution.

7 Fuller's Earth.

Procedure

Weigh or pipette an amount of sample containing not more than 200 μg creatinine-N into a 100 ml conical flask.

Add 30 ml 0.1 M HCl.

Place glass bubble in neck of flask and simmer gently on hotplate for 2 hours. Top up the level at intervals with 0.1 M HCl.

Filter hot into a 100 ml volumetric flask using a No. 541 paper.

Wash the residue and filter paper with hot water.

Dilute to volume with water when cool.

Pipette 0–5 ml of creatinine working standard into clean and dry 15 ml conical centrifuge tubes to give a range of standards from 0 to 10 μg creatinine-N.

Pipette 5 ml of the sample extract into a similar tube.

From this point treat samples and standards in the same way.

Add 0.5 ml oxalic acid solution.

Add approx. 200 mg of Fuller's Earth.

Stopper and shake for 5 minutes.

Decant and discard supernatant and drain by inverting tube on filter paper.

Add 7.5 ml alkaline picric acid reagent.

Shake for 5 minutes.

Centrifuge as before.

Leave for 20 minutes at room temperature.

Record the absorbance of the supernatant liquid at 520 nm using water as a reference.

Prepare a calibration curve from the standards and use it to obtain μg creatinine-N in the sample aliquot.

Carry out a blank determination in the same way and subtract where necessary.

Calculation

If $C = \mu g$ (creatine + creatinine)-N obtained from the graph then:

$$\text{(Creatine + creatinine)-N (\%)} = \frac{C\,(\mu g) \times \text{soln vol (ml)}}{10^4 \times \text{aliquot (ml)} \times \text{sample wt (g)}}$$

Correct to dry weight where necessary.

Note

Include the following stage for samples rich in polyphenols:

Add approximately 400 mg polyvinyl pyrrolidone to 10 ml HCl extract. Shake for 5 minutes, filter and then pipette 5 ml into the centrifuge tube and proceed as above.

Fatty acids and lipids

The lipids considered here are all saponifiable in that they form water soluble salts (soaps) on heating with alkali. They are all present in living tissue and are of considerable metabolic and structural importance. An outline classification of the most abundant types is given below.

FATTY ACIDS

Higher members of aliphatic carboxylic acid series.

Saturated $C_nH_{2n}O_2$, e.g. palmitic, stearic (n usually even).

Unsaturated $C_nH_{(2n-x)}O_2$, e.g. oleic, linoleic (x usually even).

Unsaturated hydroxy, e.g. ricinoleic.

SIMPLE LIPIDS

a Esters of fatty acids with glycerol: triglycerides (fats and oils).

b Esters of fatty acids with long chain aliphatic alcohols or with sterols: waxes, components of cutin etc.

PHOSPHOLIPIDS

Esters of fatty acids with:

a 1-glyceryl-*o*-phosphorylcholine (α-lecithin);

b 1-glyceryl-*o*-phosphorylethanolamine (α-cephalin);

c 1-glyceryl-*o*-phosphorylserine (phosphatidyl serine);

d Inositol-1,3-diphosphate;

e Sphingosine-phosphorylcholine (sphingomyelin).

GLYCOLIPIDS

Lipids containing sphingosine, a fatty acid and a sugar, usually glucose or galactose, but no phosphorus (cerebrosides and gangliosides).

Attention is given mainly to plant tissues and in particular to the methods for extraction and fractionation. Methods for estimating total lipids and total fatty acids are described here, but the determination of crude fat is described under proximate analysis.

There are a variety of other substances with similar properties that cannot be saponified. Some are present in waxes and resins and are widely distributed in plant tissue. They include aliphatic components and derivatives of aromatic groups such as quinone and phloroglucinol. Their ecological significance has not yet been explored and methods for their estimation are not given here. A bibliography of works on lipids has been published (Gunstone, 1977), and Dutton (1983) reviews chemical methods for lipids.

Total lipids

Lipids are susceptible to oxidation and molecular re-arrangements and care must be taken in the storage and treatment of samples. Fresh samples should be analysed or partially treated and purified lipid solutions should not be heated or allowed to stand too long in air. Determinations are best carried out in an atmosphere of nitrogen.

A simple method for total lipid estimation is the procedure of Bligh and Dyer (1959). They found that an optimum extraction is obtained when the final mixture contains chloroform, methanol and water in the ratio 1:2:0.8. It is necessary before extraction to carry out a separate moisture determination so that the appropriate amount of chloroform and methanol can be added to achieve the correct ratio.

Reagents

Chloroform–methanol mixture.

Mix 1 volume of chloroform with 2 volumes of methanol.

Procedure

Weigh between 15 and 25 g of fresh material.

Add chloroform–methanol mixture so that the volumes of chloroform, methanol and water in the sample are in the proportions 1:2:0.8.

Homogenize for 2 minutes, preferably under oxygen-free nitrogen (OFN) with cooling.

Add chloroform such that its proportion is doubled.

Homogenize for 30 seconds.

Add the same amount of water as the second chloroform addition.

Homogenize for a further 30 seconds.

The proportions of chloroform, methanol and water should now be 2:2:1.8.

Filter the mixture on a Buchner funnel with slight suction, pressing the residue on the paper to obtain good recovery.
Transfer the filtrate to a graduated cylinder and note the volume of the lower chloroform layer.
Discard the upper alcoholic layer.
Weigh a clean, dry evaporation dish.
Pipette a suitable amount of chloroform layer into the dish.
Remove the solvent on a warm water bath with a stream of OFN.
Place the dish in vacuum oven for 30 minutes at 40 °C.
Cool in desiccator and weigh.
Calculate weight of residue in dish.

Calculation

Total lipids (%) =

$$\frac{\text{residue wt (g)} \times \text{vol chloroform layer (ml)} \times 10^2}{\text{aliquot (ml)} \times \text{sample wt (g)}}$$

Correct to dry weight where necessary.

Note
In some cases where the sample has a low water content it may be necessary to add water to achieve the correct proportions.

Fractionation of lipids

Lipids may be partially fractionated by successive extractions with solvents of different polarities, e.g. diethyl ether, benzene, chloroform–methanol mixtures. Individual extracts may then be further fractionated by a chromatographic method.

Chromatography has now superseded most other methods of lipid fractionation and early reviews include those of Morris (1966) and Nicholls (1966). One procedure uses a column of silicic acid previously activated at 120 °C for 24 hours. The sample is applied as a solution in petroleum ether (40–60 °C boiling range) and the column successively eluted with:

1 petroleum ether (40–60 °C) (removes hydrocarbons);
2 1% diethyl ether in petroleum ether (40–60 °C) (removes sterol esters);
3 3% diethyl ether in petroleum ether (40–60 °C) (removes triglycerides and free acids);
4 10% diethyl ether in petroleum ether (40–60 °C) (removes triglycerides, free acids and sterols);
5 ether:ethanol (3:1) (removes phospholipids and non-lipids materials).

The column fractions may then be isolated and weighed or further examined by chromatography.

Thin layer chromatography is probably more suited to the separation of lipids than paper chromatography. Suitable texts covering TLC of lipids are those of Randerath (1966) and Stahl (1970) whilst reviews have been made by Murata (1980) and Nicholls (1966).

Silica gel G layers are normally used. Reversed phase partition chromatography on silanized silica gel G treated with silicones is also widely used. Solvent systems are mostly mixtures of petrol ether and diethyl ether. A higher proportion of diethyl ether is required for the investigation of substances of increasing polarity whilst the polar lipids such as phospholipids require an alcoholic solvent to bring about migration.

A full list of location reagents is given by Randerath (1966), including iodine vapour, 2:7 dichlorofluorescein and Rhodamine B, and charring by concentrated H_2SO_4.

Total lipids can be fractionated by HPLC using silica columns, but because of the wide polarity range of lipids a two or three stage gradient elution is required. Cooper and Anders (1975) have reviewed work in the use of HPLC in lipid analysis. Applications specifically for plant materials have been described by Erdahl *et al.* (1973) and Privett *et al.* (1973).

Fatty acids

Fatty acids are obtained from the parent lipids by saponification, which may be carried out on the total lipid extract or on individual fractions.

Total fatty acids

The method given here for the determination of total fatty acid is based on that of Garton *et al.* (1961).

Reagents

1 Alcoholic potassium hydroxide, 0.5 M.
Dissolve 30 g KOH in industrial spirit and dilute to 1 litre.
2 Sulphuric acid, approx. 5 M.
3 Diethyl ether.
4 Potassium hydroxide, 0.5% w/v.

Procedure

Extract the total lipids as described on p. 186.
Add excess 0.5 M alcoholic KOH.
Reflux for 2–3 hours preferably in an atmosphere of N_2.
Acidify drop-wise with 5 M H_2SO_4.
Extract with 3 × 50 ml of ether.
Wash the combined extracts by swirling with 2 × 50 ml water.
Make acid by addition of a drop of 5 M H_2SO_4 followed by a final wash with water.
Shake the extract with 3 × 50 ml of 0.5% KOH followed by one washing with 50 ml water.
Discard the ether layer which contains the unsaponifiable matter.
Acidify the alkaline solution drop-wise until no more fatty acid is precipitated.
Extract with 3 × 5 ml ether.
Wash ether extract with water until acid free.
Evaporate the ether on a water bath with a stream of oxygen-free N_2 in a tared flask.
Cool in desiccator and re-weigh.
Calculate weight of fatty acid residue.

Calculation

$$\text{Total fatty acids (\%)} = \frac{\text{residue wt (g)} \times 10^2}{\text{sample wt (g)}}$$

Correct to dry weight where necessary.

Free fatty acids

Free fatty acids in the total lipid extract may be determined by using the above extraction procedure, but omitting saponification with alcoholic potassium hydroxide. Alternatively passage through a column of silica impregnated with potassium hydroxide can be used to remove acids, which can subsequently be eluted using 2% formic acid in ether.

Fractionation of fatty acids

The fractionation and estimation of fatty acids is most readily carried out by gas liquid chromatography (GLC). Accounts of the application of GLC to lipids is given by Horning *et al.* (1964), Woodford (1964) and Kuksis (1977, 1978).

Fatty acids are usually chromatographed as their methyl esters which must be prepared beforehand. Methods available for their preparation include the use of methanol–sulphuric acid (described below), diazomethane, trans-esterification and boron trifluoride as described by Browning (1967). Trans-esterification can be carried out directly on the extracted lipids, without isolating the acids. The other three methods are carried out on the acids after isolation.

Esterification

Reagents

1 Methanol–sulphuric acid.
Add 1 ml of conc H_2SO_4 to 100 ml dry methanol.
2 Diethyl ether or petroleum ether, 40 to 60 °C.
3 Sodium sulphate, anhydrous.

Procedure

Add 15 ml methanol–H_2SO_4 to the mixture of about 0.2 g fatty acids prepared as described earlier.
Reflux gently for 1 hour.
Cool and add 30 ml water.
Transfer to a separating funnel and extract with 3 × 10 ml of ether or petroleum ether.
Wash the ether layer with 3 × 10 ml of water.
Add 2–3 g of anhydrous Na_2SO_4.
Shake and allow to stand for a few hours.
Decant or filter the solution and wash the Na_2SO_4 with a small amount of ether.

Remove the ether on a warm water bath using a stream of oxygen-free N_2.

Take up the residue in 1 ml of hexane or petroleum ether.

Apply to a GLC column with a hypodermic syringe.

Gas chromatography

The methyl esters are separated using 2–3 m columns of 4–6 mm internal diameter. Ethylene glycol polyester with succinic or adipic acids on Celite 545 (80–100 or 100–200 mesh) form suitable column packings. Other stationary phases include Apiezon M and L. Much better resolution is, however, obtained using capillary columns. The ethylene glycol polyester columns should be used at 175–195 °C whilst the Apiezon columns can be used at higher temperatures, 200–225 °C. In cases of complex fatty acid mixtures it may be advantageous to obtain chromatograms from two columns where each is packed with material containing one of the two classes of stationary phase. Short capillary columns have been found to be effective for lipids (Lercker, 1983).

Any of the standard detectors may be used including flame ionization, thermal conductivity or argon ionization. The amount of sample applied to the columns will vary according to the detector used and account must be taken of this to obtain optimum peak heights. Argon or helium are generally used as carrier gases.

Temperature programming is not normally necessary, but can be used providing the temperature does not exceed 200 °C because the polyethylene glycol adipate and succinate stationary phases are then subject to high column bleed.

High performance liquid chromatography

Although gas-liquid chromatography is still much favoured for the analysis of fatty acids, HPLC is an effective technique. Free fatty acids can be separated on an ion-exchange resin and detected using refractive index, ultra-violet detectors. Fluorescent derivatives are also popular as pre-column detectors. The derivatives used are benzyl, phenyl acyl or related esters. A reaction stage using alkaline salts to free the acid is inserted before column injection. An octadecyl or similar column packing is used and a methanol or acetonitrile water gradient mixture is used for the mobile phase. HPLC methods are given by Ryan and Honeyman (1984), Hamilton and Comai (1984). Smith (1983) reviews the use of this technique for fatty acids.

Phospholipids

A simple method of obtaining a result for total phospholipids is by selective extraction so that the phospholipids are separated from other phosphorus fractions. Total phosphorus can then be determined on the residue and a correction factor applied. Baker (1975) described such a method for phospholipids in soil. Phospholipids containing primary amino-groups have been separated after forming the biphenylcarboryl derivatives (Jungalwala *et al.*, 1975). These can be fractionated by HPLC although the use of ultra-violet absorbing derivitives increases sensitivity. The use of HPLC for the determination of phospholipids is described by Patton *et al.* (1982).

Flavonoids and related compounds

Estimation of polyphenol compounds is not often required by the ecologist, except for the somewhat arbitrary determination of soluble tannins. Occasionally, however, more complex determinations of individual components are required and these are briefly discussed later in this section. A brief summary of the various flavonoid types is given in Table 6.5. It should be noted that flavonoids occur mainly as glycosides in the natural state and it is possible for a single flavonoid aglycone to give rise to a score or more glycosides. Geissman (1962), Swain (1963) and Harborne (1964, 1967) all discuss this group of compounds in detail.

Soluble tannins

The method for the determination of soluble tannins given here is based on that of the Association of Official Agricultural Chemists (1970),

Table 6.5. Flavonoids and related compounds

Formulae	Flavonoid groups	Comments
A—*Flavonoids* (C_6-C_3-C_6 skeletons)		
	Catechins (R = H) Leucoanthocyanidins (R = OH)	Colourless Polymers include 'condensed tannin' 'phlobaphenes' (red)
	Flavanones (R = H) Flavanonols (R = OH)	Colourless or slightly yellow Occur in traces only
	Flavones (R = H) Flavonols (R = OH)	Yellow pigments
	Anthocyanidins	Red and blue pigments
	Chalcones	Only a few occur naturally
	Dihydrochalcones	
	Aurones	Yellow pigments
	Isoflavones	Yellow pigments, very few known
B—*Some related compounds*		
(1) C_6-C_1-C_6	Benzophenones	Including xanthones—yellow pigments
(2) C_6-C_2-C_6	Stilbenes	Toxic to some fauna

'Hydrolysable tannins' are not flavonoids but esters formed between phenolic acids and sugars (e.g. between gallic acid and glucose).

using a reagent first described by Folin and Denis (1912). Box (1983) reported it to be superior to all others. The method is not specific for tannins but will also measure flavonoids and other easily oxidized compounds which may be extracted by hot water. The choice of a standard is also a problem. Ideally the compound which is present in the extract as a major component should be used but in practice tannic acid or catechin are more popular. Gravimetric techniques are also available for precipitating tannins which can be weighed directly. Tannin measuring techniques suitable for plants are reviewed by Tempel (1982). Fresh or air-dry material should be used in preference to that dried at 105 °C. Waters should be filtered before analysis.

Reagents

1 Tannic acid standard (1 ml ≡ 0.1 mg tannic acid). Dissolve 0.05 g tannic acid in water and dilute to 500 ml. This solution should be prepared immediately before use.

2 Folin–Denis reagent.

Add 50 g sodium tungstate, 10 g phosphomolybdic acid and 25 ml *ortho*-phosphoric acid to 375 ml water. Reflux for 2 hours, cool and dilute to 500 ml.

(It is important to place a few glass beads in the

flask during refluxing to prevent superheating of the solution.)
3 Sodium carbonate, 17% w/v.

Procedure

1 ***Extraction procedure*** (for plant material and litter)
Weigh 0.1 g air-dry ground sample into a 100 ml conical flask.
Add 50 ml water.
Boil *gently* on hotplate for 1 hour with glass bubble in neck.
Filter whilst warm through a No. 44 filter paper into a 50 ml volumetric flask.
Wash paper and dilute to volume when cool.
(The extract will not keep overnight and must be prepared shortly before colour development.)
Take water blanks through the same stages.

2 ***Colour development***
Pipette 0–3 ml of the tannic acid standard into 50 ml volumetric flasks to give a standard range from 0 to 0.3 mg tannic acid.
Pipette a suitable aliquot of the sample extract into a 50 ml volumetric flask.
From this point treat samples and standards in the same way.
Add water until the flask is two-thirds full.
Add 2.5 ml Folin–Denis reagent.
Add 10 ml Na_2CO_3 solution.
Dilute to volume and mix.
Stand in water bath at 25 °C for 20 minutes.
Measure the absorbance at 760 nm or with a red filter using water as a reference.
Prepare a calibration curve from the standard readings and use it to obtain mg tannic acid in the sample aliquot.
Carry out blank determinations and subtract where necessary.

Calculation
If C = mg tannic acid obtained from the graph then for:
1 Plant materials and soil extracts,

$$\text{Soluble tannins (\%)} = \frac{C\text{ (mg)} \times \text{extract vol (ml)}}{10 \times \text{aliquot (ml)} \times \text{sample wt (g)}}$$

2 Waters,

$$\text{Soluble tannins (mg l}^{-1}\text{)} = \frac{C\text{ (mg)} \times 10^3}{\text{aliquot (ml)}}$$

Correct to dry weight where necessary.

Fractionation of polyphenolic compounds

Estimation of total flavonoids is difficult due to lack of specificity in the methods available. An account of these procedures is given by Swain and Goldstein (1963). Determination of sub-groups and individual compounds requires some form of fractionation.

Many polyphenols occur as glycosides the majority of which are very readily hydrolysed to the constituent sugar especially by the enzymes present in the material. A few flavonoids are very susceptible to oxidation and samples containing these substances should be analysed fresh as soon after collection as possible.

Extraction techniques
The method of extraction depends to some extent on the sample and the compounds to be analysed. In general 70–95% ethanol or methanol has been found to extract almost all flavonoids and their glycosides. The sample should be dropped into the boiling solvent to prevent enzymic action. The inclusion of a small amount of hydrochloric acid in the solvent (0.1%) may also be advantageous especially when anthocyanins are to be investigated.

The extract normally contains many other compounds besides polyphenols and purification is usually necessary. Seikel (1962) used cation exchange resins to separate flavonoids from sugars, salts and organic acids. Alternatively organic solvents may be used either to extract constituents such as chlorophyll and lipids from an aqueous–alcoholic solution of flavonoids or to transfer flavonoid aglycones to the organic layer (Quarmby, 1968). Many polyphenols may be precipitated as their lead salts. The flavonoids can then be regenerated by treatment with hydrogen sulphide or sulphuric acid in the presence of alcohol followed by separation of insoluble lead salts (Browning, 1967).

Chromatographic separation

Fractionation of flavonoid compounds is carried out by chromatographic methods and a full account of the earlier techniques is given by Seikel (1962). Large scale fractionations are usually carried out on columns of magnesia, polyamide, silica-gel or cellulose powder. For analytical work, paper and thin-layer chromatography and HPLC are more convenient. Bate-Smith (1948) applied the technique of paper chromatography to flavonoids and since then much work has been done in this field. Harborne (1958) reviewed the then available methods and an account of the paper chromatography of plant phenols is given by Smith (1969). Applications of TLC and HPLC are referred to later.

Paper chromatography

Loading

It is essential for quantitative work that spots of samples and standards are circular, similar in size and as small as possible. 10 μg of each component is a suitable amount for loading and the concentration should be such that not more than 50 μl need be applied to give the required weight.

Solvent systems

Four solvent systems are commonly used and include two for aglycones and two for glycosides. These are given below.

For aglycones:

1 HOAc:conc HCl : water = 30 : 3 : 10 (Forrestal solvent);

2 Toluene : HOAc : water = 4 : 1 : 5 (upper phase only).

For glycosides:

3 n-Butanol : HOAc:water = 4 : 1 : 5 (upper phase only);

4 HOAc, 2 to 30%.

Location techniques and reagents

1 VISIBLE AND ULTRA-VIOLET COLOURS

Many flavonoids form coloured phenolic anions when treated with alkali and also chelate with aluminium chloride. Observation of the colour shifts both in visible and ultra-violet light following such treatment often allows the compounds in a spot to be classified into one of the flavonoid groups, e.g. chalcone. A suitable procedure is given below:

Examine chromatogram in visible light and note position and colour of spots.

Examine in ultra-violet light and note position and colour of spots.

Pour 0.88 NH_3 solution into suitable tank to about 1 cm depth.

Hang chromatogram in tank for 5 minutes.

Note colour and position of spots in both visible and ultra-violet light.

Hang chromatogram in current of air until all traces NH_3 on paper are removed.

Dissolve 5 g Al_2Cl_6 in absolute ethanol (CARE—heat of solution may result in boiling).

Filter into dry flask.

Spray chromatogram.

Observe spots under visible and ultra-violet light.

Refer to Table 6.6 and identify flavonoid groups by comparison.

2 FERRIC CHLORIDE–POTASSIUM FERRICYANIDE

Dissolve 1 g $FeCl_3.6H_2O$ in 100 ml water.

Dissolve 1 g $K_3Fe(CN)_6$ in 100 ml water.

Mix in ratio 1:1 just before use.

All flavonoids give deep blue spots due to formation of Prussian Blue.

3 NEU'S REAGENT (Neu, 1957)

Dissolve 1 g diphenylboric acid ethanolamine complex in ethanol.

Spray and examine under ultra-violet light.

In general this reagent does not significantly alter the wavelength of the natural fluorescence, but increases the brightness of fluorescence of many flavonoids.

4 DIAZOTIZED *p*-NITROANILINE

Dissolve 0.3 g *p*-nitroaniline in 100 ml 8% v/v HCl.

Dissolve 5 g $NaNO_2$ in 100 ml water.

Dissolve 27 g NaOAc in 100 ml water.

Cool each of above solutions to 0 °C.

Table 6.6. Flavonoid colours on chromatograms (largely derived from Seikel, 1962)*

	No reagent		NH_3 solution		$AlCl_3$ reagent	
	Visible	UV	Visible	UV	Visible	UV
Flavone	Pale yellow	Dull brown Red-brown Yellow-brown	Yellow	Bright yellow Yellow-green Dull purple	Pale yellow	Fluorescent Green Yellows Browns
Flavonol	Pale yellow	Bright yellow Yellow-green Browns	Yellow	Bright yellow Yellow-green Green	Yellow	Fluorescent Yellow-green
Isoflavone	Colourless	Faint purple Pale yellow†	Colourless	Faint purple Pale yellow	Colourless	Fluorescent Yellow
Catechin	Colourless	Colourless†	Colourless	Fluorescent Pale blue Black	Colourless	Colourless Pale blue Yellow-white
Flavanone	Colourless	Colourless	Colourless	Colourless Pale yellow Yellow-green	Colourless	Fluorescent Green-yellow Blue-white
Leucoanthocyanin	Colourless	Colourless				
Anthocyanin	Pink Orange Red-purple	Dull red or Purple Pink Brown	Blue-grey Blue	Bluish	‡	
Aurone	Bright yellow	Bright yellow Green-yellow	Orange Orange-pink	Yellow-orange Orange Red-orange Brown	Pale yellow Orange	Fluorescent Green Green-yellow Pale-brown
Chalcone	Yellow	Brown Black Yellow-brown	Yellow Orange Red-orange Pink	Orange Red Purple Black	Yellow Orange Yellow-orange	Fluorescent Orange Brown Pink

* With $FeCl_3$ and $K_3Fe(CN)_6$ reagents all these phenolic pigments produce blue colours.
† Short wavelength ultraviolet: isoflavones, yellow; catechins, black.
‡ Only derivatives of cyandin, delphinidin and petunidin change colour.

Mix 5 ml *p*-nitroaniline and 3 ml $NaNO_2$.
Add 50 ml of the NaOAc solution.

The latter reagent is useful for the detection of simple monohydroxy phenols which frequently escape identification by other reagents, e.g. sinapic and ferulic acids give rose pink colours and are distinguished from umbelliferone and scopoletin which are yellow.

Other sprays may be used for more specific types of flavonoids; for example, Gibbs reagent (2-6-dichlorobenzoquinone-4-*N*-chloroimine) reacts with phenols unsubstituted in the *para* position (Smith, 1969), whilst leucoanthocyanidins may be detected as anthocyanidins after spraying with weak acid to bring about oxidation.

Identification

If sufficient material can be obtained at the outset, identification of individual compounds becomes easier in spite of the large number of flavonoids which are now known. Fortunately the ultraviolet spectral characteristics of all the common flavonoids have been recorded. The principle is to elute sufficient of the purified material from the chromatogram to record the absorption spectrum of the compound alone and in the presence of reagents such as sodium borate, sodium ethoxide and aluminium chloride. From observations of the resultant colour shifts it is frequently possible to identify the flavonoid without further analysis. A guide to the techniques used, and extensive tables of absorption maxima, absorption coefficients and wavelength shifts are given by Geissman (1962) and Harborne (1964).

Thin layer chromatography

Thin layer chromatography has not replaced paper chromatography for the separation of flavonoids to the same extent it has for other groups. Paper seems to be a better choice for the separation of complex mixtures where all types from simple phenols to hydrolysable tannins need to be resolved. Nevertheless TLC can be very useful in the rapid identification of flavonoids. Identification by reflectance spectra obtained directly from spots on a thin layer plate has been described (Hiermann, 1980). Quantitative results can be obtained through the use of high performance TLC (Tomimori *et al.*, 1985). Many combinations of adsorbent and solvent systems have been tried. These include:

1 Silica gel
With toluene:ethyl acetate:formic acid (5:4:1) for the separation of aglycones.

2 Polyamide
With ethanol:water (3:2) for the separation of flavonal glycosides.

3 Poly-n-vinylpyrrolidone (PVP)
With 90% formic acid in direction I and Forrestal solvent—HOAc:conc HCl:water (30:3:10) in direction II for the separation of aglycones other than anthocyanidines (Quarmby, 1968).

An advantage of this method is the greatly enhanced fluorescence and range of colours given by flavonoids on PVP. This makes the identification of spots easier even if they are not fully separated or are masked by tailing.

High performance liquid chromatography

HPLC is an effective technique for the fractionation of flavonoids. μ Bonda pak C_{18} is a favoured column packing and various solvents have been described for gradient elution followed by ultraviolet detection. Daigle and Cankerton (1983) have reviewed the analysis of flavonoids by HPLC and Wagner *et al.* (1983) compare HPLC with other methods as a technique for flavonoids. Applications for plant materials have been described by Ward and Pelter (1974), Wulf and Nagel (1976), Pietta *et al.* (1983) and Hosettmann *et al.* (1984). Kingston (1975), Hardin and Stulte (1980) and Anderson and Pederson (1982) also use HPLC for plant phenolics, whilst Cieslak (1983) reviews its suitability for polyphenols and proteins.

Other organic constituents

In this handbook it has only been possible to look at the more important organic groups. Other compounds are of occasional interest to the ecologist and some of these are considered in the following section. Included are organic acids, chlorophylls, carotenoids, phytic acid, sterols and humus residues.

Organic acids

Only the water soluble organic acids are dealt with here, the insoluble fatty acids having been discussed earlier. The water soluble acids are a chemically diverse group although many of them are found in plant tissues. They normally occur free or as inorganic salts (Browning, 1967), but some may be present as esters. The more common groups of acids are listed below.

ALIPHATIC ACIDS

1 Those which occur in the citric acid cycle, e.g. pyruvic, citric, malic and succinic.

2 The lower members of the fatty acid series, e.g. acetic and butyric acids.

SUGAR ACIDS

Ascorbic, gluconic and galacturonic acids are examples.

AROMATIC ACIDS

These include the phenolic acids (gallic, salicylic, shikimic, etc).

ISOPRENOID DERIVATIVES

These are functionalized oligomers of the

```
   C
   |
C–C–C–C unit
```

which occurs in monomeric form in isovaleric acid.

Various less common types of acids are also found in plant tissues. Many species accumulate relatively large quantities of otherwise atypical acid forms. Identification of unusual acids, together with flavonoids, alkaloids, polyacetylenes and sulphur compounds form the basis of chemical taxonomy, a useful adjunct to classical morphological taxonomy (Swain, 1963).

The main role of many of the plant acids appears to be in metabolic pathways such as the citric acid cycle. They are therefore more often investigated in biochemical rather than ecological studies. Only an outline of their extraction and preliminary fractionation is given here.

Polar solvents such as water and water–alcohol mixtures are most frequently used for extracting plant materials. To remove calcium salts an acidified (pH 1) solution will be required. Fresh plant material should be used for the extraction since drying may remove some of the more volatile acids. Extraction details are given by Paech and Tracey (1955) who also discuss purification of the extract by steam and fractional distillation. Purification using ion-exchange resins is also possible. The total acid content of the extract is then easily estimated by titrating against standard alkali.

Lower carboxylic acids are usually fractionated using a chromatographic technique. Column chromatography using ion-exchange resins is simple and convenient as shown by Samuelson (1963). The eluate can be collected with a fraction collector enabling the individual acids to be estimated. Clement and Loubinoux (1983) combine reversed-phase and ion-exchange techniques.

Paper chromatography can be used although it is necessary to prepare salts or derivatives of the volatile acids before separation. This is not necessary for non-volatile acids. The TLC separation of carboxylic acids on silica gel requires solvents that are relatively polar. Suitable basic solvents are methanol:ammonia:water or ethanol:ammonia:water mixtures. Acidic solvents include benzene:methanol:acetic acid and benzene: dioxan:acetic acid. Randerath (1966) and Stahl (1970) give details of the applicability of TLC to organic acids. Acid–base indicators, e.g. bromophenol blue or bromocresol purple, are used as spray reagents for both paper and thin layer chromatography.

Gas (GLC) and liquid (HPLC) chromatography are now widely used for the separation of organic acids. Volatile acids may be run directly on GLC using acid stationary phases (Lanigan and Jackson, 1965), but non-volatile acids normally have to be esterified first. Details are given by Greeley (1974) and by Horvat and Senter (1980). A mass spectrometer coupled with the gas chromatograph can be used for identification.

HPLC using ion exchange, ion exclusion and reversed phase have been used for the fractionation of organic acids. An acid buffered mobile phase with ultra-violet detection is generally favoured for carboxylic acids. Methods are described by van Niekerk and du Plessis (1976), Hartley and Buchan (1979) and Marsili (1981).

Chlorophyll

Plant pigments mainly consist of chlorophyll, carotenoids and flavonoids.

Chlorophyll is a green porphyrin compound containing magnesium as its central atom and is present in all photosynthetic tissues of higher plants; the pigment of lower photosynthetic organisms is closely related to chlorophyll. There are two main forms, chlorophyll *a* and *b*, normally present in the ratio of about 3:1. Other variants are not discussed here.

Paech and Tracey (1955) discussed methods for determining chlorophyll and a method review was produced by Goodwin (1965). More recent publications dealing with problems and procedures encountered in this determination include a supplement of Archiv fur Hydrobiologie (1980) and Lichtenhaler and Wellburn (1983). The use of HPLC for chlorophyll has been described by Brown *et al.* (1981).

Where possible, chlorophyll determinations should be carried out on fresh material because drying out tends to degrade the pigment resulting in colour change. If drying is unavoidable freeze-drying is probably the best method. If fresh material has to be stored, deep-freezing at −20 to −30 °C is adequate.

Total chlorophyll

The method given here is based on the original procedure of Comar and Zscheille (1942). The chlorophyll is extracted with aqueous acetone, transferred into ether and the optical density measured at 660 and 643 nm.

Fresh material should be cut finely and mixed thoroughly before weighing. Dried material should be ground to pass an 0.5 mm sieve.

Reagents

1 Acetone, 85% v/v.
2 Diethyl ether.
3 Sodium sulphate, anhydrous.

Procedure

1 *Extraction*

Homogenize or macerate 1–5 g sample by adding small amounts of 85% acetone.
Filter on Buchner funnel and wash with 85% acetone.
Repeat homogenization and filtration until filtrate and washings are colourless.
Homogenize once with a known volume of acetone and after filtration add water to adjust acetone infiltrate concentration to 85%.
Transfer combined filtrates and washings to a suitable volumetric flask and dilute to volume with 85% acetone.

2 *Colorimetry*

Add 25 ml or suitable aliquot of extract to 50 ml diethyl ether in a separating funnel.
Mix well.
Add water until the chlorophyll has passed into the ether layer.
Discard water layer.
Wash ether layer 4 or 5 times with water.
Transfer ether phase to a volumetric flask, dilute to volume with ether and mix well.
Add 2 g Na_2SO_4 and allow to stand with occasional shaking until a clear solution is obtained.
Measure the absorbance at 660 and 643 nm in 1 cm cells using ether as a reference.

Calculation

If C = total chlorophyll in ether solution ($mg\,l^{-1}$) = 7.12 × optical density at 660 nm + 16.8 × optical density at 643 nm, then:

Total chlorophyll (%) =

$$\frac{C\,(mg\,l^{-1}) \times \text{ether soln (ml)} \times \text{acetone extr (ml)}}{10^4 \times \text{acetone aliquot (ml)} \times \text{sample wt (g)}}$$

Correct to dry weight where necessary.
Chlorophylls *a* and *b* may be calculated separately as follows:
for chlorophyll *a*:
$C = 9.93 \times A\,(660\text{ nm}) - 0.777 \times A\,(643\text{ nm})$;
for chlorophyll *b*:
$C = 17.6 \times A\,(660) - 2.81 \times A\,(643)$.
(A = absorbance reading)

Notes

1 Initial water washings of ether solution should be swirled and not shaken to prevent formation of emulsions.

2 The volume used and aliquots taken should be such that an absorbance reading of about 0.6 is obtained at 600 nm. This ensures a satisfactory reading at 643 nm.

3 It is essential that absorbance measurements are made on the peaks of the absorption curve, therefore the wavelength calibration of the spectrophotometer should be checked against a suitable emission line from a mercury or hydrogen discharge lamp and if necessary a correction made to the wavelength setting. Alternatively measurements should be taken around the two peaks and the wavelengths giving highest values of absorbance taken to be the peaks.

Simplified procedure

Determine the total chlorophyll in the 85% acetone extract of the first sample in the run by the method given above.

Measure the absorbance of the 85% acetone extract at 660 nm and also of a series of dilutions with 85% acetone of this extract.

Calculate the concentration in the diluted solutions from the total chlorophyll of the undiluted extract.

Construct a curve of absorbance at 660 nm against concentration of chlorophyll ($mg\,l^{-1}$).

Measure the absorbance values of the remaining 85% acetone extracts at 660 nm.

Calculation

If $C = (mg\,l^{-1})$ chlorophyll obtained from the graph then:

Total chlorophyll (%) =

$$\frac{C\,(mg\,l^{-1}) \times \text{acetone extract (ml)}}{10^4 \times \text{sample wt (g)}}$$

Correct to dry weight where necessary.

Carotenoids

The carotenoids are an extensive group of yellow to red pigments widely distributed in plants. They are based on a carbon chain skeleton which has a cyclic structure at one or both ends. Oxygenated derivatives are known as xanthophylls. The best known carotenoid is β-carotene and is used as a standard in the method for total carotenoids given below.

Fresh tissue should be used for extraction. Drying or even exposure to strong light should be avoided since carotenoids are readily oxidized. They are extracted into a fat solvent such as petroleum ether, but the high water content of fresh tissue makes it desirable to dehydrate initially with a polar solvent such as acetone or alcohol (Goodwin, 1965). In the present method, 85% acetone as used for chlorophyll can also serve as the initial extraction for carotenoids.

Fractionation of carotenoids is best carried out by column chromatography and a review of methods is given by Goodwin (1965). Separation on Sephadex gel is described by Hasegawa (1980). Individual pigments may then be estimated on the basis of known absorption characteristics and extinction coefficients at visible and near ultraviolet wavelengths.

Paper and thin layer chromatography are useful for qualitative examination. Reversed phase partition chromatography is often used. Available systems are described by Goodwin (1965), Randerath (1966), Stahl (1970) and Quackanbush and Miller (1972). Schwartz and Van Elbe (1982) reviewed the use of HPLC.

In the method given here chlorophyll, carotenoids and xanthophylls are extracted with 85% acetone as described in the chlorophyll method. Chlorophyll is often removed on a column, but in the present method refluxing with anhydrous barium hydroxide is used. The other pigments are then extracted into ether. The xanthophylls are separated with aqueous methanol leaving the carotenoids in ether to be estimated at 470 nm.

Reagents

1 Standard solutions.

Prepare a working range of standards containing from 0 to 0.20 mg β-carotene in 100 ml petroleum ether.

2 Acetone, 85% v/v.

3 Petroleum ether, 40 to 60 °C.

4 Methanol, 90% v/v (saturated with petroleum ether).

5 Barium hydroxide, anhydrous.

6 Sodium sulphate, anhydrous.

Procedure

Extract with 85% acetone as described for chlorophyll.

Reflux 100 ml for 30 minutes with 2 g $Ba(OH)_2$.

Filter into separating funnel and wash residue with 85% acetone.

Swirl *gently* with 50 ml petroleum ether.

Run off acetone and wash with 10 ml portions of ether adding washings to original ether phase.

Wash ether with 20 ml portions of water and discard washings.

Extract with 30 ml portions of 90% methanol until the extracts are colourless.

Retain the extracts.

Wash the combined methanol extracts with 10 ml petroleum ether and add washings to original ether phase.

Wash ether phase with water and discard washings.

Filter through Na_2SO_4 to dry.

Wash with petroleum ether and dilute to 100 ml.

Measure absorbance of standards and samples at 470 nm using ether as a reference.

Prepare a calibration curve for the standards and use it to obtain mg β-carotene in the samples.

Calculation

If C = mg β-carotene obtained from the graph then:

$$\beta\text{-Carotene } (\%) = \frac{C(\text{mg}) \times \text{acetone extract (ml)}}{10 \times \text{acetone aliquot (ml)} \times \text{sample wt (g)}}$$

Correct to dry weight where necessary.

Note

Other chemicals have been suggested as substitute standards instead of β-carotene. For example, 0.025% $K_2Cr_2O_7$ has been found to be equivalent to 0.158 mg β-carotene per 100 ml petroleum ether.

Sterols

Sterols are extracted from biological material by lipophilic solvents and may therefore be determined on lipid extracts. They occur in the extract either free or esterified with fatty acids. For the determination of total sterols it may be necessary to saponify to break down the esters. This is often carried out directly on biological material. After saponification the mixture is acidified and extracted with petroleum ether and the sterols go with the fatty acids into the organic phase. Methods for extraction, purification and identification are reviewed by Thompson *et al.* (1980).

There are two common methods for the determination of total sterols; precipitation by digitonin and colour formation by the Liebermann–Buchard reaction. Procedures using both these reactions are given by Browning (1967).

Chromatography is the most widely used technique for the fractionation of sterols. A good column method for the separation of free sterols and sterol esters and other lipids is that of Garton and Duncan (1957) which uses silicic acid as adsorbent and ether–petroleum ether mixtures as eluants. Alumina is also used as a column adsorbent and the 1984 edition of Official Methods of the AOAC gives a method for the determination of cholesterol using an alumina column. Paper chromatography has been widely used for the separation of sterols. Adsorption, partition and reversed phase partition chromatography have also been tried.

Although thin-layer chromatography is now less used for most purposes it still has advantages for sterols. Quick separations may be obtained on alumina and silica gel. Silica gel impregnated with silver nitrate has also been used. Accounts of the TLC of steroids are given by Randerath (1966) and Stahl (1970).

Gas-liquid chromatography is also frequently used, and coupled with mass spectrometry provides a very powerful method of analysis. The use of mass spectrometry in natural products work has been reviewed by Games (1979). Stationary phases which have been used include Apiezon L polyesters and especially silicones. Volatile derivatives such as trimethyl silyl ethers and trifluroacetyl esters are often prepared to facilitate separation. Horning and Van den Heuval (1965) give a review of the GLC of steroids. A general discussion of the chromatographic separation of steroids is given by Neher (1967). HPLC is also suitable for steroids and its application is reviewed by Heftmann and Hunter (1979).

Phytic acid

Phytic acid is the hexaphosphate ester of myo-inositol, one of the stereo isomers of hexahydrocylcyclohexane. It is widely distributed in plants particularly in seeds where it also occurs as a calcium plus magnesium salt known as phytin. In soil, phytin is extremely resistant to degradation and contains a significant fraction of the soil organic phosphorus.

The method given below is taken from Kent-Jones and Amos (1967). An acidified extract is used to dissolve any phytin present. Phytic acid is then isolated as ferric phytate which is recovered, digested and estimated as phytate-phosphorus.

Reagents

1 Hydrochloric acid, 0.5 M and 0.17 M.
2 Sodium hydroxide, 0.5 M.
3 Ferric chloride, 0.25% w/v.
4 Digestion reagents as on p. 59.
5 Colorimetric reagents and standards as for phosphorus on p. 136.

Procedure

Extract a suitable weight of ground material for 2 hours with 100 ml 0.5 M HCl and filter.
Take an aliquot, neutralize with 0.5 M NaOH and make slightly acid with 0.17 M HCl.
Dilute to 50 ml.
Take an aliquot in a centrifuge tube, add 4 ml $FeCl_3$ solution, heat 15 minutes at 100 °C, cool, centrifuge and discard supernatant liquid.
Wash with 0.5 M and then 0.17 M acid, centrifuge and discard liquid.
Add 2 ml water to residue and heat a few minutes at 100 °C.
Add 2 ml 0.5 M NaOH, heat 15 minutes, and filter into Kjeldahl flask.
Wash with hot water and retain washings in the flask.
Add 0.5 ml conc H_2SO_4 and boil down to white fumes.
Add HNO_3–$HClO_4$ mixture and digest as on p. 59.
Dilute to 50 ml.
Take 1–5 ml and determine P colorimetrically as given on p. 136.

Calculation

If C = mg P obtained from the graph then:
Phytate-P (%) =

$$\frac{C(\text{mg}) \times \text{ext(ml)} \times \text{nsol(ml)} \times \text{dsol(ml)}}{\text{nal(ml)} \times \text{dal(ml)} \times \text{Pal(ml)}}$$

where dsol is digest solution; nsol is neutral solution; ext is acid extractant; Pal is aliquot for P; dal is aliquot for digestion; nal is aliquot for neutralization.
Correct to dry weight where necessary.

Humus and resistant residues

The methods already described in this section may be applied to soil organic matter if desired, although practical difficulties may arise in many cases. Waksman and Stevens (1930) derived a system of proximate analysis similar to that presented at the beginning of this chapter although it too has its limitations (Page *et al.*, 1982).

There is one fraction in the soil which has no direct counterpart in plant material and special techniques have been developed for its investigation. This is the so-called 'humus' fraction which is an undefined complex of substances containing resistant residues remaining from plant decomposition together with substances synthesized in the soil itself (Dubach and Mehta, 1963; Zeichmann, 1980).

Analysis has always proved difficult because the material isolated from humus seems to bear little resemblance to any structures thought to be present in the soil. Nevertheless a preliminary fractionation based on an alkaline extraction followed by precipitation stages has been widely used (Bremner, 1949b) and an outline is given below. Two of the crude fractions are normally labelled as 'fulvic acid' (alkali soluble) and as 'humic acid' (alkali soluble). Their composition is uncertain but humic acid may be partly polyphenolic in nature. Alternative methods for determining humic matter in soil have been given by Mehlich (1984) and by Blondeau (1986) who use a Sephadex column. Swift (1985) also discusses the fractionation of soil humus substances whilst Steinberg *et al.* (1984) use HPLC in the examination of degradation products. Methods are reviewed by Cheshire *et al.* (1975).

Fractionation of humus

The most widely used extractants are sodium hydroxide (0.1–0.5 M) or sodium carbonate. Prior decalcification of the soil with 0.1 M acid is

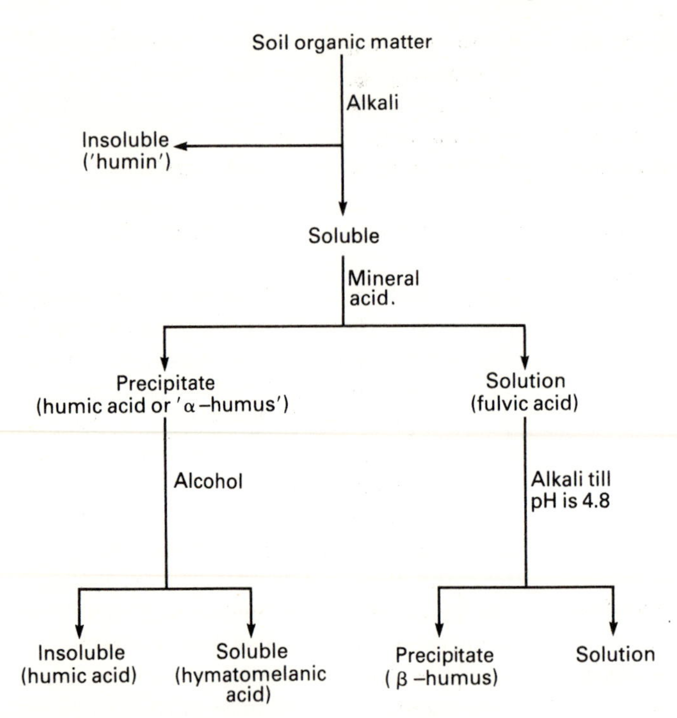

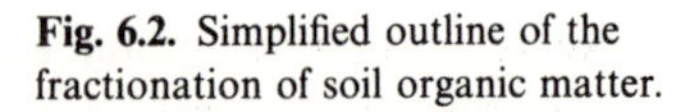
Fig. 6.2. Simplified outline of the fractionation of soil organic matter.

desirable to release any organic substances which have formed insoluble complexes with calcium. Bremner and Lees (1949) used sodium pyrophosphate which is less drastic than the other two extractants. Decalcification is not then necessary because this substance forms precipitates with calcium and other metals and liberates the organic material into solution. A pre-extraction with ether, petroleum ether or ethanol:benzene may be necessary to remove oils, waxes etc.

Many solvents alone or in mixtures have been used to remove undecomposed material (e.g. acetyl bromide, acetic acid and acetic anhydride). A two-phase system is proposed by Lindquist (1984) and pre-treatments are discussed by Russell (1961) and by Page *et al.* (1982).

A schematic diagram for the fractionation of humus is given in Fig. 6.2.

The optimum amount of soil taken for this type of fractionation ranges from 15 to 40 g in inverse proportion to the amount of organic matter in the soil. The processes involved in the various stages are straightforward and a centrifuge is generally used for separating the fractions. Any pH adjustment should be carried out using dilute hydrochloric acid or sodium hydroxide. Further examination of the fractions is possible using chromatographic, electrophoretic and gel-filtration techniques.

Additional information about humus products may be obtained from Kononova (1966), Schnitzer and Khan (1972) and Page *et al.* (1982).

7
Pollutants

This book is primarily concerned with the estimation of chemical nutrients and various organic compounds of metabolic significance. These are generally needed for the healthy development of plants and animals and are not harmful at the levels normally present. In excess, however, even an essential element may have toxic effects. Natural cases of toxicity to plants and animals occur on soils containing large amounts of a particular element and also when some plants produce metabolic by-products which are toxic to grazing fauna. A very important example of dramage to the environment caused by an excess of nutrients occurs when eutrophication follows the large discharge of fertilizer phosphates and nitrates.

In general, however, most harmful effects in the environment arise from exposure to urban, industrial and agricultural pollutants. Although some of these effects are associated with traditional industrial processes, most are of fairly recent origin. They are all of concern, not only because of their potential effect on an individual species, including man, but because they have wider implications for ecology in general and for the conservation of natural habitats.

For these reasons this chapter has been included. The analytical procedures which are presented are divided into four groups of pollutants. The first group includes the heavy metals, together with arsenic and selenium and the second concerns the synthetic, organic pesticides. A third category includes other substances not fitting into the first two groups, whilst atmospheric pollutants are dealt with towards the end of the chapter.

Heavy metals

The heavy metals considered here include antimony (sometimes called a metalloid) cadmium, chromium, lead, mercury and nickel. Many other metals are toxic to plants or animals if absorbed in excessive amounts and these include the essential nutrients copper, cobalt and to some extent manganese and zinc. In particular they are harmful to aquatic organisms. Methods for these elements are given in Chapter 5. Traces of the six listed above are present in most soils but their minerals are relatively rare. They enter the environment mostly in industrial waste products but there is also concern about the amounts in agricultural slurry and sewage sludge. The discharge of lead through vehicle exhausts has also received much publicity recently.

Most toxic heavy metals are present in the environment as cations or inorganic chelates. The exceptions include metals contained in certain pesticide formulations, mostly dithiocarbamate fungicides and the organomercury compounds. Methods for total and organic mercury are included in this section and two non-metals, arsenic, and selenium are also considered.

Other inorganic ions and compounds which may concern the ecologist and conservationist because of their toxic effects are considered later. There are a few other metals not normally present in the environment in significant amounts but which could be a danger to wildlife in particular situations. Some of these are given later in Table 7.1 together with typical concentration ranges to be found in ecological materials and a few references to analytical techniques.

In pollution studies, as in most analytical work, it is important to run replicated blank solutions prepared and treated in exactly the same way as the standard and sample solutions. All reagents used for trace element work in particular should be the purest quality available and where appropriate further purified in the laboratory. Special attention must also be paid to the possibility of external contamination during analysis. Some information about contamination sources are provided in Appendix III and there are also notes in text.

The methods in this section are almost entirely based on colorimetry and atomic absorption, techniques commonplace in most laboratories. However, some of the more expensive instrumental systems now available have distinct advantages for the determination of heavy metals. For example, emission spectrometry (p. 256), which will quantitatively scan the heavy metals, can be used for identifying the cause of pollution. By using inductively coupled argon plasma (ICP) as the emission source it is possible to use most of the sample solutions prepared for other techniques described in this book. Detection limits vary between instruments and elements but are of the order of 0.01 mg l^{-1}. The need for a concentration stage will depend on the element and sample type.

Ion chromatography (p. 280), using post column reaction and a uv/visible detector is useful for trace level water analysis. For instance cadmium, cobalt, copper, iron, lead, manganese, nickel and zinc can be separated in a typical run time of 20 minutes. Possible pollutants such as chromium and cyanide can also be quantified. Detection limits are in the range of 0.5–15 ng on the column.

Anodic stripping voltammetry also has potential for pollution monitoring, although the equipment and methods have not always proved robust. Field equipment now available provides the ability to trace the pollution discharge, or metal deposit rapidly. Further information about this technique is provided on p. 267.

Antimony

Very little attention has been paid to antimony in soils and biomaterials, mainly because it is not recognized as having nutritional significance and the amounts present in most materials are very low. Nevertheless, compounds of antimony are moderately toxic to most organisms. They are present in the waste products from a number of industrial processes, but the greatest discharge is into the atmosphere.

Some of the naturally occurring concentrations are given below.

Soils	$< 5\ \mu g\ g^{-1}$
Plant materials	$0.01–0.2\ \mu g\ g^{-1}$
Animal tissue	$0.005–0.05\ \mu g\ g^{-1}$
Freshwater	$< 0.02\ \mu g\ l^{-1}$

One of the principal difficulties in the determination of antimony is the preparation of a suitable sample solution. Excessive heating or the incautious use of powerful oxidizing agents will result in some loss of antimony. Dry-ashing procedures cannot be used and acid digestion techniques are sometimes suspect. Procedures which have been found suitable for arsenic have been used for antimony on the basis of their similarity but Jacobs (1967) warns against relying on this assumption.

It is recommended that soils be fused with potassium bisulphate as described on p. 27. The ratio of dry ground sample to the fusion salt should be 1:4. This fusion procedure can also be used for plant materials but the proportion of fusion salt must be higher. An alternative method for soils is a sublimation technique using ammonium chloride and is described by Stanton (1966). The sample is finally taken up in hydrochloric acid.

A mixed acid procedure based on that given on p. 59 is normally convenient for plant materials. 2–5 g of sample are heated gently with a mixture of 1 ml sulphuric acid and 10–20 ml of nitric acid. A further 1–2 ml of nitric acid should be added if charring occurs. When most of the organic matter has been destroyed and the solution is only faintly yellow, 1 ml of perchloric acid is added and the heat is increased until dense fumes of sulphuric acid just appear. Under these conditions antimony losses are minimized.

A colorimetric procedure is given below. Other colorimetric methods are described in Christopher and West (1966) (bromopyrogallic acid), Fogg *et al.* (1969) (brilliant green) and by Abu-Hilal and Riley (1981) (crystal violet).

Atomic absorption can be used for antimony which produces a suitable absorption peak at 217.6 nm. The detection limit in solution is not likely to be better than 0.3 $\mu g\ ml^{-1}$, although prior concentration is possible using an APDC/MIBK extraction at pH 3. Few interferences have been reported although copper and lead cause some enhancement and mineral acids are depressants. The hydride procedure, in which volatile hydrides are vaporized in the flame, is now generally preferred for antimony. This procedure is described by Godden and Thomson (1980) and Andreae *et al.* (1981) and in the latter case is used for the fractionation of antimony compounds in waters.

Rhodamine B method

This colorimetric technique is probably the most widely used for the determination of antimony. It is sufficiently sensitive for most purposes, although iron interferes and less than 1 mg should be present when the final colour is developed. A precipitation step must be introduced in which antimony is separated as the sulphate if iron exceeds this level.

Reagents

1 Antimony standards.
Stock solution (1 ml ≡ 100 μg Sb): dissolve 0.2669 g antimony potassium tartrate [$K(SbO)C_4H_4O_6$] in water and dilute to 1 litre.
Working standard (1 ml ≡ 1 μg Sb): dilute the stock solution 100 times.
2 Rhodamine B, 0.2% w/v.
3 Ceric sulphate, 0.05 M in 0.5 M H_2SO_4.
4 Sodium sulphite hydrated, 1% w/v.
5 Hydroxylamine hydrochloride, 1% w/v.
6 Hydrochloric acid, 6 M.
7 Isopropyl ether.

Procedure

Prepare sample solution and blanks as described above, finally transferring to beaker flasks, bringing just to dryness.
Pipette 0–10 ml standard into beaker flasks to give a range of 0–10 μg Sb and evaporate just to dryness.
Add 6 ml 6M HCl to standards and samples and warm to dissolve.
Add 1 ml Na_2SO_3 solution.
Add 3 ml ceric sulphate solution.
Add 6 drops 1% hydroxylamine hydrochloride.
Add 5 ml 6 M HCl.
Transfer to separating funnel with 13 ml water.
Add 10 ml isopropyl ether and shake for 1 minute.
Discard aqueous layer.
Add 5 ml rhodamine B solution and shake for 1 minute.
Measure the absorbance of the ether layer at 545 nm using water as a reference.
Prepare a calibration curve from standard readings and use to calculate the concentration of Sb.
Subtract blank values and correct to dry weight where appropriate.

Arsenic

The concentrations of arsenic normally found in soils and other biomaterials are very low but higher levels are often reported and ascribed to some source of pollution. The compounds of arsenic have long been recognized as poisonous to mammals. Although they are less toxic to vegetation they are still harmful in excess.

Pollution arises from the use of arsenic compounds in herbicides and insecticides (e.g. sheep dips) and also from its presence in certain manufacturing and mining waste products. In particular it is used in the production of glass, paints and alloys and is a by-product in several smelting processes.

The concentration ranges normally found in biomaterials are given below.

Soil	0.5–30 $\mu g\ g^{-1}$
Plant materials	0.1–1.0 $\mu g\ g^{-1}$
Animal tissue	0.1–0.5 $\mu g\ g^{-1}$ (accumulates in certain organisms)
Freshwater	0.2–1.0 $\mu g\ l^{-1}$

Most of the points made about the preparation of sample solutions for antimony also apply to arsenic. The same potassium bisulphate fusion method is suitable for arsenic and the acid digestion method given for antimony, in which care must be taken not to char the sample, can be used for arsenic. No digestion procedure using hydrochloric acid should be used since the chlorides of arsenic are readily volatilized. Dry-ashing is generally unsuitable, although Morrison and George (1969) describe an ashing procedure for arsenic in poultry tissues.

The procedure given here for arsenic is based on the formation of the molybdenum heteropolyblue arsenic complex in solution. The other method which is commonly used is derived from the classical Gutzeit procedure in which arsenic reacts with mercuric chloride impregnated on filter paper strips, with a resultant colour change of yellow through brown to black. Details are given by the Society for Analytical Chemistry (1968). Other colorimetric procedures are described by Powers *et al.* (1959), Lenstra and de Wolf (1965) and Kaneko (1982). Electroanalytical techniques, although little used, are available (Holak, 1980 and Bodewig *et al.*, 1982).

Arsenic can be determined by atomic absorption by first preparing the volatile hydride (AsH_3). A hydride method for arsenic in acid digests of soils is described by Kokot (1976). Van der Veen *et al.* (1985) look at digestion procedures for hydride generation.

Molybdenum blue method

The method is very similar to those used for phosphorus and silicon. To avoid interference from these two elements arsenic is separated by reduction and distillation and is then collected in an iodine–potassium iodide solution. Reay (1974) also describes a molybdenum blue procedure suitable for arsenic in plants, sediments and water, and discusses digestions, interferences, accuracy and precision.

Reagents

1 Arsenic standards.
Stock solution (1 ml ≡ 100 μg As): dissolve 0.416 g of sodium arsenate ($Na_2HAsO_4 . 7H_2O$) in water, add 1 ml HCl and dilute to 1 litre.
Working standard (1 ml ≡ 1 μg As): dilute the stock solution 100 times.
Include 1 ml conc HCl. Prepare fresh daily.
2 Ammonium molybdate–hydrazine sulphate reagent.
Dissolve 1 g ammonium molybdate in 90 ml 3 M H_2SO_4. Prepare separately 0.15% (w/v) hydrazine sulphate. Mix 10 ml of each solution and dilute to 100 ml.
3 Potassium iodide, 15% w/v.
4 Iodine–potassium iodide solution.
Dissolve 0.25 g iodine and 0.4 g potassium iodide in water and dilute to 100 ml.
5 Stannous chloride reagent, 40% (w/v) in conc HCl.
6 Sodium carbonate, 4.2% w/v.
Dissolve 4.2 g of anhydrous salt in water and dilute to 100 ml.
7 Zinc granules.
8 Hydrochloric acid, conc.
9 Lead acetate, saturated solution.

Procedure

Assemble apparatus as illustrated in Fig. 7.1.
Prepare sample solution as indicated above and wash into 50 ml conical flask.

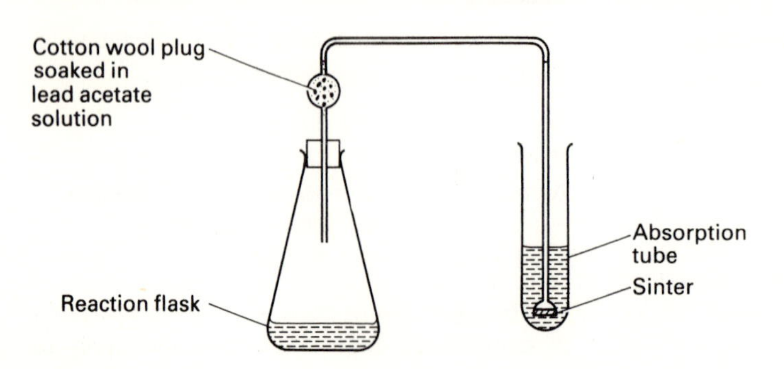

Fig. 7.1. Apparatus suitable for the separation of arsenic as arsine.

Pipette 0 to 10 ml stock solution into 50 ml conical flasks to give a range of 0 to 10 μg As.
Add 5 ml HCl.
Add 2 ml 15% KI.
Add 0.5 ml stannous chloride reagent.
Stand for 15–30 min for reduction to arsenic (III).
Prepare absorption test tube by adding 1 ml I_2–KI solution and 0.2 ml 4.2% Na_2CO_3. Connect tube to conical flask.
Add 2 g Zn granules to flask.
Connect to tube immediately and allow gases to bubble through for 30 minutes.
Immerse test tube in crushed ice.
Add 5 ml ammonium molybdate–hydrazine reagent.
Add drop(s) 0.5% $NaHSO_3$ until iodine colour disappears.
Heat for 15 minutes at 95–100 °C.
Cool and dilute to volume in 25 ml volumetric flask.
Measure absorbance at 840 nm, using water as a reference.
Prepare a calibration curve from standard readings and use to calculate the concentration of As.
Correct for blank values and to dry weight where appropriate.

Cadmium

Measurable amounts of cadmium occur in many soils and plant materials and some attention has been paid to its concentration in these materials for geochemical prospecting, in view of its association with zinc and other metals. There is much concern about the levels present in the environment, because it is a cumulative poison to mammals. It becomes concentrated in some organs where it can exceed 100 $\mu g\,g^{-1}$. It enters the environment as a waste product, especially from metal refining and electroplating works. It is also used in the chemical and paint industries. It is often discharged as an effluent into drainage courses or reaches the atmosphere through stack emissions.

Some naturally occurring concentrations are given below.

Soils	0.03–0.3 $\mu g\,g^{-1}$
Plant materials	0.01–0.3 $\mu g\,g^{-1}$
Animal tissue	0.05–0.5 $\mu g\,g^{-1}$
Freshwater	1–10 $\mu g\,l^{-1}$

Soil sample solutions may be prepared as in Chapter 2 using the sodium carbonate fusion, or the hydrofluoric–perchloric digestion. It may be necessary to take relatively large weights of sample and amounts of reagents.

The mixed acid digestion (p. 59) or nitric acid alone is suitable for biomaterials. If the former, a sample weighing from 1 to 5 g is oxidized with sulphuric and nitric acid mixture and brought to the white fume stage. The oxidation is completed by the repeated addition of small quantities of nitric–perchloric acid mixture. A dry-ashing procedure may also be used, although there is a danger of low recoveries if the combustion temperature is too high. In the case of water samples a preliminary concentration stage will be required. The formation of a metal complex followed by extraction is generally preferred. This topic is dealt with on p. 73.

Atomic absorption is the best analytical technique for this element and a procedure is outlined below. Electrothermal atomization can be used for materials very low in cadmium.

If the cadmium concentration is sufficiently high, a direct polarographic method with 10% hydrochloric acid of 0.1 M acetate solution as the support electrolyte can be used ($E_{\frac{1}{2}} = -0.65$ to -0.70). For low concentrations an initial extraction stage may be needed.

A colorimetric method based on the formation of cadmium dithizonate is available. This is fully described by Sandell (1959) and involves the measurement of the colour intensity after successive extractions from basic tartrate and cyanide solutions to remove other complexing ions. Holtz (1983) developed an automated method after forming the sulfazen complex.

Atomic absorption method

This technique is sensitive and relatively free from interferences. The principles are described in Chapter 8. The method is sensitive to about

0.02 ppm at 1% absorption. However, lower concentrations can be measured using the APDC–MIBK extraction procedure (p. 107) or electrothermal atomization.

Reagents

1 Cadmium standards.

Stock solution (100 ppm Cd): dissolve 0.2282 g $3CdSO_4 . 8H_2O$ in water and make up to 1 litre.

Working standards: prepare a range of 0–2 ppm Cd and include acid as appropriate to match the sample solutions.

Procedure

Prepare the sample solution as indicated above and dilute to volume.

Select the 229 nm wavelength and adjust acetylene and air flows and other settings as recommended for the instrument employed.

Allow the hollow cathode lamp adequate time to stabilize.

Aspirate standards and samples as given for methods in Chapter 5.

Prepare a calibration curve from the standard readings and use to calculate the concentration of Cd.

Subtract blank values and correct to dry weight where appropriate.

Chromium

Chromium is widely distributed in soils and vegetation although the concentrations are generally very low. The levels in some basic igneous soils such as serpentine are relatively high. It is toxic to animals, particularly in the hexavalent state, although less so to plants. As with other heavy metals, pollution problems can arise from the discharge of waste products from electro-plating, manufacture of alloys and other industrial processes.

The concentrations normally encountered are given below.

Soils	10–200 $\mu g\,g^{-1}$
Plant materials	0.05–0.5 $\mu g\,g^{-1}$
Animal tissue	0.01–0.3 $\mu g\,g^{-1}$
Freshwater	0.1–0.5 $\mu g\,l^{-1}$

Many of the methods described in Chapters 2, 3 and 4 for the preparation of the sample solution can be used for chromium. For soils the sodium carbonate fusion procedure is suitable and should include a little potassium nitrate. For organic samples the mixed acid digestion or dry-ashing procedures may be used. Larger sample weights may be required to offset the low concentrations of chromium generally encountered. Waters may have to be concentrated.

A colorimetric procedure (diphenylcarbazide) is described here. Atomic absorption is also suitable for chromium and the procedure is discussed below. Further information on the use of this technique for chromium is given by Feldman and Purdy (1965), Premi and Cornfield (1968) and Van Loon (1985).

Diphenylcarbazide method

This reagent provides a sensitive method for chromium and should be used for estimating low levels in soils and biomaterials. High iron levels interfere, particularly in hydrochloric acid, and so the fusate should be taken up in sulphuric acid. The procedure is adapted from that first described by Saltzman (1952).

Reagents

1 Chromium standards.

Stock solution (1 ml ≡ 1 mg Cr): dissolve 2.828 g $K_2Cr_2O_7$ in water and dilute to 1 litre.

Working standard (1 ml ≡ 10 μg Cr): dilute the stock solution 100 times.

2 Diphenylcarbazide, 0.2% in acetone. Prepare fresh.

Procedure

Prepare the sample solution as indicated above.

Filter into a 50 ml volumetric flask.

Adjust acidity to 0.1 M with dilute H_2SO_4.

Pipette 0–5 ml standard into a volumetric flask to give a range of 0–50 μg Cr.

Add 2 ml diphenylcarbazide reagent and adjust acidity as for samples.

Dilute to volume and mix.

Measure absorbance at 540 nm without delay, using water as a reference.

Prepare a calibration curve from standard readings and use to calculate the concentration of Cr.
Subtract blank values and correct to dry weight where appropriate.

Notes
1 The residual acid following fusion or dry-ashing should be sulphuric and not hydrochloric because the yellow colour of ferric chloride interferes in the diphenylcarbazide method given below.
2 The addition of a few drops of phosphoric acid will control iron interference.
3 If iron is removed as a gelatinous precipitate at the neutralization stage, chromium may be lost by adsorption on the precipitate.
4 If the digestion procedures described above are followed, some divalent chromium may be present. The acid solution should then be boiled with 0.1 M potassium permanganate to ensure all chromium is in the form of chromium (VI). 5 percent sodium azide solution, added dropwise, will destroy any residual permanganate colour.

Atomic absorption method

Atomic absorption can be used for chromium, although only the more sensitive instruments will directly measure the low levels often encountered in biomaterials.

The signal is depressed by a number of elements, notably calcium, silicon, phosphorus and iron. The inclusion of similar amounts of calcium and silicon in the standards will compensate for those two elements and also phosphorus. To a certain extent this is possible for iron but if the amount of iron exceeds that of chromium by a thousand fold it must be separated by complexing and extracting. Acids also cause signal depression but this can be controlled by matching the levels in standards and sample. Iron, cobalt and nickel interfere in the presence of perchloric acid but this can be overcome by using an oxidizing, or preferably a nitrous oxide–acetylene flame. Sensitivity is variable for different oxidation states when using an air–acetylene flame and there can be marked non-linear behaviour. It is important to ensure that the chromium in the standards and sample is in the same oxidation state.

The sensitivity is about 0.02 ppm at 1% absorption but this can be improved by using an organic carrier solvent. Methyl isobutyl ketone (MIBK) is the most suitable and can be used to extract chromium (VI) directly from an acid solution. Alternatively, diphenylthiocarbazone can be used to complex chromium which is then extracted into the MIBK phase in the presence of EDTA which will remove some interfering elements.

Electrothermal atomization
If the concentration of chromium in the residual sample solution is below the limit of detection of flame atomic absorption then electrothermal atomization will generally be sufficiently sensitive. Details of this technique and operating conditions suitable for most elements of interest to the ecologist are given in Chapter 8.

Lead

Debris from old mine workings and some industrial processes result in severe lead contamination in restricted localities, as shown by the limited array of plant species. There is concern about lead distribution in forms that could enter the human food chain, e.g. via water supplies and vehicle exhaust fumes. There is little evidence that these seriously affect wildlife, and on roadside verges de-icing salt is a more likely cause of plant damage.

Some naturally occurring concentrations are given below.

Soil	2–20 $\mu g\ g^{-1}$
Plant materials	0.05–3 $\mu g\ g^{-1}$
Animal tissue	0.1–3 $\mu g\ g^{-1}$
Freshwater	1–20 $\mu g\ l^{-1}$

To prepare sample solutions suitable for instrumental analysis many of the fusion and acid digestion procedures can be used. Soils can be fused with sodium carbonate (see Chapter 2) using a 0.5 g sample with 3 g sodium carbonate and including 0.3 g potassium nitrate. The hydrofluoric digestion procedure may also be used, though the amounts of reagents taken will have to be increased. Biomaterials should be digested as given in Chapter 3 taking 2–5 g sample with 1 ml sulphuric acid, 15 ml or more of nitric acid and 2 ml of perchloric acid. It is advisable to add perchloric acid at a late stage. Dry-ashing is not recommended, owing to the risk of loss by volatilization, although Webber (1972) obtained suitable results provided a temperature of 430 °C was not exceeded. Waters may require some concentration procedure based on evaporation, ion-exchange or preferably solvent extraction.

A colorimetric method and an atomic absorption method are given here. Electroanalytical techniques may also be used. The absence of interfering ions in most environmental samples is an advantage of polarography. Dilute hydrochloric acid or acetate solution at pH 4.5 can be used as an electrolyte. Anodic stripping voltammetry has selectivity and sensitivity which makes it even more suitable, especially for the examination of organic lead compounds. A comparative study of different methods suitable for the determination of lead in soils has been carried out by Harrison and Laxen (1977).

Particular care is needed in lead analysis to avoid contamination. All glassware should be treated with hydrochloric acid before use.

Dithizone method

Dithizone is the classical reagent used for colorimetric estimations. The method is laborious but it is sensitive, gives good results and is suitable for anyone without access to expensive instrumentation. The method given below is based on that of the Analytical Methods Committee (1965). Other colorimetric methods have been reported but most are also lengthy. Examples include the use of diethyldithiocarbamate (Keil, 1967) and 4-(2-pyridylazo)resorcinol (Dagnall *et al.*, 1965). Agrawal and Patke (1981) describe a rapid spectrophotometric method for lead based on the formation of a yellow complex with *N*-phenylbenzohydroxamic acid at pH 9.5.

Reagents

1 Lead standards.
Stock solution (1 ml ≡ 100 μg Pb): dissolve 0.1599 g $Pb(NO_3)_2$ in 20 ml 1% v/v HNO_3 and dilute to 1 litre with water.
Working standards (1 ml ≡ 10 μg Pb): prepare by dilution of the stock solution.

2 Dithizone solutions.
Stock solution, 0.1% w/v in chloroform.
Working solution. Prepare fresh by mixing 6 volumes of stock solution with 9 volumes of water and 1 volume of 5 M NH_4OH.
Reject lower layer.

3 Ammonium citrate, 25% w/v.

4 Sodium hexametaphosphate, 10% w/v.
If necessary, remove lead traces by extraction with dithizone working solution at pH 9. Remove dithizone traces with chloroform (pH 3–4) and readjust to pH 9–10.

5 Potassium cyanide, 10% w/v.

6 Nitric acid, 1% v/v.

7 Ammonia solution, SG = 0.880.

8 Chloroform.

9 Ammonium sulphite–cyanide solution.
Mix 340 ml ammonia solution, 75 ml 2% w/v $Na_2SO_3.7H_2O$, 30 ml 10% KCN and 605 ml water.

10 Hydroxylamine hydrochloride.

Procedure

Prepare sample solutions as indicated above but omit sulphuric acid from mixed acid digestion if there is any risk of calcium sulphate forming and adsorbing lead. Wash sample into a conical flask.

1 *First extraction*

Add 5 ml ammonium citrate solution.
Add 10 ml sodium hexametaphosphate solution.
Add ammonia solution until pH 9.0–9.5.
Cool and add 1 ml 10% KCN.
Add 1 ml hydroxylamine hydrochloride.
Transfer to separating funnel containing 10 ml chloroform.
Add 0.5 ml dithizone working solution and shake for 1 minute.
Add further dithizone and repeat until lower layer is purple or blue.

2 *Second extraction*

Transfer chloroform layer to another separating funnel.

Shake first funnel with 3 ml chloroform and 0.2 ml dithizone working solution.

Add chloroform layer to second funnel.

Repeat until chloroform washing is green.

Add 10 ml 1% v/v HNO_3 and shake for 1 minute.

Reject chloroform layer.

3 *Third extraction and colour development*

Pipette up to 5 ml standard into separating funnels to give a range of 0–40 μg Pb.

Add 10 ml 1% v/v HNO_3 and treat as for samples.

Add 30 ml ammonium sulphite–cyanide solution.

Add exactly 10 ml chloroform.

Add 0.5 ml dithizone working solution and shake for 1 minute.

Measure absorbance of chloroform layer at 520 nm.

Prepare calibration curve from standard readings and use to calculate the concentration of Pb.

Subtract blank values and correct to dry weight where appropriate.

Atomic absorption method

Flame atomic absorption is straightforward for lead as interference problems are not serious, although the effects of relatively high levels of elements present in the sample solutions should be checked. The sensitivity is only about 0.1 ppm at 1% absorption. Trace amounts can be estimated if the APDC–MIBK extraction procedure (p. 107) is used, but if available electrothermal atomization (see below) is preferable. Sensitivity may also be increased by using the 217 nm line but the noise level is then high due to light scattering effects. Hydride generation atomic absorption has also been tried and good sensitivity was reported (Jin and Taga, 1980).

Reagents

Lead solutions.

Stock solution (100 ppm Pb): prepare as given above.

Working standards: prepare a range from 0 to 10 ppm Pb by dilution of the stock solution and include acid as appropriate to match the sample conditions.

Procedure

Prepare sample solutions as indicated above and dilute to volume.

Select 283.3 nm wavelength with a narrow slit-width and adjust air and acetylene flows and other settings as recommended for the instrument employed.

Always allow the hollow cathode lamp adequate time to stabilize.

Aspirate standards and samples as given for the methods in Chapter 5.

Prepare a calibration curve from standards and use to calculate the concentration of Pb.

Subtract blank values and correct to dry weight where necessary.

Electrothermal atomization

If the concentration of lead in the sample solution is below the limit of detection using flame atomic absorption then electrothermal atomization will generally give a reading. Details of this technique and operating conditions for most elements of interest to the ecologist are given in Chapter 8. Chloride ions in particular cause problems, and so it is important to avoid hydrochloric acid and to use methods of standard addition.

Mercury

A number of industrial processes utilize mercury and its compounds, and mercurial waste often finds its way into the environment. Industries concerned include: chlor-alkali plants, pulp mills (although less than formerly), plastic and drug manufacturers. Mercury compounds are also used for timber preservation. Apart from these industrial uses, mercury compounds have been widely distributed through their use as fungicides, especially for seed dressings. The formulations for some pesticides include organomercury compounds together with chlorinated hydrocarbons. There have been many cases recorded where mercury compounds have resulted in serious environmental pollution, some with tragic consequences.

Naturally occurring concentrations are given below.

Soils	0.1–1 μg g^{-1}
Plant materials	0.005–0.1 μg g^{-1}
Animal tissue	0.03–0.3 μg g^{-1}
Freshwater	0.3–3 μg l^{-1}

Samples for mercury determination should be analysed as soon as possible after collection to minimize loss due to volatilization and also to prevent any changes in the relative amounts of organic mercurial compounds which may be present. Storage of fresh material at low temperatures is advisable. Another reason for mercury loss, particularly affecting water samples, was reported by Coyne and Collins (1972). They found that mercury compounds were adsorbed on to the walls of collection and storage vessels and they recommended that samples be analysed within 24 hours. Iskander *et al.* (1972) even reported losses on drying soil at 60 °C, whilst LaFleur (1973) dealt with the effects of freeze drying.

The possibility of mercury loss through volatilization, referred to above, prevents the use of fusion or dry-ashing techniques to bring mercury into solution. Acid digestion methods are therefore normally used for preparing the test solution although here also, some acid treatments result in low recoveries of mercury. The Analytical Methods Committee (1965) considered this problem and proposed a modified oxidation procedure using nitric and sulphuric acids followed by reducing and complexing stages. The method described below uses these acids and is followed by further oxidation with potassium permanganate.

Once in solution mercury can be determined colorimetrically or, probably more frequently today, by cold-vapour atomic absorption. The well-established dithizone colorimetric method, although lengthy, is straightforward and is also given below.

Solution preparation

Transfer about 1 g of soil or 2–5 g of plant materials to a digestion flask.

Add 1 ml conc H_2SO_4, 10–15 ml conc HNO_3 and digest slowly.

When most of the organic matter has been destroyed, filter and carefully add approx. 0.5 g quantities of $KMnO_4$.

Continue until a permanent precipitate of MnO_2 remains.

Cool and dilute to approx. 10 ml.

Add 50% $NH_2OH.HCl$ solution until the pink colour of $KMnO_4$ is destroyed.

Neutralize with M NaOH using bromophenol blue as indicator.

Dilute to a suitable volume.

Dithizone method

Reagents

1 Mercury standards.

Stock solution (1 ml ≡ 100 μg Hg): dissolve 0.135 g $HgCl_2$ in 1 litre.

Working standard (1 ml ≡ 1 μg Hg): dilute the stock solution 100 times. Prepare fresh before use.

2 Hydroxylamine hydrochloride, 20% w/v.

Prepare a 50% w/v $NH_2OH.HCl$ solution and shake with dithizone until no colour change takes place. Wash with $CHCl_3$, separate and filter. Dilute to give 20% strength.

3 Dithizone, 0.6% w/v in $CHCl_3$.

4 Sodium thiosulphate, 1.5% w/v aqueous.

Purify by extracting with dithizone as described under **2**.

5 Acetic acid, 30% v/v.

6 Sodium hypochlorite, 5% available Cl.

Prepare by diluting commercial NaOCl solution.

7 Hydrochloric acid, 0.1 M.

Procedure

Three separating funnels are required for each sample. These contain:

A 5 ml dithizone;

B 20 ml 0.1 M HCl + 5 ml 20% NH_2OH. HCl;

C 50 ml 0.1 M HCl.

Transfer sample to funnel A and shake.

Transfer the chloroform layer into funnel B.

Repeat twice more, then shake B.

Transfer chloroform layer to funnel C.

Wash B with about 3 ml chloroform and add to C.

Add 2 ml $Na_2S_2O_3$ solution to C and shake, then reject the chloroform layer.
Wash C with a further 3 ml chloroform and again reject.
Add 3.5 ml NaOCl solution to C and shake.
Release pressure by removal of stopper. Do not invert and open tap.
Add 5 ml 20% $NH_2OH.HCl$ to C and shake.
Carefully reject any droplets of blue organic layer.
Add 3 ml chloroform to C, shake and then reject.
Add 3 ml 30%HOAc to C, swirl to mix.
Add exactly 10 ml dithizone reagent to C and shake.
Run off first 2 ml of chloroform layer.
Filter remaining chloroform layer into cuvette through filter plug.
Measure absorbance at 490 nm.

Preparation of calibration curve
Shake 50 ml 0.1 M HCl, 5 ml 20%NH_2OH. HCl and up to 20 ml standard Hg solution with 0.5 ml chloroform.
Reject chloroform layer.
Add 3 ml 30% HOAc and shake.
Carefully reject any droplets of chloroform.
Add 10 ml dithizone reagent and shake.
Measure the absorbance as above.
Prepare calibration curve from standard readings and use to calculate the concentration of Hg.
Subtract blank values and correct to dry weight where appropriate.

Atomic absorption method

Direct application of atomic absorption for estimating trace amounts of mercury is limited because the sensitivity is only about 0.2 ppm and most of the conventional flames tend to give a somewhat noisy signal at the most intense line of 185 nm. In practice the 253.7 line is always used. The operational details are basically as given for other metals and a 0–30 ppm range of standards can be employed.

Mercury in environmental materials is best determined using the flameless cold vapour technique (Hatch and Ott, 1968). Mercury compounds are reduced to the metallic state which is then vapourized in a stream of air and swept through an absorption cell placed in the hollow cathode lamp beam. Very low concentrations of mercury can be measured by first concentrating as gold amalgam prior to cold vapour atomic absorption (Yamamoto *et al.*, 1983).

Two variants of the cold-vapour procedure are in common use. These are the open-ended and recirculating techniques. In the open-ended system the mercury vapour escapes to the atmosphere after passing through the cell giving rise to a sharp absorption peak. In the recirculating technique, the mercury vapour is continuously circulated through the test solution and cell until a steady absorption signal is obtained. In the former, air-flow and solution volume are critical and must be carefully optimized and then reproduced throughout the run of standards and samples.

Most manufacturers of atomic absorption spectrophotometers offer a cold vapour kit as an accessory for their instrument. Comprehensive operating instructions are generally supplied with these attachments. Typical conditions are:

Wavelength	253.7 nm
Reducing solution	10–20% w/v stannous chloride in 20% v/v hydrochloric acid.
Air flow	2–3 l min^{-1}

The digestion procedure given in the previous section is suitable for solution preparation for plants, animal material and soils. Water samples can often be analysed as received. The sensitivity of the technique allows levels in the solution aliquot of 1 mg or below to be determined.

Organo-mercurial compounds

Most of the organo-mercurial compounds which have been distributed in the environment are alkyl- and aryl-mercuric chlorides, although other anions sometimes replace chlorides. It has been shown by Tonamura *et al.* (1968) and others that mercury resistant bacteria are capable of breaking down these compounds to simpler forms including elemental mercury. Another process which has been suggested involves the transfer of methyl groups from other compounds to link up with mercuric ions to form the cumulative poison, methyl-mercury. It is this form which has been shown to accumulate in the marine, freshwater and terrestrial predator chains.

Although aryl-mercury (generally phenyl) is less used in fungicides than alkyl mercurial compounds, it is probably more widely applied in industry, notably as the acetate and chloride and is therefore likely to enter the environment as a pollutant. Apart from its estimation by chromatographic procedures, phenyl-mercury can be determined alone on the basis of slightly different reactions of the organic mercurial compounds with dithizone. Miller *et al.* (1958) exploited this difference to give a procedure for the direct determination of phenyl-mercury in animal tissue. Goulden *et al.* (1981) describe an automated procedure for the separation of inorganic, alkyl- and aryl- forms of mercury.

Extraction

The first stage in the examination of organo-mercurial compounds is to homogenize the fresh sample, preferably in an enclosed chamber. The sample can then be extracted with a solution of 4 M hydrochloric acid and the extract separated using a centrifuge. The mercury compounds can then be extracted into a chloroform or benzene phase.

Examination by TLC

Organo-mercurial compounds separate well on thin-layer chromatograms. A concentration stage may be required before applying the spots to the plates. Pre-coated plates may be used, or otherwise coat with 250 μ layers of silica gel and activate at 105 °C. A large number of developing solvent mixtures are available, including petroleum ether:acetone or cyclohexane:acetone (9:1). The spots can be detected using dithizone or mercuric nitrate sprays.

The organo-mercurial compounds can be converted to dithizonates during the extraction stage. This is done by shaking the 4 M hydrochloric acid extract (above) with 0.1% dithizone in chloroform. The adsorbant and developing solvents mentioned above are also suitable for the mercurial dithizonates. These appear after development as yellow spots and will have approximately the following R_f values:

Hg^{2+} dithizonates	0.1–0.2
Phenyl dithizonates	0.25–0.3
Methyl/ethyl dithizonates	0.3–0.35

Semi-quantitative estimation of the mercurial compounds is possible by scraping off the spots, taking up in chloroform and measuring the absorption at 485 nm. Alternatively, flameless atomic absorption can be used. The mercurial dithizonates are heated to generate mercury vapour which is carried through the light beam in a stream of air.

Further details on the use of thin-layer chromatography for separating organo-mercurial compounds can be obtained from Tatton and Wagstaffe (1969), Johnson and Vickers (1970), Osawa *et al.* (1980, 1981) and Margler and Mah (1981).

GLC and other techniques

Gas-liquid chromatography is more reliable than TLC for the quantitative analysis of organo-mercurial compounds. Good resolutions can be achieved, although some development work will always be needed to achieve optimum operating conditions. A suitable trial combination is given below.

Support material	Chromasorb W 60/80 mesh
Liquid phase	10% polyethyleneglycol succinate
Carrier gas	Nitrogen, 60 ml min^{-1}
Oven temperature	170 °C
Column	Steel or glass
Detector	Electron capture

The sample extract can be prepared as for TLC.

One disadvantage of GLC is that it is particularly subject to interference by organo-sulphur compounds. These procedures and further information on the application of GLC for the estimation of organo-mercurial compounds can be obtained from Westoo (1967) and Uthe (1971).

Nickel

Although levels of this element are generally less than 100 $\mu g\, g^{-1}$ in soils, it can be exceptionally high in some cases, especially if ultra-basic rocks are present. Plants appear to be more sensitive to nickel toxicity than animals, although both can be affected by gaseous discharge and solid waste from metallurgical industries. Any emission as nickel carbonyl is particularly hazardous to mammals. It is used in steel plating, and various alloys, asbestos manufacture and in some fuels.

The concentration ranges normally encountered are given below.

Soil	5–500 $\mu g\, g^{-1}$	May be higher locally
Plant material	0.5–5 $\mu g\, g^{-1}$	
Animal tissue	0.1–5 $\mu g\, g^{-1}$	
Freshwater	5–100 $\mu g\, l^{-1}$	

To bring soils into solution the fusion method using potassium bisulphate as given in Chapter 2 is basically suitable although larger sample weights should be taken. From 0.5 to 1 g of sample with 3–4 g bisulphate is generally adequate. Biomaterial may be digested using the mixed acid technique given in Chapter 3, taking 2–5 g sample with 1 ml sulphuric acid, 15 ml (or more) of nitric and 2 ml perchloric acid. The latter may be added at a later stage although there is less risk of nickel loss during this type of digestion than is the case with some other elements. Dry-ashing is also suitable for nickel in vegetation and animal materials but larger sample weights should be taken.

Three quantitative methods for the determination of nickel are mentioned here. The first is a colorimetric method using dimethylglyoxime, the second is an atomic absorption method, whilst the third uses a polarograph.

Dimethylglyoxime method

Nickel forms coloured complexes with a number of reagents but dimethylglyoxime is the most widely used. A number of metals, including copper, cobalt, iron and sometimes manganese, also react under the same conditions. Its use is described below and further information about this reagent can be obtained from Sandell (1959) and Classen and Bastings (1966).

Reagents

1 Nickel standards.
Stock solution (1 ml ≡ 0.1 mg Ni): dissolve 0.6728 g $(NH_4)_2SO_4.NiSO_4.6H_2O$ in water and dilute to 1 litre.
Working standard (1 ml ≡ 1 μg Ni): dilute the stock solution 100 times.
2 Dimethylglyoxime, 1% w/v in alcohol.
3 Hydroxylamine hydrochloride, 10% w/v.
4 Sodium citrate, 20% w/v.
5 Ammonia solution, SG = 0.880.
6 Ammonium hydroxide (1 + 50).
7 Hydrochloric acid (1 + 30).
8 Chloroform.
9 Sodium tartrate, 20% w/v.
10 Potassium persulphate, 4% w/v.
11 Sodium hydroxide, 5 M.

Procedure

Prepare the sample solution as indicated above.
Transfer the solution to a separating funnel.

1 ***Separation stage***

Add 2 ml hydroxylamine hydrochloride solution.
Add 5 ml sodium citrate solution.
Add NH_3 solution until just pink to phenolphthalein and then add a further 4 drops.
Add 2 ml dimethylglyoxime reagent and dilute to about 60 ml.
Extract 3 times with 5 ml chloroform (shake for 30 seconds) and combine extracts.
Shake combined extracts with 5 ml (1 + 50) NH_4OH and discard aqueous layer.
Shake organic layer with first 10 ml and then 5 ml (1 + 3) HCl (shake for 1 minute).
Collect acid extracts in 50 ml volumetric flask.

2 ***Colorimetric stage***

Pipette 0–30 ml standard into 50 ml volumetric flasks to give a range of 0–30 μg Ni and include 15 ml (1 + 30) HCl.
Add 2 ml sodium tartrate solution.
Add 10 ml potassium persulphate solution.
Add 0.6 ml dimethylglyoxime reagent.

Add 2.5 ml 5 M NaOH.
Dilute to volume, mix and leave for 30 minutes.
Measure to absorbance at 465 nm using water as a reference.
Prepare a calibration curve from standard readings and use to calculate the concentration of Ni.
Subtract blank values and correct to dry weight where appropriate.

Atomic absorption method

Atomic absorption provides the most straightforward method for nickel. The procedure is relatively free from interference, but it is only moderately sensitive, with a detection limit of about 0.2 ppm, so fairly concentrated solutions will be required. Solutions very low in nickel are best extracted with a chelating reagent such as ammonium tetramethylenedithiocarbamate (APDC) or diethyldiammonium dithiocarbamate (DDDC).

For very low concentrations, electrothermal atomization should be used.

Polarographic method

Probably the most convenient electrolyte to use for the polarographic determination of nickel is the ammonium hydroxide–ammonium chloride buffered solution because it can also be used for copper, zinc and cobalt. The half wave potential for nickel III is -1.1 V. Sirois (1962) describes a suitable method for plant materials. An initial dithizone extraction is normally required.

Selenium

Selenium is of interest both because of its nutritional value and because of its toxicity, and there is some justification for including it here.

It is now generally accepted that selenium is an essential micro-element for plants (Shrift, 1964). Even in excess it is not highly toxic to vegetation and some 'indicator' plants will accumulate selenium.

Grazing stock are subject to 'alkali disease' and 'blind staggers', which are disorders associated with excess selenium levels although alkaloids are also considered to be a contributory factor. Some conditions, notably 'white muscle' disease in lambs can be prevented by adding small amounts of selenium to the diet of the pregnant ewes.

The possibility of industrial pollution is sometimes referred to but there is little documented evidence. Selenium is often associated with sulphur compounds and it is widely distributed by combustion fumes especially from power stations. West (1971) evaluates selenium as an air pollutant. It is also present in paints and in waste products from a number of industries, notably metal refining, glass, rubber and pigments and is used in some insecticides.

Naturally occurring concentrations are given below.

Soil	0.15–1.5 $\mu g\,g^{-1}$ exceptionally to 100 $\mu g\,g^{-1}$
Plant materials	0.05–0.5 $\mu g\,g^{-1}$
Animal tissue	0.1–2 $\mu g\,g^{-1}$
Freshwater	0.5–10 $\mu g\,l^{-1}$

Solutions suitable for most methods can be prepared by digesting soils and plant materials with a 3+1 mixture of nitric and perchloric acids. Up to 1 g of soil and 2–5 g plant material may be digested with 10–20 ml of the acid mixture. The digest solution is taken down, almost to dryness, and dense white fumes of perchloric acid are evolved in the process. 5 ml of 25% HCl are added which is also evaporated to near dryness before diluting with water and filtering. The oxidation stage is needed, even for waters, in order to convert inorganic selenium compounds to the VI valency state. This form must then be reduced to selenium (IV) which is done by heating with HCl.

According to Shendrikar and West (1975) selenium can be lost when water samples are stored. Addition of H_2SO_4 to give a pH of 1.5 will prevent the loss. About 1 litre of water is generally adequate for analysis.

A number of colorimetric procedures have been proposed for selenium. The more successful are based on the use of 3,3-diaminobenzidine (Hoste and Gillis, 1955) or related compounds. The method was described in the first edition of this book (Allen *et al.*, 1974). Thorburn Burns discusses the principles of the reaction in some detail and gives procedures in West and Nürnberg (1988). The pH needs to be carefully controlled in this method and both oxidizing and reducing substances must be absent. Iron and copper interferences are controlled by adding EDTA but very high levels of iron have to be removed by precipitation. The absorbance of the reaction product, monopiazselenol, is measured at 420 nm after organic solvent extraction.

The sensitivity can be improved by using spectrofluorimetry. A procedure using 2,3-diaminonaphthalene as the complexing agent was described by Hall and Gupta (1969) and this is the basis of the method used by the Agricultural Development and Advisory Service (1986). In this method the complex is first extracted in decahydronaphthalene and a freeze-centrifuge process is then used to isolate the piazselenol. Its fluorescence is measured at 520 nm following excitation at 366 nm. Another way to improve the sensitivity of this reaction is to form a suitable derivative for gas chromatography (Shimoishi and Toei, 1978).

The main disadvantage of the piazselenol methods, which are in many respects ideal, is that the complexing agents are now considered to be carcinogenic. Although an experienced chemist can take the necessary precautions, partially trained assistants and students should not be expected to use these methods.

Perhaps the most effective method for selenium (and also for antimony and arsenic), providing appropriate equipment is available, is one which links hydride generation with atomic absorption measurement (the cold vapour technique). This depends on the evolution of volatile hydrides when sodium borohydride, or certain other reducing agents, react with the element in acidic solution. The hydrides are carried along in a stream of nitrogen or other inert gas into a hydrogen (diffused in nitrogen) air flame or into a heated silica tube. Both serve atomizers although the silica tube will give a steadier signal. It is difficult to produce continuous hydride generation so brief signal maxima must be measured unless integration facilities are available. A detection limit of 1 $\mu g\, l^{-1}$ is possible. Interference from other elements which produce hydrides on reduction may be serious and an internal standard will then be necessary. Much depends on the nature of the original matrix, especially with soils. Apte and Howard (1986) and Parisis and Heyndrickx (1986) describe the method, and its use with ICP is discussed by Hutton and Preston (1983).

Other methods that have been proposed for selenium include:

Turbidity	Society for Analytical Chemistry (1963)
Polarography	Nangniot (1967)
Cathodic stripping voltammetry	Adeljou *et al.* (1984), Othman *et al.* (1984)
XRF spectrometry	Robberecht & van Grieken (1980)
Electrothermal atomization (AAS)	Neve *et al.* (1980), Chiou & Manuel (1984)

Shendrikar (1974), Fishbein (1984) and the International Union of Pure and Applied Chemistry (1984) review and evaluate methods for selenium in biological materials and waters.

Some minor pollutant elements are given in Table 7.1.

Organic pesticides

The use of pesticides for agricultural, forestry and health purposes is now so extensive that it has affected almost all forms of wildlife in some way. Even in remote situations and countries the soils, plants, animals and waters are found to contain pesticide residues. Information on the spread and impact of pesticides on the environment can be obtained from White–Stevens (1971), Edwards (1974) and McEwan and Stephenson (1979).

Table 7.1. Some minor pollutant elements

Elements	Soils		Plant materials		Waters		Analytical techniques	References
	$\mu g\ g^{-1}$	Soln prep	$\mu g\ g^{-1}$	Soln prep	$\mu g\ l^{-1}$	Treatment		
Ba	100–2000	Take 0.5 g, fuse w Na_2CO_3–KNO_3	0.5–20	Take 3 g, ash 450 °C then + HCl *or* digest w HNO_3–$HClO_4$	5–100	Take 2 l, evap *or* ion-exch	At. absorption El. atomization	Stupar & Ajlec (1982) Eaton *et al.* (1982) Schramel & Xu (1982)
Be	0.2–10	Take 5 g, digest w H_2SO_4–HF	(0.02)	Take 10 g, digest w mixed acids	(0.5)	Take 5 l, evap *or* acetylacetone extr	C w Eriochrome cyanine R El. atomization	Sauerer & Troll (1984) Hurlbut (1978)
Sn	0.2–30	Take 1 g, fuse w Na_2CO_3–KNO_3	0.1–2	Take 5 g, digest w mixed acids	0.5–(100)	Take 5 l, evap *or* extr	C w catechol violet At. absorption El. atomization	Smith (1971) BSI (1980) Newman & Jones (1966) Nadkarni (1982)
Te	0.01	Take 10 g, digest w HNO_3–$HClO_4$	0.1	Take 10 g, digest w HNO_3–$HClO_4$	< 0.01	Take 20 l, part evap then extr w dithizone at pH < 1	At. absorption El. atomization XRFS	Beaty (1973) Maher (1984) Corbett & Godbeer (1977)
Tl	0.1–1.0	Take 5 g, digest w H_2SO_4–HNO_3 *or* HCl digest	(0.2)	Take 5 g, digest w mixed acids	< 0.1	Take 20 l, evap *or* dithizone extr at pH 10	C w rhodamine B El. atomization	Matthews & Riley (1964) Sagar & Toelg (1982) Kempton *et al.* (1982) Schmidt & Dietl (1983)
V	20–200	Take 1 g, fuse w $KHSO_4$	0.02–2	Take 5 g, ash 450 °C then + HCl *or* digest w HNO_3–$HClO_4$	0.1–3	Take 10 l, extr w 8-OH-quinoline	C w phosphotungston –vanadic acid Autocolorimetry El. atomization	Roberts (1971) Basson and Kempster (1980) Veno & Ishizaki (1979)

C = colorimetric, At. = atomic, El. = electrothermal, w = with, extr = extract, evap = evaporate.
Note that the solution preparations are not taken from the references quoted.

This section does not give detailed methods for every pesticide likely to be encountered. The variety and complexity of these products make this virtually impossible in a book of this nature. Guidance is given on the general procedures and techniques that can be followed and this information is supported by references to specialized works covering different aspects. However, even after following up some of these publications a certain amount of analytical development work may still be required.

Broadly, the chlorinated hydrocarbon and organophosphorus compounds are used as insecticides. Because of their environmental impact they are dealt with in more detail here than other pesticides. The chlorinated hydrocarbons (organochlorines) are generally more resistant to degradation and tend to be used rather less than the organophosphorus products. The most important herbicides belong to the phenoxyalkanoic and ester group. Some of these have a restricted toxicity and are not persistent, although others are a matter of concern. Other herbicides include the urea, carbamate and triazine compounds and these are briefly discussed later. A summary of the organic pesticides in use in the United Kingdom and their properties is given in Table 7.2. Some are now little used due to their persistence but residues may be retained in the environment.

The estimation of organic pesticides involves a number of difficulties not encountered with most other substances covered in this book. In the first place, the commercial products are often based on a mixture of different compounds. Another difficulty arises because many pesticides are readily hydrolysed or oxidized to form derivatives or metabolites. For example, of the chlorinated hydrocarbons, aldrin is converted to dieldrin, pp-DDT to pp-DDE and heptachlor to its epoxide whilst the organic phosphorus compound, demeton, is converted to its sulphoxide and sulphone. Some of these residues are actually more toxic to organisms than the parent substance. Degradation itself may involve several stages. In addition post-mortem changes can occur, such as the breakdown of pp-DDT to TDE whilst γ-BHC will eventually disappear altogether. Hence, there will always be considerable problems both on the practical side and over the interpretation of results. Many of these difficulties can only be resolved through experience.

In spite of their diversity, pesticides are almost invariably isolated by a chromatographic procedure, although quantitative measurements and final confirmation of identity can be carried out using other procedures such as infra-red and mass spectrometry. As with all chromatographic work, the preparatory stages are particularly important in order to remove interfering substances.

Pesticide analysis can be considered in four stages.

1 Initial collection and storage of samples.
2 Extraction of pesticides.
3 Clean-up of extracts to remove interfering substances.
4 Examination of cleaned-up extract.

These will be considered in turn.

For many purposes, general screening procedures to give presence or absence may be as valuable as the more specific techniques required for quantitative work. General reference books on pesticide analysis include the US Environmental Protection Agency (USEPA, 1971). Goerlitz and Brown (1972), Eagle *et al.* (1983), Zweig and Sherma (1967–86), and the Food and Drug Administration (1985). Papers of interest include Luke *et al.* (1981), Ambrus *et al.* (1981), Osselton and Snelling (1986) and Neuray *et al.* (1986).

Collection and storage

The wide distribution of pesticides may necessitate examination of a great variety of plant, animal and soil materials together with natural waters.

Problems of collection and storage were discussed for organic materials in Chapter 3 and for soils in Chapter 2. Most of the procedures described there also apply to pesticide samples but the points listed below should be especially noted.

1 Polythene, metal and paper containers are normally suitable for sample collection although Rourke *et al.* (1977) reported difficulties with

Table 7.2. Some common insecticides, herbicides and fungicides

Compound	Properties	Method	Reference
Insecticides			
Organochlorine DDT—DDE, BHC, Lindane (volatile, noxious smell), Aldrin, Dieldrin, Heptachlor→metabolites, Endrin (highly toxic)	Persistent, varying toxicity to mammals and birds. Accumulate in body fat, especially predators	See text	
Organophosphorus Disulfoton, Dichlorovos, Fonofos, Chlorofenvinphos, Bromofenvinphos, Malathion, Parathion (highly toxic), Demeton (persistent residue)	Generally low persistence. Toxicity to mammals and birds variable. Mostly systematic insecticides	See text	
Herbicides (mostly)			
Phenoxyacids Dichloroprop; Dicamba, Mecoprop, MCPA, 2,4-D, 2,3,5-T (woody spp.)	Control of broad-leaved spp. Persistent but low-toxicity	GLC & HPLC	Akerblom (1985) van Damme & Galoux (1980)
Triazines Atrazine, Isoproturon	General and pre-emergence herbicides. Persistent but low-toxicity	GLC & HPLC	Zweig & Sherma (1972) McKone *et al.* (1972) Walker (1976) Matisova *et al.* (1979) Byast *et al.* (1977)
Substituted ureas Menuron, Isoproturon	Broad-leaved spp. Persistent. Low toxicity	GLC & HPLC	Caverly & Denney (1978) Byast *et al.* (1977)
Nitriles Bromoxynil, Dichlorbenil	Contact and soil acting	HPLC	Lawrence *et al.* (1980) Pik & Hodgson (1976)
Carbamates Methomyl, Carbofuran Asulam	Aphid control. Translocated herbicide for docks & bracken	GLC HPLC Color. TLC	Hall & Harris (1979) Lauren (1984) Gustafsson & Thompson (1981) Smith & Walker (1977) Smith & Milward (1983)
Quat. ammonium compounds Paraquat, Diquat	Contact herbicides. Toxic. Inactivated by soil	GLC	Luciano & Bosetto (1968) King (1978) Lott *et al.* (1978)–Review
Other herbicides			
Glyphosate + metabolites	Translocated herbicide used on actively growing vegetation.	HPLC & UV/Color.	Rueppel *et al.* (1977)
TCA	Soil-acting herbicide	HPLC	
Fungicides Benomyl, Captan, Maneb, Zineb, Mancozeb	Mainly for horticultural and fruit treatment. Low persistence	At. Abs. after separation GLC	 Zweig & Sherma (1972) Gorbach (1980)–Review.

polythene. Hexane-washed glass containers can also be used. Other plastic materials, together with waxed containers and rubber seals should be avoided. Cardboard containers and cloth bags have been found to cause interferences (Bailey *et al.*, 1970; Levi *et al.*, 1972).

2 Apart from noting normal background details of sample origin etc., any relevant information on pesticide applications, sprays used and any likely contaminants should be recorded at the time of collection.

3 Samples should be kept as cool as possible, preferably frozen during transit to the laboratory. The use of cold packs, dry ice and insulated containers are suitable aids. Waters can also be preserved by adding 2 ml sulphuric acid per litre of sample, especially if chlorinated phenoxyacid herbicides are to be determined (Goerlitz and Brown, 1972). For field samples, and when freezing facilities are unavailable, French and Jeffries (1971) recommend the use of formaldehyde as a preservative for biological samples prior to analysis for organochlorine insecticides.

Samples should be freeze-stored (−10 to −15 °C) in the dark until analysed.

4 Plant materials and animal tissues should be analysed in the fresh state. Soils can be air-dried at a low temperature but it is preferable to treat them in the fresh state. Waters are generally examined as collected, unless suspended matter has to be separated for separate treatment.

5 In pesticide work it is particularly important that representative samples are taken, especially for soils and plant material. The distribution of the pesticide is likely to be erratic as a result of application techniques, so large samples should be taken if possible.

Extraction systems

Although a great variety of pesticides and their residues have been identified, relatively few solvents are used for their extraction, probably because most workers have followed the successful practices described in early papers. However, the extraction efficiency of any solvent or solvent mixture should always be checked for the compound or compounds being investigated.

1 All glassware should be rinsed with redistilled acetone and redistilled hexane before use. It has been recommended that all glassware used for pesticide analysis should be heated to 400 °C prior to use.

2 Plastic ware and filter paper should not be used in the analytical procedure.

3 High quality reagents, free from pesticides, must be used for the extractions. Solvents should be redistilled before use. In all cases blank determinations should be taken through every stage.

4 For the removal of moisture, granular anhydrous sodium sulphate, dried before use at 700 °C, is generally the most suitable. Sand is rendered pesticide free in a muffle furnace at the same temperature.

Hexane or petroleum ether (40–60 °C) are generally adequate for the extraction of most chlorinated insecticides. If the material is wet there is a possibility of emulsions forming, but this can be avoided by adding from 10 to 20% of a polar solvent such as acetone or iso-propanol to the extractant. Alternatively, the sample can be ground or macerated in advance with anhydrous sodium sulphate and sand to remove water before extraction.

Hexane is also a suitable solvent for extracting non-polar organophosphorus insecticides. If there is a need to examine polar compounds in the same solution this can often be achieved by using a composite solution containing equal volumes of acetone or iso-propanol with hexane.

Samples high in fat content, such as some animal tissues, seeds, etc., are conveniently extracted with hexane alone by warming on a steam bath. Two or three extractions may be necessary. Some plant cells, especially when fresh, are not easily penetrated by hexane and a pre-extraction with isopropanol may then be needed. Alternatively, acetonitrile can be used as extractant.

The phenoxyalkanoic acids and their esters may not be extracted efficiently by the above treatments and a separate extraction is required which is described later.

In general, the use of composite solvents will probably bring into solution more interfering substances, increasing the need for clean-up procedures.

Techniques and equipment

Most macerators, blenders and homogenizers are suitable for ensuring intimate contact between the sample and the extractant. High speed macerators are especially suitable for plant material. For animal material some prior chopping may often be necessary. Benville and Tindle (1970) recommend pulverizing in the presence of solid carbon dioxide for fish samples. Others have suggested shaking in a steel tube with steel ball bearings to extract samples rapidly.

Soils can be conveniently extracted by shaking on a horizontal or rotary shaker for a suitable period. One hour will generally be adequate but this should be checked.

Soxhlet extraction can be very efficient if the compounds under investigation are heat stable. Pre-wetting with extractant will improve recoveries. If paper extraction thimbles are used, these are likely to introduce interfering substances unless pre-extracted for several hours with dichloromethane, or the solvent to be used for extraction of the sample. For waters, some form of stirring or shaking in a large vessel is needed.

Extraction procedures

The methods given here should not be regarded too rigidly. The extraction conditions are flexible enough to permit variations to be made to fit the different circumstances.

SOILS

There are alternative views about the wisdom of drying soils before extraction. Although the sample is easier to handle when dry there is a risk of loss of volatile pesticides during the drying process. In addition, some of the chlorinated hydrocarbons appear to be more strongly retained by dried soil. In most cases partial drying at room temperature is a suitable treatment.

The extraction procedure given here is applicable to most pesticides except phenoxyalkanoic free acids.

Allow the soil to dry at room temperature for several hours.
Weigh about 10 g for extraction (taking a corresponding sample for moisture correction). Add roughly equal volumes of sand and granular anhydrous Na_2SO_4 and grind.
Add 100 ml 1:5 acetone:hexane mixture.
Extract with a high speed macerator for a few minutes, or with a rotary shaker for 1 hour.

PLANT MATERIAL

Finely chop and mix fresh plant material.
Weigh about 10 g for extraction (taking a corresponding sample for moisture correction).
Add 50 ml 1:2 acetone:hexane mixture.
Macerate in a high speed homogenizer.
Separate the extract by filtering or centrifuge.
Wash the residue with 20 ml solvent mixture and combine with first extract.
Shake with anhydrous Na_2SO_4 to remove water.

Procedure for phenoxyacids and ester herbicides

Weigh about 40 g chopped and mixed plant material.
Add water, if necessary, to about 50% concentration and mix well.
Add in succession:
10 ml 10% H_2SO_4
20 ml ethanol,
40 ml petroleum ether (40–60 °C),
100 ml diethyl ether.
Macerate well for about 15 minutes.
Filter with suction.
Wash the residue with a mixture of equal volumes of diethyl ether and petroleum ether.
Combine extract and washings in a separating funnel.
Add 100 ml 4% w/v Na_2SO_3 solution.
Shake for 1 minute.
Allow phases to separate, then run off aqueous layer into another separating funnel.
Wash with 25 ml diethyl ether and combine with earlier ether extract.
Use this for the separation of herbicide esters.
Wash the aqueous layer three times with 50 ml $CHCl_3$.

Acidify the aqueous layer with 10% H_2SO_4 and re-extract with 2×50 ml $CHCl_3$.
Combine $CHCl_3$ washings and dry with anhydrous Na_2SO_4.
Use for examination of acids and salts.

ANIMAL TISSUE
Finely chop the fresh sample.
Weigh out about 2 g for extraction.
Grind with a roughly equal volume of sand and 5 g anhydrous Na_2SO_4 in a 100 ml beaker until a granular, friable mixture is obtained.
Add 50 ml distilled hexane, mix, and then decant the hexane extract.
Repeat the process with alternate 50 ml aliquots of distilled acetone and hexane until 250 ml extract is obtained.

WATER
Measure out an appropriate volume of the water (at least 1 litre, often much more).
Extract with 100 ml $CHCl_3$ by vigorous stirring or using a large separating funnel to separate the organic layer.
Repeat with 2×50 ml $CHCl_3$ and combine all extracts.
Shake with anhydrous Na_2SO_4 to remove water.

Compared with most other classes of material, waters are relatively free of the organic substances which have to be removed in the clean-up stage. It is therefore often possible to pass directly to chromatographic separation after extraction and sample volume reduction.

Clean-up stage

The stage between the extraction and final separation of the individual pesticides can broadly be termed the clean-up stage. This involves the removal of interfering substances whilst keeping losses of pesticides to a minimum.

Many compounds in natural biomaterials can interfere with pesticide estimations, including lipids, pigments, polar organic compounds and industrial pollutants. Among the latter are polychlorinated biphenyls which are now recognized as a class of pollutants in their own right. A large number of methods have been proposed for clean-up, but the two most commonly used are based on adsorption chromatography and liquid–liquid partition. Other chemical processes have been used for the separation of specific pesticides. These processes include oxidation, pH adjustment, hydrolysis, acid reaction and fractional distillation. Saponification of lipids using alkali has been used by several workers as the basis for fat separation. In recent years, solid phase extraction with chemically bonded silicas has been used for the clean-up of pesticide extracts. Food and Drug Administration (1985) give clean-up methods for organochlorines, Van Horne (1985) for atrozine and PCBs, and Wilson and Phelan (1986) for carbamates.

Liquid–liquid extraction

This form of partition is used mainly for separating pesticides from fats and waxes and is particularly important for the clean-up of extracts of animal tissue and products, providing that the fat content is not excessive. It can also be used to give some preliminary separation of different pesticides and some workers have developed successive liquid–liquid partition to give extensive pesticide separations together with evidence of identification.

One of the most important partition pairs is acetonitrile as the polar solvent and hexane as the non-polar solvent. Most pesticides will appear in the acetonitrile phase. Alternatively petroleum ether can generally be substituted for hexane and acetone for acetonitrile.

The partitioning of pesticides from lipid products is reported to be effective using *N-N*-dimethyl formamide. This can be paired successfully with hexane. A liquid partition system using dimethyl sulphoxide with hexane was found to give more advantageous partition coefficients for most common pesticides than acetonitrile with hexane.

Sorption methods

Adsorption techniques are probably the most widely used of the clean-up systems employed for the removal of interfering substances. They are particularly suitable for removing pigments and general organic contaminants as opposed to fats and waxes, which can be more easily dealt with using liquid–liquid partition, or by a freeze separation technique. Sometimes column adsorption may be used as a secondary clean-up procedure following initial fat separation, and is often necessary even after liquid–liquid partition. Some unsaponifiable lipids are also more conveniently removed by adsorption.

If the concentration of the interfering substances is not too high, batch shaking may be adequate to remove them. In this case an adsorbent is added to the extract solution and the mixture is well shaken for several minutes. About 10 g adsorbent to 100 ml solution is a suitable amount.

In most cases, however, a column leaching system is generally preferred as being much more effective for removing interfering substances. Chromatographic separation of pesticides of interest may also take place to some extent. Like liquid–liquid extraction, some systems have been highly developed for specific separations, but in general a great deal of care and experimental work is necessary to establish optimum conditions.

Column design

The basic procedure used with all adsorption columns is much the same. A column approximately 30 cm long and 1.5 cm in diameter, having a tap or other suitable control at the bottom, and containing a sintered glass disc or glass wool plug, is filled with a slurry of adsorbent in a suitable solvent. It may be advantageous to include a plug of anhydrous sodium sulphate at each end of the column. The column should be irrigated well with the eluting solvent before use.

In recent years the use of the microcolumn has increased, where a Pasteur pipette of I.D. 0.7 cm drawn to a fine tip is filled with about 4 cm adsorbent. These are disposable and hence reduce the risk of cross-contamination.

Reduction of sample volume

Before transferring to the column it will generally be necessary to reduce the volume of the extract. This may be carried out in various ways:

1 Evaporation on a steam bath using a stream of warm inert gas or clean air. Flammability of the solvent plus stability and volatility of the pesticide have to be taken into account before using this method.

2 Evaporation with a rotary film evaporator (although this apparatus may result in losses of the more volatile compounds).

3 Concentration in the Kuderna–Danish evaporator. The apparatus is described by Thornburg (1963). It enables the solvent to be evaporated but a partial reflux allows any pesticide left on the sides of the vessel to be rinsed down into a small tube.

In most cases evaporation to less than 2 ml will be needed. For some of the column clean-up procedures it may be necessary to evaporate the extracting solvent completely and then take up the residue in 1–2 ml of petroleum ether. A small volume of the eluate should be used to wash out the sample vessel and the remaining eluant (about 250 ml) used for the column separation.

Adsorbents

A large number of different adsorbents are described in the literature. Some of these are complementary, that is they are suitable for different groups of pesticides or will remove different interfering substances. Accordingly, more than one column is often necessary (Law and Goerlitz, 1970). Some adsorbents vary from batch to batch so recovery checks are advisable before running the sample solutions. Some of the more common adsorbents are:

1 *Activated charcoal*

This has been used alone and mixed with cellulose (McLeod *et al.*, 1969), with celite and other adsorption materials. It can be used for chlorinated hydrocarbons and non-polar organic phosphorus compounds, although prolonged elution may be necessary to obtain good recoveries of some compounds. Hexane is suitable as an elution solvent.

2 *Alumina*

This is usually activated by prior heating at 600–800 °C. Some consider it to be more efficient than silica as an adsorbent for pesticide clean-up. It is particularly suitable for removing interferences prior to the determination of organophosphorus compounds (Aharouson and Resnick, 1972).

3 *Silica gel*

This adsorbent has a chromatographic effect on the pesticide mixtures being passed through it. Kadoum (1967) describes a rapid clean-up procedure using micro-columns of the adsorbent. Aharouson and Resnick (1972), however, state that removal of interfering plant extractives by silica gel is not always satisfactory in the case of organophosphorus insecticides. It is particularly suitable though for the clean-up of solutions prior to the determination of polychlorinated biphenyl compounds.

4 *Florisil*

Florisil is a synthetic magnesium silicate product. It may be purchased in the activated state but it is more usual to heat the commercial product to 650 °C before use. It is particularly important to standardize each column first. For complete recovery, about 25% ether in petroleum ether is a suitable solvent. Other solvent mixtures of increasing polarity may also be used.

5 *Magnesium oxide*

Although it is used as a column clean-up material in its own right, magnesium oxide is probably more popular as a secondary column.

Columns for phenoxy acids and esters (anionic pesticides)

Liquid–liquid partition is generally used for separating most impurities prior to the determination of anionic pesticides. If column adsorption clean-up is necessary this usually follows partition. Florisil is commonly used for this purpose. Shafik *et al.* (1971) describe a method using silica gel after hydrolysis and ethylation.

Minor clean-up techniques

Thin-layer chromatography

This technique is used both as a clean-up procedure prior to gas chromatography and as a method in its own right for the determination of pesticides. For clean-up using TLC, silica gels are effective adsorbents and 250 μ layers are suitable. A general review of the use of TLC for the determination of pesticides has been prepared by Sherma (1986).

Low temperature precipitation

During extraction of high fat content materials for pesticide analysis most solvents will also remove large amounts of crude fat and other lipid residues. Although these may be reduced to acceptable levels by successive liquid–liquid partitions, low temperature precipitation can be more effective. The method involves cooling to temperatures ranging from -10 to -70 °C. Under these conditions the fat will precipitate and can be removed by filtration. For large scale routine work on pesticides it is possible to purchase suitable equipment to facilitate this operation.

Examination of extracts

After removal of interfering substances, the final examination of the compounds or groups of compounds of interest is carried out. This may be just a qualitative or semi-quantitative screening of the extract or the determination of a specific compound or compounds may be required. Instrumental techniques suitable for this purpose are described in the following pages.

Gas chromatography

This is the most widely used technique for the separation and estimation of organic pesticides, especially chlorinated hydrocarbons. General reviews on the application of GLC for pesticide residue analyses include Leathard and Shurlock (1970), Luke *et al.* (1981) and the reviews in Analytical Chemistry and chromatographic journals are a valuable source of information on new developments. Most of the references quoted on p. 218 provide details of the application of gas chromatography for the examination of organochlorine and organophosphorus pesticides. Other works giving methods for pesticides include McLeod *et al.* (1969), Zweig and Sherma (1967–86) and Standard Methods for the Examination of Waters and Wastewaters (1985). Waliszewski and Szymczynski (1982) describe a simple method specifically for organochlorines and the use of GLC for organophosphorus compounds is reviewed by Greve and Goewie (1985). Zweig and Sherma give considerable information on retention data under different operating conditions for both classes of compounds.

In order to achieve good separations by GLC it is essential to develop efficient clean-up procedures. One of the principal difficulties in the determination of pesticide residues is the presence of extraneous organic compounds which lead to confusion in the interpretation of chromatograms.

The operational techniques used in the examination of chlorinated hydrocarbons are broadly similar to those used for organophosphorus compounds. The main difference is the need for specific detectors and these are considered later.

Operational conditions

The notes given here should be used in conjunction with those given on GLC techniques in Chapter 8, together with the instructions in the manuals issued by the instrument manufacturers. It is impossible to set out detailed instructions to meet all circumstances and a certain amount of development work will always be needed for this type of investigation.

Many of the recommended materials and instrumental conditions are listed in Table 7.3.

The support materials are normally diatomaceous earth products. It has been reported that silanized supports improve efficiency by allowing certain components, which tend to decompose and give more than one peak on untreated supports, to be eluted as a single peak. The addition of Epikote resin in small amounts with the stationary phase has been found to have a similar effect.

The stationary phases are based on silicone oils and gums. As with the support materials, they are available from most suppliers of chromatographic materials.

Columns should be made of borosilicate glass since hot metals can result in decomposition of some pesticides (Beckman and Bevenue, 1963). Before use, the columns generally have to be conditioned at 250–300 °C for 48 hours. The use of capillary columns has become more widespread in recent years. These may be made of glass or silica and are coated with the stationary phases used in conventional columns. The main advantage is better resolution due to sharper peaks.

For screening purposes one column is adequate, but for analytical operations two columns, one polar, the other non-polar, are essential.

Table 7.3. Materials and settings used in the GLC determination of organic pesticides

Support materials	Celite 545, Chromasorbs P, W, G, Gas Chrom Q	Diatomaceous earth products improved by acid and base washing and silanizing treatment.
Particle size	80–100 mesh	Smaller sizes improve resolution Larger sizes give shorter retention times.
Stationary phase		
– non-polar	DC-200, SE-30 OV-1	5–10% w/w
– polar	QF-1, GE-XE-60, OV-210, OV-225	4–6% w/w Mixed non-polar/polar are often used
Coating solvents	Chloroform, methylene dichloride	
Columns	Borosilicate glass	2 m × 4–6 mm
Column temperature	185–250 °C	Temperature programming sometimes used
Carrier gas	N_2, He or Ar	40–100 ml min^{-1}
Detector temperature	200–250 °C	

Standard solutions must be injected frequently to check retention times.

Improvement in the resolution of more volatile mixtures is obtained through the use of temperature programming. However, temperatures over 150 °C are normally required for adequate retention characteristics.

Detectors

Electron capture detector

In this device an inert gas such as argon or nitrogen passes through the cell which contains a radiation source, usually tritium or nickel-63. The inert gas becomes ionized and electrons are released which migrate to the anode under a fixed voltage, causing a current to flow. The current falls when a compound which has an affinity for electrons enters the cell, and the reduction is proportional to the amount of compound present. If tritium is the radiation source the detector cannot be used above 250 °C. The use of nickel-63 enables the maximum operating temperature to be raised to 350 °C.

The detector is very sensitive to halogen compounds. It is less sensitive for nitrogen compounds and insensitive to hydrocarbons and oxygenated compounds. Few organophosphorus compounds are detected although malathion and parathion are exceptions.

Microcoulometric detector

This detector consists of a pyrolysis unit and a titration cell. When organic compounds containing halogen and/or sulphur pass through the pyrolysis unit they are decomposed. The gases produced are then bubbled through the titration cell which contains a Ag/Ag^+ measuring electrode, a Ag anode, a Pt reference electrode and a Pt cathode. The electrolyte is $AgNO_3$ and the reaction removes silver from solution with a change of potential between the measuring electrode and the reference electrode. This change in potential is used to operate a servomechanism which passes a current through the other two electrodes allowing silver to be dissolved from the anode to exactly replace the portion removed. This current is recorded and from it the amount of the original compound can be assessed.

The detector is specific for halogen and sulphur compounds. However, a modification of the furnace will allow organophosphorus compounds to be measured. Its principal disadvantage is its high cost.

Thermionic detectors

Giuffrida (1964) modified a flame ionization detector by coating the collector electrode with a sodium salt and obtained very high responses to chlorine and phosphorus compounds, especially the latter. This principle forms the basis of the modern thermionic detector in which a pellet of caesium bromide with a hole bored centrally is placed over the FID jet so that the hydrogen flame burns above it (Hartmann, 1966). This type of detector is supplied by most instrument manufacturers specifically for organophosphorus pesticide analysis. Later developments have been described by Kolb *et al.* (1977), Burgett *et al.* (1977) and Patterson and Howe (1978).

Flame photometric detector

Phosphorus compounds can be detected using the flame photometric detector originally described by Brody and Chaney (1966). In this device, the column effluent is passed into an oxyhydrogen flame. Light emitted from the flame passes through an interference filter (526 nm) and its intensity is measured with a photomultiplier. Flame background is shielded from the photomultiplier by a cylindrical housing surrounding the lower part of the flame. Emission due to phosphorus occurs in the higher region of the flame beyond the background shield and therefore the detector is specific for phosphorus.

Hall detector

This detector carries the name of the man who developed and patented it. It has become popular with many analysts, especially with those involved in environmental monitoring. The detector was developed from the microcoulometric device mentioned earlier. The effluent from the separation column reacts to form simple acids or bases which are then transferred to an electrolyte system, the difference between the reference and the analytical electrodes being the output of the detector. The main advantage of this detector is its ability to be tuned to pick up certain groups of compounds only. It has four specific modes: halogen, nitrogen, sulphur and nitrosamines. The change from one mode to another is determined by the choice of reaction gas, temperature, electrolytic composition and other basic properties and can be easily carried out.

Gas chromatography of phenoxyalkanoic acid and ester herbicides

For analytical purposes these compounds may be considered as a subdivision of the organochlorine pesticides.

The esters may be examined by GLC immediately after clean-up. However, if acids or salts are present they must first be esterified. The most common derivatives used are the methyl or ethyl esters. Esterification may be carried out using diazomethane or diazoethane. Alternatively, the BF_3–methanol or BF_3–ethanol reaction may be used. Both methods are described by Browning (1967) for fatty acids. Once esterified the acids may be separated chromatographically in the same way as for organochlorine compounds. Baur *et al.* (1971) describe the gas chromatography of phenoxy acids using their trimethyl silane derivatives.

Separations of phenoxyalkanoic esters may also be carried out with the GLC systems used for other fatty acid esters in which the stationary phases used are Apiezon L or M and polyethylene glycol succinate or adipate.

Methods for the determination of phenoxy acid pesticides and their residues are given by Shafik *et al.* (1971), Goerlitz and Brown (1972) and Akerblom (1985).

Quantification and interpretation of chromatograms

Only a gas chromatograph fitted with a microcoulometer detector will give an absolute measure of the amount of substance passing through. Other detectors have to be calibrated. This is carried out by preparing standard solutions of individual pesticides and injecting different amounts to obtain a calibration curve. The standards may be made up as mixtures, providing the system chosen gives good resolution of the components of the mixture. At least two runs should be carried out and the mean used for calibration.

For results of the highest accuracy the area under the peak should be measured, but for most purposes it is adequate to use peak height. If only one or two individual compounds are to be measured, the method of standard addition may be applied.

Identification of individual peaks may present problems, especially in the examination of residues as opposed to original pesticides. The method usually employed is to compare retention times with those of standard solutions on at least two different columns. In difficult cases, responses to different selective detectors may also have to be compared. The addition of known pesticides to the sample ('spiking') is a valuable method for identification.

Even if clean-up procedures have been efficient, interfering substances may be introduced at a later stage. Croll (1971) reported the presence of extraneous peaks derived from injection septa.

As an absolute method of peak identification the mass spectrograph can be used to monitor column effluent, although the high cost puts this technique beyond the scope of some laboratories. Zweig and Sherma (1972) discuss the use of this instrument.

High performance liquid chromatography

The emergence of HPLC as an important analytical technique has resulted in the development of a wide range of applications in the pesticide and residue field. A number of pesticide groups which had been difficult to analyse by GLC were found to be more easily separated using HPLC. In particular, non-volatile compounds or those not readily converted into volatile derivatives comprised a field to which the technique was generally applicable. Notably, carbamates, triazines, substituted urea herbicides and phenoxyacid herbicides have been widely analysed by HPLC, along with warfarin-type rodenticides and synthetic pyrethroid insecticides.

The determination of pesticides and residues by HPLC has been described by Cotteril and Byast (1984). Reviews on pesticide and residue separations have been published by Lawrence and Turton (1978), Ivie (1980), Moye (1981) and Greve and Goewie (1985). Cochrane (1979) looks at chemical derivative techniques using both GLC and HPLC for some of the more common pesticides. Further information on the application of HPLC in the examination of specific compounds or pesticide groups can be obtained from the many papers that are now published on this subject in the chromatographic journals.

As with GLC, efficient clean-up procedures are required to obtain good separations and simplify chromatogram interpretation. In fact, HPLC itself has often been used as a clean-up procedure.

Operational conditions

General principles of HPLC are given in Chapter 8. The technique is a general one and detailed instructions cannot be given to meet all circumstances. The notes given here are for guidance only and development work will be needed in any specific application.

For most pesticide separations, columns containing a silica support with a bonded stationary phase of octadecylsilane (C_{18} bonded column) will normally be suitable. Similar column packings from different manufacturers may have different retention characteristics (De Stephano *et al.*, 1980), and this should be borne in mind when choosing. A reversed-phase system using C_{18} columns with polar solvents such as water, methanol or acetonitrile (or mixtures of these) will separate most pesticide mixtures encountered. However, non-polar mixtures like organochlorine insecticides are better handled using polar columns ($-NH_2$ or $-CN$ bonded silica or silica alone) with non-polar eluents. Hexane and octane are commonly used.

The most popular detector used in pesticide applications is ultra-violet absorbance. The refractometer is generally not very satisfactory due to the low levels of individual constituents likely to be found in environmental samples. The ultraviolet detector is non-destructive, and effluent fractions may therefore be collected for further examination if necessary.

Confirmation techniques in HPLC are almost identical to those of GLC discussed earlier and the same precautions apply.

Minor techniques

Various other techniques have been used in the analysis of biomaterials for pesticides and their residues but none have the overall advantage of GLC and HPLC. They include thin-layer chromatography, electro-analytical methods and total element determinations. The latter approach is non-specific but is useful for simple survey purposes. The total amounts of both organochlorine and organophosphorus compounds can be obtained after suitable separation and oxidation stages and measurement of the appropriate element. Enzyme inhibition, biological assays and spectrophotometry have also been tried. A description of minor techniques used for pesticides is given by Zweig and Sherma (1973, 1977).

Substituted urea and carbamate herbicides

The substituted urea herbicides have been used for the control of grasses on agricultural crop land. In general they are of low toxicity to mammals. They are, however, resistant to attack by micro-organisms and are relatively stable and therefore tend to be very persistent. This is especially true of monuron and diuron. Carbamate herbicides are also active against monocotyledons but relatively ineffective against dicotyledons. They are, however, more readily degraded by micro-organisms and also tend to be hydrolysed in soil or aqueous media. Some carbamates are also used as insecticides.

The usual method for determining urea herbicides is to extract the sample with a suitable solvent such as acetone or iso-propanol. Clean-up is by liquid–liquid partitioning followed by hydrolysis to the corresponding amine which is determined colorimetrically or by TLC or GLC. Alternatively the hydrolysis may be carried out before the clean-up stage. Accounts of the determination of substituted urea herbicides are given by Zweig and Sherma (1972). A TLC method for urea and carbamate herbicides in natural waters is described by Frei *et al.* (1973). Finocchiaro and Benson (1967) give a TLC method for the estimation of these compounds in other sample materials. Oswiecimska and Golz (1969) give a paper chromatographic method for the separation of urea herbicides.

Extraction methods for carbamate pesticides depend on the type of sample to be analysed. For non-fatty samples methylene chloride is the most widely used solvent. For fatty samples such as animal tissue, hexane–ether mixtures, acetonitrile or methylene chloride have been used. Clean-up is by liquid–liquid partition and/or Florisil column adsorption. The extracts may then be examined by TLC, GLC or colorimetry. In some cases hydrolysis can be employed to facilitate determination of individual carbamates such as carbaryl. The procedures for carbamate residues are reviewed by Van Middelem (1971). Zweig and Sherma (1972), Hall and Harris (1979), Abdel-Kader *et al.* (1984) and Standing Committee of Analysts (1988a) all describe methods for carbamates and the metabolites either by GLC or TLC. Lauren (1984), and Gustafsson and Thompson (1981) apply HPLC.

Other pesticides

In general, ecologists are less interested in other pesticides (mostly herbicides) because they are mostly less persistent and toxic to animals than the ones already considered. However, there are exceptions and some ecologists would maintain that all unnatural products affect the environment. There are also new products that need to be monitored. Some of the other pesticide classes are mentioned in Table 7.3 with a few analytical method references. Further information on pesticide methods can be obtained from the Food and Drug Administration (1985), Zweig and Sherma (1967–86) and from specialist journals and abstract publications.

Polychlorinated biphenyl compounds

Polychlorinated biphenyl compounds (PCBs), although not pesticides, are mentioned here because they are closely related chemically to the chlorinated hydrocarbon insecticides. They are also widely distributed in the environment due to their use in industry as a plasticizer in paint, resin and plastics manufacture and as hydraulic fluids and transformer oils. They are particularly resistant to degradation. They have been found to accumulate in animals, birds and fish in relatively large amounts.

They were first found to be present in wildlife when a number of extraneous peaks in organochlorine pesticide analysis by GLC were identified (Holmes *et al.*, 1967). The PCBs behave in a similar way to the organochlorine insecticides and this has resulted in erroneous interpretations of chromatograms. Since the proportions of the two classes of compound are so closely allied, a very efficient clean-up sequence is required before examination of either group using gas chromatography.

Reynolds (1969) separated PCBs from some chlorinated pesticides using a Florisil column. The former were eluted with hexane and the pesticides eluted afterwards with a mixture of diethyl ether–hexane. Bevenue and Ogata (1970), however, stated that the behaviour of this type of column had to be carefully checked before this system was used. Armour and Burke (1970) used a silicic acid–celite column, the PCBs being eluted with light petroleum and the pesticides with acetonitrile hexane–dichloromethane mixture whilst a silica gel column was used by Snyder and Reinert (1971) for the rapid separation of PCBs from DDT and its analogues. Schneider *et al.* (1984) describe the use of capillary column gas chromatography for separating PCBs from organochlorine residues. Additional references on the use of gas chromatography for PCBs can be obtained from the review of Creaser and Fernandes (1987). Thin-layer chromatography has also been applied for separating PCBs from the organochlorines. A two-dimensional system found to be successful uses silica gel–silver nitrate layers with *n*-heptane as the solvent in one direction and *n*-

heptane:acetone (98:2) for the other direction (Fehringer and Westfall, 1971).

One of the biggest problems in the examination of PCBs stems from the lack of suitable standards. The industrial preparations which are the source of PCB pollution are complex mixtures and most of the individual components are not available. As many as twenty peaks have been found on gas chromatograms but often they cannot be identified. In pollution investigations it may be necessary to match the entire chromatogram against others obtained from industrial plants and processes, to identify the source of the pollution.

Other pollutants

There are some chemical pollutants which do not fit into the two main groups already covered in this chapter. These substances mostly affect water courses and lakes, so the general topic of water pollution is discussed first. Later in the section, five specific pollutants, namely cyanides, fluorine, phenols, sulphides and detergents, which have been known to affect natural habitats of conservation interest, are dealt with in more detail. Finally, in this part there is a discussion on atmospheric pollutants.

Many synthetic substances which have no commercial value are produced as side products by chemical industries. Little is known of the toxic effects on wildlife of the substances which are often dumped or discharged in the countryside. One example of this class of material which has had considerable recent publicity are the PCBs. The methods for analysing this group are discussed at the end of the section on organic pesticides because of their similarities with chlorinated hydrocarbon pesticides.

Water pollutants

Various Acts of Parliament have gradually brought under tighter control the discharge of objectionable industrial wastes into water courses. Even so, some loopholes remain and various pollutants find their way into rivers and lakes. The standards used to control pollution are based on human health requirements and amenity considerations, as well as the needs of wildlife conservation. Clearly, for most cases the standards are similar but there may be instances where a particular water treatment is not to the liking of ecologists. For some communities the pH or nutrient balance may be critical and the optimum conditions are easily disturbed. This is a factor which must not be overlooked when using the criteria given in Table 7.4 where field observations are given in addition to some chemical parameters.

Some of the more simple tests which are often carried out on water samples are: suspended solids, total dissolved matter, pH, alkalinity and colour and turbidity. These tests are described in Chapter 4 which also deals with preservation and initial treatment procedures for water samples.

There are a number of tests which are commonly used for the estimation of the general level of pollution in water bodies. These include:

1 Permanganate value (PV).
2 Chemical oxygen demand (COD); see p. 78.
3 Biochemical oxygen demand (BOD).
4 Methylene blue stability.

To a certain extent the results from these tests are assessed in conjunction with each other.

Permanganate value gives a measure of oxidizable organic matter and some of the inorganic substances which are subject to oxidation. The method is based on reduction of acid permanganate solution to the manganous form and the unused permanganate is determined by titration with thiosulphate solution. Different reaction times are specified but the ones most commonly used are 3 minutes, for a quick test, and 4 hours.

The COD also gives a measure of oxidizable organic matter and is generally preferred to PV because the oxidation is more efficient and the test is less subject to errors. The water sample is treated with hot acid dichromate solution and the excess dichromate is determined by titration with ferrous sulphate.

The other two tests in this group are more closely analogous to biological processes. In the BOD procedure the micro-organisms are allowed to decompose the organic pollutants using up

Table 7.4. Generally accepted standards for the purity of river waters (derived from official reports)

	Very clean	Clean	Fairly clean	Doubtful	Bad
Suspended matter	None	None	Fairly clear	Turbid	Turbid
Smell	Odourless	Faintly earthy	Strong earthy	Strong earthy or wormy	Soapy Faecal
Appearance	Clear	Clear	Slightly opalescent	Dark and opalescent	Brown or black and soapy
Delicate fish–trout, grayling, etc.	Plentiful if river appropriate	Scarce	Probably absent	Absent	Absent
Coarse fish–roach, gudgeon, etc.	Plentiful	Plentiful	Plentiful	Scarce	Absent
Stones in shallows	Clean and bare	Clean	Light coating brown fluffy deposit	Coating brown fluffy deposit	Grey growth
Stones in pools	Clean and bare	Fine light brown coating	As above	As above	Brown and black mud
Water vegetation	Scarce	Plentiful; fronds clean	Plentiful; fronds brown in places	Plentiful; fronds covered with light deposit	Scarce
Green algae	Scarce	Moderate in shallows	Plentiful in shallows	Abundant	Abundant
PV (ppm)	0–3	3–5	6–8	8–10	>10
BOD (ppm)	0–1	2	3	5	>10
Dissolved O_2% saturation	100–90	90–80	80–65	65–50	<50

dissolved oxygen in the process. The oxygen left in the water is determined after an incubation period. The time allowed for this process is usually five days, which is a disadvantage. Another difficulty arises out of the use of some of the oxygen for oxidizing ammonium and organic nitrogen to nitrates and nitrites.

The methylene blue stability test is a much simpler and somewhat arbitrary test in which the stability of the oxygen level is measured, this being dependent on the amount of oxidizable substances present in the water. The test can easily be carried out by adding 0.5 ml of 0.035% w/v methylene blue solution to a bottle of about 100 ml capacity and carefully filling with the water sample, taking care to exclude air bubbles. The sample is then incubated for 5 days at 20 °C and the time taken for the blue colour to disappear is noted. The bottles should be checked frequently the first day, and thereafter only two or three times a day.

In view of the complexity of the first three tests in this group the procedures are not given here and the reader is referred to the two principal texts on this subject which give full details. These are Standard Methods for the Examination of Waters and Wastewaters (1985) and Department of the Environment (1972). These books are also valuable reference works for other water pollutants. Leithe (1973) specifically deals with the analysis of organic pollutants in water.

Some of the procedures given elsewhere in this

book are also of value in assessing the nature and extent of any pollution, in conjunction with those discussed above. In particular, dissolved oxygen is an important characteristic and a procedure for its determination is given in Chapter 4. The levels of ammonium and nitrate-nitrogen, and phosphate in waters have relevance in problems of eutrophication.

Cyanide

Large amounts of cyanide are used in industry, especially for electroplating, for case hardening of steel and for purifying various metals. Generally, the residues are discharged as liquid waste into sewers or even drainage channels. If the concentration is high the bulk of the cyanide is removed before discharge as an insoluble ferricyanide which is normally reprocessed for further use. In some cases both ferricyanide or spent cyanide residues have been dumped as solid waste.

Cyanide is extremely toxic to higher animals but many bacteria can oxidize it fully to carbon dioxide and ammonia whilst others may convert it to non-toxic cyanate or thiocyanate.

Apart from isolated incidents involving solid dumping, most cyanide pollution of concern to the ecologist or conservationist will be associated with an aquatic habitat. The biological system of the water can be disrupted or even destroyed by cyanide before the latter is broken down by natural organisms.

Both titration and photometric techniques can be used for estimating cyanide in solution. The titration is only suitable if the concentration of CN^- is above about 1 mg l^{-1}. A suitable method is given by the Department of the Environment (1972). Colorimetry is more sensitive and a method is given below. An ion-selective electrode can be used for the estimation of cyanide if sufficient is present. For rapid determinations a simple detector tube procedure is described by Kobayashi *et al.* (1966).

Samples should be analysed soon after collection but if storage is unavoidable the pH should be raised to at least 11 by addition of sodium hydroxide.

A problem might arise in the determination of total cyanide when some of the cyanide complexes are present. A few of them are very stable and special treatment is needed to break these down, for example with cuprous chloride in hydrochloric acid (ABCM-SAC, 1957). This increased resistance to oxidation can extend the area affected by a pollution incident before the compounds involved are harmless.

Pyridine–pyrazolone method

This procedure involves conversion of CN^- (following distillation) to CNCl by reaction with chloramine-T before development of the colour. The colorimetric method given here is derived from Standard Methods for the Examination of Water and Wastewater (1985). It was developed by Kruse and Mellon (1952) and gives total cyanide.

Reagents

1 Cyanide solutions.

Stock solution (1 ml ≡ 100 μg CN^-): dissolve 0.125 g KCN in water and dilute to 500 ml. Prepare fresh.

Working solution (1 ml ≡ μg CN^-): dilute the stock solution 100 times.

Caution: take great care in handling this toxic substance.

2 Pyridine–pyrazolone reagent.

A 1-phenyl-3-methyl-5-pyrazolone, 0.5% w/v. Dissolve at 75 °C, cool and filter.

B Bis-pyrazolone solution.

Dissolve 0.025 g bis-pyrazolone in 25 ml pyridine and filter.

Mix solution B with 125 ml solution A. Prepare fresh.

3 Chloramine-T, 1% w/v; prepare fresh.

4 Sodium hydroxide, M.

5 Acetic acid, 1 + 4.

6 Cuprous chloride, 2% w/v in 5 M HCl.

Procedure

Set up distillation apparatus preferably on a small scale. Include splash head and vertical water condenser (see Note **1**).

Take an aliquot of sample containing not more than 2 or 3 mg CN^-.
Add 2 ml acid cuprous chloride and dilute to about 80 ml.
Heat and distil HCN into 5 ml M NaOH until volume is 10 ml (condenser tip below surface) (see Note **2**).
Collect blank distillates.
Take aliquot of distillate in 25 ml volumetric flask and dilute to about 15 ml.
Bring to pH 6 to 7 with 1+4 acetic acid.
Pipette 0–10 ml standard into 25 ml volumetric flask to give a range of 0–10 μg CN^-.
Add NaOH as for samples and then neutralize as above.
Add 0.2 ml chloramine-T to standards and samples.
Invert three times and leave 2 minutes.
Add 5 ml pyridine–pyrazolone reagent.
Dilute to volume, mix and leave for 20 minutes.
Measure the absorbance at 620 nm using water as a reference.
Prepare a calibration curve from standard readings and use to calculate the concentration of CN^-.
Subtract any blank values if present.

Notes

1 The distillation apparatus *must* be operated in a fume cupboard because of HCN toxicity.

2 Some complex cyanides are slow to break down. If it is suspected that these are present, collect further distillate.

Fluorine

Fluorine appears to be essential for most fauna and some flora although conclusive evidence is not yet available for some groups. Its best known function is in mammalian tooth development. The addition of fluorine as fluoride to drinking water supplies to lower the incidence of dental decay is a widespread though controversial practice.

Some naturally occurring concentrations are given below.

Soils	10–250 $\mu g\,g^{-1}$
Plant materials	0.5–10 $\mu g\,g^{-1}$
Animal tissue	1–50 $\mu g\,g^{-1}$ (can accumulate in bones and teeth)
Freshwater	0.1–10 $mg\,l^{-1}$

Fluoride is moderately toxic at higher concentrations. It enters the environment in this form and there has been increasing concern because of its widespread use in industrial processes. Fluoride may be discharged in solid, liquid and gaseous forms which enhance the widespread nature of its distribution. In some specific cases heavy contamination of the area immediately surrounding an industrial plant has been reported. This is especially true close to brickworks, aluminium smelters and some pottery factories. Steel works and many coal burning plants also contribute to fluoride pollution. Levels in agricultural soils and near fertilizer plants have increased because of relatively high amounts in many phosphate fertilizers.

Some organofluorines can be extremely toxic to fauna (fluoroacetate and fluorocitrate). However, the mechanism of fluorosis is not fully understood. The effects of fluoride on vegetation are described by the National Research Council Committee on Biological Effects of Atmospheric Pollutants (1971).

Solution preparation requires careful control in the early stages. In general, high temperature fusion and acid digestion result in loss of the element. It is recommended that both soils and plant materials are ashed and fused. If there is the least danger of acid products being formed during the ashing, a base should be mixed in with the sample as a fixative. Potassium carbonate is recommended here, as it was found that potassium hydroxide prevented complete combustion of some organic samples. Excess calcium present in bone or egg shells produces solubility problems which can be overcome by using only 0.3 g or less of the sample.

Alternative rapid acid extraction methods with either sulphuric acid (Jacobson and Heller, 1971) or perchloric acid (Agricultural Development and Advisory Service, 1986) have been used in the estimation of fluoride in plant material. Soils may be ashed and fused, or extracted with water (10 g to 50 ml for ½ hour). Soil extracts or water samples should be analysed directly for fluoride using the ion-selective electrode after mixing 30 ml with 20 ml acetate and 10 ml citrate buffer.

The method based on the ion-selective electrode is now the most popular for fluoride determination. There is no other straightforward instrumental technique available so most other methods are relatively laborious and, in some cases, insensitive. An ion-selective electrode procedure is described below. If this equipment is not at hand then the following alternatives can be considered:

Titration	Analytical Methods Committee (1972)
Spectrophotometric	Hall (1968)
	Thomas and Amtower (1969)
	Banerjee (1975)
Radiometric	van der Mark and Das (1975)
Ion chromatography	(other ions co-elute with fluoride)

A review of techniques for fluoride in vegetation has been carried out by Cooke *et al.* (1976), and Boniface *et al.* (1975) compare methods for its determination in waters.

Ion-selective electrode method

The principles of selective ion electrodes are discussed in Chapter 8. The manufacturers' instruction manuals will give details on how to use individual electrodes and meters.

Reagents

1 Fluorine standards.
Stock solution (100 ppm F): dissolve 0.221 g NaF in water and dilute to 1 litre. Store in polythene bottle.
Working standard (10 ppm F): dilute the stock solution 10 times.
Store in a polythene bottle.
2 Potassium carbonate, anhydrous.
3 Potassium hydroxide, pellets.
4 Sodium citrate, M.
Dissolve 294 g $Na_3C_6H_5O_7.2H_2O$ in water and dilute to 1 litre.
5 Sodium acetate, 2 M.
Dissolve 272 g $CH_3COONa.3H_2O$ in water and dilute to 1 litre.
6 Hydrochloric acid, 50% v/v.

Procedure

Weigh 2 to 5 g sample into a nickel crucible.
Cover with 1 g K_2CO_3 (see Note **1**).
Ash at 600 °C for 2 hours.
Add 5 g KOH and fuse until all the contents are dissolved.
Allow to cool.
Add 20 ml M sodium citrate.
Add 10 ml 2M sodium acetate.
Dissolve by gentle swirling.
Transfer to 250 ml beaker and dilute to 50 ml.
Add 17.5 ml 50% HCl (see Note **1**).
Allow to cool when effervescence subsides.
Dilute to approx. 90 ml.
Adjust to pH 5.5 with 50% HCl.
Transfer to a volumetric flask and dilute to 100 ml.
Set up the meter and prepare the electrodes as given in the manufacturers' instructions (see Note **3**).
Prepare three standards containing 0.1, 1.0 and 10.0 ppm F. Include fusion blanks, sodium citrate and sodium acetate to match the sample solutions.
Use the standards to calibrate the meter according to the manufacturers' instructions.
Read the concentration of F in the sample solutions directly and from this calculate the amount of the original samples (see Note **2**).
Subtract blank values and correct to dry weight where appropriate.

Notes

1 It is essential to cover the whole sample with a thin layer of K_2CO_3 to prevent loss of fluorine during the ashing stage. If more than 1 g is required (e.g. when fresh material is analysed) the weight should be recorded and a further 2.5 ml of 50% HCl be added at the neutralization stage for each extra gram taken.

2 The calibration is not linear near the detection limit and extra standards should be included if sample concentrations are at the lower end of the range.

3 Citrate causes a slow electrode response and it is desirable to use a chart recorder to indicate the approach of equilibrium. Alternatively cyclohexane diaminotetracetic acid may replace citrate (Harwood, 1969). Very high levels of Al and Mg may cause interference with this stage in which case they must be separated first.

Sulphide

Sulphide ions are present in soils and waters when anaerobic bacteria decompose organic matter. They are found especially in bog waters, waterlogged peats and similar stagnant situations. It is often introduced into water courses as a pollutant from paper mills, gas works, tanneries, sewage works and other chemical plants.

Apart from investigations into pollution incidents there is some ecological interest in the sulphide content of bog waters. In the section dealing with redox systems there is reference to a method for indicating anaerobic conditions through the reaction of sulphides.

Since sulphides are so readily oxidized, particular care has to be taken at the time of sampling to exclude air. Air can be displaced by using inert gases such as nitrogen or carbon dioxide but perhaps the most appropriate treatment is to 'fix' the samples immediately on collection. This can be done by adding a small volume of acetate solution to the sample (about 10 ml of the Cd–Zn acetate solution described below will be suitable). Acid samples should be first neutralized by adding a small excess of alkali. Once the sample is 'fixed' in this way it can be stored for two to three days before analysis.

No one procedure is used for sulphide to the exclusion of all others. Much depends on the sample state. A titration method is given here. Ion-selective membrane electrodes are suitable if the sample is in aqueous solution; a procedure is given by Baumann (1974). Gas chromatography can be used if hydrogen sulphide is first generated from the sample (Sugino, 1976). Further information on methods for sulphide can be obtained from Beaton *et al.* (1968), Hesse (1971), Golterman *et al.* (1978) and Standard Methods for the Determination of Water and Wastewater (1985).

Reagents

1 Sodium thiosulphate, 0.05 M.
Dissolve 12.410 g $Na_2S_2O_3.5H_2O$ in CO_2-free water and dilute to 1 litre.
For maximum accuracy standardize against KIO_3.

2 Iodine, 0.025 M.
Dissolve about 10 g KI in 30 ml water. Add 6.3454 g I_2 and stir to dissolve. Dilute to 1 litre with water. Standardize against 0.05 M $Na_2S_2O_3$.

3 Starch indicator.
Prepare a paste by mixing about 0.5 g starch powder with water and then add to about 100 ml boiling water while stirring. Cool before use. Prepare fresh daily.

4 Cadmium–zinc acetate.
Dissolve 50 g cadmium acetate and 50 g zinc acetate in 1 litre of water.

5 Hydrochloric acid, 1 + 1.

Procedure

1 Soils.

Use a distillation apparatus of the type illustrated in Fig. 7.2. Connect inlet tube to a source of nitrogen and the outlet to 2 gas bubbling bottles in series, and containing 20 ml of Cd–Zn acetate solution.

Weigh from 2 to 10 g fresh soil into the distillation flask and mix with about 10 ml water.

Connect up the apparatus.

Add 25 ml of 1 + 1 HCl slowly through the tap funnel.

Pass nitrogen slowly through the mixture and boil for 1 hour, allowing the H_2S evolved to be 'carried' through the acetate bottles via the nitrogen gas stream.

Combine the acetate solutions for the next stage.

2 Waters and extracts.

Take from 10 to 80 ml of water.

Add 20 ml Cd–Zn acetate solution and sufficient water to give a total volume of about 100 ml.

Transfer all samples to titration vessels.

Add 20 ml 0.025 M I_2 solution.

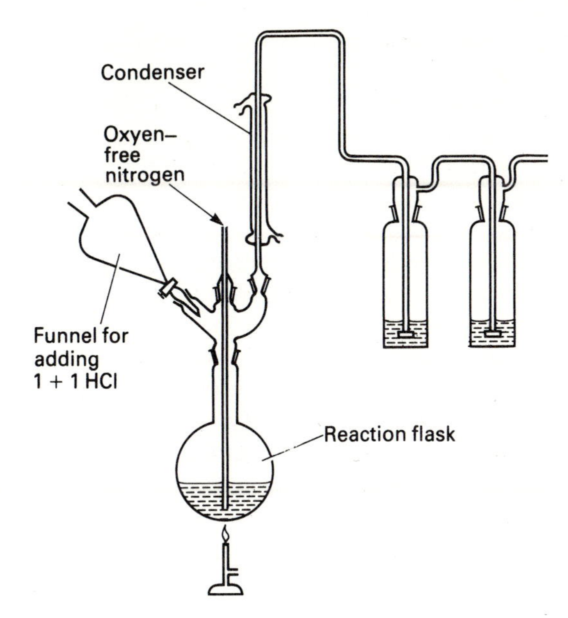

Fig. 7.2. Apparatus for the evolution and trapping of hydrogen sulphide during the determination of sulphide.

Immediately add 15 ml 1 + 1 HCl and mix.

Rapidly titrate excess I_2 against 0.05 M $Na_2S_2O_3$, adding starch solution towards the end point.

Calculate the sulphide in the original samples from the amount of I_2 used in the reaction with the H_2S.

Subtract blank values if present.

Surfactants

Surfactants, the active ingredients of detergent formulations are mainly synthetic compounds which are widely used in industry and in the home for washing, cleaning and rinsing, and in various other processes where wetting properties are of importance. They have the ability to lower the surface tension of water and depend for their properties on the presence of hydrophilic and hydrophobic groups on the same molecule.

The extensive frothing caused by surfactants, which used to disfigure the water courses and affect aeration, is now a thing of the past. The so-called 'hard' detergents containing aromatic rings and branched-chain hydrocarbons are now replaced by formulations containing straight-chain alkyl groups which are readily oxidized ('soft' compounds). Regulations on biodegradability are now strictly applied. There is, however, still concern about the effects of surfactants on nutrition, toxicity and in some instances on the acid–base status of waters. It is these properties that the ecologist will generally want to measure, rather than the surfactant concentrations, when studying the effects of these compounds in the environment.

Apart from the active ingredients, the detergent formulations contain various additives. The most important of these are the 'builders' which are included to prevent re-deposition of the dirt particles. These often contain phosphates which contribute to eutrophication which is so often a matter of concern. Many substances have been tried in place of phosphates. Some proved to be toxic themselves or had the ability to sequester heavy metals (e.g. nitrilotriacetic acid). Others have been less effective or are expensive to produce so phosphates are still included. Other additives in the detergent formulations include bleach, water conditioners, brighteners, and anti-foam agents.

There are three classes of surfactants in general use:

1 Anionic, e.g. alkaline salts of alkyl sulphates and sulphonates.

2 Cationic, e.g. salts of quaternary organic bases incorporating long-chain hydrocarbons.

3 Non-ionic, e.g. polyglycol ethers of alkyl phenols or fatty alcohol.

Identification of the class of detergent can be carried out after extracting the sample residue with an ethanol–acetone mixture, filtering, evaporating to dryness and then dissolving in a 1% aqueous ethanol solution. The addition of an equal volume of 1% cationic surfactant to this sample solution will produce turbidity when an anionic surfactant is present. Conversely, the addition of anionic surfactant will give a turbid solution if the sample contains a cationic detergent. Non-ionic surfactants may also react under these conditions, but can be separated from the others by passing an aqueous solution through a mixed bed ion-exchange column which will retain the charged ions.

Most of the surfactants in general use and contained in domestic detergents are the anionic type. Almost all methods for their determination are derived from a procedure first described by Longwell and Maniece (1955). This depends on the formation of large hydrophobic ion-pair complexes when the surfactant reacts with methylene blue or other dyes.

A procedure for estimating the linear alkyl sulphonates (LAS), the most common of the anionic surfactants, was described by Bailey *et al.* (1971) and reproduced in the first edition of this book (Allen *et al.*, 1974). This involves shaking about 100 ml of the water sample for 1 minute with 2 ml pH 6 phosphate buffer, 2 ml 0.025% copper sulphate solution, 2 ml 0.01 M 1.10—phenanthroline, 2 ml 0.001 M EDTA (added to complex the iron) and 25 ml chloroform. The chloroform layer is separated, then extracted for 1 minute with 10 ml 0.1% erythrosine, 2 ml pH 8 phosphate buffer and 10 ml water. The absorbance of the organic phase is measured at 545 nm. Solutions containing up to 50 μg Manoxol OT are used as arbitrary standards.

The non-ionic surfactants, some of which include natural products (e.g. the saponins), are now being used increasingly because they are more acceptable ecologically. Most contain alcohol- and alkyl phenol ethoxylates. A complexometric titration method (Wickwold, 1971) or a colorimetric method based on a reaction with cobalt thiocyanate (Anderson and Girling, 1982) are generally preferred for the non-ionic surfactants. These methods are precise but lack sensitivity and thin layer chromatography is sometimes used as an approximate test if levels in natural waters are low.

Cationic surfactants are used much less than the other two types and are generally limited to special applications as in fabric softeners and disinfectants. They can be determined by methods analogous to those used for anionic surfactants but with the substitution of an anionic dye. A booklet by the Standing Committee of Analysts (1982) deals with the analysis of surfactants.

Polycyclic aromatic hydrocarbons

Polycyclic aromatic hydrocarbons (PAHs) are produced as a result of the incomlete combustion of various fuels. Among the more common of these compounds are pyrene, benzo(a)pyrene, benzo(e)pyrene, benz(a)anthracene, fluoranthone and chrysene. Many of them are considered to be carcinogenic, so their presence in the environment, especially close to busy roads is a matter of concern. Other forms of animal life, apart from man may be affected but there is little or no evidence on this or about damage to plant life. However, there is the possibility that wildlife and soils may be of importance in trapping, recycling or even concentrating these compounds. Total PAH concentrations in soils can exceed 10 $\mu g\,g^{-1}$.

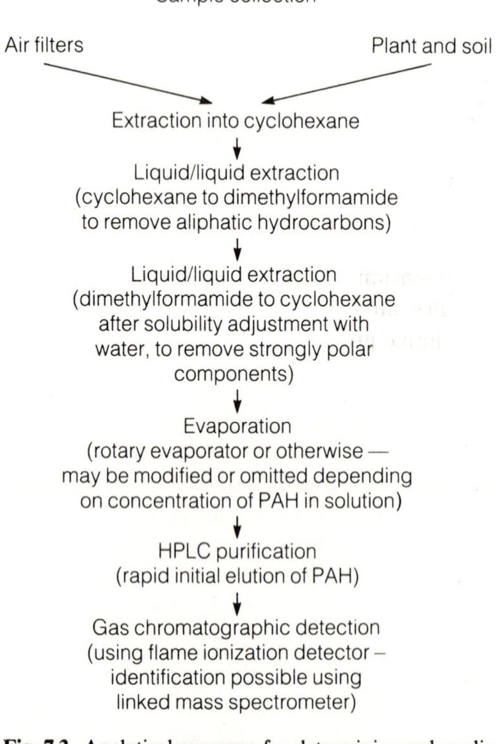

Fig. 7.3. Analytical sequence for determining polycyclic aromatic hydrocarbons.

Most of the published methods are concerned with ambient air filters. It is necessary to collect the air through multi-component filters so that volatile and particulate material can be retained. The analytical sequence is given in Fig. 7.3.

Because of PAH losses during the clean-up stages (especially liquid/liquid extraction) internal standards are essential, although the variable losses of different PAHs pose a problem. Good laboratory practice is essential if meaningful results are to be obtained. Further information about the chemical analysis of PAHs can be obtained from Jones and Leber (1979), Lawrence (1984), Tong and Karasek (1984), Jaklin and Kronmayr (1985), Hagenmaier *et al.* (1986) and the Standing Committee of Analysts (1986).

Atmospheric deposition

Over recent years there have been numerous studies and methods developed to investigate atmospheric deposition processes, and there are many publications providing details (e.g. Perry and Young, 1977, Cheremisinoff and Morresi, 1978, Scheider *et al.*, 1979). Apart from the more obvious pollution interest, it should not be overlooked that atmospheric deposition may be of importance in the supply of nutrient elements. Marine influence is also a feature of some deposition studies.

The effects of acidic deposition from anthropogenic sources are considered to have drastic consequences on vulnerable ecosystems and so methods for fractionating oxides of nitrogen, sulphur dioxide and ozone will be considered. It may not always be feasible to relate measurements of pollutants to observed effects in the field, so simulation studies are sometimes used to investigate mechanisms and critical concentration. It is then possible strictly to control experimental conditions.

For the present purpose it is convenient to divide atmospheric materials into two classes, wet deposition and dry deposition, the latter including gases, aerosols and dust.

Wet deposition

Apart from the interest in the chemical composition of rainwater there is sometimes a need to determine how the precipitation is modified as it passes through the vegetation canopy. When collecting canopy leachates (throughfall) and stemflow, debris must be excluded from the collection vessel. Coarse filters (bolting cloth) or inert funnel plugs can be used for this purpose but there is a disadvantage because the deposits are continually leached by incoming precipitation. There is no ideal solution apart from frequent attention. To obtain the concentration of the truly soluble components in the water it is necessary to filter immediately using a 0.45 μm membrane filter. A 'total income' result may be required but this to some extent depends on what is excluded by the coarse filter. It is difficult to differentiate between wet and dry deposition using standard rain gauges since the funnels may adsorb particulates and gases in dry periods and then be washed off.

Collectors that are sensitive to precipitation can be constructed, (e.g. Benham and Mellanby, 1978), but it is essential for these to respond to the early part of a rainfall event when the chemical composition is much greater. In high rainfall areas, large volumes of stemflow are produced which soon fill collection vessels, and subsampling is necessary using stream splitters (see Perry and Young, 1977). Regular and frequent sampling is essential to minimize the changes in the samples, and in some cases 'event' sampling may be required. Cloud deposition in upland areas is significant and not captured in standard rainfall collectors and therefore suitable occult collectors are required to simulate canopy capture.

In many respects ion chromatography (p. 280) is the ideal technique for anion determination in 'deposition' waters, but alternative colorimetric methods are available for most ions. In the case of phosphate the colorimetric (molybdenum blue) technique is preferable because of its greater sensitivity. Cations can also be determined by ion chromatography but in general atomic absorption methods are more convenient for the metallic elements. The treatments and procedures are given in Chapters 4 and 5.

Dry deposition

Gases

Dry deposition is a continuous process with gases being deposited onto vegetation and natural surfaces. Sulphur dioxide and nitrogen dioxide are reactive gases which are readily deposited and they, together with ozone, are now the subject of much research. In cerrtain cases there may be a local need to examine other common toxic gases such as hydrogen chloride, hydrogen sulphide and fluorine. Other gases of possible interest including ammonia and carbon dioxide are produced by natural processes. Methods for the determination of these are given in Stern (1968), and Perry and Young (1977).

Continuous field monitors are available for most of these gases, especially the pollutants. For a detailed analysis of atmospheric gases, adsorption techniques commonly using non-active polar or non-polar gas chromatographic support phases are much employed. For individual gases, absorption by a chemically active medium involving the passage of air through filter traps and flow meters, is much simpler. Gas bubbling bottles are often preferred as gas trapping media. Dreschel bottles are relatively inefficient in comparison to those fitted with fritted glass diffusers. Efficiency and saturation values should be determined.

Sulphur dioxide is the most common gaseous pollutant and is widely distributed, being a product of combustion processes. Atmospheric concentrations can be estimated by passing air (about 5–8 l min^{-1}) through 1% hydrogen peroxide solution. The sulphuric acid formed may be determined by either the ion chromatography (p. 280) or turbidity methods (p. 151). Significant amounts of nitrogen dioxide will also be absorbed, not in quantifiable amounts, but enough to enhance sulphur dioxide values if these are determined by titration against an alkali.

The amounts of nitrogen oxides (NO_x) in the atmosphere are generally low, but these species play an important role in atmospheric chemistry. A variety of methods have been used to absorb the nitrogen oxides on to cellulose filters impregnated with sodium carbonate or sodium chloride but these methods may be subject to interferences from particulate material (Appel and Tokiwa, 1981, Forest *et al.*, 1982). Diffusion tubes (Palmes *et al.*, 1975) enable nitrogen dioxides to be collected on a triethanolamine-coated gauze. The gauze can be extracted and nitrite determined colorimetrically.

Ozone is a natural constituent of the earth's atmosphere, but high concentrations are found in urban industrial environments resulting from the photochemical reactions of nitrogen oxides and unburned hydrocarbons emitted principally from vehicle exhausts. No truly specific chemical method has yet been developed for ozone. Indirect methods involving the liberation and detection of iodine (Lindquist, 1972) are available but other pollutants need to be first eliminated. Other methods depend on the bleaching properties of ozone (e.g. Guicherit *et al.*, 1973). However, chemoluminescent methods which rely on its reaction with an organic dye (rhodamine B) to produce sensitive and selective light emissions are now widely used and form the basis of continuous monitors (Perry and Young, 1977). These and other methods are reviewed by Stevens and Hodgeson (1973).

Carbon dioxide is not a pollutant, but its estimation is often required in connection with soil and plant respiration studies. There is now much concern about the effects of combustion and forest felling on atmospheric carbon dioxide concentrations so there may be more demand in future for monitoring studies. The simplest analytical procedure is to trap the carbon dioxide in an aqueous solution of alkali and then determine either gravimetrically, conductimetrically or by titration. Infrared gas analysers provide a sensitive and reliable method. Further information can be obtained from the method review of Anderson and Girling (1982).

Ammonia occurs in the atmosphere as a result of natural processes and also as a pollutant from industrial and agricultural practices. Atmospheric concentrations can be readily estimated by passing air through 1% sulphuric acid solution followed by a distillation procedure (as on p. 122).

Aerosols

Aerosols are considered to be solid or liquid particles suspended in a gaseous media and require special consideration. They are of particular interest in plant nutritional studies because foliage is able to retain aerosols directly.

Various sampling devices can be used for the collection of aerosols, varying from simple filter traps, to relatively complex electrostatic separators. Because of their simplicity, filter and impact devices are the most widely used in the field. The inert filters commonly used for sample collection are either fibrous (cellulose or glass) or the membrane type (PTFE, plastic or metal). Glass fibre filters are recommended for universal application. The filters are mounted on a disc or plug through which air can be drawn. It is important to determine background levels in the filters since they may contain significant and variable concentrations of certain elements. The size of the particles retained will depend on the nature of the filter. Normal laboratory filter papers cannot be used for particles less than about 1 μm but some glass fibre filters can be used down to 0.2 μm. The pressure drop developed as the deposition builds up must be considered, and Stern (1968) compared the characteristics for common filter types.

Impact collectors depend on a stream of air being forced through a jet and the particles (>0.5 μm) are allowed to impinge on a wet or dry surface.

Sampling instruments suitable for aerosols and other atmospheric samples are described by Stern (1968), Lee *et al.* (1972), Dzubay (1977) and Perry and Young (1977).

X-ray fluorescence spectrometry and neutron activation are particularly suitable techniques for the analysis of atmospheric deposits. In XRF, filter discs can be placed directly into the X-ray beam, but it is essential first to calibrate the instrument using a chemical method and to ensure that all the impacted sample falls within the X-ray beam. If this type of equipment is not available, it is usual to treat the filters and deposits with suitable reagents to prepare a sample solution. The acid digestion mixtures described in Chapter 3 are generally adequate for this purpose. In view of the small amounts present, some concentration stages may be required before the analytical techniques given in Chapter 5 can be used.

Dust

Dust is considered to be the fraction which is small enough to pass through a 1 mm sieve, but large enough to settle. Standard deposit gauges suffer from losses in high winds and therefore a directional dust gauge is preferred (British Standards, 1972). This type of gauge is also useful for pinpointing the source emission of the particles. Plastic funnels or cylinders may be adequate in some cases. Most of the aerosol collectors also include the dust component.

Once collected the dust samples can be analysed using the analytical techniques, given in Chapters 5 and 7. Prior acid oxidation treatment will generally bring all the dust samples into solution although in special cases fusion may be necessary.

8
Instrumental Procedures

In this chapter, various instrumental techniques available at the present time for the analysis of ecological materials are reviewed. Although some theoretical information is given, more attention is paid to practical aspects such as the applications and limitations of the techniques. References to specialist works dealing more fully with instrumental design and theory are given later in the text.

The first three instrumental groups (colorimetric, atomic absorption and electroanalytical) are mostly used for the instrumental methods given in Chapters 5 and 7. X-ray fluorescence spectrometry can also be used for the determination of many of the elements and is especially useful for multi-element assays. In the first edition of this book (1974), polarography was discussed and many of the analytical procedures were based upon its use. Since then, polarography has declined in popularity although other branches of analytical electrochemistry have evolved.

The chromatographic techniques considered here are mostly used for the organic determinations. Paper and column chromatography, discussed in the first edition, are now little used but thin-layer chromatography still has much support because it is a relatively simple, low-cost method with many applications. However, over the last 10–15 years instrumental techniques have been employed much more in the examination of organic components and two are dealt with in this chapter. Gas chromatography is of particular interest because of its suitability for pesticide analyses as well as for natural process studies. High performance liquid chromatography (HPLC), essentially an extension of column chromatography, is becoming perhaps the most widely applied of the organic analytical tools. Ion chromatography is a particular application of HPLC but is usually employed for the estimation of anions in waters.

Bomb calorimetry is only of interest to those ecologists concerned with nutrition or energy transfer studies. The technique is still included in this edition because there are relatively few sources of information about its use elsewhere.

Many analytical systems are now partly or fully automated so any analytical publication has to take these developments into account. Continuous flow colorimetry was one of the earliest of the automated processes to be accepted and with the availability of cheaper modules it is still probably the most used. Some continuous flow methods are given in Chapter 5 and general information about the technique is given in this chapter. Other automation processes are briefly discussed, and data acquisition systems are dealt with in Chapter 9.

This book has concentrated on procedures using equipment which is now widely available. There are other, more powerful analytical appliances installed in some laboratories. These instruments often have full time operators who are fully conversant with the technical features. Examples include spark source spectrometry (see Ure and Bacon, 1978) and neutron activation (see Filby and Shah, 1974; Campbell and Bewick, 1978) both used for elemental analyses, and mass spectrometry (see Williams, 1981, and Chemical Society Reviews of Mass Spectrometry) used in association with gas chromatography for the identification of unknown organic compounds. One of the more recent of the relatively expensive analytical systems, which is gaining in popularity, is inductively coupled plasma spectrometry. It is particularly appropriate for ecological research which generates large sample numbers because the equipment can be programmed to examine selected elements in rapid succession. It is briefly discussed in this chapter under Emission Spectrometry and references to additional information are provided there.

Colorimetry and related techniques

The absorption of electromagnetic radiation by chemical substances and the measurement of this absorption is the basis of colorimetry and spectrophotometry. Visible spectrophotometry extends from approximately 380 to 750 nm and is concerned with light absorption by solutions which the eye sees as coloured. The principles involved in ultraviolet and infra-red spectrophotometry are basically similar.

The technique was first used for quantitative purposes in the last century and is now very widely applied following the development of sensitive instruments.

Principles

The degree of absorption of monochromatic light passing through a solution is related to the number of molecules or ions in its path. It is often expressed as the fraction of light transmitted and is logarithmically related to both the solution thickness (Lambert's Law) and the concentration of solute in solution (Beer's Law). It is usual to combine these two relations to form the Beer–Lambert Law whereby:

$$\frac{I_t}{I_0} = e^{-kct}$$

where I_0 = incident light, I_t = emergent light, c = concentration, t = solution thickness and k = constant; or alternatively,

$$\log_e \frac{I_0}{I_t} = kct.$$

Colorimeters are normally calibrated in absorbance (formerly optical density) which is expressed as follows:

$$\text{absorbance} = \log_{10} \frac{I_0}{I_t} = k'ct$$

where k' is a new constant.

For a range of concentrations obeying the Beer–Lambert Law the plot of absorbance vs. concentration will be linear. This assumes a constant solution thickness and is applicable to nearly all the concentration ranges recommended for the colorimetric methods given in this book. However, some deviation from linearity may arise through chemical effects which include association and disassociation of molecules, interaction between solvent and solute and also pH changes. Physical effects including change of refractive index or temperature changes may also be important.

The ultraviolet region below 190 nm is less often used for analytical purposes. This is because atmospheric gases absorb ultraviolet light at these wavelengths and it is necessary to replace these with a non-absorbing gas or even to evacuate the monochromator. The spectrum usually covered in infra-red spectrometry ranges from 0.8 to 300 μm. There are relatively few infra-red applications in ecological analysis and it is not considered here although it is a valuable research tool, particularly for investigating and characterizing organic molecules.

Instruments

Instruments used in colorimetry and spectrophotometry consist of four main parts:

1 A source of radiation.

2 A means of isolating the required wavelength.

3 Compartments with suitable sample containers.

4 A means of detecting and measuring the intensity of radiation.

Sources

For wavelengths in the visible range, a source of white light such as a tungsten filament bulb is used and is often fed by a stable voltage supply to avoid fluctuations in light intensity. The common sources of ultraviolet radiation are deuterium or mercury vapour discharge lamps. In modern instruments, both the visible and ultraviolet sources are housed in the same compartment, or a single source used for both ultraviolet and visible regions.

Wavelength selection

In simpler instruments, filters are used to give a waveband covering the peak absorption. Coloured filters transmit complementary wavelengths but the bandwidth of the filters is quite wide. Interference filters have a much narrower wavelength range and consist of a thin layer of dielectric spacer material of low refractive index sandwiched between two cover glasses of optical quality. The transmitted wavelength depends on the thickness of the spacer material.

Prisms or gratings are used in spectrophotometers whose basic component is the monochromator. If used for ultraviolet work as well as visible, the prism must be made of quartz since glass absorbs ultraviolet light. Rotation of the prism or grating allows any part of the spectrum to be focused on the exit slit of the monochromator. The dispersion of a prism increases with longer wavelength so that the band-spread at the ultraviolet end of the spectrum is much less than that at the red-end where a filter may be incorporated to reduce stray light. Gratings give uniform dispersion throughout the spectrum. The use of an adjustable slit-width also reduces the passage of unwanted radiation. Excessive slit-widths are undesirable because stray light may be admitted which has not been absorbed by the sample. In some cases, e.g. the estimation of silicon by molybdenum blue, the wavelength peak is very sharp and care is needed to select and maintain the maximum absorbance. It may in these circumstances be preferable to use a narrow-band filter.

Cells

Sample cells (or cuvettes) are made from glass or quartz and used according to the wavelength of the transmitted light. When working in the visible or near-ultraviolet, the cell faces are usually made from soft optical glass which needs careful handling. Only clean soft tissues or lint should be used for drying or wiping. A mild detergent is generally adequate for cleaning dirty cells, but more resistant stains can be removed by chromic acid or an acid permanganate–periodate solution in which they may be warmed up to about 40 °C. Grease can often be removed by organic solvents.

Cells of various path lengths enable solutions with a wide range of colour intensities to be measured. For any particular run of samples it is important for the reference and sample cells to be well matched. The cell faces must always be inspected and wiped clean between readings.

Special flow cells are available for use in automated systems. These differ slightly in design depending on whether they are to be used in continuous flow systems or with 'sipper' units used in batch analysers.

Detectors

Instruments in the past used the barrier-layer cell but these have long since been superseded by phototubes and photomultipliers which are more sensitive. The phototube consists of a vacuum tube containing a cathode plate coated with an alkali metal compound which emits electrons when illuminated. The electrons pass to the anode and give a current which may be amplified. Some instruments contain both blue and red phototubes to cover a wide range of wavelengths. A more sensitive development of the phototube is the photomultiplier which, in addition to the two electrodes, has a number of dynodes enabling very small light intensities to be detected.

Most instruments give digital absorbance values, with facilities for output in concentration units, or incorporate microprocessors. Most detectors are also compatible with automatic samplers to provide fully automated analysis.

Single beam instruments are set up by using water or solvent as a reference solution to obtain zero absorbance (100% transmittance). The sample is then placed in the light path and the reading taken. Fluctuations in the intensity of the incident light will affect readings although voltage stabilizers will minimize this problem. Alternatively a double beam instrument can be used, in which the light beam is split after leaving the monochromator so that it can pass through both the reference and sample cells. In some systems the beam alternates rapidly between the reference and sample cells using a vibrating mirror. In others, two detectors are employed and signals are compared directly or the reference beam is reduced mechanically by a wedge until it is the same intensity as the sample beam. Infra-red instruments generally use double-beam scanning systems designed so that the spectra may be recorded.

Techniques

Few constituents of interest to the ecologist contain sufficient intrinsic colour to be measured directly. Exceptions include humus in natural waters and solutions containing chlorophyll and carotenoids. For most constituents it is much more usual for the colour to be developed by reaction with a chromogenic reagent. The colour intensity usually increases with the concentration of the constituent being determined, but in a number of cases an inverse relationship occurs. To be suitable for colorimetry the reaction should ideally:

1 be stoichiometric, fairly simple, and free from interferences,
2 give a colour which is stable with time and relatively insensitive to pH and temperature changes,
3 not be critically affected by slight changes in the wavelength of the incident light,
4 produce a reasonably intense colour at low concentrations,
5 be specific for the constituent being measured,
6 conform to the Beer–Lambert Law over a wide range of concentrations, and
7 be sufficiently reproducible to allow the use of permanent standards.

The conditions are never fully achieved in practice because some of the reagents used are unstable and it is recommended that fresh sets of standards be prepared with every batch of samples. Most of the reactions utilized in the methods given in Chapter 5 conform reasonably well.

Absorbance measurements are made by comparing the sample with a reference solution. In the methods described, water or occasionally the solvent is used. Some workers use a reagent blank but since this may have a measurable absorbance of its own it is preferable to use it as a zero standard and include it in the calibration curve. Absorbances of the samples and standards must be measured under the same instrumental conditions.

Construction of a calibration curve from the standard readings provides the simplest method of converting sample readings into concentrations. The graph will be linear if the Beer–Lambert relationship is followed. Departures from the law are acceptable if sufficient standards are included to plot the extent of the deviation. It should not be assumed that dilution of a sample or standard after the colour has been developed will produce a proportional meter reading. Caution is also needed over extrapolation and interpolation since the curve may not follow its predicted form. The use of permanent graphs is not recommended because of reagent instability as mentioned above.

If a computer is accessible the graphical stage can be omitted and the sample concentrations may be directly obtained. This is discussed further in Chapter 9. In some circumstances, e.g. for interference checks, it might be preferable to use the method of standard addition. In this case a known amount of the standard is added to the sample solution and is therefore subject to any interferences which may be present.

For samples which give a measurable absorbance at the wavelength of interest, a correction value for the colour will be required. This correction is determined by proceeding with the method as described but excluding the chromogenic reagent. The absorbance value obtained should be converted into a solution concentration before subtracting the correction. However, this can lead to error if the chromogenic reagent alters the background colour intensity.

Applications

A large number of the chemical estimations required by the ecologist may be carried out using colorimetric procedures. Some of these are the most convenient and sensitive available. Chromogenic reagents are available for almost all elemental and some organic constituents.

Many colorimetric procedures are described in the text and references to other procedures of possible interest are included for most constituents. Further information on colorimetric methods of analysis may be obtained from authorities such as Sandell (1959) and Boltz and Howell (1978). Some of the colorimetric techniques lend themselves to automation and these are discussed later in this chapter.

Turbidimetry and fluorimetry

Gravimetric methods are tedious and time consuming when applied to a large number of samples. It is possible to adapt some of them to a spectrophotometric system by working with very dilute solutions. If the reaction conditions are carefully controlled, the precipitate can be held in suspension long enough to measure its absorbance and the conditions of the reaction can be adjusted to control the size and shape of the particles. The relation between concentration and absorbance can be expressed by an equation analogous to the Beer–Lambert Law.

Although some instruments are specifically designed for turbidity measurements, colorimeters and spectrophotometers may be used. The choice of wavelength is not usually critical and most operating conditions are similar to those for colorimetric reactions although more stringent. A turbidimetric method for sulphur is given on p. 151.

Nephelometry is similar to turbidimetry but measurements are made using scattered as opposed to transmitted light. It has little application in the analysis of ecological materials.

Fluorescence is the radiant energy emitted when an atom returns to its ground state after being excited by absorption of energy. If the emission continues after removal of the exciting source, even for a fraction of a second, the phenomena is called phosphorescence. The Beer–Lambert Law is obeyed when the fluorescing substance is in dilute solution, but at higher concentrations the incident light is unable to penetrate the solution.

The technique has the advantages of being sensitive, precise and simple to use. Nevertheless, the number of ecological applications is limited, mainly due to a lack of suitable fluorescent derivatives for the constituents of interest. Interference from organic substances must be controlled. Some fluorimetric methods of interest have been described for aluminium (White *et al.*, 1967), boron (Parker and Barnes, 1960), fluoride (Guyon *et al.*, 1969) and selenium (Olson, 1969). Fluorescent dyes are also used in tracer studies of pesticides. Reviews on the general application of fluorimetric techniques are published in alternate years in the review editions of Analytical Chemistry and include White and Weissler (1970).

The instruments used for measurement of fluorescence are similar to those used for nephelometry since both depend on the measurement of scattered light at right angles to the incident beam. The fluorimeters generally use a high pressure mercury lamp and particular care is needed to shield against stray ultraviolet radiation.

Continuous flow technique

This technique was first described by Skeggs (1957) and subsequently further developed by Technicon Instruments with their AutoAnalyzer system. Despite the development of techniques with greater sample throughput, such as flow injection systems and discrete analysers, continuous flow analysis (CFA) is still widely used for the analysis of ecological materials. Comprehensive texts on CFA include Furman (1976), Snyder *et al.* (1976) and Coakley (1981), whilst Smith (1983) considers its theory and application for soil analysis.

Principles

The analysers consist of separate modules linked to form one flow system. The basic modules are the sampler, peristaltic pump, colorimeter and recorder. The flow of the system is controlled by the peristaltic pump in which a number of plastic tubes are squeezed between a system of rollers on a platen. The volume of fluid pumped per unit time is dependent upon the bore of the tubing used and thus reagents and sample are introduced into the system in the correct proportions. Systems of T- and Y-pieces bring together the various streams which are then mixed in horizontal glass coils. The system of pump tubes, T- and Y-pieces and mixing coils is termed a manifold. At least one tube in each manifold introduces air bubbles into the flow system (air segmentation). These facilitate mixing in the coils and sweep the liquid along the tubes thereby minimizing contamination between samples which are already separated by water introduced from the sampler. In addition to the modules listed above, others can be obtained for heating, dialysis, filtration, digestion, distillation, flame photometry and electrode measurement. Dedicated microprocessor systems are also available to control the equipment and process the data.

Many basic colorimetric reactions have been adapted to continuous flow analysis, but in some cases, e.g. when concentrated acids are involved, it is more difficult to modify manual methods. On the other hand, other methods such as the indophenol-blue reaction for nitrogen are more suited to continuous flow than manual operation.

Equipment

It is normal but not essential to purchase complete working systems from one manufacturer because manifolds can be specifically designed for the analytical requirement. Most systems are modular and expandable, and usually four channels can conveniently be housed together.

Peristaltic pumps now have the rollers close together to give pulse-free flow. Older designs, with rollers spaced well apart tend to suffer from pulsations although these may be reduced by injecting the air bubbles in phase with the pump. For the applications covered in this chapter standard PVC tubing is suitable. If solvents or concentrated acids are used then special purpose tubing must be used. Manifolds for the original AutoAnalyzers (AAI system) originally needed fairly high flow rates (typically 3–4 ml min^{-1}), but capillary connections and small coils are now available which allow lower flow rates and better control and mixing of the liquid (AA II). The connectors and coils of the manifold are mounted on plates to fix them firmly in position. Digestor modules have been used for the determination of nitrogen and phosphorus, but results tend to be low even from easily decomposed biological materials. It is also possible to incorporate distillation into continuous flow modules and this procedure may be used to overcome interferences from soil samples. Dialysers may also be included in continuous flow systems to remove colloidal interferences before the colour is developed. For many of the reactions, it is necessary to add a heating stage to speed up the rate of reaction. Rapid circulation heating baths can be used which are filled with either water (for lower temperatures) or mineral oil. Water baths are useful to supply heat to jacketed coils or for immersing mixing coils

Colorimetry is the most common detection system for continuous flow and all the main applications covered in this text use this form of detection. Filter detectors are adequate for this purpose and have the advantage of compactness. It is important to be able to accommodate flow cells up to 50 mm in order to obtain adequate sensitivity for some elements. Either tubular or square flow cells are used, and usually the sample stream needs to be debubbled before the flow cell although some flow cells do include integral debubblers. The latest generation of detectors allows bubbles to pass through without interference. Flame photometry and atomic absorption detection may be used in continuous flow whilst electrode detectors have been applied in the measurement of pH and fluoride.

Samples are presented to the continuous flow analyser from an autosampler which is controlled with timers to regulate the sample and wash cycle. The small air bubble introduced between the sample and wash solutions may be troublesome on some manifolds but can be removed by debubbling.

Techniques

Working instructions for operating the instruments will be given in the manual supplied with the equipment. The purpose of this section is to supplement this information with practical hints derived from experience.

Checklist

Problems can be avoided by carrying out a number of simple checks before the analysis is commenced. Before starting the pump check that:
1 the correct manifold is being used,
2 the pump tubes are in good order,
3 there is an adequate supply of pure water,
4 all other modules are switched on and operating,
5 heating bath solution levels are adequate and sufficient time is allowed for the bath to reach its operating temperature,
6 the wash receptacle on the autosampler is full.

Following these preliminary checks, the pump tubes should be located in position with all tubes parallel and the pump switched on initially to pump water through the manifold. Then the operator should ensure that:
7 all connections and tubes are intact and there are no obvious leaks,
8 the bubble pattern is regular,
9 the flow cell is correctly aligned and free from leaks,
10 an appropriate filter or the correct wavelength settings have been selected for the detector,
11 the reagents are prepared freshly as indicated on the method sheet.

Once it has been established that the hydraulic system is free from problems, place the tubes in the appropriate reagents and allow the system to equilibrate. Once a stable baseline reading has been established, check:
12 the zero and 100% settings on the detector,
13 the recorder is set on the appropriate range,
14 the gain on the detector using the highest standard solution,
15 the sample and wash times on the sampler and ensure the sample probe is in the correct position and exactly vertical.

Fault finding

Despite these routine checks a number of unexpected faults are bound to occur periodically. Some of these faults together with suggestions for their correction are given in Table 8.1.

Table 8.1. Continuous flow technique—faults that may arise with the equipment

Fault	Cause	Correction
1 Suck-back of reagents	**a** Blockage in coil system	Locate and clear blockage
	b Pump tubes not compressed on to platen due to:	
	i old tubing	Replace
	ii mixture of small and large size	Group small and large tubes separately
	c Flow resistance on long line	If practical to shorten, insert debubbler and resample
	d Obstruction in sample line (often due to fluff-like material in water reservior)	Examine frequently and clean out as required
2 Tendency to blow joints	Tubing blocked or restricted near joint due to foreign matter or use of excess cyclohexanone	Locate and clear blockage. Progressively uncouple joints until a liquid jet appears

Table 8.1. (*Cont.*)

Fault	Cause	Correction
3 Baseline noise	**a** Minute air bubbles or particulate matter	(i) Clear flow cell by squeezing pull-through tube or by removing flow cell and invert while running. (ii) Remove dirt from bottom of debubbler
	b Poor pump tube performance	Locate and replace defective tube
	c Detector cell or light source defective	Check and repair
	d Detector or recorder gain too high	Optimize instrument settings
	e Syphoning on waste line from debubbler to colorimeter (slow baseline oscillations and uneven bubble pattern)	Avoid long sections of waste tubing coupled to debubbler. If necessary insert small waste funnel near colorimeter
4 Baseline drift or gradual loss of sensitivity	Flow cell coated or blocked	Clean manifold and cell by washing through with a recommended reagent
5 Failure of peaks to appear	**a** Pump tube failure	Replace tube
	b Reagents incorrectly prepared	Check and remake
	c Reagent decomposed or degraded	Prepare fresh reagent. Some reagents must always be fresh
6 Pointed peaks (Fig. 8.1a)	Back pressure	See **1(c)**
7 Double peaks (Fig. 8.1c) Pointed peaks with dialyser (Fig. 8.1b)	**a** Failure of air segmentation	Replace air pump tube
	b Backpressure on long line causing suck-back	Adjust platen spring tension, insert pulse chambers or suppression
	c Optimum conditions not achieved especially in pH dependent reactions	Modify reagents to give optimum conditions
8 Broad based peaks with shallow troughs (Fig. 8.1d)	**a** Poor wash out conditions on manifold, poor bubble pattern or incorrect debubbling	Adjust pull-through pump tube to take 2/3 of total liquid volume
	b Poor joints	Check manifold
	c Dirt in system	Clean with recommended reagent
9 Sharp peaks	Peaks not reaching a steady signal	Increase sample time
10 Broad peaks that return to baseline	Sample time too long	Decrease sample time
11 Signal off-scale and not returning	Air in flow cell caused by reagent running out	(i) Squeeze-pull through tube (ii) Check reagent vessels
12 Spurious spikes on baseline and peaks or baseline shift	Air bubble in cell	Squeeze-pull through tube or remove flow cell and invert while running
13 Poor bubble pattern or bubble pattern breaking up	Grease in tubing	Wash through manifold and add detergent to one of the reagents
14 Carry over	**a** Coating on manifold	Wash out manifold
	b Poor joints	Check manifold and remake defective joint

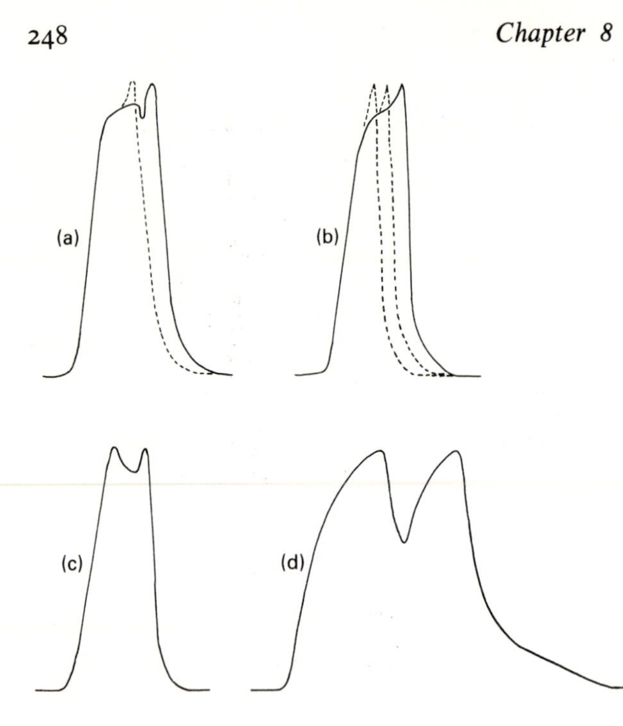

Fig. 8.1. Irregular peak patterns associated with continuous flow faults.

Hints for assembling manifolds

When assembling a manifold from a flow diagram it is important to keep the following points in mind:

1 Keep all tubing as short as possible.

2 Avoid using glass transmission tubing.

3 Use appropriate fittings. Capillary joints should be used to introduce sample streams of low flow rate. Older methods use AA1 coils and fittings which have a larger internal diameter than the AA2 type which require lower flow rates.

4 Connections must be snug and without spaces. Cyclohexanone solvent may be used for joints, but should be used sparingly.

5 The sample line should be kept short and the internal diameter must not exceed the pump tube.

6 Group all large and small tubes separately on the pump to prevent interference in the flow of the small tubes.

7 When assembled, pump water through the manifold to check the performance for leaks, bubble pattern, back pressure etc. Pump a detergent solution to clean any grease from the tubes and establish a good bubble pattern.

8 Label all reagent lines (e.g. Dymo labels) to simplify the change over from reagents to wash.

Applications

A number of continuous flow methods are described in Chapter 5. Included are methods for chloride (p. 99), iron (p. 109), ammonium-nitrogen (p. 125), nitrate-nitrogen (p. 130), phosphate (p. 137) and silicate (p. 147). These have all been well tested in the authors' laboratory and are suitable for routine purposes.

Many other methods have been published and equipment manufacturers also have standard manifolds which are suitable for a large range of analytical requirements. The review here concentrates on some of the chemical constituents of ecological importance.

Although atomic absorption is the generally favoured method for aluminium, there are some advantages in using a colorimetric technique, e.g. for sensitivity and speciation. Lancaster and Balasubramaniam (1974) developed a method based on alizarin red S to examine soil and plant digests and an alternative method suitable for water analyses was described by Henriksen *et al.* (1975). Basson *et al.* (1969) proposed a method to determine trace levels of boron using azomethine-

H as the chromogenic reagent and this has been successfully applied in the analysis of plant tissues and soil extracts.

Most of the published sulphate methods have limitations, especially for low level determinations. Indirect methods based on methylthymol blue, chloranilate or thoron suffer from interference from cations and so adaptations to continuous flow require an ion-exchange column in the manifold to remove them. The most common continuous flow method for sulphate is based on barium sulphate turbidity, e.g. Sinclair (1973), and Wall *et al.* (1980), but this method is far from ideal and suffers from deposition of the precipitate on the tubes. A wash with EDTA between samples prevents build up of the deposition, but also complicates the procedure.

Organic carbon is an important parameter in natural waters and recently a continuous flow method based on ultraviolet oxidation with persulphate and colorimetric detection at 550 nm has been produced by one manufacturer (Skalar).

Although methods based on concentrated acids are difficult, one application for soluble carbohydrate analysis of herbage (Thomas, 1977) uses anthrone in 76% H_2SO_4. This simple manifold overcomes some of the control problems encountered in the manual method. An alternative method for total sugars (Gaines and Gascho, 1985) with potassium ferricyanide has recently appeared.

Modules are available for electrode measurement, and there are several references to pH measurement in soil solution. It is important to check the continuous flow methods against manual methods as electrodes tend to be slow to reach equilibrium for some samples which may cause bias (Grigg *et al.*, 1980).

Atomic spectrometry

The development of flame techniques, using emission and absorption spectra has made sensitive and highly specific methods available to the analyst. For many elements, in particular the alkali and alkaline earth elements, these methods have superseded most others. Estimates take only a few minutes providing certain precautions are taken regarding interferences, flame conditions, etc. Initially instruments depended on direct emission but in ecological studies were only suitable for sodium, potassium and sometimes calcium. Now atomic absorption can be used for most metals of interest. More recently resistively heated graphite rods or furnaces have been used in place of flames for atomization, especially in the measurement of trace levels of heavy metals. All three techniques are considered in the ensuing discussion.

Flame emission

Principles

When atoms of an element are heated in a flame, some of the heat energy is absorbed by a few of the atoms which become excited, i.e. there is a transition by one or more electrons from the ground state to higher energy levels. On reverting to the ground state the electrons lose this energy which is emitted as electro-magnetic radiation. The wavelength of the emitted radiation is governed by the energy change involved in the transition. For each element there are certain permitted shifts giving rise to a series of lines, each series being characteristic of the element. The intensity of any one line is governed principally by flame temperature, other atomic species present and the number of atoms of the element in the flame at any one instant. If operating conditions are kept constant, the intensity of radiation will be a measure of this number. This is the basis of quantitative flame photometry.

Instruments

Instruments used for flame emission consist of three main parts: the burner-sampler system, a means of isolating the required emission line and the detection measuring system.

The design of the aspiration system which introduces the sample into the flame has been an important factor in the success of flame instruments. Most widely used is the nebulizer-spray chamber system. In this the flow of oxidizing gas through a tube draws the solution through a concentric inner capillary. The solution leaving the inlet jet is partly disintegrated into a fine spray by the gas stream. This process is called nebulization and the term 'atomization' which was formerly used is now restricted to the reduction of elements to the atomic state. Only a small proportion of the solution aspirated passes into the flame and the remainder condenses in the spray chamber and is drained away. Both the size of this proportion and the overall rate of aspiration are related factors which must be controlled to achieve maximum sensitivity.

Some instruments are designed to monitor internal standards. These contain two detectors, one of which monitors the emission line of the internal standard and the other the line of the element being determined. Changes in the emission of the internal standard cause the instrument automatically to adjust the apparent emission of the element line to compensate for variations in flame background and in solution uptake rate (see Feldman, 1970).

The emission line can be isolated by a filter, prism or grating as described for colorimetry. Simple flame photometers employ a filter, but for sensitive work a spectrophotometer is desirable. The detectors and measuring system are basically the same as discussed under colorimetry but can be elaborated as suggested later under atomic absorption. Multi-element instruments are now available, usually for sodium, potassium and lithium, but sometimes including calcium. These have a series of filters and detectors around the flame, and measure all the elements simultaneously. The lithium channel can be used for an internal standard or for determination of the element. These instruments were developed for clinical use but are suitable for ecological analyses with little or no modification.

The addition of a sequential sampler and chart recorder results in an automated system which can be further enhanced by monitoring the analogue output by interfacing with an external computer, hence allowing automated data processing. Modern equipment often incorporates a microprocessor which allows various degrees of sophistication from simple monitoring of instrumental parameters to complete control of the analysis and internal data processing.

Techniques

Nebulizer

Much of the success in achieving good results with flame instruments depends on the design and correct use of the nebulizer-flame system (see Smith and Bremner, 1984). Solutions must be free from suspended matter to avoid blocking the nebulizer. This is a frequent cause of instability and may be checked by spraying standard solutions regularly. Dust from the surroundings and the fibres picked up from filter papers used during prior preparation of the sample solution are often troublesome. It is often preferable to allow a sample solution to settle and decant off a suitable volume for aspiration. Many blockages can be cleared with the fine wire supplied for this purpose or back flushing by placing a finger over the outlet end with the oxidizing gas flow on. The flame must be extinguished before removing the nebulizer for back flushing. Regular flushing with solvent and checking of the zero standard should always be carried out and at the end of a run thorough flushing with the solvent (usually water) is essential.

Changes in the spray rate from sample to sample resulting from variations in viscosity can sometimes be troublesome but may be controlled by adding an internal standard such as lithium (see below). The dual detector instrument mentioned above is then needed.

Flame conditions

There are great differences in the ease with which various elements are excited in a flame: e.g. sodium and potassium are more readily excited than calcium and magnesium. For emission in particular the available energy is therefore important and varies from the very hot cyanogen–oxygen flame (4550 °C) to the relatively cool propane–air system (1920 °C). Metals such as calcium which tend to form heat stable oxides in the flame are better determined at the higher temperatures obtained with air or oxy–acetylene flames.

It is essential to maintain stable flame conditions. Small changes in temperature can lead to considerable instrumental instability. The flame and burner temperatures should be allowed to come to equilibrium by aspirating the solvent into the flame for a few minutes before taking any readings. The flow rates of the fuel and oxidant gases are extremely important in this respect and should be carefully set and maintained throughout, the run to achieve maximum sensitivity and stability. The height of the burners is a contributing factor which must also be regulated. Manufacturers usually supply details of optimum flame conditions for their instruments but these should be checked independently.

Sudden changes in gas pressure must be avoided. A large fall in air pressure may result in a minor explosion caused by the flame 'sucking-back' into the chamber. For this reason it is important to turn the fuel supply off first at the end of the test. When using acetylene the cylinder pressure should not be allowed to fall too low or the acetone vapour from the solvent in these cylinders may be aspirated into the flame and affect the readings.

As mentioned above, flame and nebulizer errors can be controlled by the use of internal standards to compensate for the background variations. Feldman (1970) discusses the importance of controlling flame conditions for both emission and absorption techniques.

Interferences

A number of interferences occur which are associated with atomization conditions. These can usually be reduced by careful control of the flame conditions or solution preparation and have largely been taken into account in the methods given in this manual. The more important classes of interferences are given below.

1 *Radiation effects*: resulting from band spectra of other elements or molecules and also from an intense adjacent emission line.

2 *Self absorption*: this is a related effect whereby unexcited atoms absorb emitted radiation resulting in reduced sensitivity.

3 *Formation of refractory molecules*: this reduces the number of atoms available for excitation. An example of this is the interference of aluminium and phosphate on calcium emission. It may be suppressed by the addition of lanthanum or strontium as a releasing agent (see p. 89). The formation of oxides in lower temperature flames, already referred to is another example of this effect. The use of a nitrous oxide flame often overcomes this difficulty.

4 *Ionization*: this occurs most readily with the alkali metals in hotter flames and is significant because ionized atoms do not emit radiation at the wavelength normally selected for the analysis. The effect increases with temperature and is greater for a metal present in low concentrations. However, when other ionizing species are present these furnish electrons which tend to inhibit ionization. An example of this is the enhancement of alkali metals by other alkalis. This type of interference can usually be controlled by the addition of an ionization buffer to both standard and sample solutions. These are solutions of salts of easily ionized elements, such as potassium chloride or lithium sulphate in sufficient concentrations to swamp inter-element effects due to sample composition.

The concentrations at which inter-element interferences occur differ with each type of instrument and should be checked before bringing the instrument into routine use.

Choice of spectral line

In general the spectral line chosen for the determination of an element is that which gives the greatest sensitivity and will usually be the most intense line. However, if there are interferences with this line it may be necessary to choose another. The elements determined by flame methods in this handbook each have one particularly intense line which is utilized, although others are available in some cases.

If a prism instrument is used, temperature changes may alter the refractive index of the prism and hence the scale reading of peak wavelength.

To avoid this possibility the instrument must be allowed to come to equilibrium before use.

The wavelength scale should only be used as a guide. The actual position of the spectral line can be found by scanning the wavelength control until maximum transmission is obtained for a standard solution of the element being tested.

Other features

The standard range chosen should allow the sample readings to be in the optimum range of the instrument employed. If scale expansion is unavoidable, only the minimum amplifier gain should be used. It is better to obtain maximum readings through burner and flame control rather than the use of scale expansion because this amplifies any background noise.

A calibration curve constructed from standard solutions is generally used to determine the concentration of the test solution. Some of these graphs are permanent, although they should be checked at intervals.

Organic solvents such as iso-propanol may be used to enhance the sensitivity. The optimum conditions for this effect need to be found for each system (background noise or fuel conditions). Any dilution required to introduce the solvent will partly offset any enhancement effect.

Applications

Methods of determining sodium, potassium and calcium by flame photometry are given in the appropriate sections of this text and techniques for overcoming chemical interferences are suggested where necessary. Reference works dealing with flame photometry include Ramirez–Munoz (1968) and Alkemade and Herrmann (1979).

Atomic absorption

Many of the points dealt with under flame emission concerning nebulizer design, flame conditions and interferences also apply to atomic absorption. Some of the more important differences are dealt with below.

Principles

The use of flame absorption spectra instead of emission spectra was first proposed and developed by Walsh (1955). Atomic absorption spectra are formed by the absorption of radiation of certain wavelengths by atoms whose electrons are in the ground state. On absorbing this energy the atoms become excited. The extent of absorption is dependent on the number of atoms in the ground state in the path of the radiation beam at any one time and can thus be used as a quantitative method of determining this number.

Atomization can be achieved in a number of ways, the most common being the use of the air–acetylene or nitrous oxide–acetylene flames, the sample being introduced by means of a nebulizer-spray chamber system. Radiation of a characteristic wavelength from a hollow cathode discharge lamp is passed through the flame and the decrease in intensity is measured using a monochromator and detector system. This decrease is related to the concentration of the element in solution. Electrothermal atomization is dealt with later.

Other atomization techniques in common use are the cold vapour technique for mercury determination and hydride generation for elements such as arsenic, antimony, tin and bismuth. The former depends on the fact that mercury has an appreciable vapour pressure at room temperature. Mercury compounds are reduced to the metal using stannous chloride, and transported into the tube in the radiation beam by passing nitrogen or argon through the solution. The hydride technique depends on the ability of some elements to form volatile hydrides on reduction with sodium borohydride or lithium aluminum hydride. These can be swept by a stream of nitrogen into a silica tube mounted in the radiation beam above a conventional air–acetylene burner. This gives an increase in sensitivity of 100–300 times for many of the elements.

Instruments

Atomic absorption instruments are basically spectrophotometers with a burner compartment instead of a cell compartment. They consist of a source of radiation, burner plus sampler compartment, monochromator, and a detection and measurement system.

Radiation source

Light sources such as tungsten filament and hydrogen discharge lamps give a continuous spectrum and so are unsuitable for atomic absorption work. The usual source is a hollow cathode lamp. This is a discharge lamp which has a hollow cylindrical cathode made from a material which contains a substantial proportion of the element to be measured. The radiation produced corresponds to the emission spectrum of that element and so the required line may be readily isolated by the monochromator. Individual lamps are available for a large number of elements. Lamp turrets which allow advance warming-up to take place are a feature on most instruments.

An alternative to the hollow cathode lamp is the electrode-less discharge lamp. This type of source is not so widely used, but, due to its more intense spectral line emission, can give increased sensitivity for some of the more volatile elements such as arsenic, antimony, bismuth, selenium and tellurium.

Burner system

Most instruments incorporate a spray chamber as part of the nebulizer system. The essential features are the same as those discussed for flame emission.

Many manufacturers supply interchangeable burners so that different fuel systems can be used or to enable the instrument to be used in emission or absorption modes. Multi-slot burners specifically designed for use with sample solutions of high solid content (Boling burner) greatly reduce the problem of burner blockage. An auxiliary air supply is incorporated for use with these burners and allows a fuel-rich flame to be achieved without a large increase in flame background effects. Another burner in frequent use burns a nitrous oxide–acetylene mixture enabling high temperatures to be achieved with a normal flame velocity. This is particularly useful for estimating aluminium, but may be employed for the estimation of calcium without the addition of a releasing agent. The interferences expected in this type of flame were discussed by Marks and Welcher (1970).

Instrumental features

Some instruments use a double beam arrangement as already mentioned in the colorimetry section. The radiation from the source is split by means of a vibrating mirror so that the beam alternately passes and by-passes the flame. This has the effect of modulating the flame beam enabling the detector circuit to adjust for changes in source intensity.

In single beam instruments changes in source intensity have to be allowed for by adjustment of the gain control and by frequent checking of the standard readings, although automatic zero control is a feature of most modern equipment. In addition improvements in power supply design have reduced source drift considerably.

Background absorption can be troublesome, especially in the ultraviolet part of the spectrum, i.e. below 300 nm, and many instruments offer correction systems for this kind of interference. The traditional and simplest method is to monitor the continuum output of a deuterium lamp passed through the flame alternately with the hollow cathode lamp beam and to correct the analyte signal for any absorption detected. The Zeeman effect may also be utilized, in which atomic emission or absorption spectral lines are polarized by a magnetic field, but not the background, thus offering a method of correction for the latter. A recently introduced procedure for background correction uses the hollow cathode lamp as its own background corrector. Each pulse at normal lamp current is followed by a short pulse of much higher current which broadens the emission line sufficiently for background absorption to be dominant during this short period. The analyte signal detected during the normal current pulse can then be corrected accordingly.

The detection arrangements in atomic absorption are essentially the same as those described under colorimetry and flame emission. The addition of a sequential sampler and chart recorder forms the basis of an automatic system.

Initially, extensive developments in atomic absorption instrumentation occurred. However, latterly, these developments have slowed somewhat, especially as regards the flame-based systems. Improvements in recent years have mainly been seen in the increased use of microprocessors for instrument control and data processing (Chapter 9) and also in flameless atomization systems, particularly electrothermal atomization.

Techniques

One of the most important points in good atomic absorption practice is the need for correct treatment of the source lamps. Hollow cathode lamps must be run at their specified currents. Too low a current may give insufficient sensitivity but too high a current will shorten the life of the lamp. As a general rule, best sensitivity is obtained between 50 and 70 per cent of the maximum current, and best precision between 75 and 90%. The output of hollow cathode lamps can drift, producing a gradual shift of the zero standard reading. Frequent checking on the instrument controls is the most effective way of counteracting this. Alignment of the lamp is critical and this should be checked periodically.

In general the design of and conditions for using the nebulizer burner and detection system are very similar to those discussed under flame emission. Fuel and oxidant supplies are especially critical and should be measured as flow rates rather than gas pressures. Long multislot burners with an auxiliary air supply have already been mentioned. The flow rate of the auxiliary air must be carefully adjusted so that the pressure drop across the nebulizer is not reduced, thus causing a decrease in the aspiration rate.

The flame itself will emit some light of the wavelength in use but in practice this is eliminated by using a modulated source and an amplifier in the detector circuit tuned to the frequency of modulation. Background correction may also be necessary.

In general there are few interferences in flame AAS and techniques to overcome these are noted for the individual elements. After the elimination of flame and chemical interferences the most important causes of error in atomic absorption are:

1 nebulizer blockage,
2 changes in air and fuel flow rate,
3 wavelength drifting off peak,
4 very low acetylene cylinder pressure,
5 hollow cathode lamp drift.

Burner height should be optimized for each element analysed. The problem of wavelength drift is sometimes accentuated by the heat absorbed by the monochromator compartment when the burner is first switched on. In such cases the instrument should be allowed to stabilize. Most modern instruments are sufficiently well-insulated to avoid this problem.

Calibration with standard solutions to produce a graph is similar to the practice followed in emission and colorimetric techniques. Standard graphs may be permanent, but are often slightly curvilinear since resonance broadening and flame absorption variations tend to bend the calibration lines towards the concentration axis.

Applications

Methods for the determination of aluminium, calcium, copper, iron, magnesium, manganese and zinc are given elsewhere in this text and include details for controlling interferences. Sodium and potassium are readily determined by atomic absorption, but flame emission is still commonly used because it is just as sensitive for these elements and the extra cost of the lamps is avoided.

Techniques to concentrate trace metals, involving complexation and solvent extraction are often used in association with atomic absorption. Examples of these are given in Chapter 4 and 5 and further details are provided by Subramanian and Meranger (1979). A guide to literature on this subject has been prepared by Wilson (1979).

Publications dealing with atomic absorption spectrometry and its applications include Christian and Feldman (1970), Price (1979), Slavin (1979) and Isaacs (1980). Flame techniques (emission and absorption) suitable for elements of ecological interest are summarized in Table 8.2.

Table 8.2. Summary of flame techniques applicable to the analysis of ecological materials

Element	Wave length (nm)	Mode	Flame	Suitable ranges (ppm)	Potential interferences	Control of interferences	Comments
Al	309.3	A	N_2O–C_2H_2	0–20	Fe, Ti, Ca, HOAc	Include Fe in standards	Very hot flame essential EA possible
Ca	422.7	E	Air–C_2H_2 O_2–C_2H_2	0–40	Al, P, Si, S	Add releasing agent (Sr or La salt)	A more sensitive than E
		A	Air–C_2H_2	0–10 0–40	As above	As above	
		A	N_2O–C_2H_2	0–40	Na, K	Include ionization buffer	
Co	240.7	A EA	Air–C_2H_2	0–50 0–1	Fe if high	Extract	Prior concn normally reqd. EA preferable
Cr	357.9	EA	—	0–0.1	Ca, Fe	Standard compensation	EA usually essential
Cu	324.8	A	Air–C_2H_2	0–5	None significant		Prior concn desirable. EA preferable
Fe	248.3	A	Air–C_2H_2	0–20	Si	Add $CaCl_2$	
Hg	253.7	A	Flameless	0–1	None		Hg vapour is produced which is passed into an abs cell in the light path
Mg	285.2	A	Air–C_2H_2	0–3	Ca, Al, P, Si	Add releasing agent (Sr or La salt)	
Mn	279.5	A	Air–C_2H_2	0–2 0–20	None significant		Samples and standards to be in same valency state
Mo	313.3	A	Air–C_2H_2	0–3			Prior concn with organic extraction
K	766.5		Air–C_2H_2 Air–propane	0–10 0–100	Na, Ca, H_2SO_4, HCl	Include in standards Dilute sample	A has no marked advantage over E
Na	589.0		Air–C_2H_2 Air–propane	0–5 0–20	Ca, H_2SO_4, HCl	Include in standards Dilute sample	E preferable
Ni	232.0	EA	—	0–0.2	None serious		EA essential
Pb	217.0	EA A	— Air–C_2H_2	0–0.5 0–10	None serious, Cl possible		EA desirable Prior concn necessary
Zn	213.9	A	Air–C_2H_2	0–1	None significant		

E = Emission, A = Atomic absorption, EA = Electrothermal atomization.

Electrothermal atomization

The short time that the excited atoms within the flame stay in the radiation path limits the sensitivity of conventional atomic absorption. The development of a small, high-temperature graphite furnace has enabled the atoms to be contained within the beam much longer, resulting in higher absorbance values, and gives a better signal to noise ratio, thus reducing detection limits. These are in the picogram range in many cases, which makes the technique of particular value for the determination of trace metals, for example in natural waters. Early development of the technique was carried out by L'vov (1970) and most commercial units are based on the design of Massman (1967).

The furnace consists of a graphite tube clamped between two direct current (dc) terminals. The geometry of the tubes varies between manufacturers but, in general, the length is 3–5 cm and the internal diameter 5–8 mm. A 2 mm diameter hole is located midway along the tube to allow sample introduction. In practice the arrangement is such that the furnace can be water-cooled and also surrounded by an inert atmosphere of nitrogen or argon to prevent oxidation of the graphite at the high temperatures used. The furnace is mounted in the burner compartment of the spectrophotometer so that the tube is along the optical axis of the hollow cathode lamp beam. The dc power supply must be capable of high currents at low voltages allowing the tube to be heated to about 3000 °C in less than a second. Temperature control circuitry is incorporated as different analyses require different temperature regimes.

A drop of sample is introduced into the tube using a micropipette. The volume used is usually 50 μl or less. The analysis then proceeds using a four stage cycle:

1 Drying—to remove solvent.
2 Ashing—to remove organic matter.
3 Atomization—analysis stage.
4 Burn-off—to clean tube.

Each stage is programmable for temperature and time and, in the case of drying and ashing, most modern instruments also allow the rate of temperature increase to be specified.

The drying temperature employed depends on the solvent. In the case of aqueous solutions 100–150 °C is normally used with a rate of temperature increase chosen to avoid spitting within the tube. The ashing phase takes place 200–300 °C below the atomization temperature. Optimum rate of temperature increase and final temperature and time will depend on the sample matrix and requires investigation for each sample type. The atomization temperature depends on the element being determined. The optimum value can be found by experiment, but most manufacturers recommend values for their equipment. The atomization phase usually lasts 3–5 seconds, the analytical signal being a pulse displayed on a flat-bed chart recorder or detected and processed by a computer. Finally the graphite tube is heated to about 3000 °C to burn-off any residues and provide a clean tube for the next analysis.

Quantitative analysis may be carried out by comparison with a standard curve in the normal manner or by standard addition. Graphite furnace atomic absorption is particularly suited to the latter technique, and, since matrix problems can be more acute than in flame atomic absorption, standard addition is often the method of choice. The ashing and atomization stages are extremely important in overcoming matrix interferences and it is worth spending some time optimizing these aspects to avoid errors later. Background correction is essential in furnace AA to overcome non-specific matrix effects and the addition of a matrix modifier, e.g., nitric acid may be necessary. The introduction of pyrolytically-coated and totally pyrolytic graphite tubes by manufacturers has reduced some interference effects encountered when using normal graphite.

Reviews that deal with electrothermal atomization include L'vov (1978) and Sturgeon and Chakrabarti (1978) and conditions for the determination of elements by this method are discussed by Fuller (1977). Recent developments, especially 'atom trapping' techniques are summarized by Brown and Morton (1985).

Emission spectrometry

Emission spectrometry with dc arc excitation has been used for many years in the analysis of soils and vegetation but this excitation procedure limits its use to solid materials. Recently though, the development of the inductively coupled argon plasma source (ICP), which requires the sample to be in a liquid form has increased the scope of the technique. The availability of relatively cheap and powerful microcomputers for spectrum evaluation has also contributed greatly to the effectiveness of the instrument and it is increasingly being used in larger laboratories.

A detailed description of the technique is beyond the scope of this book. Briefly however, the sample is introduced into the plasma using a conventional nebulizer-spray chamber arrangement. The high temperature maintained in the plasma (8000–10 000 °C) gives rise to a composite emission spectrum due to all the elements present in the sample. This is separated into its component lines using a high resolution optical system. The intensity of a particular line is proportional to the concentration in the sample of the element producing it. Optical systems may be simultaneous or sequential. The former contain a number of detectors each permanently positioned to monitor a given line. Intensities of the lines are measured together. In a sequential system the spectrum is scanned by a single detector and the required lines examined as they are encountered.

The efficient atomization and detection system makes ICP spectrometry one of the most powerful of all analytical systems. As a technique it is preferable to most others for the determination of many of the elements in ecological materials but its cost is high. Books describing the technique include Thompson and Walsh (1983) and Boumans (1987). Further information may be obtained from articles and Atomic Spectroscopy Updates in the Journal of Analytical Atomic Spectroscopy and the Annual Reviews in Analytical Chemistry.

X-ray fluorescence spectrometry

Principles

When an atom of an element is bombarded with particles or radiation of sufficient energy, an electron from one of the inner shells of the atom may be ejected. The resulting vacancy is then filled by an electron from a higher energy level. The excess energy released by this electronic transition is given off as fluorescent X-rays. A number of different transitions are possible, giving rise to a series of X-ray lines of different energies, and the energy spectrum obtained is characteristic of the element concerned. The energy of a particular X-ray line is inversely proportional to the wavelength and proportional to the square of the atomic number of the excited element, thus heavier elements emit higher energy and shorter wavelength lines than light elements. The desired line is isolated using either a wavelength or an energy dispersive system. In the former case a diffraction crystal is used and the intensity is measured at a known angle. In the latter the detector is a high resolution semiconductor and the intensities of the various lines in the spectrum are stored in a multichannel analyser. For a given element line, intensity is proportional to the amount present in the sample.

Instruments

Equipment used for XRF measurements range from manual to fully automated systems interfaced directly or indirectly to a computer. These include wavelength and energy dispersive instruments. Both types have similar excitation sources and sample compartments.

Excitation source

The traditional excitation source is an X-ray tube. This is operated at high voltage and low current, provided by a generator with a very stable output. The anode, or target, of the X-ray tube is coated with a suitable metal, the choice of which depends on the application. The most useful target materials for ecological work are chromium (for the lighter elements, Cu to Na) and rhodium or molybdenum (for the heavier trace elements). The tube is operated at up to 100 kV and up to 50 mA with total power output of up to 4 kW. X-ray production efficiency is about 1%, the remaining energy being dissipated as heat which is removed by a water cooling system.

Most spectrometers employ X-ray excitation, but in recent years, other excitation techniques have become popular for some applications. The use of high energy electrons for this purpose has been known for some time. Electron excitation results in higher fluorescent yields especially for light elements and can be used down to atomic number (Z) = 6 (carbon). X-ray excitation is only effective down to $Z = 9$ (fluorine), although for practical purposes, in biological material the limit is $Z = 11$ (sodium). Excitation may be achieved using an electron gun or, more usually, in association with an electron microscope. In the latter case extremely small specimens can be examined, and the technique is particularly useful in geological work for the examination and characterization of minerals. Further information on electron excitation may be obtained from Reed (1975) and Heinrich (1981).

Particle induced X-ray emission (PIXE) spectrometry, in which excitation is achieved by bombarding the sample with high energy protons or α-particles rather than electrons is becoming increasingly prominent. Its main advantages are the very low backgrounds from the source itself, resulting in lower detection limits, and the high fluxes available. However, in the case of high energy protons, access to a suitable accelerator is required, which means the technique is unavailable for many routine analytical laboratories. α-particle excitation involves the use of fairly active americium or curium sources and the hazards associated with the handling of these materials make it unattractive to many users. Nevertheless, PIXE is now a well-established technique.

Radioactive sources emitting γ- or β-rays can also be used for excitation and are particularly useful where power and weight restrictions may occur, for example, in portable X-ray analysers.

Sample compartment

The sample compartment allows introduction of the specimen into the exciting radiation beam. Most modern commercial X-ray spectrometers use an evacuated X-ray path to avoid attenuation of the longer wavelength, less energetic, X-radiation obtained from lighter elements. An airlock arrangement is therefore usually incorporated to avoid continually releasing the vacuum within the spectrometer itself. In addition, facilities may be available for use with an automatic sequential sampler, allowing unattended operation. This is an advantage in routine applications where large numbers of samples are handled.

Spectrometric systems

Wavelength dispersive spectrometers

Wavelength dispersive spectrometers employ a crystal to diffract the X-ray beam according to Bragg's Law:

$$n\lambda = 2d \sin\theta$$

where λ = wavelength, θ = angle of incident radiation, d = distance between crystal planes and n = order of reflection.

It follows from this that for first order reflections ($n = 1$), which are the most intense and therefore normally used, $2d$ must be greater than the required wavelength. However, it also follows that angular dispersion is inversely proportional to $2d$ for a given wavelength. Therefore, in practice it is preferable to employ a series of crystals with different $2d$ spacings and to select the one which gives optimum angular dispersion of the spectral line of interest.

An outline of a wavelength dispersive spectrometer is shown in Fig. 8.2 and instrumental conditions are summarized in Table 8.3. The fluorescent X-rays from the sample pass through a primary collimator to give a parallel beam at the correct take-off angle relative to the primary beam (usually 90°). After diffraction by the analysing crystal, the beam passes through a second collimator to the detector. This secondary collimator increases resolution by excluding unwanted radiation scattered from the surface of the crystal. The angles of the crystal and detector relative to the primary collimator are controlled by a goniometer, the geometric arrangement being such that the detector arm always moves through exactly twice the angle moved by the crystal. Thus the goniometer is calibrated in terms of 2θ. Tables giving 2θ values of X-ray lines for the analysing crystals in common use are readily available (e.g. Powers, 1960) and can be used to obtain the position of the detector arm for any particular line of interest. Alternatively 2θ values

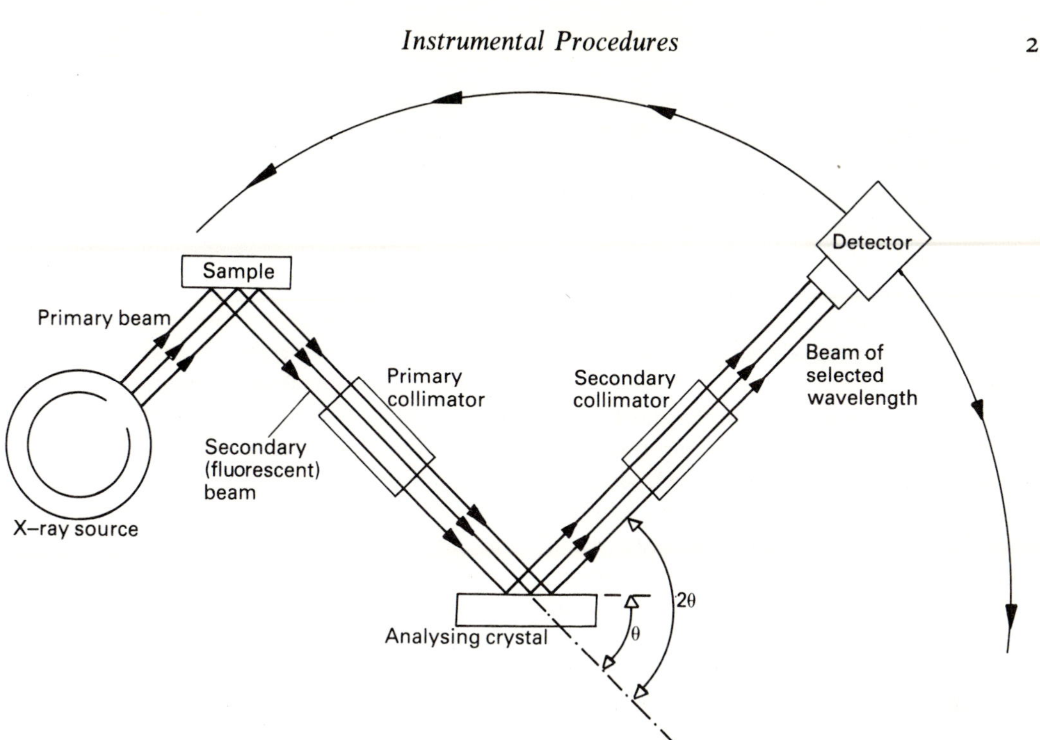

Fig. 8.2. Outline of a wavelength dispersive XRF spectrometer.

Table 8.3. Suggested instrumental conditions for wavelength dispersive XRFS

	Wavelength[1] (nm)	Tube, anode material	Crystal	Counter
Aluminium	0.8339	Chromium	PET	Gas flow proportional
Calcium	0.3360	Chromium	LiF	Gas flow proportional
Cobalt	0.1790	Gold or tungsten	LiF	Gas flow proportional
Copper	0.1542	Gold or tungsten	LiF	Scintillation
Chlorine	0.4729	Chromium	PET	Gas flow proportional
Iron	0.1937	Gold or tungsten	LiF	Gas flow proportional
Magnesium	0.9889	Chromium	ADP or KAP[2]	Gas flow proportional
Manganese	0.2103	Gold or tungsten	LiF	Gas flow proportional
Molybdenum	0.0710	Gold or tungsten	LiF	Scintillation
Phosphorus	0.6155	Chromium	PET[3]	Gas flow proportional
Potassium	0.3744	Chromium	PET	Gas flow proportional
Silicon	0.7126	Chromium	PET	Gas flow proportional
Sodium	1.1909	Chromium	KAP[2]	Gas flow proportional
Sulphur	0.5373	Chromium	PET	Gas flow proportional
Titanium	0.2750	Chromium	LiF	Gas flow proportional
Zinc	0.1437	Gold or tungsten	LiF	Scintillation

KAP—Potassium acid phthalate, ADP—Ammonium dihydrogen phosphate, PET—Pentaerythritol, LiF—Lithium fluoride (200). [1] First order $K\alpha$ lines for all elements; [2] Rubidium acid phthalate (RAP) may be used instead of KAP; [3] Germanium (Ge) which eliminates even order reflections may be advantageous in this case.

can be calculated and tabulated quickly using a computer if the $2d$ spacing of the crystal is known.

The two types of detector employed in wavelength dispersive spectrometry are the gas flow proportional counter and the scintillation counter. Both depend on the ionizing properties of X-rays.

The gas flow proportional counter is used for

longer wavelengths and is particularly useful in biological applications. The ionizing medium is a constant flow of argon containing about 10% methane. Electrons produced when an X-ray photon causes ionization are attracted to an anode wire held at a positive potential. Each photon thus causes a voltage pulse, the amplitude of which is proportional to the energy of the photon and hence inversely proportional to the wavelength.

The scintillation counter, which is more efficient for shorter wavelengths, contains a crystal of thallium-activated sodium iodide which acts as a phosphor. Interaction with an X-ray photon causes a flash of light which is detected by a photomultiplier tube and converted into a voltage pulse. As before, the amplitude of the pulse is proportional to the energy of the X-ray photon.

Pulses from the detectors are amplified and shaped, subjected to pulse height analysis and finally counted. Pulse height analysis allows only a limited range of pulse amplitudes to be counted by excluding voltages above and below certain operator-defined thresholds. The thresholds are set so that the 'window' between them just encloses the pulse height distribution of the spectral line of interest. This is particularly useful for eliminating interference due to high order lines. Pulse height selection has been fully discussed by Heinrich (1981) and is covered in most text books devoted to X-ray Spectrometry (e.g. Bertin, 1978; Herglotz and Birks, 1978 and Jenkins *et al*., 1981).

If several elements are to be determined on the same sample, tedious adjustments to the pulse height analyser, amplifier gain and detector bias on changing analyte lines, can be avoided by the inclusion of a sine θ potentiometer, coupled to the goniometer. This ensures that the mean pulse amplitude at the pulse height analyser remains constant for lines of the same order for all values of 2θ for a given crystal.

An alternative to the sequential type of spectrometer described above is the simultaneous instrument. This consists of a number of separate spectrometers grouped around the specimen, allowing simultaneous multi-element analysis. This is an expensive arrangement but has throughput advantages if large numbers of repetitive analyses for the same range of elements are required.

Energy dispersive spectrometers

Energy dispersive X-ray spectrometers (EDXRF) take advantage of the superior resolution offered by semi-conductor detectors compared with the gas-flow proportional and sodium iodide scintillation type. The detectors are lithium drifted silicon, Si(Li), or germanium, Ge(Li), diodes and both have sufficient energy resolution to resolve adjacent $K\alpha$ lines. In practice the Ge(Li) detector has an effective lower limit of about 6 keV and therefore, for environmental purposes the Si(Li) detector with an optimum range of 1–40 keV is preferred.

A schematic diagram of an energy dispersive spectrometer is shown in Fig. 8.3. The mechanical arrangement is much simpler than in wavelength dispersive systems. The detector is at a fixed angle (usually 90°) to the primary exciting beam. The X-ray spectrum from the specimen can then be collected and processed via the Si(Li) detector, and a multichannel pulse height analyser.

The Si(Li) detector consists basically of a crystal of p-type silicon. To compensate for impurities which impair the electrical charge generation properties, the crystal is drifted with lithium ions which allow the crystal to act as a semi-conductor device. When an X-ray photon enters the crystal, ionization occurs, the resulting charge is collected rapidly, due to the high voltage bias applied, and integrated. The voltage pulse obtained is very weak with a poor signal to noise ratio. To reduce the noise to an acceptable level, and to minimize loss of lithium ions, the detector must be stored and operated at very low temperatures. In practice the detector and its pre-amplifier are operated in a cryostat, the reservoir of which must be kept topped up with liquid nitrogen. Similar Ge(Li) detectors are used to examine gamma emitters in environmental materials.

Pulses from the pre-amplifier are further amplified and shaped, using a high performance amplifier, before being passed to the multichannel pulse height analyser. Pulses passing through the

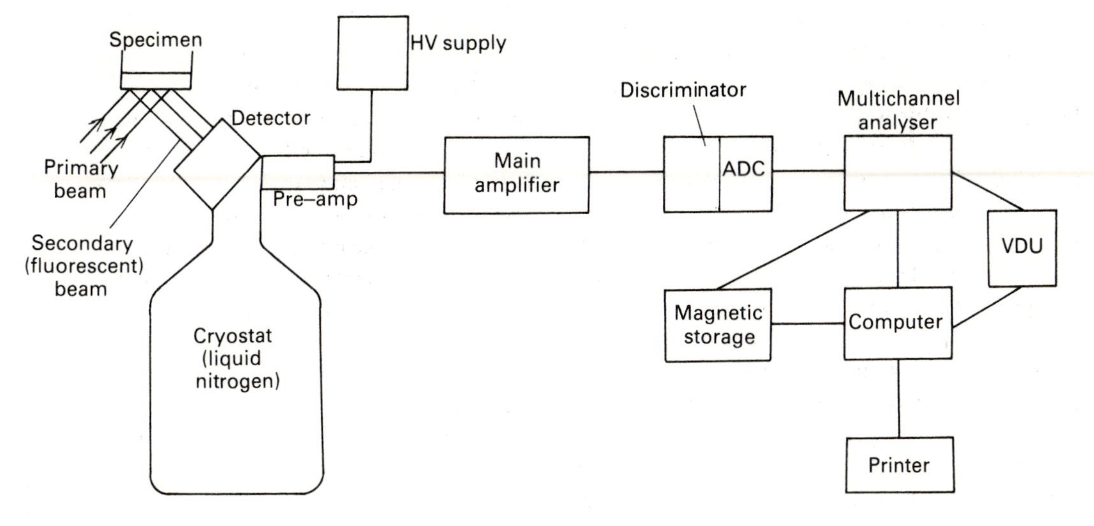

Fig. 8.3. Outline of an energy dispersive XRF spectrometer.

window of the pulse height selector are digitized using an analogue to digital converter.

Every channel of the multichannel analyser corresponds to a particular pulse amplitude. Each time a pulse is processed the value stored in the relevant channel is incremented by one, so that a frequency distribution of pulse amplitudes is built up as the count progresses. This frequency distribution is a representation of the spectrum of X-ray photon energies falling on the Si(Li) detector.

In modern instruments the multichannel analyser is part of a larger computer system which can then be used for qualitative and quantitative analysis of the spectrum.

Techniques

Sample preparation

Sample preparation depends on the nature of the sample and the analyses required. In general, vegetation and similar material can be analysed in powder form, usually after being pressed into a disc-shaped briquette or pellet. Particle size must be uniform to obtain reproducible results, particularly in analysis of lighter elements. Grinding to less than 0.5 mm is often sufficient (Ball and Perkins, 1962; Mudroch & Mudroch, 1977). In the authors' laboratory, results obtained for sulphur in vegetation, litter and peat using samples of this kind pressed at 20 tons into 40 mm diameter briquettes show good agreement with values obtained by wet chemical techniques. However, if a number of elements are required on the same sample it is better to pulverize to pass 100 mesh (about 0.15 mm) or finer (Evans, 1970; Norrish and Hutton, 1977) as particle size effects are wavelength dependent.

Briquettes should be thick enough to maximize the count-rate for the shortest wavelength line. If samples do not easily bind into a stable disc, cellulose or acrylic powder may be added as a binder, either as a backing or mixed directly with the sample before compression. The latter introduces a dilution effect which is not always desirable. The use of collapsable metal cups to contain the sample during compression usually results in durable pellets. A useful technique for small samples, such as insects, is to prepare a thick layer of cellulose. A depression is then made in the centre into which the sample is placed and the whole is then compressed as normal (Dempster *et al.*, 1986). Other procedures include supporting the loose powder on a thin film of polyester or polycarbonate. However, this is not considered very satisfactory for quantitative work (Jenkins *et al.*, 1981).

In general a simple grinding and pelleting technique is unsuitable for quantitative analysis of soils and rocks (Adler, 1966). Fusion and preparation of a glass disc is necessary to remove particle size and heterogeneous effects. The fusion fluxes most suitable for ecological materials are based on sodium or lithium borate. Many recipes involving these two compounds have appeared in the literature over the years, with sample/flux ratios ranging from 1:1 to 1:20 or greater. Many involve the addition of a heavy absorber such as lanthanum oxide, to minimize matrix effects, and include small amounts of various other salts to reduce the melting point of the flux and/or aid dissolution of the sample. Typical procedures are those of Padfield and Gray (1971) for sodium borate and Norrish and Hutton (1969) for lithium borate. Soils containing more than 10% organic matter are best ashed prior to fusion.

Liquid samples such as soil extracts and waters require little preparation providing analyte concentrations are high enough for direct analysis. Prior degassing of the sample may be required if a vacuum spectrometer is to be used, to prevent rupture of the supporting membrane of the special cell required for liquid samples. These cells must always be filled completely excluding air bubbles before capping, for the same reason. Alternatively, for lighter elements, measurements can be made in an atmosphere of helium to reduce attenuation difficulties.

For trace level analysis, a concentration step is required. Vegetation, soils and rocks may be digested, dry-ashed or fused by the methods given in Chapters 2 and 3. This can then be followed by a suitable chemical concentration step. Waters and soil extracts can be directly evaporated or treated with ion-exchange resin (Marcie, 1967; Cowgill, 1968). Blount *et al.* (1973) describe the application of chelating ion-exchange resins in the analysis of geological samples. Absorption on activated carbon has been used for trace element enrichment from water samples (Van der Borgth and Van Grieken, 1977) and Harris & Baines (1976) determined copper after solvent extraction.

Filters used in air pollution studies may be analysed directly or following special treatment. Dzubay (1977) covers this subject in detail.

Instrument calibration

Standards used for instrument calibration may be samples whose elemental composition is known or synthetic mixtures prepared from known amounts of compounds mixed to simulate the type of sample being analysed. In the latter case, the composition is known with more certainty, although practice is required in the preparation to obtain reproducible results. For solid mixtures it is important to ensure the material is finely ground and adequately mixed. Simulated vegetation is often made by adding solutions of analyte elements to cellulose, followed by drying and grinding. In this case it is better to carry out the whole operation in the barrel of the grinder to avoid losses due to evaporation residues deposited on the walls of the drying vessel. Previously analysed samples are more convenient but may be limited by the availability of specimens having a suitable composition. Vegetation and rock standards are not usually a problem in this respect, but soils may be more difficult. A number of certified reference materials (CRMs) are available which include a variety of vegetation and rock types but not many soils. These can be used as primary standards. Secondary standards, carefully analysed using CRMs can then be used as day to day working standards.

Calibration curves are normally rectilinear. Any curvature will be due to sample matrix effects or counting equipment dead time. The latter will only be significant at rates in excess of 5000 cps and can be corrected to the true count mathematically if the dead time is known. Matrix effects and their correction are dealt with below.

Matrix effects and their correction

Matrix effects due to physical heterogeneity and poor presentation can be largely eliminated by careful and reproducible sample preparation. Inter-element effects are not so easily controlled and serious errors can occur if they are not taken into account.

Quantitative analysis by X-ray fluorescence spectrometry is normally carried out in one of four ways (Jenkins *et al.* 1981):

1 Ignoring matrix effects.

2 Minimization of matrix effects.

3 Compensation for matrix effects.
4 Mathematical correction for matrix effects.

1 *Ignoring matrix effects.* This is the simplest procedure and involves the use of standards whose composition closely matches that of the samples. This is relatively simple for vegetation where the bulk of the matrix is carbon, hydrogen and oxygen from cellulose. Soils, however, can have widely differing compositions according to the minerals present and a large number of standards would be required to cover all applications. Soils are therefore best handled using one of the other methods.

2 *Minimization of matrix effects.* This is usually achieved by diluting the sample and standard so that absorption and enhancement effects become negligible. Often a 'heavy absorber', an element which absorbs the analyte line strongly, is included with the diluent and this additive then becomes the dominant matrix factor. The need for the standard composition closely to resemble that of the sample thus becomes less important. A good example of the use of this method is the glass disc fusion procedure mentioned earlier. In this case there is the additional benefit of creating a homogeneous specimen.

Double dilution (Tertian, 1972) and standard addition methods also come into this category and in these cases the use of a standard is not necessary.

3 *Compensation for matrix effects.* This may be achieved by the use of an internal standard, by monitoring scattered source radiation or by measuring the degree of absorption of the analyte line by the sample matrix. The internal standard procedure involves the addition of an element having a spectral line which is affected by the sample matrix in the same way as the analyte line. This can be cumbersome if the sample is to be analysed for more than one element, as different internal standards will be required for each. It is preferable to use scattered source radiation as the internal standard. This is highly matrix dependent and can be utilized via the characteristic lines of the anode target material or via the spectral background in the region of the line of interest.

Anderman and Kemp (1958) first proposed the use of scattered target lines. The procedure has since been refined by Champion *et al.* (1966), Taylor and Anderman (1973) and many others but the method is somewhat empirical and the choice of variant for any particular application is probably best found by experiment. The use of the background intensity as internal standard is common and in many cases improved calibrations are obtained by using the ratio of net counts to background counts rather than net counts alone.

4 *Mathematical correction for matrix effects.* In simple terms, for a complex sample, the concentration of each element in the sample matrix may be expressed as:

True concentration =
measured concentration × correction factor.

The correction factor includes components due to each of the other elements present in the sample. Each component consists of an influence (or alpha) coefficient multiplied by the weight fraction in the sample of the interfering element. Hence, if the concentrations of the interfering elements and the alpha coefficients are known, the correction factor can be calculated by summation of the components. In practice an approximate composition of the sample is determined by comparison with standards and a new composition then computed using the approximate weight fractions and alpha coefficients. The process is then iterated until convergence to a constant result. Many models and algorithms for the determination and application of alpha coefficients have been published. The most popular are those of Lachance and Traill (1966), Claisse and Quintin (1967) and Raspberry and Heinrich (1974). The book by Lachance (1982) covers the subject in detail. All are impractical for routine use without access to a computer and therefore the increased availability and low cost of small computers has increased the popularity of the procedure.

Reproducibility and sensitivity

Reproducibility is largely dependent on the homogeneity of the specimen and counting statistics. The latter is discussed thoroughly in most text books on X-ray fluorescence. Day-to-day instrument variation can be monitored using a permanent reference disc (Evans, 1970).

Sensitivity is affected by many factors. Any dilution introduced during sample preparation will obviously increase detection limits. Sensitivity is also element dependent. For plant materials, detection limits of around 1 ppm are obtainable for heavier elements such as iron, copper and zinc but rise to several hundred ppm for the lighter elements, magnesium and sodium. Limits of detection have been discussed by Pantony and Hurley (1972).

In general wavelength dispersive instruments are more sensitive than energy dispersive instruments as the overall resolution is superior. However, excitation techniques which reduce the background, such as the use of radioisotopes, can result in dramatic improvements in detection limits and these techniques are normally more applicable to energy dispersive spectrometers.

Applications

The use of XRFS for rock and mineral analysis has frequently been described and most of the techniques are also applicable to soils (Jones, 1982; Wilkins, 1983). The procedures of Brown and Kananis-Sotiriou (1969), Norrish and Hutton (1969) and Williams (1976) are widely used for the total and trace analysis of soils. For accurate quantitative work fusion is recommended, however in semi-quantitative and other less demanding applications the use of pressed powders may be adequate. The initial treatment of plant materials is summarized in Table 8.4.

A number of workers have applied XRFS to plant materials including Evans (1970), Hutton and Norrish (1977), Mudroch and Mudroch (1977) and Norrish and Hutton (1977). Fieldes and Furkert (1971) describe a direct procedure using small weights (50 mg) of plant material. Bowden *et al.* (1979) also examine small samples. Two examples of the use of the energy dispersion technique are for the analysis of insects (Turnock *et al.*, 1979), and for trace metal determinations (Elder *et al.*, 1975). Small samples can be examined by first digesting followed by a concentration

Table 8.4. Treatment of plant materials for analysis by X-ray fluorescence spectrometry

	Atomic number	Minimum sample weight (g)	Sample preparation
Aluminium	13	0.5	Grind and press into discs
Calcium	20	0.5	Grind and press into discs
Cobalt	27	5–20	Dry-ash and concentrate at low levels (1-nitroso-2-naphthol)
Copper	29	5–20	Dry-ash
Chlorine	17	0.5	Grind and press into discs
Iron	26	1.0	Grind and press into discs
Magnesium	12	0.5	Grind and press into discs
Manganese	25	1.0	Grind and press into discs
Molybdenum	42	5–20	Dry-ash and concentrate at low levels (α-benzoinoxime)
Phosphorus	15	0.5	Grind and press into discs
Potassium	19	0.5	Grind and press into discs
Silicon	14	0.5	Grind and press into discs
Sodium	11	0.5	Grind and press into discs
Sulphur	16	0.5	Grind and press into discs
Titanium	22	5–20	Dry-ash and concentrate at low levels (cupferron)
Zinc	30	2.0	Grind and press into discs

procedure using co-precipitation or chelation (Knapp, 1985).

The direct analysis of water is generally difficult due to concentration limitations. However, a number of procedures are available, (Peter and Tuchscheerer, 1966, Van Grieken *et al.*, 1976, Kalam *et al.*, 1978 and Sichere *et al.*, 1978). Pre-concentration methods are generally preferred and a large number of procedures have been published. Nevertheless, no single procedure appears to be suitable for all cases and the reader is referred to the reviews of Van Grieken (1982) and Leyden and Wegscheider (1981) for specific applications. Suspended matter in waters may be estimated after filtration, preferably using membrance filters, the amount of pre-concentration depending on the volume of water used. The filters are dried and analysed directly. Particle size effects may be troublesome in this application and, if these cannot be overcome, fusion into a glass bead may be necessary.

Atmospheric particulate matter and aerosols may be handled in a similar way. Many applications in this field are given by Dzubay (1977) and are covered along with water applications in the review by Van Grieken and La Brecque (1985).

Electroanalytical techniques

All electrochemical methods measure either the current or potential which results when two or more electrodes are immersed in solution. For analytical purposes the different techniques can be divided into two broad categories. Those that apply a voltage and measure the resultant current are collectively called voltammetric methods and include polarography. Potentiometric methods, which measure the potential between two electrodes when they are immersed in solution, include the use of glass electrode for measuring pH as well as general ion selective electrode measurements.

Voltammetric methods

General texts dealing with this topic include Whitfield and Jagner (1981), Bard and Faulkner (1980) and Bond (1980). Davison and Whitfield (1977) list applications of modern instrumental techniques.

Principles

Ions in solution may be oxidized by removing electrons or reduced by adding electrons e.g.

oxidation: $Fe^{2+} \rightarrow Fe^{3+} + e$
reduction: $Fe^{2+} + 2e \rightarrow Fe$

The electrical potential at which these redox reactions occur depends on the ion in solution, some being reduced or oxidized more easily than others. By inserting electrodes into solution and applying an appropriate potential some ions will be either oxidized or reduced, the resultant transfer of electrons causing a current to flow. This current, according to Faraday's Law, is directly proportional to the concentration of the ion in solution. By using three electrodes and appropriate electronics the potential is accurately applied between the working electrode, where the reactions take place, and a reference electrode which maintains a constant potential. The current which flows between the working electrode and a third, inert, auxilliary electrode is measured.

Many techniques have evolved from this basic voltammetric procedure. They are classified and named according to the type of working electrode which is used and the manner in which the potential is applied. Classical polarography, the forerunner of modern voltammetric methods, uses a dropping mercury electrode and linearly increases the potential with time. The word polarography is now commonly used synonymously with voltammetry, and modern polarographs are usually multifunctional instruments offering a comprehensive range of voltammetric techniques.

Techniques

1 Without pre-concentration
Although any conducting material can, in principle, be used as the working electrode, mercury is still the most popular. Being a liquid, the surfaces

can be easily renewed to keep them fresh and clean. A series of mercury drops are usually employed, but instead of using free falling drops from a capillary, as in classical polarography, the drops are formed mechanically. Each drop is allowed to grow quickly and then maintained at a constant volume and surface area while the current is measured. The drop is then dislodged and discarded and a new one formed. This type of electrode is known as a static mercury drop electrode. The potential is supplied as a series of pulses rather than linearly changing with time, and each pulse is synchronised with the formation of a new drop. The current is only sampled for a short time, a few milliseconds, while the drop is stationary, but as the drops are re-formed at typically 0.5 s intervals, many current measurements are made as the potential is scanned for several minutes.

Many different techniques are available, their names being derived from the shape of the potential–time curve used to supply the pulses and the way in which the sampled current is processed (Turner and Whitfield, 1974). 'Differential pulse', 'square wave' and 'staircase' are the most common. The final output with all these methods is a series of peaks at different potentials, corresponding to different ions in solution (Fig. 8.4). Although the theoretical best possible detection limit is 10^{-8} mol l^{-1}, a practical limit for routine work is 10^{-7} mol l^{-1}. The peak current which is measured depends on the electrode reaction, as well as the concentration, and so calibration curves may be non-linear. The current in conventional polarography, is independent of the electrode reaction and so calibration curves are always linear. This less sensitive technique may be preferable if concentrations exceed 10^{-6} mol l^{-1}.

The most recent polarographs use microprocessor technology to simulate digitally the imposed potential and to sample and process the current. The output is displayed on a VDU or a graph plotter and the instrument is programmed to measure and print out the current, and if necessary to convert it to concentration. They usually also offer the facility of performing cyclic voltammetry, whereby a rapid linear scan is applied to a single mercury drop or to a solid electrode such as glassy carbon. This traditional technique, which is less sensitive than the modern

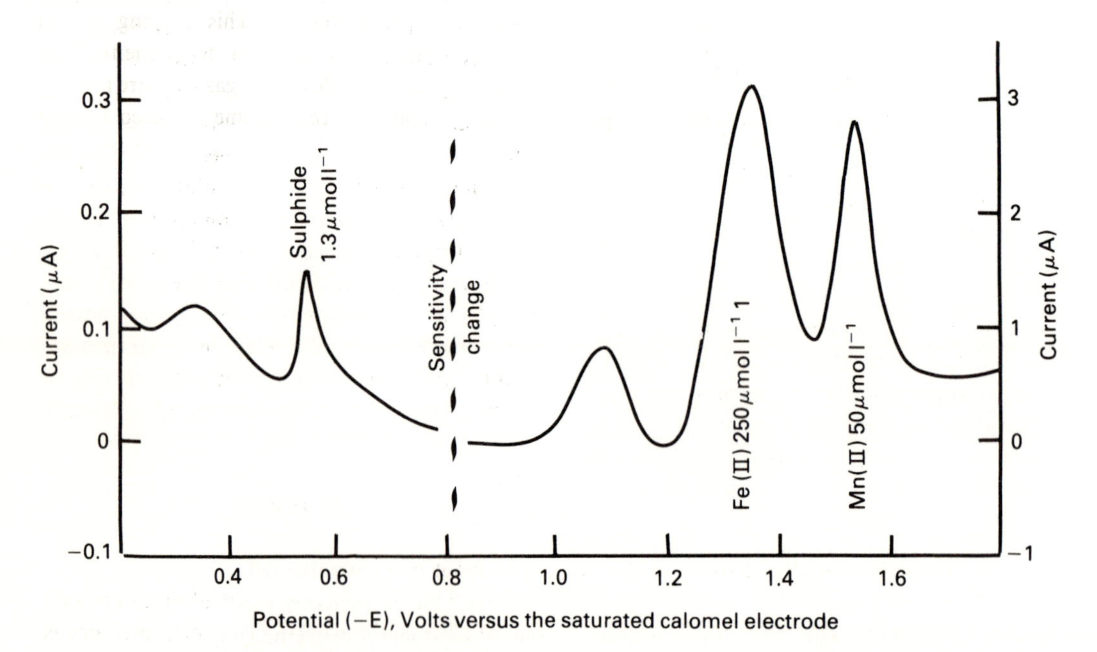

Fig. 8.4. Example of peaks at different potentials corresponding to different ions in solution.

pulse procedures, is being superceded by 'staircase' or 'square wave' for analyses at stationary electrodes.

2 Anodic stripping

Anodic stripping voltammetry (ASV) is widely used because it can determine metals at concentrations as low as 10^{-9}–10^{-11} mol l^{-1}. This sensitivity is achieved by incorporating a preconcentration step into the electrochemical procedure. A stationary mercury electrode is held at a negative potential for a fixed time (usually 1–5 minutes) so that metal ions in solution will be reduced to elemental metal which then dissolves in the mercury:

$$M^{2+} + 2e \rightarrow M$$
$$M + Hg \rightarrow M(Hg)$$

As the mercury electrode has a very small volume the metal quickly reaches a high concentration compared with the concentration of its ion in solution. When the electrolysis is complete the potential is scanned in a positive (anodic) direction, and at the appropriate potential the metal in the mercury is re-oxidized to its ion which reenters the solution and is stripped out of the mercury. The current that flows is measured using one of the direct voltammetric techniques.

The same static mercury drop assembly used in direct voltammetry can be used for ASV, but only one drop is required instead of a series of drops. Most modern polarographs are capable of performing ASV and those that are microprocessor controlled will execute all steps automatically. To improve sensitivity to 10^{-9} mol l^{-1}, one of the pulse techniques can be used in the stripping step. During the deposition step the solution must be stirred to maintain the flow of ions to the electrode, but this too can be automated. Thin film electrodes provide enhanced sensitivity, to 10^{-11} mol l^{-1}. They use a carbon electrode plated with a thin film of mercury giving even greater pre-concentration.

ASV is limited to those metals which dissolve in mercury, but cathodic stripping voltammetry can also be used (van den Berg, 1985). Glassy carbon or an inert metal such as platinum, gold or silver can replace mercury as the working electrode. Some metals, or organics in solution will spontaneously adsorb on to the electrode surface or react with the electrode. Others can be encouraged to do so by polarizing the electrode at a positive potential.

Instruments

Modern instruments are capable of performing a wide range of techniques. They are designed to be used with a dedicated electrode assembly and the computer controlled versions can be instructed to perform repetitive tasks. Metrohm and Princeton Applied Research Corporation are the two major manufacturers.

There are no limits to the types of cells which can be used and flowing versions are becoming more popular for research purposes and as electrochemical detectors for HPLC. However, most cells are basically simple in design. A silver chloride electrode usually serves as the reference and platinum wire is used as the auxilliary electrode.

Dissolved oxygen can be reduced at an electrode and so can be measured polarographically. Generally, however, it is regarded as an interference and so is removed from the solution by bubbling with an inert, oxygen-free gas (Ar or N_2) prior to the measurement. This purging step is also automated on modern instruments. Although three electrodes and a gas tube are immersed in the solution, the volume required is only about 5 ml. Automated sample presentation modules are available.

Applications

Voltammetric methods measure those ions in solution which can be oxidized or reduced at an electrode. However, most solutions contain complicated mixtures of ions and a metal may be present in the free ionic state and also as inorganic and organic complexes. Usually only one ion is reduced at the electrode and this is often the free uncomplexed aquo-ion. However, voltammetry is not an instantaneous technique and throughout the period that current flows, ions continue to be reduced. If some of the inorganic or organic complexes can be rapidly converted to the aquo-ion which is being removed from solution, they too will be effectively reduced. Thus voltammetry measures the aquo-ion and labile complexes in solution, but excludes complexes in which the metal is tightly bound (Davison, 1978). Simple inorganic complexes such as metal chlorides or carbonates are usually measured, as are weakly bound organic complexes such as acetates. Metals bound to particulates or in mineral lattices, or present in solution as another oxidation state or as very stable organic complexes, are not measured.

Electrochemical methods, being non-destructive, and sensitive to the form of the element are not particularly suited to total analysis which usually requires prior digestion of the sample. Moreover, only those elements which are electroactive can be determined (Table 8.5). However, voltammetric methods do have complementary analytical roles and some examples are listed below:

1 *Extreme sensitivity.* Anodic stripping voltammetry, which is extremely sensitive, can determine

Table 8.5. Approximate peak or half-wave potentials for inorganic substances at a mercury electrode with a saturated calomel electrode as reference

Ion	Supporting electrolyte	Potential	Reaction	ASV
Al(III)	0.4 M acetate + Solochrome violet	−0.46	R	
As(III)	M HCl	−0.42, −0.84	R,R	
Bi(III)	M HCl	−0.09	R	✓
Br^-	0.1 M KNO_3	+0.12	O	
Cd(II)	0.1 M acetate at pH 4.5	−0.65	R	✓
Cl^-	0.1 M KNO_3	+0.25	O	
CN^-	0.1 M NaOH	−0.36	O	
Co(II)	M NH_3/0.1 M NH_4Cl	−1.30	R	
	M NH_3/M NH_4Cl + dimethylglyoxime	−1.12	R	
Cr(VI)	M NH_3/0.1 M NH_4Cl	−0.30	R	
Cu(II)	M NH_3/M NH_4Cl	−0.43	R	✓
	0.1 M acetate at pH 4.5	−0.07	R	
Fe(II)	0.1 M KNO_3	−1.35	R	
Fe(III)	0.1 M oxalate at pH 4.0	−0.23	R	
I^-	0.1-M KNO_3	−0.03	O	
Mn(II)	0.1 M KNO_3	−1.55	R	✓
Mo(VI)	0.3 M HCl	−0.26, −0.63	R,R	
Ni(II)	M NH_3/0.1 M NH_4Cl	−1.10	R	
N_3^-	0.1 M KNO_3	+0.25	O	
NO_3^-	0.2 M KCl/0.1 M acetic acid + U(VI)	−0.98	R	
O_2	0.1 M KNO_3	−0.05, −0.90	R,R	
Pb(II)	0.1 M acetate at pH 4.5	−0.50	R	✓
S^{2-}	0.1 M NaOH	−0.78	O	
Sb(III)	M HCl	−0.15	R	✓
Se(IV)	M HCl	−0.10, −0.40	R,R	✓
Sn(II)	M HCl	−0.10, −0.47	O,R	✓
SO_3^{2-}	0.1 M acetate at pH 5	−0.62	O	
Ti(IV)	0.2 M acetic acid	−0.37	R	
V(VI)	0.1 M HCl	−0.18, −0.94	R	
V(V)	0.1 M H_2SO_4/0.1 M KSCN	−0.52	R	
Zn(II)	M NH_3/0.1 M NH_4Cl	−1.24	R	✓
	0.1 M acetate at pH 4.5	−1.10	R	

O and R indicate whether the electrode reaction is an oxidation or reduction. Ions suitable for determination by anodic stripping volammetry, ASV, are indicated. More comprehensive tables are available in Princeton Applied Research Corporation Application Note H-1, Sawyer and Roberts (1974) and Turner and Whitfield (1981).

ultra-trace levels of metals in natural waters (e.g. Vos *et al.*, 1986).

2 *Speciation.* Voltammetric methods, including ASV, automatically exclude particulate forms without recourse to filtration. They also distinguish between labile and strongly bound complex species.

3 *Redox specific.* Voltammetric methods measure a single oxidation state, for example they determine Cr(VI) as opposed to Cr(III) and As(III) rather than As (V). In each case the toxic form of the metal is measured.

4 *Multi-element analysis.* As up to 5 or 6 elements can be determined in a single potential scan which takes only a few minutes, voltammetric techniques offer rapid analysis. Gustavsson and Hansson (1984) and the Standing Committee of Analysts (1987) describe ASV methods.

5 *Organics.* Organic compounds as well as metal ions may be oxidized and reduced and so determined using voltammetric methods (Smyth and Healy, 1984). Important metal organic pollutants such as methyl mercury and triethyl lead may also be measured (Birnie and Hodges, 1981). The tendency of organics to adsorb on to mercury has been exploited by using voltammetric methods to estimate the humic acid content of waters (Cosovic, 1985).

Supporting electrolytes

The potential at which an ion is oxidized or reduced depends on the solution matrix. Ionic strength, pH and the presence of complexing agents can all have an effect. Some solutions are particularly suitable for determining certain species. However, the list of solutions in Table 8.5 should only be used as a guide because others may be more suitable for particular applications. In some cases, as for seawater, no supporting electrolyte may be necessary.

Ion-selective electrodes

General guidance concerning ion-selective electrodes is available in Durst (1969) and Bailey (1980), whilst Yu (1985) has reviewed applications in soil science. Bates (1964) and Westcott (1978) are two useful standard works on the measurement of pH, the most common example of the application of the ion-selective technique. Some aspects of soil pH have already been discussed in Chapter 2.

Principles

When an ion-selective electrode and a reference electrode are immersed in solution the potential, E, between the two electrodes is related to the activity of the 'selected' ion by the Nernst equation.

$$E = E_0 \pm \log a_X (2.3\,\boldsymbol{RT}/nF)$$

E_0 is a constant, E is the standard potential of the cell, a_X is the activity of the ion X, n is the charge of the ion. In a potentiometric measurement there is effectively no flow of current; the solution remains unperturbed and the electrode responds to the activity of the free uncomplexed ion.

Equipment

Sensing electrodes

There are four types of electrodes commonly used. Three of them use a membrane with the test solution on one side, and on the other, an internal reference solution which provides the electrical contact.

1 *Glass electrodes.* Glasses used for pH measurement are particularly sensitive to hydrogen ions, but other glasses are manufactured for measuring Na^+, K^+, NH_4^+ and some other cations. The determination of pH is particularly free from interferences and discrimination to 0.01 pH can easily be attained.

2 *Inorganic salt electrodes.* The membranes of these electrodes have been produced in a variety of homogeneous and heterogeneous forms ranging from single crystals to dispersion of the active material in an inert matrix such as silicone rubber or polythene. Lanthanum fluoride, silver halides and metal sulphides are the most commonly used, chiefly for measuring fluoride, chloride and sulphide in fresh and waste waters.

3 *Organic ion-exchangers and neutral carriers.* These electrodes either have an active material which is a charged ion-exchanger made up from large organic molecules, or a neutral matrix containing cavities of discrete size. Although there is a wide variety of electrodes of this type only a few perform well enough for routine use. These include those sensitive to Ca^{2+} and NO_3^-.

4 *Gas sensing probes.* These are complete electrochemical cells which depend on gas diffusing through a membrane, modifying the composition of a solution and in turn a potential sensed by an internal electrode. CO_2, NH_3 and SO_2 can be determined in this way.

Reference electrodes

The potential developed at an ion-selective electrode is measured with respect to a reference electrode which ideally provides a constant potential independent of any changes in solution. This reference electrode is commonly made of Ag/AgCl or Hg/HgCl immersed in a solution of KCl and provides a very stable potential. The solution of potassium chloride is usually brought into contact with the solution being measured in a ceramic frit which is incorporated in the electrode. There is a potential associated with the junction between the two liquids. If the liquid junction potentials, which develop when the electrode is immersed in the standard and test solutions, are the same, they cancel one another and so introduce no errors. Any difference between these two potentials constitutes a measurement error. This problem is most likely to arise when measurements are made in low-ionic strength solutions, such as freshwaters. For many ion-selective electrode measurements a common supporting medium is used for both standard and test solutions and so the problem does not occur. However, addition of an inert electrolyte would change the pH of a natural water, and so this remedy is not available. The best solution is to use reference electrodes with free-flowing junctions and to use quality control procedures to ensure that they work well (Davison and Woof, 1985).

If the solution being measured contains something which will react with the potassium or chloride ions, or be contaminated by them, a double junction reference electrode is used. The electrode is immersed in a sheath containing a non-reacting solution, such as KNO_3, which contacts the solution being measured via a further ceramic frit.

Combination electrodes, which incorporate sensing and reference elements in the same body, are commonly used for measuring pH. Although these are very convenient, their fritted junctions often have a very low flow rate and are prone to clogging.

Meters

Recent advances in electronics have ensured that most meters are reliable and easy to use. For ion-selective measurements the meter should discriminate to 0.1 mV. Microprocessor based instruments can incorporate calculation factors for direct display of concentration or activity. Portable meters suitable for field use are freely available and usually reliable. They are, however, susceptible to humidity problems because their low power consumptions provide insufficient heat to prevent condensation. Slight shorting of a high impedance circuit will cause the potential to drift.

Techniques

A calibration curve of the measured potential plotted against the logarithm of the activity or concentration is constructed by using a series of standard solutions. The slope of the curve, which is linear over several decades, should theoretically be $59/n$ mV at 25 °C, but in practice may be less. As the concentration decreases the plot starts to curve and the mV change per unit log concentration change declines. This performance threshold denotes the limit of detection of the electrode. Some electrodes are highly selective to one ion, but others are prone to interferences from other ionic species.

Attention to quality control is as important in ion-selective electrode measurements as any other analytical procedure. If the electrodes are to be dipped into a beaker with a magnetic stirrer their geometry should be fixed so that the hydrodynamics are constant. As the response depends on RT/nF the temperature should be controlled to at least $\pm 2\,°C$. A pre-treatment reagent is normally added to ensure that samples and standards have similar composition. Electrodes may be incorporated in flowing systems which help to ensure well defined procedures.

Measurement of pH differs slightly because only two standard solutions, or buffers, are usually employed. It is also more customary to calibrate and measure in the pH mode using the dedicated buffer and slope adjust controls of the meter. In this mode the mV reading is simply converted to pH within the meter using the buffers to define the multiplication factor and an intercept. However, simple calibration in mV is quite acceptable and in some cases may be preferable.

Seldom will ion-selective electrodes give a precision better than $\pm 2\%$, even at high concentrations of determinand. This is partly due to the logarithmic calibration curve because $\pm 2\%$ corresponds to ± 0.01 pH or ± 0.6 mV.

Applications

Direct potentiometry

Apart from pH measurements, ion-selective electrodes are routinely used for measuring F^- and some of the other ions listed in Table 8.6. They are often used for research purposes when the activity of a single species needs to be determined, or when it is desirable to monitor continuously the composition of a solution. However, it must be recognized that they are not specific, but merely select toward a particular ion. Although they can be used in complexation studies their response to single species is a disadvantage for most routine analytical work. Another disadvantage is that their response depends on the pH of the solution and so routine measurements must be performed in buffered solutions. Also the response characteristics and selectivity of each electrode is different and may change with time.

Potentiometric titrations

End-point detection in volumetric titrations has traditionally used indicator solutions. To an increasing extent electrodes which follow the change in activity of either the added or titrated species are being used for this purpose. The end-point, which often coincides directly with the equivalence point, can be determined very accurately.

Chromatography

Chromatography is a versatile, rapid and basically non-destructive technique which enables a wide variety of organic compounds to be separated and detected. Martin and Synge (1941) were the first to use partition chromatography, for

Table 8.6. Typical properties of some ion-selective electrodes and a gas sensing electrode

Ion/ gas	Electrode type	Concentration range (M)	Principal interferences	pH	Reference
Ca^{2+}	Ion-exchange	10^{0}–10^{-5}	Zn^{2+}, Fe^{2+}, Mg^{2+}	6–10	Craggs *et al.*, 1979
Cl^-	Ion-exchange	10^{-1}–10^{-5}	ClO_4^-, I^-, NO_3	3–10	Langmiur and Jacobson, 1970
CO_2	Gas sensing	10^{-1}–10^{-5}	NO_x, H_2S	<3	Midgley, 1975
F^-	Inorganic salt	10^{0}–10^{-6}	OH^-	5–8	Bailey, 1980
NO_3^-	Ion-exchange	10^{-1}–10^{-5}	ClO_4^-, I^-, Br^-	4–10	Ross, 1969
Na^+	Glass	10^{0}–10^{-5}	K^+, Ag^+	>4	Wilson *et al.*, 1975
S^{2-}	Inorganic salt	10^{0}–10^{-7}	Hg^{2+}	>13	Mor *et al.*, 1975

amino-acids, and Consden *et al.* (1944) later used a filter paper instead of a column. Paper chromatography reached its peak of popularity during the fifties and early sixties but declined following the introduction of thin-layer chromatography and the development of instrumental procedures. It has the advantage of being a low cost technique and, although its principles are not discussed here, examples of its use are included in Chapter 6. Additional information about paper chromatography can be obtained from the first edition of this book (1974). The work of Stahl (1970) demonstrated the value of thin-layer chromatography. Despite competition from the instrumental procedures, thin-layer chromatography is still often used because it is a simple, low cost, versatile and rapid method suitable for many organic separations.

Partially automated column chromatographic equipment was introduced quite early but has now been superseded by high performance liquid chromatography (HPLC). This is probably the most powerful and versatile of all the instrumental systems. Ion-exchange chromatography is a branch of column chromatography and is the basis of ion chromatography which itself is a particular example of HPLC. Ion chromatography differs from other chromatographic techniques in that it is mostly used for the separation of inorganic ions in waters. Another instrumental technique of importance is gas-liquid chromatography (GLC). This is especially valuable for the separation of organic pesticides and their residues although it is also of benefit for gas analyses and for the resolution of other compounds giving volatile derivatives. Most of these techniques are discussed below and references are given in the text. A comprehensive reference work on chromatography is the CRC Handbook of Chromatography (Zweig & Sherma).

Adsorption chromatography depends on the attraction which a support material exerts on a solute through weak-bonding forces. This retards the progress of the solute in relation to the solvent as the solution passes across the surface of the support. The magnitude of the bonding forces varies between compounds so that mixtures may be fractionated as the solution flows across the support material. Normal phase partition chromatography is based on the relative distribution of compounds between a polar stationary phase and a non-polar solvent (mobile phase). Different compounds have specific distribution ratios for a particular system which forms the basis for the separation of the components in the mixture. On the other hand, reversed phase systems use a stationary phase, with a non-polar functional group chemically bonded to the surface, and a polar solvent. This is more versatile than normal phase chromatography and is well suited to the separation of the water soluble components in ecological materials. Ion-exchange separations are achieved through competition of the ionic species in the solvent and sample for ion-exchange sites on the stationary phase.

Thin-layer chromatography

Although, to some extent thin-layer chromatography (TLC) has been superseded by instrumental techniques it is still widely used especially in exploratory work. Some of its limitations have been minimized by developments which have produced high performance thin-layer chromatography (HPTLC). TLC is also of value for quick testing of methods for standard HPLC.

Principles

Liquid samples are applied as spots to plates which are spread with thin layers of adsorbent, and the components are separated as they are transported along the plate with a suitable solvent. The component separated is assigned a constant value (R_f) for that system:

$$R_f = \frac{\text{Distance travelled by spot from origin}}{\text{Distance travelled by solvent from origin}}$$

The R_f value depends mainly on the composition of the developing solvent and the adsorbent used.

Adsorbent layers are normally spread uniformly thinly on glass or aluminium backed plates 20×20, 20×10 or 20×5 cm, but commercially available plates are generally preferred to laboratory prepared plates. Microscope slides are particularly useful for rapid qualitative separations. The high performance technique adapted for thin-layer chromatography (HPTLC) is more sensitive and precise than normal TLC. It also gives improved sensitivity in a shorter time. It is especially suitable for separating simple mixtures but has no advantages over TLC for complex samples.

Techniques

Adsorbents

The adsorbents in common use include silica gel, alumina, cellulose and polyamide, although others are available for special cases. If desired most adsorbers can be mixed with a binder, usually calcium sulphate. Prepared plates are available containing fluorescent indicators and concentration zones. For certain applications, an additional reverse phase layer is added to the basic adsorbent. In general adsorbents used for HPTLC have a smaller mean particle size with a much narrower size range than for TLC and the layer thickness used is slightly less than 0.20 mm as opposed to 0.25 mm.

Spreading the layer

The layers are usually spread in the form of a slurry with water, although other solvents may occasionally be used. The solvent chosen should not swell the adsorbent so much that on drying cracks appear in the layer. In general 25 g of adsorbent will require 50–80 ml of solvent to provide a suitable slurry. This should be used within 2 minutes if a binder is present. Before spreading, the plates should be wiped with a tissue or cotton wool soaked in acetone or ether to remove the grease. Layers of 0.25 mm are generally used except in preparative work when a greater thickness is needed. Gradient mixed layers may be prepared with the aid of a special applicator.

Application of spots

Sample application needs much care and it is important not to handle the plate surface and to ensure that the needle of the applicator does not touch and damage the surface of the adsorbent. Care is needed to ensure that the solution does not 'creep back' up the needle or be allowed to evaporate from the tip. For qualitative separations a fine tipped Pasteur pipette may be adequate for spotting, otherwise a calibrated microsyringe is essential. Automatic applicators are available for routine and precision work. Contact spotting may be used for HPTLC when the spot is introduced on to a treated fluopolymer film, and once dried, a slight pressure is applied to transfer the spot to the plate. For HPTLC, up to 30 samples may be applied along one edge of a 20 cm plate and as opposite edges can be used, up to 60 samples can be accommodated on one plate. On TLC plates the number of sample spots is limited to about 10, which should be placed approximately 2 cm apart and 2 cm from one edge when using a 20 cm plate. It is most important that the sample spot should be small, which usually requires repetitive spotting, in which case it is essential to ensure that the spot is dry before the next application. Larger size spots generally result in a reduction in sensitivity and overloading may produce tailing of the faster moving components.

Solvents and development

Ascending chromatography has traditionally been used in TLC. The solvent is placed in a glass tank with a flat lid. The principal precaution required is to ensure saturation of the tank atmosphere by lining the tank with filter paper soaked in solvent and ensuring the tank is sealed. Silicone grease is preferable to petroleum jelly for sealing the lid to the tank. Temperature changes do not seriously affect saturation, but may alter the phase composition in solvent mixtures. Two-dimensional development can be applied with loadings of 4 to 10 times the amount for single dimension TLC. All traces of the original solvent should be removed prior to the second run and it is useful to scribe a line immediately below the front of solvent 1 to ensure the even flow of solvent 2. In multiple development, the same solvent is used in successive runs, drying the plate after each run. It is particularly suitable for separating substances with low R_f values.

Similarly, with HPTLC, separation of components may be accomplished by standing the plate vertically in a few millilitres of solvent. Better performance can be achieved if the plate is horizontal and the solvent fed by capillary action. This also enables the solvent to be fed to opposite edges of the plate at the same time. In circular HPTLC the solvent is fed to the centre which produces an improvement in resolution of low R_f components in a shorter time. If the solvent is fed to the periphery of the circle (anticircular), the resolution of high R_f value components is improved also in a shorter run time. Development chambers for HPTLC are much smaller to enable more reproducible conditions to be obtained than in TLC.

Detection and quantification methods

Spots may be detected by colour, spraying to produce a visible spot or using instrument densitometry with absorbance or fluorescence detection. A useful spray reagent for the detection of organic compounds on inorganic adsorbents (e.g. silica gel, alumina) is concentrated sulphuric acid. On subsequent heating the organic compounds become charred and appear as dark spots. Spots may be scraped off plates and eluted with a solvent and estimated by spectrophotometry, but care must be taken to avoid mechanical losses. Densitometry is the most practical method of quantifying spots although it may, in certain circumstances, be subject to error. Variable wavelength instruments overcome many of the drawbacks of single beam scanning densitometers and give a more realistic background value.

Equipment

Pre-coated plates are available for TLC and HPTLC and are usually preferred. These are not necessarily supported on glass, but may also be on plastic sheets, fibre sheets or aluminium foil.

Equipment for spreading plates is available commercially. There are three main types.

1 Those for plates of equal thickness.

2 Those which allow plates of different thickness to be coated together.

3 Automatic spreaders.

In addition kits are available for spreading microscopic slides. These are very useful for technique development.

Densitometers, which scan both absorbance and fluorescence, measure spectra and take both standard and HPTLC plates, are available.

Applications

Thin-layer chromatography may be used for many of the organic fractions covered in this manual, e.g. amino-acids, proteins, lipids, environmental pollutants and pesticides. It may also be used as an aid in HPLC development work. Many applications are covered in the bi-annual reviews in Analytical Chemistry, by Stahl (1970), and in the CRC Handbook of Chromatography (Zweig & Sherma). The technique and its applications are also described by Götz *et al.* (1980) and Touchstone and Rogers (1980).

Gas chromatography

Gas chromatography is required for separating mixtures of volatile substances and has several advantages including speed and good resolution as well as being quantitative for very small samples. The most widely used applications employ an absorbent coated with a relatively non-volatile liquid to form gas-liquid chromatography (GLC). Absorbents without a liquid coating are used in gas-solid chromatography (GSC). The principles of gas chromatography are discussed here with emphasis on GLC but are also relevant to GSC. The applications mentioned are confined to GLC.

Principles

Both adsorption and partition may be involved in gas-liquid chromatography according to the nature of the separation. Most methods of interest to the ecologist depend on partition.

The stationary phase generally used in partition is a non-volatile liquid, coated on to an inert support material. An inert carrier gas acts as the mobile phase. Components are continually absorbed and released according to the equilibrium between the carrier gas and the stationary phase. Separation is achieved by differences in the partition coefficients of the components of the mixture, those which favour the mobile phase being eluted first. The time taken for a compound to pass through the column is known as its retention time.

The column is a long thin tube in the form of a coil or loop. This is placed in a thermostatically controlled oven, since temperature is an important variable in gas-liquid chromatography. The components of the mixture after separation are detected by a suitable device which transmits a signal to a chart recorder or integrator which records the series of peaks. Each of these represents a separate compound or group of compounds.

Techniques

A schematic outline of a GLC system is given in Fig. 8.5. Some of the more important stages are discussed in the rest of this section.

Columns

The traditional columns used in GLC are packed glass or stainless steel tubes, and these are still widely used. Although glass is resilient it is fragile when compared with stainless steel and is also less suitable for high temperature work (over 250 °C). Stainless steel is much more robust but unlike glass it is not possible to see the packing. Both column length and diameter affect performance. In general the resolution is proportional to the square root of the length. The diameter influences the length of the zones in which equilibrium is established between the gas and stationary phase. For many purposes an internal diameter of 3–4 mm is used. Columns are usually used as concentric coils or are U-shaped with a total length of up to 2 m.

Packed columns are either filled with a solid adsorbing stationary phase or an inert support material on which is coated a liquid stationary phase. Solid adsorbants include activated charcoal, silica gel and alumina. In addition many cross-linked polymers based on polystyrene, acrylics and PTFE are available as well as a number of molecular sieves. A large number of liquid stationary phases are available, the most popular being silicone oils of various kinds and high molecular weight esters of polyalcohols such as ethylene glycol. A newer development involves the liquid stationary phase being chemically bonded to the support material. This type of packing is mainly used in high performance liquid

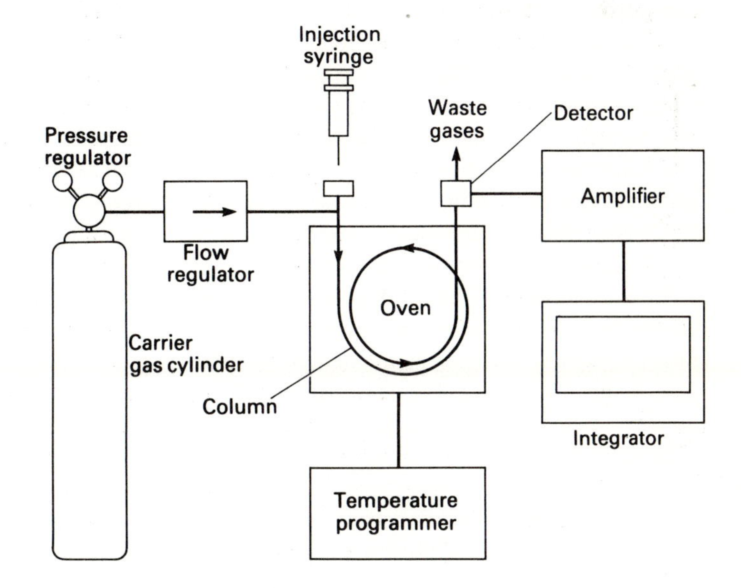

Fig. 8.5. Outline of a gas liquid chromatograph.

chromatography but has found some application in GLC. The bonded phases result in more even coating and give rise to sharper, more symmetrical peaks, with less column bleed.

The most significant advances in column technology in recent years have occurred in the field of capillary columns which are operated on the 'open tubular' principle. These have been reviewed by Lee *et al.* (1984) and Karasek *et al.* (1984). The capillaries are made of glass or quartz and the liquid phase is coated onto the inside wall in one of two ways. In 'wall coated open tubular' (WCOT) columns, the liquid phase is uniformly coated or bonded directly on to the wall of the capillary. Alternatively a 'support coated open tubular' (SCOT) column may be used, where the liquid phase is coated on a porous supporting layer on the inside of the capillary.

Because of the open tubular arrangement, capillary columns are generally of much greater length than packed columns as pressure drop symptoms along the length of the column are much less. The increase in length results in better separation characteristics. The disadvantages are that extremely small sample aliquots must be used to avoid overloading and the difficulties of injecting the sample into columns with internal diameters of 0.5 mm or less.

Analytical packed columns are relatively easily prepared by the user, although they are available commercially ready for use. The support material must be of uniform mesh size and well mixed. It is normally coated with up to 20% of the liquid phase which is chosen according to the mixture to be separated. Some supports will hold more liquid which increases capacity but may reduce flow efficiency. The coating procedure usually consists of stirring a slurry of the support material with the liquid phase or a solution of it.

Ideally the packing should flow freely into the column tubes but adsorbents are sometimes rather sticky and can only be introduced into the tube in small amounts. The material should be poured slowly through a funnel and the column continually tapped to give an even packing density. It can be helpful to use a close fitting flexible rod to push gently the adsorbent through the column tube. If glasswool plugs are used to retain the column packing these should be silanized by treating with 10% dimethyldichlorosilane in toluene for 10 minutes.

Capillary columns are much more difficult to prepare and unless the user is experienced it is probably better to buy ready for use columns which are commercially available.

Production of chromatograms

Before use the column is placed in an oven maintained at a suitable temperature and the carrier gas allowed to pass through. This must continue for 24 hours before sample injection to remove contamination and stabilize the instrument. The carrier gas must be inert as far as the separation is concerned. The choice of carrier gas can influence the performance of the gas chromatograph since the size and weight of the molecules are related to diffusion characteristics. Argon is most frequently used for routine purposes although helium, hydrogen and nitrogen are also employed. The carrier gas cylinder should be replaced when the pressure drops to 200 psi.

The partition coefficient is influenced by temperature and it is necessary to work at the optimum for each separation. In general higher temperatures will give sharper peaks but this may be at the expense of accuracy. When a mixture of components having a wide boiling range is to be separated the first peaks to appear may be narrow and tall whereas later components with a long retention time give broad, low traces. Accurate results cannot be obtained from either extreme, and programmed temperature control is used to maintain ideal conditions. The oven temperature is allowed to increase at a pre-determined rate to give ideal separation conditions. A linear temperature control is probably of most general use. In some cases it may be advantageous to have the injection head enclosed in an oven.

When optimum operating conditions have been established they may be maintained by:

1 checking for gas leaks regularly,

2 changing the injection septum daily,

3 replacing gas cylinders before the pressure drops too low, and

4 regularly checking on the injection block, column oven and detector temperatures.

Given these conditions, any changes in elution pattern, relative peak proportions and peak geometry indicate a deteriorating column. The column should then be replaced immediately.

The sample mixture is injected with a microlitre syringe into the entrance of packed columns. The volumes generally range from 1 to 50 μl. The most suitable syringes are those with a plunger made from stainless steel wire. Injection septa may be conditioned by heating under vacuum at 250 °C for 2 hours.

When using capillary columns sample injection is much more specialized and may be accomplished in a number of ways. The commonly used procedures have been summarized by Schomburg (1983). In the 'split injection' technique, the sample is vaporized in a stream of carrier gas and a small proportion then split off and passed to the analytical column. Other methods involve concentration of the sample solvents by condensing the solvent at the head of the column or by absorbing the solvents on carbon or porous polymers before backpurging into the column. The choice of procedure will depend on the application and the amount and concentration of the sample solution.

Detectors

The detector is placed in the carrier stream at the outlet end of the column where it monitors the gas flow from the column for the component bands. A good detector must have the characteristics listed below.

1 Wide range of application.
2 High sensitivity.
3 Quick response.
4 Simple calibration with linearity against standards.
5 High signal-to-noise ratio and freedom from drift.
6 Freedom from external variables.

Many different detector systems are available, including some employed for specific applications. Some of the latter which are used in pesticide analysis are described in Chapter 7. The detectors usually used for general work are as follows:

Thermal conductivity (Katharometer)

This system records the changes in the resistance of a hot wire as the separation products flow over it. The signal is independent of flow rates. Its main advantages are simplicity and reliability. It is subject to noise and is less sensitive than an ionization detector.

Flame ionization

This device depends on the increases in electrical conductivity of a hydrogen flame in the presence of an organic compound. The carrier gas is mixed with air and hydrogen and the mixture burnt in a jet. Organic compounds entering the flame furnish electrons which are collected by one of a pair of electrodes thus generating an electric current. This ionization current is very small and a sensitive amplifier has to be included in the circuit. This detector is applicable to nearly all organic compounds but is unsuitable for the inorganic gases. It has the advantage of having a linear response over a wide range.

Radiation ionization

This system uses sources of ionizing radiation such as α or β rays to produce changes in conductivity according to the components in the carrier gas. There are several types of radiation detector, perhaps the most common being the argon detector. The radiation detectors are the most sensitive of the detecting systems mentioned here, although the health hazard necessitates careful handling.

As the separated components pass sequentially through the detector the amplified current changes are fed into a recorder or integrator where a peak is produced for each chromatographic zone.

Identification and quantification

Individual compounds may be identified by running synthetic mixtures of known compounds or by their addition to the sample mixture.

The area under the individual peaks is used as a measure of the amounts of the compounds present. The detector must be calibrated for each compound by injecting known amounts of reference standards or alternatively using an internal standard.

Most chromatographs use a computing integrator to collect data. This computes areas and then calculates solution concentrations using either internal or external standardization techniques.

Provided the peaks are relatively thin an adequate approximation can be obtained from measuring the peak height. Peak measurements are discussed in more detail by Giddings and Keller (1968).

Skewed peaks may occur if the rate of adsorption or gas concentration changes during a run. Column packing and operating conditions may have to be modified if irregular or broad peaks are produced.

Equipment

Many companies manufacture gas chromatographs. Most are modular in construction so that the various facilities such as temperature programming and injection heaters may be added or removed as appropriate.

Automatic integrators and peak sensing devices are also used. In conjunction with automatic sample injectors and computing equipment it is possible to run instruments unattended for long periods.

The use of gas-liquid chromatography has been extended by linking with other analytical systems, such as mass spectrometry.

Applications

Gas-liquid chromatography can be used for tests of purity, identification and for quantitative analysis. The ecologist is usually interested in the latter although, in view of the complex nature of organic separations, identification is sometimes required. Ideally a mass spectrograph is then coupled to the gas chromatograph. There have already been references to the use of gas chromatography for the examination of natural compounds and pesticides in Chapters 6 and 7.

It is possible to introduce many of the organic extracts directly on to the columns but some mixtures of non-volatile compounds such as carbohydrates and amino-acids have first to be converted to volatile derivatives. A large number of references to the use of GLC for the analysis of soils, plant materials and waters are given by Chesters *et al.* (1971). Hunt and Wilson (1986) describe its use in water analysis and Smith (1983) discusses the examination of soil atmospheres using gas chromatography. Other information is given by Tranchant (1969) and Leathard and Shurlock (1970).

High performance liquid chromatography

High performance liquid chromatography (HPLC) developed in recent years from traditional column chromatography, has become as important as gas chromatography for the separation of mixtures of organic fractions. The main advantage compared with gas chromatography, is that it is not necessary first to prepare volatile derivatives. Only a brief review of the HPLC is given here, but for further in-depth treatment of the subject there are many books available dealing with the subject in general, e.g. Parris (1976) and Poole and Schuette (1984), and its application to environmental analysis (Lawrence, 1984). A series of volumes by Colin *et al.* (1984–1985) provides a good guide to HPLC literature.

Principles

In HPLC, the liquid mobile phase (solvent) is pumped through a column (typically 5 × 300 mm) filled with a solid stationary phase. A liquid sample is injected into the solvent and washed through the column, individual components being separated on the basis of their relative affinities between the mobile and stationary phases. The time for a particular component to reach the detector is used in identification. Separation of the components on the stationary phase may be due to adsorption, partition, ion-exchange or size exclusion mechanisms or combinations of these, and so the column should be selected according to the application. Usually the system is operated with a solvent of constant composition (isocratically), but when retention times are excessive or resolution needs improving the composition may be gradually changed (gradient elution).

Equipment and techniques

A schematic outline of a simple manual high performance liquid chromatograph is given in Fig. 8.6. However, gradient elution facilities, multiple column techniques and the associated more complex sampling and switching valves may be incorporated. McNair (1982) reviews equipment used for HPLC. The mobile phase pump is an important component in the production of good chromatograms and must provide a pulseless, reproducible flow at pressures up to 6000 psi. Most commercially available pumps are based on reciprocating pistons with the flow-rate being controlled by increasing or decreasing the speed of the drive motor. Where gradient elution is used an additional pump (high pressure mixing) or liquid valves (low pressure mixing) will be required.

Liquid samples are injected on to the column using a valve to avoid perturbation of the solvent flow and column pressure. The basic arrangement includes a six-port valve, fitted with a suitable fixed volume sample loop, which can be switched from the sample input (stand-by) circuit to the solvent circuit for injection on to the column (Fig. 8.6). The sample loop can usually be removed to allow the use of different volume loops, or the loop can be partly filled although the injection precision may be worse. Harvey and Stearns (1984) describe sample injection techniques in detail.

Developments in column packings in recent years have probably contributed most to the

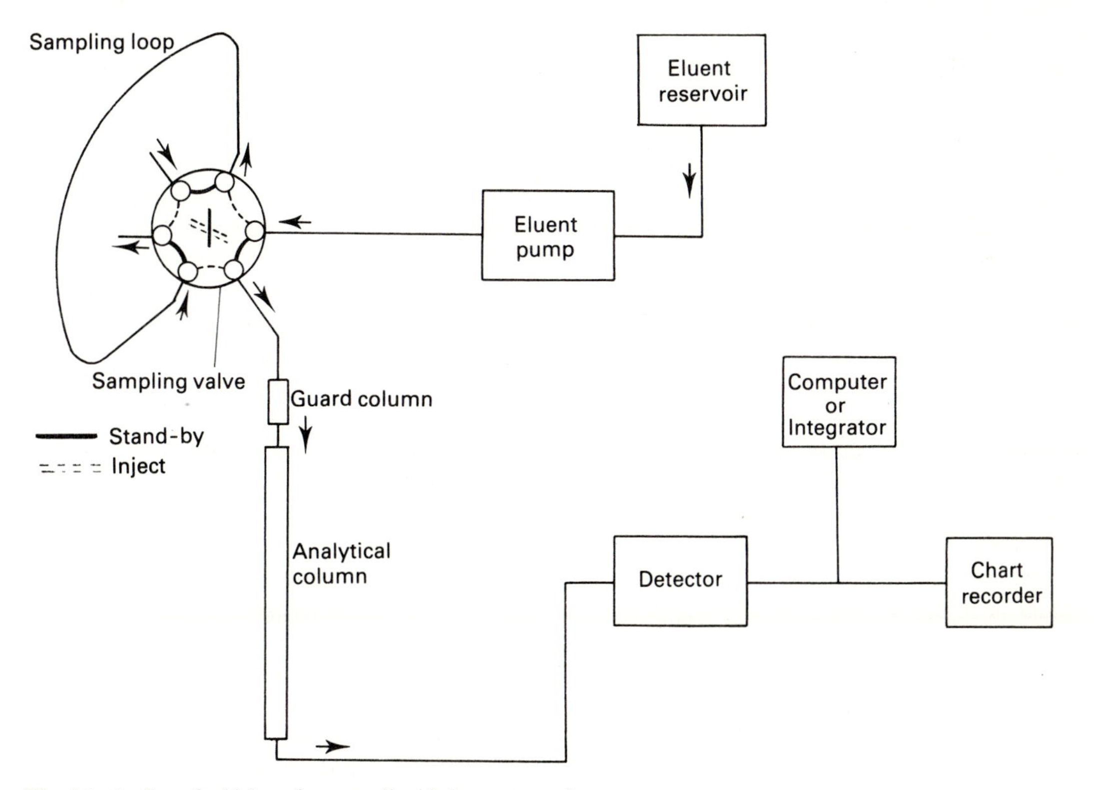

Fig. 8.6. Outline of a high performance liquid chromatograph.

quality of chromatograms now obtainable using HPLC, particularly in adsorption chromatography and bonded phase partition chromatography. A wide range of packings are commercially available. Most are based on silica support material, unbonded for adsorption applications or bonded with silane-based stationary phases of various polarities for partition work. Partition chromatography is the most widely used HPLC technique and can be either normal-phase, in which the stationary phase is more polar than the mixture to be separated, or reversed-phase in which the stationary phase is less polar than the mixture. For normal-phase work the stationary phase consists of cyano- or amino-substituted silanes and in reversed-phase HPLC octyl- or octadecyl-silanes are used. The octadecyl silane packings (ODS C18 columns) are the most versatile, and by careful choice of eluant, can be used for most separations of organic mixtures. Columns and packings have been reviewed by Majors (1980).

Solvent systems for the mobile phase are generally based on hexane, alcohol, acetonitrile or water, or mixtures of these to obtain the required polarity.

The most popular general purpose detector for HPLC is the ultraviolet photometer. There are many variants of this detector from single wavelength filter photometers to scanning spectrometers and diode array detectors which allow the full ultraviolet spectrum of each detected peak to be examined. The latter is useful for identification of unknown compounds. Other detectors in common use are the refractive index monitor, the conductivity meter or electrochemical detectors. In addition HPLC can be interfaced to atomic absorption or mass spectrometers.

As with other instrumental chromatographic techniques, peaks may be collected on a chart recorder or processed automatically by an integrator or computer.

Applications

HPLC has considerable potential in ecological chemistry and has yet to be fully exploited. Many of the organic fractions considered earlier in the manual can be quantified using HPLC. Applications have been developed for studying organic pollutants, notably polycyclic aromatic hydrocarbons, pesticide and herbicide residues, and details of these applications are given by Lawrence (1984). It is possible to use HPLC to investigate soil decomposition products, e.g. humic acids or amino-acids, and there have also been some recently reported applications for trace metal speciation and organo-metallic complexes in soil. HPLC can be used to characterize the structural components in vegetation such as carbohydrates, and could be applied in the examination of plant lipids. Its use in the determination of carboxylic acids, alcohols, phenols and phytoplankton pigments in waters has been described by Hunt and Wilson (1986). Other applications are given by Pryde and Gilbert (1979).

Ion chromatography

Ion chromatography is in general a specialized development of high performance liquid chromatography, but because so many applications are applicable in ecological research, a separate section has been devoted to this topic. The earliest developments in the separation of anions (Small *et al.*, 1975) were commercially exploited by Dionex, and later the technique was extended to cations (e.g. Fritz *et al.*, 1980). For anions especially there are significant advantages in this multi-ion technique in terms of rate of sample throughput, ease of automation, fewer interferences, lower detection limits and especially in providing additional information when compared to alternative methods.

Principles

A liquid sample is introduced into a pumped carrier stream (eluent) which is transported through an ion-exchange column (separator column). Ions compete for the active exchange sites and this results in the separation of the ions into discrete bands (Fig. 8.7). The retention time can be

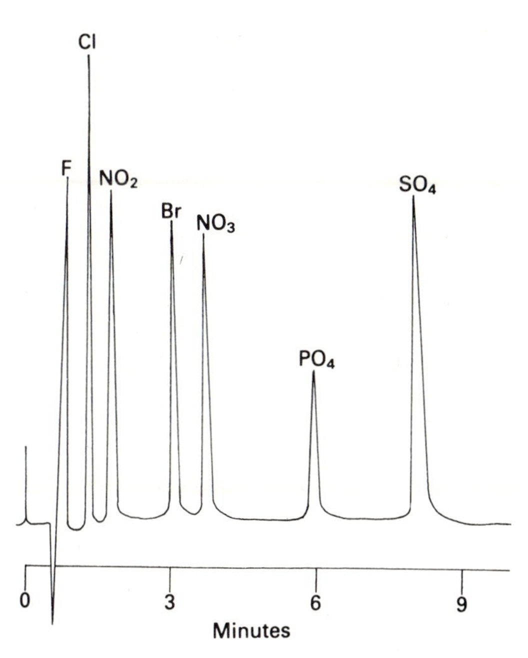

Fig. 8.7. Ion chromatographic separation of anions in water.

used to identify the fraction which flows through the detector. Ions are separated on the basis of size and charge, monovalent ions being eluted first, and within the same valency the smaller ionic radius ions come through quicker. Eluent pH affects the retention times as both the nature of the eluent and the form of the sample ion may be altered.

In general, low capacity ion-exchange columns are required with a wide range of selectivity coefficients and which are stable over a wide range of pH. Commercial columns (e.g. Vydac, Biorad, Dionex) are available which can be selected for the application required. Columns specifically designed for anionic separations (Dionex) are based on styrene divinyl benzene polymers on to which are bonded latex beads. These columns are suitable for separating seven major anions (F^-, Cl^-, NO_2^-, NO_3^-, Br^-, ${SO_4}^{2-}$ and ${PO_4}^{3-}$). Other anion columns which are silica-based can also separate these anions. Monovalent and divalent cations may be determined separately using different eluents but with one column. However, a high-efficiency cation separator based on cation latex exchanges bonded to an anion substrate (Dionex) can be used to quantify mono- and di-valent cations in one run (Na^+, K^+, NH_4^+, Ca^{2+} and Mg^{2+}). A dual channel system has been used for the simultaneous analysis of anions and cations (e.g. Cheam and Chan, 1987).

Some columns have a higher tolerance to fouling and this has to be considered when analysing soil solutions or extracts. Manufacturers also produce shorter versions of the separator columns (guard column) which are placed in line before the separator column to protect it from contamination.

Conductivity is a universal property of ionic species and is the most common detection system for both anions and cations. However, the background signal from the eluent may be a limiting factor. One commercial system utilizes an additional suppression system to reduce chemically the conductivity of the eluent, which results in improved sensitivity. Alternatively the eluent background signal may be subtracted electronically. Ultraviolet detection has been successfully applied with an absorbing eluent (e.g. phthalate), and as the non-absorbing ions pass through the detector the signal decreases. Species which absorb ultraviolet light (e.g. NO_3^-) may be directly quantified in the presence of large excesses of other ions (e.g. in seawater or soil extracts). For certain anionic species such as CN^-, S^- or I^- electrochemical detection can be used.

The choice of eluent is dependent on the type of column, detector and also whether chemical suppression is used. In addition, eluents can be modified to a certain extent to alter the elution order of the peaks for particular applications. Phthalate is a popular eluent for non-suppressed anion chromatography as it has a low background conductivity and also absorbs in the ultraviolet region. Sodium carbonate/bicarbonate has been widely adopted for routine conductivity detection of anions when chemically suppressed. If iodide is required, then stronger eluents such as kyrosine/sodium hydroxide with a micro-membrane suppressor are used. Either nitric or hydrochloric acids are recommended for the detection of monovalent cations by conductivity. Divalent cations may be eluted from the column if an organic modifier (e.g. ethylenediamine) is added to the acid, but under these conditions monovalent ions are eluted as one peak. With a micromembrane suppressor and a high efficiency column, hydrochloric acid with a histadine modifier separates mono- and di-valents (Na^+, K^+, Ca^{2+} and Mg^{2+}). Gradient elution or eluent step change improves the efficiency and speed of the chromatography and also enables ammonium ions to be quantified.

In chemically suppressed ion chromatography (Dionex) using conductivity detection, the eluent is converted to a less conductive species, and the sample ions modified to more highly conductive species. For example, in anion separations using sodium carbonate/bicarbonate, the sodium ions are exchanged with hydrogen ions to form carbonic acid whilst the sample ions are converted to strong acids. Earlier suppressor columns were based on strong acid exchange resins which required frequent regeneration.

This method was succeeded by the fibre suppressor and later by a micro-membrane system. These allow sodium and hydrogen ions to be exchanged through a fibre wall or membrane. A much wider range of eluents at higher concentrations can be used with the membrane suppressor which also offers the potential for gradient elution. Similar chemical suppressors for cations permit the exchange of hydroxide ions with the eluent anion. Non-suppressed ion chromatography is in general less sensitive and the conductivity detection is more sensitive to changes in ambient temperature. However, analyses tend to be quicker and there is a greater flexibility with ultra-violet detection.

Interferences are in general not a major problem in ion chromatography and fall into three categories:

1 Substances whose retention time coincides with the ion of interest, e.g. low molecular weight organic acids interfere with fluoride.

2 Presence of very high concentrations of other alien ions, e.g. interference of potassium by high levels of sodium and other ions in seawater.

3 Column contamination.

Techniques

Analysis

Ion chromatography has become established for anion analysis because of its significant advantages compared to other techniques. In addition to the considerable savings in laboratory time, ion chromatography offers good precision, improved detection limits and is easily automated. Although all peaks which appear on the chromatogram may be of interest, it is not always practical to quantify them all as the response factor is different for each element. In natural solutions, chloride, nitrate and sulphate are easily quantified, but the amounts of nitrate in the samples are usually lower than the others. It may be necessary either to attenuate or amplify the peaks to operate under optimum conditions. Phosphate in natural unpolluted solutions is too low to detect in most instances. Cation methods initially did not compare favourably with the alternative flame methods which were well established, but the technique has now developed so that all the major nutrient cations can be determined from one sample injection.

Sample preparation

Column contamination can be minimized if caution is exercised when handling samples of unknown origin. All samples need to be filtered, preferably through a membrane filter ($<0.45\,\mu$) and in addition an on-line filter can be fitted. This will exclude particulate matter from the system. Humic and fulvic acid fractions from soil solutions, which affect the exchange properties of the columns, can be removed using clean-up cartridges. Some of the ions in solution may change with time, and therefore it is important to minimize the delay between sampling and analysis.

Reagents and standard preparation

In comparison to the small volumes of sample passing through the columns (50–200 μl), much larger volumes of eluent (typically 2 ml min^{-1}) are used and it is therefore most important that high purity is maintained. HPLC grade purity water ($<0.05\,\mu$S) which is essential for all reagents passing through the column, may be prepared using reverse osmosis, a carbon filter and an ion-exchange system. Colloidal matter, which may remain in laboratory grade distilled or de-ionized water, can drastically reduce the useful life of a column and very pure grades of water are essential.

Multi-ion standards should also be prepared using high quality water. Concentrated stock solutions are stable if stored separately for each element, and then mixed freshly to provide a working multi-ion solution which can be further diluted to prepare a range of standards. Anion standards provide an ideal growth medium for algae and should be replaced weekly (daily if PO_4^{3-} or NO_2^- are included). In addition, a large volume of one standard solution should be prepared regularly to monitor drift.

Columns

Polymer or silica based columns are suitable for ion chromatography. The instrument manufacturer's specifications generally provide sufficient information to enable the user to choose the appropriate column. Styrene-based polymer columns are made for use with chemically suppressed systems. The columns are expensive, and therefore the addition of a guard or protection column is strongly recommended. Monitoring of the pressure, separation and detector background signals are essential in order to check when performance begins to deteriorate. A clean-up procedure is then possible. Frits at the ends of columns should be replaced when pressure in the system starts to build up and voids at the top of columns may need to be filled with old column material.

Operating procedures

Conditions and procedures vary according to the configuration of the system, but in general an eluent flow rate of around 2 ml min^{-1} should be selected. For chemical suppression using a hollow fibre, the flow rate of the sulphuric acid regenerant needs to exceed the eluent (approx. 3 ml min^{-1}). Air bubbles trapped at the top of the suppressor column occasionally restrict the flow and may be removed by gently tapping the column. The size of the sample loop fitted to the injection valve (typically 50–200 μl) will govern the sensitivity of the system. If the sample matrix is likely to contaminate the column, then a smaller injection loop is preferable. It is most important that equilibrium conditions are established before commencing analyses, especially after columns have been reconditioned.

Detection limits are not usually a constraint in ion chromatography as the system may be set on the most suitable detection range for the analysis. Exceedingly low levels can be determined if a concentrator column is added. In practical terms, however, the sensitivity not only varies from element to element but the relative concentrations found in natural solutions vary considerably. For instance, ion chromatography is also particularly sensitive for chloride which occurs in relatively high concentrations (up to 100 mg l^{-1}) whereas nitrate is less sensitive and is usually much lower (up to 5 mg l^{-1}). In order to overcome this problem, the detector range can be switched during the run, a signal amplifier used over part of the run or a dual channel recorder used which is set for differing sensitivities.

Calibration may be based on either peak height or area measurement. It is necessary to use area measurements for monovalent cation analysis as the retention time changes as the divalent ions build up on the column. Calibration curves are usually linear, but there may be some deviations from linearity at the lower concentrations due to contamination. Matrix matched standards should be used and a single standard run every 5 samples to monitor fluctuations in performance.

Ion chromatography equipment requires little maintenance except for general routine cleaning of tubing and injection lines with phosphate-free detergent in order to prevent algal growth. Occasionally the eluent pump requires lubrication and the calibration of the conductivity cell needs to be checked.

Equipment

HPLC systems for ion chromatography applications are usually sold with a pump, injection valve, columns and detector (either conductivity or ultraviolet). Alternatively a system using chemical suppression is available. The choice between either type of system will be dependent on the applications and conditions involved in the analysis.

Applications

Ion chromatography has many applications in ecological studies. Examples include the examination of nutrients and pollutants in waters, the analysis of soil extracts and air pollution investigations where sulphur dioxide and nitrogen oxides can be trapped in solution. There are other separations apart from anion and cation chromatography which have been developed, although to date there have been few reports of their use in ecological studies. They include organic acids, alcohols, carbohydrates, amino-acids and proteins, polyvalent anions and transition metals. Examples of its use for the measurement of ions of ecological interest include Smee *et al.* (1978) and Tabatabai and Dick (1983).

Bomb calorimetry

Investigations into the efficiency of ecological processes often require a measurement of the energy utilized in various stages of the system. This applies to studies involving single organisms as well as to those concerned with entire communities or ecosystems. Some of the basic concepts in bioenergetics are discussed in Appendix II.

The most convenient way to measure the energy content is by converting it to heat. This is carried out by complete combustion in the presence of oxygen in a bomb calorimeter. The instruments and techniques used in bomb calorimetry were largely developed many years ago for the fuel industries, but since energy measurements are in absolute terms they are suitable for biomaterials. The most significant of recent innovations has been the introduction of microprocessor controlled instruments.

It is possible to calculate an approximate result for calorific value from the total carbohydrate, fat and protein contents (see Appendix II) but this is unsatisfactory for most purposes.

Principles

The heat of combustion is referred to as a calorific value (CV) which is expressed as joules per gram ($J\,g^{-1}$) or kilo-joules per gram ($kJ\,g^{-1}$). Some older publications express CV as calories per gram ($cal\,g^{-1}$) where 1 calorie = 4.184 joules.

To determine the calorific value, a known amount of sample is burnt in a steel container (bomb) in the presence of oxygen under pressure. The bomb is immersed in a known quantity of water during the combustion and the temperature increase due to the release of heat is measured by an immersion thermometer. The heat released in joules by combustion of the sample (sample weight × unknown CV) is equated with the heat gained (water equivalent of system × temperature rise) ± corrections. This allows the CV to be derived, as shown later. The water equivalent of the system is the heat capacity of a known weight of water plus that of the bomb, thermometer, stirrer and fittings. This value is obtained from the combustion of benzoic acid of known calorific value. This procedure standardizes the calorimeter but both the weight of water and oxygen pressure must be kept constant for subsequent samples.

When using static or isothermal calorimeters, corrections for the cooling rate of the system must be concluded in the calculation. These corrections are based on Newton's law of cooling with allowance for any initial heat gain or loss. Most calorimeters now in routine use have an adiabatic system and this dispenses with the need to estimate heat loss. Other corrections which have to be applied are discussed later.

Instruments

Although static isothermal bomb calorimeters are available, they are rarely used in ecological studies because the need for detailed corrections of

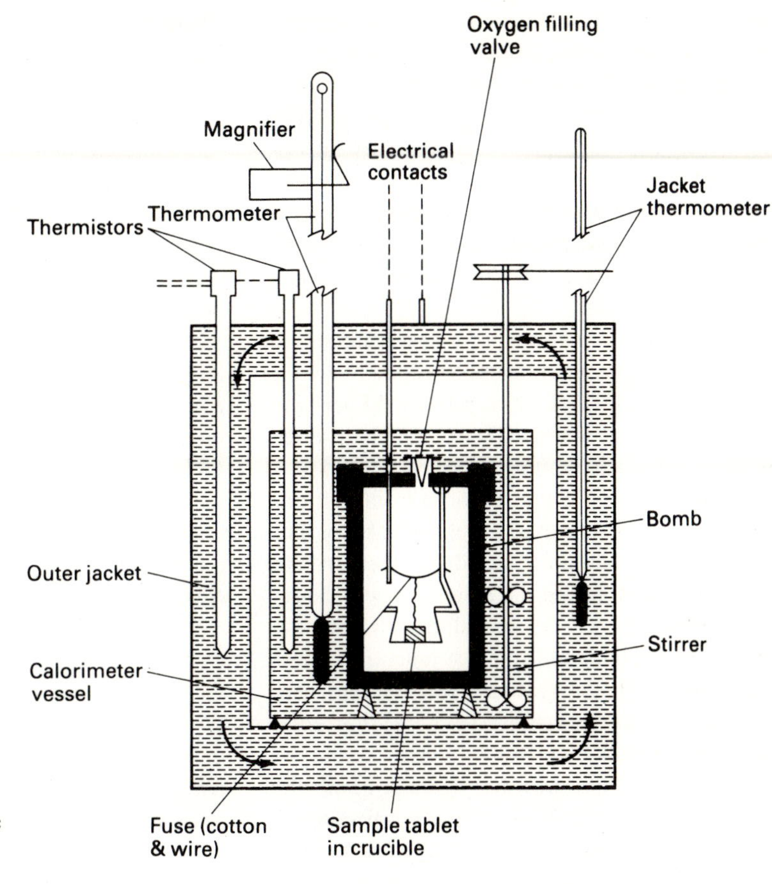

Fig. 8.8. Outline of an adiabatic bomb calorimeter.

heat losses makes them rather slow to use. Manually operated adiabatic instruments are also available but most adiabatic calorimeters make use of thermistor controlled heaters.

The basic components are illustrated in Fig. 8.8. A balanced pair of thermistors is used, one being inserted in the calorimeter vessel alongside the bomb whilst the other is in the outer jacket. When a temperature differential occurs between the water in the inner vessel and that in the outer jacket the thermistors become unbalanced. A relay then switches on compensating heaters in the outer jacket until the balance is restored. The surrounding water is circulated through the jacket and lid by a pump inside the jacket. Passage of mains water through a coil inside the jacket enables rapid cooling to be achieved.

Most adiabatic calorimeters operate in this way but differ in minor details. For example, one model incorporates thermometer vibration and a quick release system. A valuable feature on another instrument is the inclusion of a more powerful heater which can be switched on during periods of maximum temperature rise. This is additional to the smaller compensating heaters controlled by the thermistors. Another way of adjusting the heat compensation in the inner jacket is to include in the circuit a variable transformer which can be adjusted during the test to alter the current passing through the heater.

Adiabatic instruments can often be adapted for use with a 'mini bomb'. A small calorimeter vessel will also be needed and this should have as narrow a diameter as possible to allow for maximum immersion of the thermometer. In this way it is possible to reduce the total water equivalent from about 10 kJ to between 1 and 2 kJ. This makes it possible to burn sample weights as low as 100 mg with little loss of precision and smaller weights still with a proportional loss of precision.

The addition of microprocessor controllers to adiabatic systems enables high throughput to be combined with high precision. Similarly an isothermal system has been developed which maintains the jacket at a constant temperature, thus reducing the delay between samples to a minimum whilst giving results comparable with adiabatic methods.

Routine determinations have been speeded up by attaching the thermocouples directly to the bomb. A model first described by Miller and Payne (1959) has a single thermocouple inserted in the bomb and is commercially available. However, dependence on the response from just one thermocouple results in poor reproducibility. The use of several thermocouples distributed around the bomb and linked in series gives a larger integrated response, but it is difficult to achieve good surface contact for the hot junctions. These methods require calibration with standard benzoic acid and are less precise ($\pm 2.5\%$) than adiabatic instruments ($\pm 1.0\%$).

Phillipson (1966) describes a micro-bomb developed for small amounts of biological samples. The bomb is supported by a copper ring which also serves as the hot junction for a set of thermocouples arranged in series. These are connected to a chart recorder and since the test conditions are standardized it is possible to calibrate the instrument using benzoic acid of known calorific value. The instrument is suitable for biological materials weighing from 5 to 100 mg although having a lower precision than is obtainable even with the 'mini' adiabatic bomb. However, the micro-bomb is very rapid in use.

Techniques

Only a few of the more important considerations with particular reference to the adiabatic bomb colorimeter, are dealt with here. Further information on the adiabatic and other bomb systems can be obtained from the series of papers by Barker *et al.* (1955), Mott (1958) and Nijkamp (1961).

Thermometer calibration

Since the problems of external heat loss have been largely overcome the main limitations to high accuracy are the erratic nature of combustion and the inherent limitations of a mercury-in-glass thermometer.

Either a fixed range or an adjustable Beckman thermometer can be used. These are graduated to 0.01 °C and it is possible to estimate to 0.002 °C using a lens. However, calibration changes of 0.005 °C may occur over about a year, particularly with a new thermometer. Consequently it is advisable to have the thermometers tested at least once a year against a standard temperature source preferably at the National Physical Laboratory if in the UK. A small correction may then need to be applied to each temperature reading. In general, errors may be reduced by starting and finishing as close as possible to the same temperature.

Various electrical devices (thermistor, thermocouple and resistance) can be used in place of mercury-in-glass thermometers for temperature rise measurement. In older instruments careful balancing and calibration is necessary before the temperature increase can be measured. Control in later solid state systems is more effective but sensitivity is not much better than that achieved by mercury thermometers.

Microprocessor monitors and controllers are available to display the temperature with high precision. They may also control the operating conditions and later compute the corrected calorific value for the sample.

Limitations to accuracy and precision

About 1 g is the optimum sample weight for most bomb calorimeters. At this weight there should be little difficulty in reproducing results to within 1.0%. Limitations in the reliability of mercury-in-glass thermometers are mainly responsible for the loss in precision with smaller sample weights, 0.1–0.2 g being the lowest acceptable range for most purposes. Below this the micro-bomb calorimeter, previously mentioned, becomes more applicable.

Since the calorific values for most biological materials fall within a limited range (17–23 $kJ\,g^{-1}$) there is a need for the utmost operational care if meaningful differences between samples are to be recorded. Apart from the errors caused by the thermometer, others are associated with initial sample drying (see below), inefficient heat balancing systems and unsuitable heat corrections.

Initial sample preparation

Although it is possible to fire powdered material successfully, a better combustion generally occurs when the sample has first been compressed into tablets or pellets. Air-dried material should be used and a separate sub-sample must be taken for air-dry moisture and ash determination. The tablets are formed by compressing the powdered sample into a cylindrical mould using a screw press.

Misfiring, and even ejection of the sample from the crucible occurs in some cases and litter samples are especially vulnerable in this respect. The use of crucibles which taper inwards towards the top will result in a better combustion. A metal disc suspended over the sample crucible or the addition of an inert powder to the sample is also effective in certain circumstances. Other methods for facilitating smooth combustions include the use of gelatin capsules and the addition of benzoic acid to the sample. Corrections must then be applied to allow for the calorific value of gelatin or benzoic acid. Most of the special items mentioned above can be obtained from the suppliers of bomb calorimetric equipment.

Drying

Apart from the unsuitability of using dried material for making pellets there is also the danger of energy losses during oven drying. This is largely due to the volatilization of the higher energy oils and fatty constituents. Komor (1940) reported that the calorific value of dried plant material could be as much as 8% less than that of fresh material. When oven-drying is unavoidable, the sample should be thinly spread and rapidly dried in a well-ventilated oven at about 40 °C. Losses are then unlikely to exceed 1 or 2% for most species of vegetation.

Losses of labile, energy-rich materials from animal tissue can be much greater. Colovas *et al.* (1957) reported energy losses of up to 20% upon drying faeces although Flatt (1957), who used higher drying temperatures, obtained a 3% loss for similar materials. In general, it is preferable to freeze-dry this type of sample. Alternatively a low-temperature vacuum oven or a strong desiccant such as phosphorus pentoxide can be used. If it is not practicable to use any of the drying systems, the determinations can be carried out after homogenizing the fresh material with a combustible additive to produce an emulsion.

Corrections

Corrections have to be applied during the calculation stage to compensate for incomplete combustion (leaving unburnt carbon), ignition heat (from fuse wire and cotton) and heat released during acid formation. Sulphur in the sample is oxidized and then absorbed in the water giving sulphuric acid. Similarly nitric acid is formed but both the sample nitrogen and atmospheric nitrogen in the bomb contribute to this reaction. The importance of these corrections was discussed by Barker *et al.* (1955). Some faunal samples have a high content of carbonates which break down during combustion in the bomb. This is an endothermic reaction and a correction of 0.6 $J\,mg^{-1}$ $CaCO_3$ should be allowed (Paine, 1964). This correction is negligible for most samples (containing less than 5% $CaCO_3$).

Water equivalent

The water equivalent should be determined before using the bomb calorimeter for biological samples. It is determined by burning a known weight (between 0.7 and 1.0 g) of pure benzoic acid in the bomb and measuring the rise in temperature of the water as described in the procedure below. The weight of benzoic acid taken should release an amount of heat closely comparable to that released later by the selected weight of sample.

Benzoic acid is available as a thermochemical standard from B.D.H. The calorific value varies slightly with each batch but is about 26 455 $J\,g^{-1}$. Benzoic acid is non-hygroscopic and contains less than 0.1% water in the air-dried state.

The calculations for both the water equivalent and the samples are shown later. About ten separate determinations should be carried out so that a reliable working average can be obtained. This value needs to be checked regularly, some consider that checks every few days are desirable. It is essential to check after equipment modifications. The calculations for both water equivalent and sample determinations are shown later. Further information concerning water equivalent and factors affecting its determination are given by Barker *et al.* (1955).

Application

The bomb calorimeter may be used for the determination of calorific value on almost all biological materials. The nature of the material determines the actual technique to be used in the initial preparation or in handling the bomb. Allowing for slight deviations, some of which have already been discussed, the following basic procedure is suitable for adiabatic bomb calorimeters.

Procedure

Weigh a pellet of about 1 g dried ground sample into a previously weighed bomb crucible.

(Refer to earlier discussion about pelleting, drying, and other initial treatments.)

Pipette 1.0 ml of water into the bomb.

Fit known, predetermined lengths of ignition wire (platinum, diameter 0.0035 mm) and cotton thread (No. 40 or similar) to the crucible support.

Transfer crucible + sample on to support and place cotton thread under sample.

Place inside the bomb, seal and fill with 25 atmospheres of oxygen.

Weigh into the calorimeter vessel sufficient water (about 2 kg) to immerse the bomb. It is essential to use the same weight of water for the samples and for water equivalent determination. Running time can be reduced by preheating the water until the temperature is close to the initial test value.

Place the bomb in the vessel, then transfer both to the water jacket and fit the terminals, stirrer, thermistors and thermometers. Do not proceed further if oxygen bubbles are leaking from the bomb.

Allow water to run slowly through the cooling coil.

Connect the adiabatic circuit and allow the calorimeter temperature to come to equilibrium (usually about 5 minutes).

Tap the calorimeter thermometer lightly for a few seconds and record the initial reading.

Depress the ignition switch immediately.

(If the sample fires correctly the temperature will rise and within about 30 seconds the external compensating heater will come on.)

Allow the temperature to reach a maximum (about 8–12 minutes).

Tap the thermometer gently and take a final reading.

Dismantle the connections and remove bomb.

Pass water through the outer cooling coil to reduce temperature of jacket for the start of the next test.

Release the gases (where corrections for H_2SO_4 and HNO_3 formation or any analyses of the bomb gases are to be carried out, delay release for approx. 8 minutes).

Abort the test at this stage if unburnt material has been ejected.

Remove the crucible and dry for 30 minutes at 105 °C, then weigh.

Ignite in a muffle furnace at 550 °C then weigh again.

Corrections

a Unburnt carbon.

If the weight of carbon residue does not exceed 50 mg multiply the loss-on-ignition weight loss (in mg) by 35.04 (*CV* for carbon).

b Acid formation.

For most vegetation samples (with a heat release of approx. 20900 J g^{-1}) the corrections are of the order of 17 J for H_2SO_4 and 25 J for HNO_3 and separate estimations are not necessary.

However, for any critical work these corrections should be estimated and applied.

Treat the solution in the bomb as follows:

Titrate with 0.05 M $Ba(OH)_2$ (A),

Add excess 0.05 M Na_2CO_3 (B),

Titrate with 0.1 M HCl (C).
HNO_3 correction = 5.99 (B − C) J.
H_2SO_4 correction = 15.07 [A − (B − C)] J.
(The approximate sulphur content of the sample may also be obtained in this way.)

c Ignition corrections.
Deduct a predetermined value for the heat released by the platinum wire.
Deduct similarly for the known length of cotton thread.

d Dry weight and ash-free weight corrections.
Determine moisture on a separate sample and apply correction in usual way.
An ash-free correction can also be applied, if desired, after a separate ash determination. The residue left in the bomb after combustion should only be used for an approximate ash result.

Calculation
In the discussion of the principles of bomb calorimetry, it was mentioned that the sample weight × unknown *CV* was equated with [water equivalent (*WE*) × temperature rise (*T*°C)] ± corrections, hence:

$$CV(\mathrm{J\,g^{-1}}) = \frac{(WE \times T) - (\text{all corr.}) + \text{unburnt C}}{\text{sample wt (g)}}$$

where the corrections to be subtracted (heat of formation of H_2SO_4, heat of formation of HNO_3, ignition wire, cotton thread) and the correction to be added (unburnt carbon) are all expressed as joules (J).

When determining *WE*, the benzoic acid weight × known *CV* is equated as above so that by rearrangement:

$$WE\ (\mathrm{J\,°C^{-1}}) = \frac{(\text{benz. acid wt (g)} \times CV) + (\text{all corr.}) - \text{unburnt C}}{T(°\mathrm{C})}$$

where the three corrections to be added are heat of formation of HNO_3 (from nitrogen already in bomb), ignition wire and cotton thread and the correction to be subtracted is unburnt carbon.

A constant weight of water in the calorimeter and oxygen pressure in the bomb are assumed.

The final result is often expressed as kJ g^{-1} and should be on a dry weight (or ash-free dry weight) basis.

Notes
1 Match the thermistors before use by immersing in a beaker of water and adjusting the appropriate instrument control.
2 If nickel crucibles are used in the bomb the ignition temperature during subsequent ashing for unburnt carbon correction must be kept below 500 °C to prevent oxidation of the metal.
3 Consult physical tables to check for variation of specific heat of water with temperature.
4 The solution in the bomb may be used for the determination of sulphur using the procedure on p. 151–152.

Automation

Automation in this instance refers to the analysis of samples without the continuous attention of the operator. Strictly, most procedures are only partially automatic since they require the involvement of the operator in at least one step in the analytical procedure. There are some fully automatic analysers, such as those used for monitoring water flow but these are specialist instruments and are not considered here.

When dealing with heterogeneous materials such as those encountered in ecological studies it is advisable to process large numbers of samples, so some automation is desirable. Apart from facilitating the processing of the samples, the use of such equipment can lead to:
1 better accuracy and precision,
2 improved sensitivity,
3 lower analytical costs.

The following paragraphs will briefly discuss various aspects of laboratory automation and how they can be applied to some of the techniques mentioned in this book. Further information on the application of automation in chemical analysis can be obtained from Hunt and Wilson (1986) (waters) and from the Commonwealth Bureau of

Soils (1975) (soils and plant materials). A review of trends in laboratory automation was given by Stockwell (1980).

Applications

Sample preparation

Preparation of solid material using grinding or sieving remains perhaps the most labour intensive process because automated equipment has so far not been developed for laboratory scale use. This stage can be time consuming because the entire sample collected in the field will need to be handled before representative sub-samples can be taken for analysis. In addition it is important that the equipment is cleaned between samples to prevent cross-contamination and no suitable procedure has so far been devised for this stage.

Solution preparation

A large range of liquid handling equipment is available for dilution and dispensing of reagents. The calibration of these devices should be regularly checked by weighing the deliveries of water. Solution preparation from solid material initially involves weighing the sample and then transferring to a suitable container, generally the extraction or digestion vessel. Robotic devices have been described for these operations but are as yet only in use in larger laboratories with good support facilities. Dissolution procedures involving acid digestion can be speeded up using aluminium digestion blocks with programmable units to control each phase of the reaction.

Instrument control

Mechanical autosamplers are generally used to present the prepared sample solutions to the instrument. For atomic absorption or colorimetric procedures, relatively low cost devices are available which enable the sample and wash time to be set. Automation of liquid chromatographic processes (e.g. HPLC and IC) needs control of the sampler, pump and injection valve in order to propel the sample to the top of the column. The microprocessors used to collect instrumental data may also offer facilities to control the autosampler or instrument parameters. In addition the widespread use of computer and built-in microprocessing chips has increased the degree of automation of some techniques, although the set-up procedures may then be more complex.

Quantification and report preparation

The traditional instrumental signal output was in analogue form, but most microprocessor controlled units now also give digital information through a data port. Even if the instrument is interfaced to a computer, it is important to use an analogue recorder to provide a back-up in case of power failure or software problems. Data processing facilities are usually available for modern instruments, but these vary from software packages for a micro-computer, to a fully integrated data-station. The data system should be fully evaluated before purchase and installation. Co-ordination of data from the analytical instruments should also be considered in order that the analytical report can be prepared. Laboratory Information Systems (LIMS) are now becoming available to provide comprehensive management facilities.

Colorimetry

Full details of automated continuous flow analysis are given in Chapters 5 and 8. Alternatively, the manual methods given in Chapters 5 and 7 can be automated using discrete, flow injection or centrifugal analysers. In discrete analysers, samples are held in containers and transported on a moving belt or turntable. Reagents are then added and the reaction allowed to proceed before the solution is fed into the detector. Cross-contamination is virtually eliminated with this technique which is suited to high sample throughput. Flow injection analysis has developed rapidly over the last few years, but its adoption for routine analysis has been much slower. Sample solutions are injected into an unsegmented flowing carrier stream which is merged with the reagent stream. The reaction is controlled by dispersion before the solution arrives at the sensor unit which is usually a colorimeter. Centrifugal analysers are used to automate colorimetric methods for clinical applications, but the high capital cost and lower sensitivity due to small sample volumes have hindered its wider acceptance.

Atomic spectrometry

Flame techniques are readily automated by the addition of an autosampler and recorder-data unit. In general, instrumental performance is sufficiently stable to present few problems in automation. The addition of computer control enables the operator to select the operating parameters from a screen menu or facilitates automatic set up which enables the instrument to be used for sequential elemental analysis. Control of the sample input using an auto-sampler and injector probe is particularly important to obtain adequate precision and accuracy for electrothermal atomization.

Chromatography

Fully automated gas and liquid chromatography systems are available, but it is most important to examine the need for automation as sample preparation and preliminary concentration steps may be complex. Automation of liquid chromatography is readily achieved, and it is possible to incorporate column switching procedures.

9
Statistical Analysis and Data Processing

Ecological work generates both field and laboratory data when (i) information about a habitat or population is required (including surveys and monitoring) and (ii) an hypothesis is tested (an experimental approach which may also include survey and monitoring). However, these data will not give valid information unless certain aspects of the problem are adequately defined as below.

1 The objectives of the work should be clear and realistic, and any hypothesis tested should form part of a theory or model constructed to include any basic assumption underlying the work. Many workers have written on these aspects of scientific methodology including Mead and Curnow (1983) and Ridgman (1975).

2 The data collected must be relevant to the objectives of the work.

3 Sufficient data must be collected to allow an adequate statistical analysis and the collection itself must not be biased. The sample collection, survey techniques and/or experimental design must allow for these constraints (see p. 308). The concepts of statistical analysis are discussed later.

4 In the laboratory the accuracy, precision and sensitivity of any physical and chemical methods used should be known. These terms are defined and discussed later. The calculation procedures involved in converting instrument readings to sample concentrations are also discussed in this chapter.

Certain aspects of statistics which are relevant to this framework are covered here although not to the extent of providing a mini-course in statistics.

Instead the basic tests are described, supplemented by background material not always found in elementary texts. Attention is also given to assumptions and limitations of the tests. Special topics such as the design of experiments can be treated only briefly here, but others such as calibration, accuracy and precision are highlighted because of their particular relevence in chemical analysis. Several workers quote examples using analytical data including Caulcutt and Boddy (1983) and Youmens (1973).

Basic statistical concepts

The word *statistics* is not used here as a synonym for data but rather in its methodological sense and may be defined as the science of collecting, analysing and interpreting data. It finds very frequent application in many spheres of research, industry and economics although only a brief outline can be given here. Two main types of statistics are considered.

1 *Descriptive statistics*—dealing with the presentation and description of data. This includes graphical methods and the classification of data in tables. Numerical measures of location and variation could also be included here but are more conveniently discussed under statistical inference.

2 *Inferential statistics*—involving the assessment of sample data to infer conclusions about a population. In addition a few brief notes on probabilities are included with some discussion of frequency distributions. Many statistical textbooks are available including a number devoted more specifically to the biological field.

Graphs and tables

Graphs and tables enable data to be presented so that the frequency distribution and other gross features become clear. Both discrete data (obtained by counting a variable) and continuous data (obtained by measuring a variable) may be handled in this way. The use of tables to portray data needs no discussion here and several examples are given in the Appendices. Several graphical methods are in common use, including those outlined below. Many of the larger computing systems have 'graphics' facilities which not only provide clean copies of diagrams ready for public display but are more objective to use than manual methods.

1 *Bar chart.* This is used for discrete data. One example is the comparison of numbers of shoots of certain herb species in a quadrat (see Fig. 9.1).

2 *Histogram.* This is used for continuous data and is the best method of illustrating frequency distributions. A typical example in forestry is selected breast-height girths in an even-aged single species stand. The distribution of girth sizes in the stand is clear and may be further clarified by linking the mid-points of the tops of the rectangles to form a frequency polygon. It is also helpful to compute the appropriate normal curve (see p. 294) and superimpose it on to the histogram. Major departures from normality in the histogram can then be seen.

3 *Scatter diagrams.* Very frequently graphs are used to portray the relationship between two sets of continuous data. Measurements of biological variables often show a considerable scatter and mathematical tests are needed to see if there is a statistical relationship between the variables. Examples are given under regression and correlation.

4 *Other two-dimensional presentations*—include block graphs and circular graphs (pie charts). More complex is the three-dimensional histogram enabling three variables to be presented together.

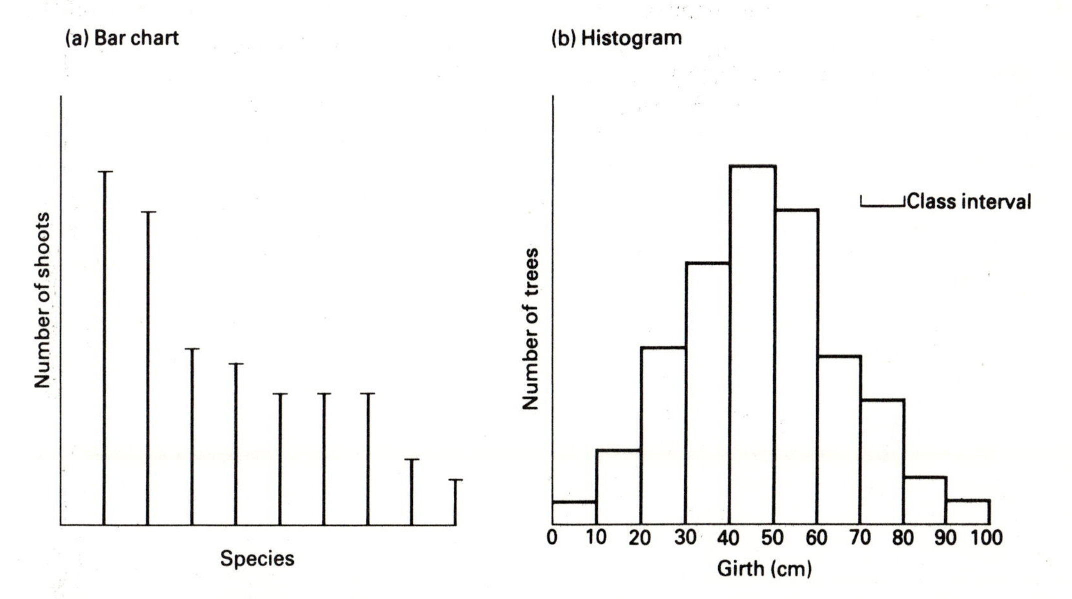

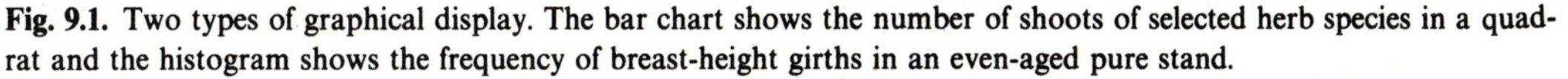

Fig. 9.1. Two types of graphical display. The bar chart shows the number of shoots of selected herb species in a quadrat and the histogram shows the frequency of breast-height girths in an even-aged pure stand.

Examples of all these are available in statistical textbooks. See also Chapman and Mahon (1986).

Elementary probability

The magnitude of a natural phenomenon can never be predicted in advance with absolute certainty and the concept of probability is essential for understanding such phenomena. Intuitive ideas about probabilities are readily obtained from games of chance involving coins, dice, etc. (Huff, 1965). Probability theory makes it possible to calculate the probability of an event occurring or a specific result being obtained. The probability of a simple event (successful result) illustrates this. If an experiment can result in n equally likely, mutually exclusive results, r of which are successful, then the probability of success is given by:

$$\frac{\text{successful results}}{\text{total results}} = \frac{r}{n}$$

A widely quoted example is the binomial experiment in which only two results (*success* and *failure*) are possible for each trial, with the trials being independent, the probability of success constant, and the successes binominally distributed.

Natural phenomena are more complex than this and the likelihood of an event or result is inevitably conditioned by other factors. Hence probability theory has to allow for the permutations and combinations arising when two or more events are likely and which may not be independent or in a particular order. Most statistical texts derive the appropriate formulae and these are not repeated here. The concepts of probability, however, form the basis of statistical theory and so if a biological population has n individuals, r of which are sampled at random, then probability statements can be made about those samples and conclusions inferred about the whole population.

Frequency distributions

Continuous and discrete data follow frequency distributions. Each is described by three parameters (location, scale and shape) although two may be sufficient.

Continuous data

1 *Normal distribution* (Fig. 9.2). This important symmetrical distribution is described by two parameters only (arithmetic mean and variance) with the shape constant. This property is the basis of all parametric tests which are not applied to other distributions because of the effects of shape on mean and variance. However, a log-transformation sometimes restores approximate normality if asymmetry is slight.

2 *Other distributions.* Many other types of distribution are known, including exponential, but

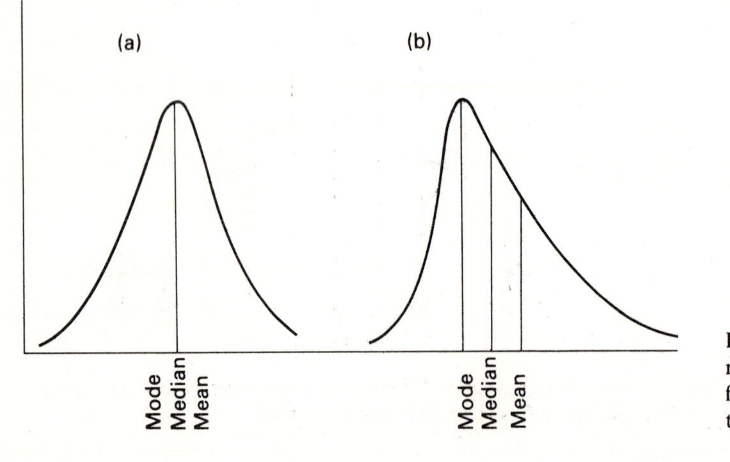

Fig. 9.2. Diagram illustrating the mode, median and arithmetic mean for (a) normal and (b) skewed distribution.

direct applications in ecology have been uncommon. Distributions of t, F and χ^2 belong here but their use in parametric tests depends on a special property (shape related to sample size) allowing small samples to be tested (see later).

Discrete data

Various binomial distributions can be fitted to discrete data (counts) and are important in population studies. The binomial itself arises where there is a probability of two alternative results for a trial. This distribution is roughly symmetrical for nearly equal probabilities and approaches normality for large numbers of trials. The related Poisson distribution is skewed and is a special case applied when the probability of one result is very low.

Departures from normality

Parametric tests require normal data but nutrient values sometimes display non-normality (Grimshaw and Allen, 1987). This is assessed by measuring the two shape parameters (skewness and kurtosis) and comparing with the normal values. Estimates are derived from the moments (m_r) about the mean where

$$m_r = \frac{(x-\mu)^r}{n}, \text{ where } r = 2, 3 \text{ or } 4.$$

The skewness coefficient is given by

$$\gamma_1 = \sqrt{\beta_1} = \frac{m_3}{m_2(m_2)^{\frac{1}{2}}} = \text{zero (normal)}$$

whilst the kurtosis coefficient is

$$\gamma_2 = \beta_2 - 3 = \frac{m_4}{(m_2)^2} - 3 = \text{zero (normal)}$$

Coefficient values are assessed by tables (Pearson and Hartley, 1966) and a significant difference from zero indicates non-normality. Large samples are needed to apply these tables.

Estimation

The main concern here is to obtain information from a sample about the population from which the sample was taken and thence to set up a mathematical model describing the population. Statistical inference concerns two closely related problems.

1 The estimation of the unknown parameters of the population (from sample data).

2 The testing of an hypothesis about the model including assessment of the significance of the sample data.

The main features of sample data become clear in a table and/or histogram. These features can be further summarized by obtaining two estimates which are needed for parametric tests and which locate the central value of the data and show how the data are scattered or dispersed.

Measures of location

Three measures should be distinguished although numerically equal for normal data (Fig. 9.2).

1 *Mode*—the value occurring most frequently. There may be more than one mode or none at all. It is seldom used and is not tested.

2 *Median*—the middle value dividing a distribution into two halves of equal area. Whilst it is unaffected by outlying values it is not tested by parametric methods and is used only for scanning data or in some non-parametric tests.

3 *Arithmetic mean*—this value is given by

$$\bar{x} = \frac{\sum_{i=1}^{n} x_i}{n}$$

This measure is affected by outliers but is always used because it is applicable to parametric tests.

Two central values which are more specific include the geometric mean (obtained from logarithms) and the harmonic mean (from reciprocals). These are not tested but can aid data understanding.

Measures of variation

Four measures are employed to show the dispersion of the data.

1 *Range*—the difference between the least and greatest values. Ranges are commonly quoted in the literature but have the drawback of ignoring most of the sample data.

2 *Mean deviation*—the arithmetical mean of the deviation of the values from the mean of the sample. This value has limitations and is little used for inferring the dispersion of the population.
3 *Variance*—the arithmetical mean of the *squares* of the deviations.

$$\text{Variance } (s^2) = \frac{\sum_{x=1}^{n} (x_i - \bar{x})^2}{n-1}$$

This measure is always used for inference and in parametric tests. It is necessary to use $n-1$ degrees of freedom (unrestricted and independent variables) instead of n values to ensure an unbiased estimate particularly for small samples (see also below).

A form suitable for computation is given by:

$$s^2 = \frac{\Sigma x^2 - \frac{(\Sigma x)^2}{n}}{n-1}$$

4 *Standard deviation*—the square root of the variance. The variance (s^2) and standard deviation (s) are the accepted sample measures for inferring σ^2 and σ of a normal population but it should be noted that they are valid for inference only if they are unbiased. The conditions for avoiding bias can be restated:
a The samples are taken at random (see p. 308).
b The variance and standard deviation are calculated from $n-1$ values and not from n (particularly for small samples where $n \leq 30$).

The relative dispersion is a widely used quantity when expressed as the coefficient of variation or (standard deviation/arithmetic mean) × 100. Other measures of relative dispersion exist but are little used in practice.

So far, it has been indicated that a random sample of n measurements with mean $\bar{x}$ and variance s^2 can be used to infer the mean μ and variance σ^2 of a normal population of an infinite number of measurements. However, inferences from samples are only approximate (have random error) even when the measures are unbiased. Hence to infer μ from $\bar{x}$ requires limits to be set about μ.

These are given by the 'central limit' theorem which states that if n random values are distributed about their mean ($\bar{x}$) with variance s^2 then $\bar{x}$ itself will be distributed about μ with variance s^2/n.

This value leads to the standard deviation or *standard error* of the population mean and is important in significance tests (see below).

Standard error of the mean =

$$\sqrt{\frac{\text{variance}}{n}} = \frac{\text{standard deviation}}{\sqrt{n}}$$

Standard errors of the variance (σ^2) itself are not needed for tests but will be mentioned later with regard to non-normality.

Confidence limits

The limits set by the standard error are attached to the population μ whose inferred value is $\bar{x}$. However, this does not guarantee that μ will fall within these limits and the uncertainty can be expressed by assigning confidence limits. These limits are chosen so that the frequency with which they enclose the true mean are satisfactorily high. 95% limits (probability ≤ 0.05) are often used, being the limits ($\mu \pm 1.96\sigma$) enclosing 95% of the area under the normal curve. The exact expression depends on whether σ is known. To summarize for $p \leq 0.05$:

If σ is known

$$95\% \text{ confidence limits} = \mu \pm 1.96 \frac{\sigma}{\sqrt{n}}$$

If σ is not known

$$95\% \text{ confidence limits} = \mu \pm 1.96 \frac{s}{\sqrt{n}} \quad (n > 30)$$

$$= \mu \pm t_{0.05} \frac{s}{\sqrt{n}} \quad (n \leq 30)$$

This format is further explained below under significance tests. Meanwhile Table 9.1 gives a typical calculation for $n \leq 30$ taken through to the confidence limits. Table 9.2 gives a program listing and print out of this calculation from SAS.

Outlying values

Outlying values affect the arithmetic mean and its apparent significance and may have an environmental interest of their own. Various tests for

Table 9.1. Calculation of variance, standard deviation, standard error and confidence limits for percentage total nitrogen (x) in 10 replicated soil samples

Sample	x	x^2
1	0.87	0.7569
2	0.83	0.6889
3	0.79	0.6241
4	0.92	0.8464
5	0.84	0.7056
6	0.97	0.9409
7	1.03	1.0609
8	0.92	0.8464
9	0.89	0.7921
10	0.90	0.8100
	$\Sigma x = 8.96$	$\Sigma x^2 = 8.0722$
	$\bar{x} = 0.90$	

$$\frac{(\Sigma x)^2}{10} = \frac{(8.96)^2}{10} = 8.0282$$

$$\text{Variance } s^2 = \frac{8.0722 - 8.0282}{10 - 1}$$

$$= 0.004889$$

Standard deviation $s = 0.06992$

Standard error $s/\sqrt{n} = 0.02211$

95% Confidence limits $= \pm 2.26 \times 0.02211$ (t for 9 degrees of freedom)

$= \pm 0.0499$

i.e. % nitrogen $= 0.90 \pm 0.05$

Table 9.2. SAS program listing for the measures in Table 9.1

```
DATA SOISAM;
  INPUT SAMPLE X;
  LIST;
  CARDS;
 1    0.87
 2    0.83
 3    0.79
 4    0.92
 5    0.84
 6    0.97
 7    1.03
 8    0.92
 9    0.89
10    0.90
PROC MEANS SUM MEAN VAR STD STDERR;
VAR X;
RUN;
```

VARIABLE	SUM	MEAN	VARIANCE	STANDARD DEVIATION	STD ERROR OF MEAN
X	8.960 000 00	0.896 000 00	0.004 893 3	0.069 952 36	0.022 120 88

Minor differences between the 'manual' and computational method are due to 'rounding' errors.

identifying outliers have been proposed and assume the population is normal. The easiest to apply are 'cut-off' limits. For example, 99% of the values under a normal curve are contained within $\mu \pm 2.58\,\sigma$, so that 3σ is a useful although arbitrary criterion. Some workers have used 4σ. However, probability tests are preferred and require a statistic similar in type to that used for comparing mean values (see p. 299). One of the simplest statistics is; (extreme value – overall mean)/ overall standard deviation. The extreme value is an outlier if this statistic exceeds the quantity tabulated for given size (n) and probability level. Davies and Goldsmith (1972) and others provide the necessary table. Snedecor and Cochran (1980) discuss similar statistics and offer appropriate tables whilst a detailed review of outlier tests was given by Hawkins (1980).

The subsequent treatment of outliers is a separate problem. In statistical terms an outlier may:

1 Belong to the same population as the rest of the data. In this case the outlier is part of the natural variation of the material and should *not* be rejected even if inconvenient for assessing the arithmetic mean. As discussed on p. 309, sample replication should be adequate to allow for high natural variation. Meanwhile possible distortion of the mean can be checked by computing the median. The difference between these two values should be trivial if the sample is approximately normal. If not then non-normality is suspected and can be confirmed from the coefficients of skewness and kurtosis.

2 Belong to a different population from the other data. This type of outlier usually indicates sampling or lab contamination or analytical error and should be rejected provided there is sufficient evidence (including repeat analyses). The dismissal of unusual values as 'fliers' should always be avoided when this evidence is lacking. A subsidiary problem arises if acceptance of the repeat value in the data set allows other outliers to emerge under test and which were initially 'masked' by the discarded value. If contamination/error is again confirmed, the whole set may now be suspect but if not, the values should remain. As noted earlier, positive skewness can be expected in field data and provides an excess of values in the extended positive tail. Whilst the probability tests discussed above are not applicable to non-normal data, an arbitrary cut-off limit will highlight the more extreme values which might have environmental implications for the particular sampling points or sites.

Other types of 'outlier' problem can arise in the lab. For example, a faulty standard solution can distort a calibration line. However, the validity of a standard is quickly checked if tolerance limits for each standard are inserted in the calibration–prediction program as appropriate for the variability of the method. This is part of quality control (see p. 314).

Significance

Significance forms another aspect of statistical inference and is important if conclusions inferred about a population are to be accepted as valid. In general if a study is made to see if a certain treatment has a specific effect, random errors may lead to an apparent effect even when none exists. Significance testing assesses the probability that the apparent effect could have arisen by chance. The lower the probability the more likely the effect was real. (See also comment below on errors.)

For all significance testing it is necessary to:

1 construct a mathematical model of the situation of which the hypothesis to be tested forms a part,

2 state precisely the hypothesis to be tested,

3 set up this as a null hypothesis (H_0) for the population, i.e. set it up as being true. Any other hypothesis is the alternative hypothesis (H_1).

4 define some measure of an unsatisfactory fit,

5 find the probability with which (on the hypothesis H_0) the observed value could by chance be equal to or more extreme than this measure,

6 consider what evidence this probability provides against the hypothesis and what action should be taken next.

It is usual to accept some arbitrary level of probability as a criterion for making a decision about the hypothesis. The most commonly accepted probability (p) is 1 in 20 or 0.05. This is often termed the 5% level because only 5% of the area under the normal curve (2.5% of each tail) is excluded. (Note that if a single tail test is appropriate then p is 0.025.) On this basis a result attaining a 5% level is considered 'significant' whilst a 1% level is 'highly significant'.

This means there is a probability of only 1 in 20 (or 1 in 100) that the observed value will by chance be equal to or more extreme than the measure representing an unsatisfactory fit. Hence H_0 can be rejected. However, there is a 1 in 20 chance of it being rejected incorrectly. In effect the probability of making errors of judgement is inherent in tests based on significance levels and must be borne in mind. The two types of possible errors are I (H_0 rejected when true) and II (H_0 accepted when false). The significance level of a test is the probability of type I occurring. See texts for further details regarding these two errors.

It should be noted here that only statistical significance is involved and not practical or causal significance. This distinction is particularly important in biological studies. A typical example might be the hypothesis that the incidence of a particular disease in an area is related to the level of element X in the crops grown in that area. Showing that X is significantly higher in the crops there than in areas free of the disease in no way proves there is any biological relation between X and the disease. External evidence is needed to affirm a decision on H_0.

Comparison of mean values

Arithmetic mean values are additive so differences are tested for significance using a statistic of known distribution. A key property of the normal curve provides a statistic. If random samples of size n are drawn from a normal population of mean μ and variance σ^2 then $\bar{x}$ is distributed normally about μ with variance σ^2/n. The square root ($\sigma/\sqrt{n}$) is known as the *standard error* of the mean and

$$\text{the test statistic} = \frac{\text{difference}}{\text{standard error}}$$

This is used in comparing two mean values as follows:

1 *Comparison of $\bar{x}$ and μ*

a If σ is known (or estimated but *not* from data under test)

$$\text{statistic} = z = \frac{\bar{x} - \mu}{\sigma}$$

and follows the standard normal curve (mean zero and unit variance) with tables in the texts. H_0: $\bar{x} = \mu$ is rejected (difference significant) if $z \geqslant 1.96$ ($p \leqslant 0.05$) or 2.58 ($p \leqslant 0.01$).

b If σ is *not* known (or is estimated from data under test)

$$\text{statistic} = \frac{\bar{x} - \mu}{s/\sqrt{n}}$$

but its distribution depends on sample size. If $n > 30$ the statistic is normal (z) and is applied as above. However, if $n \leqslant 30$ the statistic follows the t distribution whose table is entered at $n-1$ degrees of freedom (see Table 9.3).

2 *Comparison of $\bar{x}_1$ and $\bar{x}_2$ of unpaired sets*

The standard error is now drawn from both samples.

a If σ is known (or estimated but *not* from data under test)

$$z = \frac{\bar{x}_1 - \bar{x}_2}{\sqrt{\dfrac{\sigma_1{}^2}{n_1} + \dfrac{\sigma_2{}^2}{n_2}}}$$

Use standard normal tables as in **1a**.

Table 9.3. Comparison of the mean ($\bar{x}$) of the data in Table 9.1 with a specified value (μ) of 0.84% nitrogen

$$t = \frac{\bar{x} - \mu}{s/\sqrt{n}}$$

$$= \frac{0.90 - 0.84}{0.02211}$$

$$= 2.71$$

enter t table at $n-1=9$ degrees of freedom
then $p < 0.05$

hence the difference between $\bar{x}$ and μ is significant at the 5% level and the null hypothesis can be rejected.

b If σ is *not* known (or estimated from data under test) and $n > 30$ use z as given in **2a** but replace σ^2 by s^2.

If $n \leqslant 30$

$$t = \frac{\bar{x}_1 - \bar{x}_2}{\sqrt{\frac{s_1^2}{n_1} + \frac{s_2^2}{n_2}}}$$

enter t table at $n_1 + n_2 - 2$ d.f.

This statistic is often needed because biologists usually work with small samples. (see Table 9.4 and Table 9.6 for SAS listing). Random, normal data are assumed as before but 't' has further constraints.

i The variances must be equal under the F test. If not, the statistic does not follow t and cannot be used. It is possible to recalculate the degrees of freedom and enter the t table (Snedecor and Cochran, 1980) but this is approximate and little used.

ii If sample sizes are not equal, a common (weighted) variance is needed (see texts). This is applied as $s\sqrt{1/n_1 + 1/n_2}$ and used in statistical packages but is not valid if the two variances are not equal. When t is not valid, non-parametric methods are available for comparing mean values although less powerful. Simple ranking tests are widely used and given on p. 307.

Tests for comparing several mean values require the F test with the ANOVA format (p. 304).

3 *Comparison of paired sets*

As noted above, the comparison of independent sets of data is complicated by variation within

Table 9.4. Comparison of the mean percentage total nitrogen contents of 10 random soil samples taken at the same depth in each of two plots: x_1 (treated) and x_2 (control)

Sample	x_1	x_1^2	x_2	x_2^2
1	0.56	0.3136	0.50	0.2500
2	0.53	0.2809	0.49	0.2401
3	0.57	0.3249	0.42	0.1764
4	0.56	0.3136	0.53	0.2809
5	0.61	0.3721	0.49	0.2401
6	0.61	0.3721	0.43	0.1849
7	0.52	0.2704	0.50	0.2500
8	0.60	0.3600	0.42	0.1764
9	0.51	0.2601	0.53	0.2809
10	0.52	0.2704	0.41	0.1681
	$\Sigma x_1 = 5.59$	$\Sigma x_1^2 = 3.1381$	$\Sigma x_2 = 4.72$	$\Sigma x_2^2 = 2.2478$

$\bar{x}_1 = 0.559$ $\qquad$ $\bar{x}_2 = 0.472$

$(\Sigma x_1)^2/10 = 3.1248$ $\qquad$ $(\Sigma x_2)^2/10 = 2.2278$

$$\text{Variance } s_1^2 = \frac{3.1381 - 3.1248}{10 - 1} = 0.001\,478 \qquad s_2^2 = \frac{2.2478 - 2.2278}{10 - 1} = 0.002\,222$$

and $s_1^2/n_1 = 0.000\,1478$ $\qquad$ $s_2^2/n_2 = 0.000\,2222$

$$t = \frac{\bar{x}_1 - \bar{x}_2}{\sqrt{\frac{s_1^2}{n_1} + \frac{s_2^2}{n_2}}} = \frac{0.559 - 0.472}{\sqrt{(0.000\,1478 + 0.000\,2222)}} = 4.52$$

$F = s_1^2/s_2^2 = 1.50$ and is not significant at $p \leqslant 0.05$ for $n - 1 = 9$. At $t = 4.52$ $p < 0.01$ for $n_1 + n_2 - 2 = 18$. Therefore the null hypothesis can be rejected and indicates a significant difference between treatment and control values.

each set. This influences the t value and even its validity. However, if the comparison is based on pairs of sub-samples drawn from a single set of homogeneous parent samples (stable solutions or fine powders), the effects of variation are lessened. When sets are paired, the mean difference ($\bar{d}$) within pairs is tested by t for a small number of pairs (n). To test there is no difference between two paired sets (i.e. $H_0: \mu_d = 0$)

$$t = \frac{\bar{d} - \mu_d}{s_{\bar{d}}}$$

and the standard error $s_{\bar{d}} = s_d / \sqrt{n}$

where $s_d = [\Sigma(d_i - \bar{d})^2 / n-1]^{1/2}$. The t table is entered at $n-1$ degrees of freedom.

Opportunities for pairing arise in the laboratory as, for example, when a new analytical method is developed. The performance of a new method will need to be compared with that of an existing standard procedure. Where convenient, a single set of solutions or finely ground materials will be used for the comparison and pairing will then be acceptable (see Table 9.5). A significant difference between the methods might indicate that bias is present but further evidence is needed to confirm this. The analysis of one or two certified samples (p. 316) will provide evidence.

Table 9.5. Comparison of dissolved organic carbon ($mg\,l^{-1}$) by ultraviolet oxidation and combustion methods

Combustion	UV oxidation
26	24
10.2	9.5
34	28
10.5	13.6
1.8	0.80
22	20
7.2	6.3
13	13
2.0	1.85
5.0	5.2

$\bar{d} = 1.459$
$\Sigma(d_i - \bar{d})^2 = 28.294$
$s_{\bar{d}} = 0.5607$
$t = 2.602$

H_0 = null hypothesis of no difference between the two methods $t = 2.262$ at $p \leqslant 0.05$ for $n-1=9$ degrees of fredom but this value was exceeded therefore H_0 is rejected with a significant difference found between the methods.

Comparison of variances

The variance (σ^2) measures scale and is multiplicative so ratios rather than differences are tested. The test is always applied to estimates (s^2) in the form F = greater variance/lesser variance. These estimates are χ^2 variables, particularly in small samples, and their ratio follows a related though specific distribution known as F. This, like t and χ^2 is affected by the shape parameter (kurtosis) and hence by sample size, and the F table is entered by degrees of freedom $v = n-1$. However, unlike t and χ^2, the F table is a two-way structure with v_1 and v_2 entered separately. For clarity separate tables are prepared for each of the usual probability levels (designated 'upper percentage' points).

The F-statistic tests H_0: $\sigma_1{}^2 = \sigma_2{}^2$ and, like t, can form a one- or two-tailed test. However, the F-test is one-tailed in the analysis of variance where, as shown later, H_1: $\sigma_1{}^2 > \sigma_2{}^2$. The F table gives one-tailed percentage points to meet this situation and $F_{0.05}$ is read from the 5% points table. If two tails are appropriate (as prior to the t test, where H_1: $\sigma_1{}^2 \neq \sigma_2{}^2$) read $F_{0.05}$ from the table of 2.5% points. The contrast between tests with t (usually two-tailed) and F (usually one-tailed) should be kept in mind.

Values of s^2 are relatively unstable because the distribution of s^2 about σ^2 is itself skewed. It is further affected by non-normality in the populations and normality is an important assumption for the F-test. The F-test is also somewhat insensitive and an alternative is given by Bartlett's test which uses a logarithmic statistic and requires the χ^2 table. Details are given in the texts. In practice F continues to be used in statistical packages for the analysis of variance (ANOVA).

Covariance

In field experiments the measured parameters can be affected by factors operating independently of the treatments themselves. An example is the dry weights of oak seedlings which are affected by acorn weight as well as by the treatment. The seedling weights can be adjusted to remove the

Table 9.6. SAS program listing for the procedure of Table 9.4

```
DATA SOISAM;
  INPUT X $ SAMPVAL @@;
  LIST;
  CARDS;
X1 0.56   X1 0.49   X1 0.53   X1 0.57   X1 0.56   X1 0.61   X1 0.61   X1 0.52   X1 0.60
X1 0.51             X1 0.52
X2 0.50             X2 0.42   X2 0.53   X2 0.49   X2 0.43             X2 0.50   X2 0.42
X2 0.53             X2 0.41
;
PROC TEST;
 CLASS X
  VAR SAMPVAL;
RUN;
```

T-TEST PROCEDURE

VARIABLE: SAMPVAL

X	N	MEAN	STD DEV	STD ERROR	MINIMUM	MAXIMUM
X1	10	0.559 000 00	0.038 427 42	0.012 151 82	0.510 000 00	0.610 000 00
X2	10	0.472 000 00	0.047 093 29	0.014 892 21	0.410 000 00	0.530 000 00

VARIANCES	T	OF PROB	>|T|
UNEQUAL	4.5263	17.3	0.0003
EQUAL	4.5263	18.0	0.0003

FOR $H0$: VARIANCES ARE EQUAL, $F = 1.50$ WITH 9 AND 9 DF OF PROB>$F = 0.5542$

effect of varying acorn weight. An analysis of covariance of the data allows this to be carried out and details are given in statistical textbooks.

Regression and correlation

Relationships and processes are fundamental topics in the study of natural systems and are modelled by related groups of functions. Relationships (regressions) between variables are expressed as polynomial functions and are described here because of their relevance to chemical analyses as well as to ecology. In contrast, major topics such as growth and decay (modelled by exponential functions and their transforms) together with rates of change (differential equations) cannot be adequately covered in this chapter and the reader should consult the extensive literature which is available.

Regression itself raises a number of issues and even the terminology can be misleading. For example 'linear' (commonly understood to imply a straight line) has a specific meaning in statistics. In the 'general linear model' (GLM) it is the *coefficients* (b_0, b_1 etc.) which must be first-order whilst the variables can be of any order. The GLM has received much attention because it forms the basis of many simple models. Within its framework the straight line (with first-order variables) is a 'linear regression' whilst a curve (with higher order variables) is a 'curvilinear regression'. The use of 'non-linear' to describe a curve is best avoided. The true non-linear model with coefficients such as b^2, b^3 etc. plays little part in elementary modelling.

The use of 'dependent' (random) and 'independent' (non-random) to describe a variable is another example. These terms are avoided by some writers who may use 'response' and 'regressor' instead.

Choice of equation

Even within the GLM many types of equations are possible but for simplicity only bivariate data are discussed here. Prior knowledge of a relationship between two variables affects the choice of equation and two contrasting situations are considered.

1 The relationship may be known to be causal and follow a well established principle ('law'). The desired equation is that which best expresses the principle. The laws of physics and chemistry provide many proven examples of equations which are fitted to experimental data solely for making predictions. The Beer–Lambert law in colorimetry is a typical example. Data violating well-known laws for no apparent reason should be checked before being accepted for making predictions. In ecology the field situation is complex with few general laws, but in areas such as forestry simple though empirical regressions are commonly employed (Prodan, 1968).

2 The relationship may not be understood and causality cannot be assumed. In this case the 'best' equation is usually that accounting for the greatest proportion of the variation of the data. This situation is very common in ecology and the equation is fitted initially only to describe the data. The equation may be:

a First order (straight line) where $y = b_0 + b_1 x$. This line is often fitted to experimental data although true linearity in biology is usually confined to short ranges of values.

b Second order (parabola) where $y = b_0 + b_1 x + b_2 x^2$. The parabola is often a legitimate choice although x is positive only between limits. These limits can be calculated but are not often a constraint because many data sets utilize only a segment of the parabola.

c Higher order models where $y = b_0 + b_1 x + b_2 x^2 \ldots + b_n x^n$. The number of possible equations rises sharply with the order. This can lead to a situation where several equations account for similar proportions of the variations and cannot be further distinguished. This is a well-known problem in ecology particularly when causal evidence is lacking (Mead, 1971).

In chemical work, poorly understood phenomena may arise in certain areas such as atomic absorption where complex calibration curves have been recorded for certain elements such as chromium. In this case a change of flame conditions may be a better option than fitting a higher order equation. This illustrates the fact that a statistical treatment of chemical data will not always aid its interpretation if the data were obtained under non-optimal conditions.

The choice of equation for fitting multivariate data is more complex and the number of variables (x_1, x_2 etc.) within an order is a decisive factor (Draper and Smith, 1981).

Fitting an equation

Two main techniques are used to fit equations to data. These can be illustrated for the straight line where, for simplicity, only one variable (y) has random error (i.e. is dependent).

1 *Least squares*

For a straight line regression of y upon x there exists, for a given data set, one fitted model $y = b_0 + b_1 x$ whose coefficient values are such that the sum of squares (SS) of the deviations of the points from the line is a minimum. The method sets $\delta(\text{SS})/\delta(\text{coeff.})$ to zero for each coefficient thus forming two simultaneous equations whose solution gives

$$b_1 = \frac{\Sigma(x - \bar{x})(y - \bar{y})}{\Sigma(x - \bar{x})^2}$$

and $b_0 = \bar{y} - b_1 x$.

The formula for b_1 needs to be modified for ease of computation (and software) as given on p. 312. Least squares gives unbiased values for the coefficients, provided assumptions about error (such as x being fixed) are not violated. Values of y are then predicted for given x except following calibration. This procedure is applied to multiple regression but the computations increase with the number of coefficients. However, this problem is eased by using matrix methods which are often preferred to the formulae recommended on p. 312 (Draper and Smith, 1981). Curvilinear regressions can be similarly fitted but the simultaneous equations yield intractable expressions for all but the second order (parabola) whose formulae are on p. 312. Cubic and higher orders are fitted by numerical methods such as iteration which gives a sequence of estimates for each coefficient converging to the least squares value.

Although least squares gives exact values it has two general limitations in that the data should be normally distributed with variations about the line being uniform and independent. Major failures require the data to be transformed or other steps taken (Snedecor and Cochran, 1980).

Non-uniform variances can be allowed for by weighting the random variable (y) with $1/s^2$ where s^2 is the estimated variance. Weighted regression is an option in many software packages (see also Miller and Miller, 1984; Ryan *et al.*, 1986).

2 Maximum likelihood

For this method the coefficient values for the fitted straight line are such that the sum of the logarithms of the probability density functions for the observations (y_i) is a maximum. This sum relates directly to the 'likelihood' value of the parameter, which, in turn, is theoretically equivalent to the unbiased least squares value. However, small samples give only approximate estimates and least squares is preferred as in analytical calibration. Nevertheless, maximum likelihood is less restricted by assumptions and has wider applications than simple least squares including cases where both variables have random error. Ripley and Thompson (1987) have compared the two methods for analytical data.

Random and systematic variation

The validity of a fitted model is assessed by analysing the variation, particularly random 'error'. Only principles are given here and are illustrated by the straight line regression with x free of random error. Although least squares minimizes the scatter of points about the fitted line, it does not eliminate it. The minimal scatter is therefore the residual variation and is the difference between observed (y_i) and predicted ($\hat{y}_i$) values. In the ANOVA format

total SS = regression SS + residual SS

i.e. $\Sigma(y_i-\bar{y})^2 = \Sigma(\hat{y}_i-\bar{y})^2 + \Sigma(y_i-\hat{y}_i)^2$

If the degrees of freedom (v) are partitioned as

$(n-1) = 1 + (n-2)$

then in terms of the mean square (MS) where $\text{MS} = \text{SS}/v$

total MS = regression MS + residual MS

These quantities allow standard errors to be derived for predictions of y and for b_0 and b_1 (as estimators of β_0 and β_1). Another useful expression is R^2 = reg SS/total SS which is commonly used to measure the variation explained. The MS values are required for significance tests. For example, the significance of F = reg MS/residual MS is evidence for the validity of the model. Similarly t tests are helpful including the test for $H_0{:}b_1=0$ whose rejection is further evidence for the validity. In multiple regression any decrease in residual MS due to the inclusion of an extra variable can be tested. All the relevant formulae are given in the main texts including forms needed for calculations.

The residuals themselves can be examined (via plots) for evidence that the data sets are suitable for the least squares method. The residuals should be approximately normal in their distributions and independent of both x values and predicted ($\hat{y}_i$) values if the variances are uniform. Other plots are sometimes used, as shown by Draper and Smith (1981). However, Framstad *et al.* (1985) note that the plot of residuals and observed (y_i) values is not conclusive.

Data can be subject to systematic error (bias) independently of random variation. Bias is present if the fitted line does not coincide exactly with the population line ($\bar{y} \neq \mu_y$). However, μ_y is usually unknown and the statistical treatment less simple than for random variation. Sources of bias in field and lab should be eliminated beforehand where possible. Standard solutions must be correct if the formula for b_1 assumes x is free of error.

Random error in the variables

The distribution of random error amongst the variables affects the least squares estimates and these aspects are summarized here.

1 If each variable is chosen in turn to be dependent (have error) whilst the other is independent (without error) then two separate regressions can be fitted.

a y upon x with $y = b_0 + b_1 x$,
where $b_1 = \Sigma xy / \Sigma x^2$ as before.
b x upon y with $x = b_0 + b_1 y$,
where $b_1 = \Sigma xy / \Sigma y^2$.
These lines cross at $(\bar{x}, \bar{y})$ but are otherwise distinct with separate formulae for the coefficients. It is usual to fit y upon x unless the other is clearly appropriate.
2 If, in practice, the independent variable (x) has error then the least squares estimate of b_1 will be biased. Two situations are common.
a In calibration, x represents the standard values. If some of these are not what they are assumed to be then b_1 will be biased and sample predictions affected (Meites *et al.*, 1985). This may pass unnoticed unless the standard values are predicted as a check or a reference material is included (see p. 316). This underlines the need to prepare standard solutions with care.
b In ecology if two sets of random data are regressed then b_1 is again biased. However, this may be tolerable if prediction is not the purpose of the regression or the errors in x are known to be small compared to their range (Daper and Smith, 1981). Nevertheless a least squares fitting raises problems for these cases and alternative approaches may be required (Madansky, 1959).
3 Calibration—prediction sequence.

Calibration involves fitting $y = b_0 + b_1 x$ to the standard data where y is random (reading) and x fixed (concentration). It is then necessary to predict x for a given sample y value instead of y from x. For this purpose

$$x = (y - b_0)/b_1$$

although Snedecor and Cochran (1980) note that this prediction is itself biased and its confidence limits are less easy to calculate than for prediction of y from x. However, some packages do not use this formula for x and appear to assume that because

$$\text{reading} = b_0 + b_1 \times \text{concentration}$$

for the standards, it is then acceptable to predict

$$\text{concentration} = b_0 + b_1 \times \text{reading}$$

for the samples. Although such predictions may seem to agree with graphical results in a given case, the practice has been criticized on theoretical grounds by several workers (Morris, 1983) and care is needed to avoid making incorrect predictions.

Multiple linear regression

The simple regression assumes only one variable (x) affects y. This is adequate for routine calibration where other factors (such as light path length in colorimetry) are constant. However in the field, y is usually affected by several variables (x_1, x_2, etc.) so that for the first order (straight line) or 'multiple linear regression':

$$y = b_0 + b_1 x_1 + b_2 x_2 + \ldots b_n x_n.$$

Individual coefficients are derived by least squares as before where

$$b_1 = \frac{\Sigma(x - \bar{x})(y - \bar{y})}{\Sigma(x_1 - \bar{x}_1)^2}$$

although matrix methods are preferred for the computations (Draper and Smith, 1981). It is seen that y is determined by *all* the variables so it is less easy to interpret b_1 than for the simple equation. ANOVA and significance tests are applied as shown earlier and the results of both are printed out when standard packages are used.

Application of multiple regression is related to the initial purpose which, as discussed earlier, may be:
1 *Descriptive.* Here the full equation is built up stepwise by entering every variable in the decreasing order of correlation with y (see later). However, whilst such an equation is 'complete' because it uses all the data, it may be far from being a *causal* description of the behaviour of y.
2 *Predictive.* As noted earlier empirical relationships are often accepted in forestry and other areas for predictive purposes. However, the complete equation is not necessarily the best predictor of y because
a Some variables may be poorly correlated with y or not at all.
b Some variables may be sufficiently intercorrelated to form a subset accounting for most of the variation of y. This effectively makes the other variables redundant. In this case variables are entered stepwise, as above, until an entry fails

significantly to increase R^2 (as tested by F). Alternatively variables can be removed from the full equation until the decrease in R^2 becomes significant. Although stepwise regression is easily applied using packages, biased or spurious predictions are possible if, for example, many variables are entered when data sets are small (Verbyla, 1986).

A further problem is that the assumption of a straight line relationship with y may fail for a given variable as the range of values increases. However these deviations can often be corrected by transformation or another device and some packages allow for these options. If linear regression is not applicable higher orders of some variables will be needed but the new equation is. not just an alternative to multiple linear regression. Apart from any problems in fitting the equation, there may be a need for further evidence before the model is used for prediction. In this way multiple regression leads to the subject of model building which is not covered here but is well treated in many texts.

Correlation

As shown above, prediction needs regression equations because they show the nature of the xy relationship. However for a general description it is often sufficient to know if a relationship exists and how strong it is, i.e. how x and y vary together (correlate). This involves products of deviations from the mean and in particular:

1 the mean product of the deviations (or covariance).

2 the square root of the product of the variances.

The correlation coefficient (r) is based on **1/2** which leads to

$$r = \frac{\Sigma(x-\bar{x})(y-\bar{y})}{\sqrt{(\Sigma(x-\bar{x})^2\Sigma(y-\bar{y})^2)}}$$

This value is best computed using the expressions on p. 313. A straight line relationship is assumed together with bivariate normality and random sampling which are needed for parametric tests.

Coefficient values fall between + or -1 (full correlation) and 0 (no correlation). The most common test of r is $H_0: \rho = 0$. The r distribution is symmetrical about ρ when $\rho = 0$ which allows testing with a (t-type) statistic based on r and n. Critical values of r are given in a table (see texts) which is entered at $v = n-2$ degrees of freedom for a given probability level. The table assumes a two-tailed test (H_1: $\rho \neq 0$). The r distribution is *not* symmetrical about any other value of ρ (see below) and other tests of r require a logarithmic statistic (see texts).

Correlation matrices are readily computed for large blocks of data using packages. This is the first stage for certain multivariate methods which examine the variables and texts should be consulted for details.

In spite of its convenience there are problems in using this coefficient:

1 If r is not 'significant' ($H_0: \rho = 0$ is accepted) then the interpretation is uncertain, for whilst complete independence of x and y ensures $\rho = 0$ it does not follow that accepting $\rho = 0$ implies full independence.

2 If r is significant ($H_0: \rho = 0$ is rejected) some statistical relationship is implied, at least in the sample, but note:

a this does not guarantee a relationship in the population because as sample size falls, r is a much less reliable estimate of ρ due to increasing skewness in the r distribution as $\rho \to 1$.

b it does not guarantee a valid regression between x and y because as size rises, low r values can be significant yet correspond to a low R^2 (x accounting for little of the variation in y).

c regardless of sample size, significance never implies *causality* for which external evidence is always needed.

3 The r value is sensitive to outliers in the sample and if their presence reflects significant skewness r will not be valid. However a coefficient (Spearman's) based on ranking is available for non-normal samples (see later).

Non-parametric tests

Parametric tests require normal samples whilst less restricted methods are needed for non-normal data. Various non-parametric tests are available for random samples and assume only that the distributions are of a similar form. The null hypothesis states that probabilities (i.e. distributions or populations of the data) are equal. Some tests (replacing t, F and r) which are based on ranking are described below along with the main non-parametric distribution (chi-square). These tests generally have less power than parametric methods so that the probability of assessing H_0 wrongly is moderately increased. Some tests require extra tables apart from chi-square and these are found in various texts. See also Siegal (1956) for further details.

Chi-square

If in a population a normal deviate is $z = x - \mu/\sigma$ then the squares of z follow the chi-square (χ^2) distribution where

$$\chi^2 = {z_1}^2 + {z_2}^2 + \ldots {z_v}^2$$

where v = degrees of freedom

All values are positive and the distribution is asymmetric for small v (except at $v = 1$). Critical values of χ^2 are tabulated up to $v = 30$ (or more) for selected probability levels. The table is entered at $v = k - 1$ *classes* (sometimes $k-2$, $k-3$) and not by sample numbers. For two classes $v = 1$ (where χ^2 reduces to a single z^2 value) the normal table can be used provided the value is corrected for the continuity of the normal distribution. However when $v > 1$, uncorrected values of χ^2 are used. The table is only approximate for small samples and not reliable when $n \leqslant 5$ in the classes.

For tests, chi-square is usually expressed as

$$\chi^2 = \frac{(\text{observed} - \text{expected})^2}{\text{expected}}$$

Formulae for computation are found in the texts.

Comparison of two samples (classes)

The methods given below replace t and test that probabilities are equal (H_0:$p = 0.5$). The use of ranking implies that sample *medians* and not arithmetic means are compared.

1 *Ranked-sign test for paired samples (Wilcoxon)*

Absolute values of differences are ranked where the lowest = rank 1, zero differences are omitted and 'tied' values take an average rank. A sign (+ or −) is applied to each rank and ranks summed for each sign. The *smaller* rank total (T) is a test for H_0 as follows.

a Small samples ($n \leqslant 25$ pairs). The test is binomial and T is entered directly into the Wilcoxon table of probabilities provided $n \geqslant 7$.

b Large samples ($n > 25$ pairs). The test requires the corrected $z = (|\mu - T| - 0.5)/\sigma$, where μ and σ of the ranks are binomial parameters whereby $\mu = n(N+1)/4$ and $\sigma = [(2n+1)\mu]^{1/2}/6$. If, for a two tail test, $z \geqslant 1.96$ at $p \leqslant 0.05$ then H_0 can be rejected.

2 *Mann–Whitney test for independent samples*

For this test all observations (including negative values) are ranked with the lowest = 1.

The ranks are summed for each sample and H_0 tested as follows:

a Small samples (if within limits of the Mann–Whitney table).

i $n_1 = n_2$ The smaller rank total is entered into the table.

ii $n_1 \neq n_2$ If n_1 is the smaller sample with rank total T_1 then $T_2 = n_1(n_1 + n_2 + 1) - T_1$ (where $T_2 \equiv U$ in older texts).

b Large samples. This test needs the corrected $z = (|\mu - T| - 0.5)/\sigma$, as before but $\mu = n_1\ (n_1 + n_2 + 1)/2$ and $\sigma = (n_2\mu/6)^{1/2}$. If $z \geqslant 1.96$ ($p \leqslant 0.05$) then reject H_0.

Comparison of several independent samples (classes)

Here the methods replace ANOVA and test that every class comes from the same population. The best known method is the Kruskal–Wallis test.

Kruskal–Wallis test

All observations are ranked and ranks summed as for the Mann–Whitney test and if samples n_1, $n_2 \ldots$ correspond to rank totals R_1, $R_2 \ldots$ then H_0 is tested by the statistic

$$H = \frac{12}{N(N+1)} \sum_{i=1}^{k} \frac{R_i^2}{n_i} - 3(N+1)$$

where N = total number.

This value is entered into the χ^2 table at $k-1$ degrees of freedom. However, if $k \leqslant 3$ and $n \leqslant 5$ in the classes then H requires a more exact table of probabilities (Siegal, 1956). Packages are available to aid computation. The corresponding method for several matched samples is Friedman's test which uses a modified form of H and the χ^2 table. Details are available in various texts.

Rank correlation coefficient (Spearman)

In a bivariate non-normal sample, the correlation between the *ranks* of each variable provides an alternative to the parametric r value. The best known is Spearman's rank correlation coefficient (r_s). The data are ranked separately for each variable with the lowest = 1 and 'ties' take an average rank. The difference in rank for each x, y pair is obtained and squared (d^2). The test of $H_0: r_s = 0$ requires

$$r_s = 1 - \frac{\Sigma d^2}{n(n^2-1)}$$

which is entered in the Spearman table provided $n \geqslant 4$ (pairs). A version giving exact probabilities for small samples ($n \leqslant 10$) is preferred (Snedecor and Cochran, 1980). At $n > 10$ the usual table for r is entered at $v = n-2$ degrees of freedom.

The methods given above seem to be the most widely used especially for small samples of non-normal data. However, alternatives have been proposed (with their tables) and are sometimes employed.

Design of experiments

An experiment is commonly defined as the test of a hypothesis under controlled conditions. The necessity for clear objectives with hypotheses set up was mentioned at the beginning of this chapter. The aim of an experiment is to allow one or a few selected variables to vary significantly in space and/or time, whilst other factors remain constant or are measured separately.

The chosen variables are measured and the resultant data handled to see if valid conclusions about the hypothesis can be made. However, the significance tests depend upon the arithmetic mean and variance, which necessitate minimizing bias and poor precision respectively. The main ways of achieving this are discussed below. They relate closely to problems of sampling which have received much attention including general texts on basic theory (Barnett, 1974), a monograph on soil sampling (Webster, 1977) and a detailed discussion regarding waters (Hunt and Wilson, 1986). Three introductory points are made here.

1 The choice of experimental area is closely related to the objectives of the experiment. However, excessive variations of topography, soil type etc. and the presence of natural boundaries such as streams can induce sampling problems and may have to be avoided unless they are the subject of study.

2 Samples taken in a subjective manner may not be representative of the population and can introduce bias. Simple random sampling minimizes bias and is discussed below. Stratified random sampling may be preferable for very variable sites but is not considered here (see texts).

3 Practical considerations often limit the number of samples taken but if replication is not sufficient, estimation of the precision will not be adequate. Replication is discussed below and includes reference to the problems induced by bulking.

Four ways of avoiding bias and poor precision are considered briefly here but for further details many texts are available including Davies (1956), Cochran and Cox (1957), Cox (1958) and Snedecor and Cochran (1980). Sampling procedures suitable for surveys are not considered in this section but descriptions are available in the literature.

Randomization

Randomization is one of the most important methods of avoiding bias in experimental data. It involves selecting some proportion of a population so that:

1 It is impossible to deduce which particular part has been taken. This is important because probability theory underlies the statistical tests used.

2 there is an equal chance of every part selected being representative of the whole (with regard to variability), so that random errors of a given variable will be independent of each other.

There are two aspects of experimental planning which are relevant here. These are the spatial design of the experiment and the method of collecting the samples.

These are often treated separately but may be included here in one illustration.

A typical simple field experiment is one in which the effects of different unrelated treatments on a crop are compared. For this purpose dry matter production or nutrient concentration are often measured. In a completely randomized experiment the various treatments are assigned at random to the plots (often by using a table of random numbers), random samples taken and analysed and the treatment means computed and compared statistically. The two advantages in this arrangement are that the design is flexible and the statistical analysis is straightforward.

Unfortunately if large (unknown) variations are present in the site (especially in soil fertility) differences due to treatment will tend to be obscured. For this reason randomized blocks are usually recommended (see below).

It may be noted that although random sampling is adopted in most experimental designs, a systematic procedure may be acceptable in specific cases. However, all the tests described in this chapter assume random sampling.

Replication

The need for replication arises because of natural variation especially in biological material. Only rarely is it justifiable to draw conclusions from one observation since there is usually no information about the precision of such a value. Replication provides both a measure of variation and a better estimate of the precision. Sufficient replicates should be provided to allow a satisfactory statistical analysis. Excess data should not, however, be collected as this will yield no further information. Again there are two aspects of experimental planning which are relevant.

1 *Spatial design*

Field experiments should contain more than one replicated block of plots (see below) although practical considerations often limit the number to three. This would ensure three plots for each treatment.

2 *Number of samples*

If the variance of the population is known it is possible to calculate the minimum number of samples required to infer valid information about that population. Usually, however, the population variance is unknown. There may also be a practical limit to the number of samples that can be taken. Statistical tests for handling small samples were discussed earlier and the collection of five or ten samples per plot is often acceptable under these conditions.

In chemical analysis, variability in the laboratory is usually much less than in the field. Even so, the precision of each analytical method should be determined beforehand. This information can then be used as a guide when the methods are applied to a given set of experimental samples. Then it is generally sufficient to carry out duplicate or triplicate analysis on any one sample, and isolated samples should always be handled in this way. However, if sufficient samples are collected to account for field variation (as above) there will be more than enough to account for the known laboratory variation. In the case quoted above, 5 samples will yield 5 values of each element for the plot which can be meaned directly. It is not desirable to combine such samples for chemical analysis as this conceals variability. However, practical considerations may necessitate some bulking as discussed in the first edition of this book (Allen *et al.*, 1974).

Blocking

Randomized blocks

The complication of soil variations in assessing treatment effects was mentioned above. Combinations of plots into replicated blocks greatly reduce the effect of inherent plot differences and improve the estimates of precision. The treatments are allocated at random as before and comparisons are made within blocks.

Typical blocks for four treatments might be as shown below.

A	B
C	D

A	B	C	D

A	B	C	D

Block shapes and sizes are not critical but should allow for spatial trends such as slope. One plot in each block should be a control. These blocks are recommended for simple experiments but are limited in the number of treatments easily handled.

Latin squares

The Latin square is a type of randomized block design in which variation between blocks as well as between plots is reduced. A typical square might have four treatments in four rows with each treatment occurring only once in a row. The rows can be regarded as randomized blocks.

B	C	D	A
D	A	C	B
C	B	A	D
A	D	B	C

The Latin square gives the most precise results of any simple design and systematic errors too are minimal. A disadvantage is that the number of treatments must equal the number of rows or columns. Hence the number of treatments easily handled is restricted.

Factorial experiments

Simple experiments are concerned only with unrelated treatments or factors. When related treatments are being compared, more complex designs or factorial experiments are required. This situation arises very frequently in some studies, especially those involving fertilizers. Treatments then consist of combinations of two or more fertilizers and perhaps varying levels of each. Another example is the effect of temperature, pressure and concentration of reactants on the yield of a chemical reaction. Essentially factorial experiments show:

1 how the treatment response is modified by changes in other factors,

2 the interrelations between other (selected) factors, and

3 the overall combination of factor levels which gives the maximum or minimum treatment response.

Further details of this type of experiment cannot be given here but most statistical books for biologists devote a section to them.

Calibration and calculation procedures

This section concerns the procedures for calculating the amount of element or constituent present in the sample once the chemical stages are complete. The basic procedure is straightforward and involves converting an instrument reading or other measurement into the concentration of element present in the original sample. All the flame and colorimetric methods in Chapter 5 require calculating stages as follows:

sample reading	calibration → curve of standard readings	concn in sample solution	subtract blank ↓
corrected concn →	factor → (incl. wt, vol, diln)	concn in original sample →	concn on dry basis

The formula for deriving the factor is given at the end of each method. A few specific methods require their own calculations which are outlined in the text as necessary.

The individual stages are considered here. It is assumed that the digital readout of colorimetric and atomic absorption instruments is in absorbance (logarithmic) units whilst FES intensity values are linear. Modern instruments include circuitry to convert the output to be linear in absorbance. This output is usually accompanied by chart recorder output. The early stages are handled automatically in equipment with built-in calibration facilities or which is linked to a data processor. Meanwhile analytical performance must be considered before data handling.

Analytical performance

An optimal performance is sought because the calibration line is determined by chemical and physical processes and not by statistical methods. The practical steps required (also known as quality control) are best illustrated by continuous flow colorimetry for which:

1 The sample batch must include adequate numbers of standards, drift markers, sample replicate(s), blank solutions, water cups and a reference material.

2 The chart output should be examined for poor reproducibility of standards, irregularly shaped peaks, carry-over between peaks (poor wash-out), baseline noise and/or drift particularly if erratic, poor replicates, inconsistent blanks, anomalous reference peaks and sample peaks outside the optimal part of the standard range. Although this initial scan is subjective (statistical quality control is discussed on p. 316) gross symptoms indicate a fault and usually require a repeat run of the samples. Minor effects such as a small and constant drift can often be allowed for in data handling. If computer control is adopted an 'early warning' detection of some of these effects can be devised.

Calibration

Graphical methods

Although computer systems have largely displaced graphical methods in handling flame and colorimetric data they may still be needed on occasions. In addition they show features of the data which may be overlooked on a print-out. The classical practice for calibration is to plot standard readings against concentration. The 'best' line is then fitted subjectively with a ruler or 'flexicurve' to simulate the unknown mathematical line. This simulation is reasonably close if the scatter of points is small and the line is not forced through a given point (see below). The line is straight (for colorimetric values obeying the Beer–Lambert law) or slightly curved (flame methods) and allows a sample concentration to be read from its reading using the graph.

Chart traces are handled by measuring peak height (or area—see below) or by marking the peak tip on a translucent chart reader (Technicon system). The line is drawn as before. For detectors with logarithmic absorbance output, the calibration curvature will be pronounced unless semi-logarithmic graph paper is used. This is not a problem except that anomalous standards are harder to detect in a curve compared to the straight line. If desired the peak heights can be recorded in absorbance units using a special ruler (Technicon) or by computation (Allen *et al.*, 1974).

Although graphical methods are simple they raise various problems:

1 Peak height is acceptable only if the peak is symmetrical and carry-over small in relation to height. Otherwise peak area should be used. A computer-based integrative device attached to the instrument is preferable to manual 'triangulation' or 'counting squares' on the chart. Chromatographic charts can be more complex and computer methods are essential.

2 A gross error in one standard is usually noticed on a graph but a wide scatter of points can reduce graphical methods to guess-work. The scatter remaining after fitting the curve can be tested by computer. The problem of an outlying standard is also discussed on p. 298.

3 The practice of forcing a straight line through a given point (usually the zero) should be avoided unless there is prior justification. In general forcing techniques make it harder to simulate the mathematical line. A contrasting problem arises when a straight line fit misses the zero by a substantial margin thereby distorting the blank values. This can occur in flame work if the higher standards fall on a slight curve whilst the others form a straight line. If the curve is genuine (and not due to a faulty standard) the calibration line should incorporate the curve. (If the least squares method is applied the two segments should be treated separately.)

4 Drift can be allowed for if it is uniform and not excessive. It is assumed standards were run before and after samples and provide two similar shaped curves. Computer techniques for drift correction are more precise.

a If the calibration lines are obtained separately and two values are obtained for each sample, the latter can be averaged if judged to be sufficiently close.

b If a single line is obtained from two sets of standards then one sample value is provided. However, absorbance readings should not be meaned in this way except via antilogarithms unless numerically very close.

c If the batch is divided into two halves the samples can be allocated to the appropriate set of standards. However, this is arbitrary and leaves samples near the division carrying more error than those nearer the standards.

These three approaches are only acceptable if the difference between the two sets of standards is small, for example, if the peak heights of the second set are within 10% of the first set and all are well clear of the detection limit. A greater difference indicates a drift or other problem needing attention and possibly a repeat analysis of the samples.

In spite of these limitations graphical methods can, with care, be as acceptable as computer techniques (Galen *et al.*, 1985).

Computer methods

Computer-based methods are normally required for fitting equations. Suitable formulae for programming are given here and can be used if access to SAS or other software packages is not available. Formulae derived by least squares are given here for the straight line and parabola which are the commonest regressions required in routine chemical analysis. As before y is the dependent variable and x (independent) is without random error.

1 ***First order (straight line)***

On p. 303 the fitted equation was given as

$$y = b_0 + b_1 x$$

$$\text{where } b_1 = \frac{\Sigma(x-\bar{x})(y-\bar{y})}{\Sigma(x-\bar{x})^2}$$

$$\text{but for programming, } b_1 = \frac{\Sigma xy - \dfrac{(\Sigma x)(\Sigma y)}{n}}{\Sigma x^2 - \dfrac{(\Sigma x)^2}{n}}$$

and subsequently $b_0 = \bar{y} - b_1\bar{x}$.

2 ***Second order (parabola)***

The parabola is given by $y = b_0 + b_1 x + b_2 x^2$ (see also p. 303). The least squares formulae for the coefficients can be expressed in a reduced form where $X = (x-\bar{x})$ and $Y = (y-\bar{y})$ so that

$$b_2 = \frac{\Sigma X^2 \Sigma X^2 Y - \Sigma X^3 \Sigma XY}{\Sigma X^2 \Sigma X^4 - (\Sigma X^3)^2}$$

$$b_1 = \frac{\Sigma XY \Sigma X^4 - \Sigma X^3 \Sigma X^2 Y}{\Sigma X^2 \Sigma X^4 - (\Sigma X^3)^2}$$

and finally

$$b_0 = \bar{y} - \left(\frac{b_2 \Sigma x^2}{n}\right) - b_1 x$$

See also Steel and Torrie (1960).

The individual components should be programmed as follows.

$$\Sigma(x-\bar{x})^2=\Sigma x^2-\frac{(\Sigma x)^2}{n}$$

$$\Sigma(x-\bar{x})^3=\Sigma x^3-\frac{(\Sigma x)(\Sigma x^2)}{n}$$

$$\Sigma(x-\bar{x})^4=\Sigma x^4-\frac{(\Sigma x^2)^2}{n}$$

$$\Sigma(x-\bar{x})(y-\bar{y})=\Sigma xy-\frac{(\Sigma x)(\Sigma y)}{n}$$

$$\Sigma(x-\bar{x})^2(y-\bar{y})=\Sigma x^2y-\frac{(\Sigma x^2)(\Sigma y)}{n}$$

An alternative form used in some AAS packages is

$$\frac{x}{y}=b_0+b_1x+b_2x^2$$

but zero and negative values cannot be accepted.

Computer methods have many advantages but certain points should be noted.

a Only the straight line and parabola are simple enough to be fitted by the type of formulae given above. Complex regressions are fitted by matrix and numerical methods and appropriate packages are available.

b Many functions (but not the parabola) can be transformed to a straight line form. Packages usually allow for this option.

c One pitfall is the fitting of one function to data whose range is sufficient to encompass two functions. This often indicates an inappropriate choice of analytical range. For example, AAS and FES data are often 'linear with a curvilinear component' which is easily detected by a graph. Such data should be divided into sections for processing unless a specific polynomial is already known to fit the data. If the latter is used, weighting will be needed if the deviations $(y-\bar{y})$ differ greatly between the two components.

d Extrapolation is best avoided when packages are used and sample readings should always be within the standard range. Extrapolation errors may be serious when curves are fitted.

e Standard packages such as SAS, print out sufficient information (such as R^2) to evaluate the regression. Large packages may include a general 'polynomial regression' program allowing various regressions to be compared. These are also valuable features for handling ecological data.

f Correction of drift between two check standards can most conveniently and precisely be performed by computation. In this case, each sample concentration has a drift factor applied based on the drift between standards and the distance of the sample from them.

Calculation

Once the sample solution concentration has been derived, the content in the original sample is obtained by subtracting a blank value and then correcting for weight and volume together with dilution and moisture content where appropriate.

Blank values

Blank determinations are a measure of the reagent background (also seen in the 'zero' standard when standards are compensated) together with any laboratory contamination which has occurred. If the concentrations of the blank solutions are low and similar, then a mean value may be subtracted from the sample concentration. (Instrument readings should not be subtracted because this may introduce an error.) The operator should beware of subtracting high and/or dissimilar blanks particularly when they exceed some of the sample values. Conversely, care is needed before negative blanks (less than the 'zero' standard value) are accepted although these may occasionally be valid.

Calculation stages

Formulae to derive the concentration of element in the original sample from the solution concentration are given with each method. Manual calculating is much easier if a consistent weight, volume and dilution are used for a given batch. However, if a computer with file handling is used, consistency has no advantage although it is retained for analytical reasons when, for example, soils are extracted. Some data will need further correction to a dry (105°C) basis which requires the factor, 100/% dry. All the final data can be stored on disc and later recalled to compile the analytical report.

Expression of results

Rounding off must be avoided during a calculation until the final stage or bias may be introduced. The rounding off of the final value to an acceptable number of 'significant' figures is then governed by two considerations:

1 If data are to be processed statistically, several significant figures may be needed to prevent rounding off errors being carried through to values computed for testing hypotheses, etc.

2 If nutrient data are to be published then readers can be misled regarding the overall precision of a study if data (with standard errors) are presented to more significant figures than are warranted by the methodology used. If samples are analysed for a customer then the analyst will need to find a compromise and 'two significant figures' is widely adopted. This is acceptable because the relative precision of the laboratory methods described in this book is rarely better than 2–5% unless all stages are run under optimal conditions. However, flexibility is needed and a third significant figure will sometimes help to meet a customer's needs. Rounding off itself should be consistent and mathematical so that, for example, 2.65 becomes 2.7 whilst 2.648 becomes 2.6.

Analytical errors and quality control

Some knowledge of the accuracy and precision of a method is essential for assessing the real value of the results obtained. An understanding is also required of the sensitivity, selectivity and limits of various methods in order to select an appropriate procedure. Analytical errors are classified as either random or systematic (bias) and both groups include gross error. Through the use of quality control procedures it is possible to accumulate sufficient data on the accuracy and precision of the method to indicate when errors may have occurred. This section briefly considers the two types of error and examines facets of quality control procedures.

Systematic errors

For the purpose of this manual, *accuracy* is defined as the closeness of the result to the *true* or absolute value. The difference between the *mean* experimental result and the true value (if known) represents a systematic error or bias which may be inherent in the chemistry of the method employed or in the equipment used. Faulty or uncompensated standards also impart bias. Chemical interference from samples and general contamination are two further sources.

Results that are reproducable are not necessarily free from bias and specific checks are required. Method bias can be detected by analysing a certified reference material of similar matrix to the samples of interest. Suitable materials are available from the National Bureau of Standards (USA), Community Bureau of Reference (EEC) and elsewhere, including geological institutes. Rock and soil materials are obtainable for total elemental analysis, whilst 'citrus leaves', and 'tomato leaves' (NBS) and 'Bowen's kale' (UK) are widely used as plant material standards. Water references are being developed (Koch *et al.*, 1984). In general a wide range of materials may be needed for ecological work. The use of these substances is discussed by Uriano and Gravatt (1977), Wolf (1985) and Alvarez (1986). However, for routine quality control it may be better to prepare an internal reference sample for this purpose. This is discussed further below.

There is no simple way of expressing systematic errors in the absence of the true value which is usually unknown in biological materials. Methods are sometimes quoted as giving recoveries that are '5% low' or '10% high'. These indicate a systematic error but too much significance must not be attached to the figures themselves. The practice of quoting methods as being 'accurate to 2%' is to be avoided particularly when *precision* is intended (see below).

Although the procedures given in this book are sufficiently accurate for most ecological purposes, no method should be regarded as totally free from bias. Different methods offered for a particular element will not necessarily give identical values even when known interferences have been corrected. Hence the chosen method must be appropriate for the study and the type of sample and should not be changed during an experiment or some bias may be introduced.

A method for detecting bias due to sample interference is to add known amounts of the test element to the sample solution (standard addition) and measure the recovery. Contamination is revealed by high blank solution values or anomalies in reference sample values (see discussion below).

Random errors

All individual sample values have random errors associated with both the individual worker and external factors operating in the laboratory. They arise mainly through insufficient preparation of the sample, which may be too heterogeneous (pp. 12 and 50), incorrect or careless use of equipment and spurious contamination of glassware and solutions. They can be greatly reduced by improved laboratory facilities and personal technique, but cannot be eliminated entirely. For this reason experimental data should not be quoted without some indication of the errors involved.

The scatter of results due to random errors is expressed as the *precision*, which is the measure of agreement between replicated determinations made at any one time under identical conditions. It is independent of accuracy but becomes poor as the detection limit of the method is approached. The term 'reproducibility' extends the concept of precision to incorporate both within- and between- batch variation and, where appropriate, variation over a period of time.

Precision may be expressed as the standard deviation of individual results from their mean or more commonly as a percentage of the mean (relative standard deviation or coefficient of variation), but care is needed in expressing such values. For example, a result quoted as '10.0 ± 0.2 mg calcium' is clear because only absolute units are used. However, statements claiming that 'the method gave results to within 2%', or 'x% calcium was determined to within 2%' are widespread in the literature but they are ambiguous because absolute and relative errors are not distinguished. For example, the statement '10% calcium was determined with a precision of $\pm 2\%$, would usually be intended to mean that the result fell between

9.8 and 10.2% (*relative* precision $\pm 2\%$),

but it could also mean that it fell between

8 and 12% (*absolute* precision is $\pm 2\%$).

In this manual the precisions given are relative precisions (coefficients of variation) applicable to results obtained under good conditions within an optimal range of standards. If it is defined as

$$\text{relative precision} = \frac{\text{absolute precision} \times 100}{\text{mean result}}$$

then 1.00% calcium determined with a relative precision of $\pm 2\%$ implies an absolute precision of $\pm 0.02\%$, so that 1.00% calcium falls between 0.98 and 1.02%. This implies that 0.10% calcium would fall between 0.08 and 0.12% but this deduction needs checking. As noted above, error limits widen as the detection limit is approached and the relative precision should be re-determined at low levels of the element to get a realistic value.

This discussion has been confined mainly to the precision of results, though Hunt and Wilson (1986) emphasize its importance in characterizing the performance of a method generally, and they outlined the steps required.

Random variation in the field is generally greater than in the laboratory. This fact must be allowed for in planning ecological studies (see p. 308).

It is always preferable to take a number of samples in the field and analyse them separately rather than rely on several determinations on one sample or even a combined sample. Replicates of a specified sample are, however, useful for monitoring method precision as part of 'quality control' in the laboratory. This aspect is considered below.

Sensitivity and detection limit

These two concepts are important when selecting appropriate analytical methods.

'Sensitivity' expresses the response–concentration relationship but working definitions vary with the type of method. For example, light path geometry is used in colorimetry so that for a 1 cm cuvette (or 1 cm flow cell) the sensitivity is the molar concentration (c) giving unit absorbance, say 0.001 (sometimes the absorbance corresponding to unit concentration is quoted). Absorbance $= 10^{-\varepsilon bc}$, when $b = 1$ cm and $\varepsilon =$ molar absorptivity and if ε is known for a given reaction and wavelength then the sensitivity can be predicted. In flame work the sensitivity has been defined more simply as the concentration giving a 1% change in transmission which is equivalent to an absorbance of 0.0044. The sensitivity values give an indication of the lower limit to the concentrations which can be determined with a desired precision.

The detection limit has, in the past, represented an arbitrary signal to noise ratio. However, quoted limits were often unrealistically low because pure solvents were used to establish the limits. Current definitions relate to blank solutions carried through the system (including preparation stages) and utilize the standard deviation of the blank values (s_{bl}) so that the minimum sample-blank difference is now in ks_{bl} units. Many workers use $k = 2$ or 2.33 (Hunt and Wilson, 1986) but the Analytical Methods Committee (1987) strongly recommend $k = 3$. Hence a realistic sample detection limit is $3s_{bl}$ units above the blank value. If the sensitivity varies as noted above, the limit will need to be adjusted.

Control procedures

Once a method (including solution preparation) has been selected and the accuracy and precision quantified, there is a need to establish control procedures to detect when it is operating outside specified limits. To assess systematic and random errors within the batch, and long term reproducibility, suitable control samples should be run with each batch. Vegetation is used to illustrate the principles but soils and waters are mentioned.

Reference samples

The monitoring of bias, random error and contamination is considered here in turn. As discussed above, systematic errors (bias) are best detected using certified reference materials. However, only limited supplies are available and for daily monitoring over long periods it may be better to collect and prepare large quantities of an 'internal' reference (control) sample. This can be calibrated against certified reference materials. The chemical composition of the internal reference sample must be well established by repeated analyses to quantify between and within batch variation. For ecological work Rowland (1987) recommended preparing several control samples to cover a range of nutrient levels. He examined several species and found that leaves of alder (*Alnus glutinosa*), hazel (*Corylus avellana*) and meadowsweet (*Filipendula ulmaria*) and leaflets of rose-bay willowherb (*Chamaenerion angustifolium*) contain levels of macro- and micro-nutrients which would be optimal for labs handling wide ranges of species. More localized studies might require other plants. For example, current years growth of heather (*Calluna vulgaris*) would be suitable if most samples come from nutrient poor sites, whilst first-year pine needles (*Pinus sylvestris*) would be preferable for conifer samples.

Reference soils are similarly prepared. The choice is difficult because soils vary greatly and each type may need its own control sample. Relatively few certified soils are available although many rock minerals can be obtained. Waters are a problem because of instability and natural waters are unsuitable for control purposes. However, synthetic solutions have been proposed by NBS and others (Koch *et al.*, 1984) whilst sea water has long had a standard recipe (Strickland and Parsons, 1960). A synthetic stock solution can be diluted weekly to provide a control sample for fresh waters. These solutions should be stored under the same conditions as the samples and the stock renewed at intervals (see also Koch *et al.*, 1984). As noted earlier, a control sample monitors method bias only. Systematic errors in the samples (interference) are detected by standard addition tests.

Random error (precision) can also be monitored but it is important to distinguish between variations inherent in the method and those from the sample material.

Method variations

The precision of the method as applied to the control is assessed by running several replicates as if they were a batch ('within batch' variation). If this is repeated several times the 'between batch' variation is obtained. These data can be supplemented by a long term record of the control sample value for each nutrient (see control charts below). This indicates the reproducibility of the method.

Sample variations

Standard deviations obtained above must not be attached to other sample values. These may require their own replicates. In practice replicates may need to be confined to one or two samples in the batch. This is acceptable for a reproducible method if the scatter in the replicates is consistent with the method (control) precision and sample matrices resemble the control. If these conditions do not apply all samples may need to be run in duplicate or triplicate. Again it is stressed that these replicates are not a substitute for field replicates. The latter are essential to obtain a valid standard error for field nutrient data.

Contamination is the final aspect of quality control affecting both types of error. Three overlapping categories can be recognized.

1 Very low levels of contamination are present in all labs, forming the laboratory component of the sample random error.

2 Systematic background contamination is revealed by measurable and consistent blank values. This often arises during solution preparation (e.g. from filter papers) and contributes to bias so the mean blank must be subtracted from the sample values. Systematic contamination also arises from reagents but they will not be a problem provided identical amounts of a given reagent are added to all standard, sample and blank solutions. A related problem arises when a single set of standard solutions contain two or more elements. 'Multi-element standards' of this type should be made up only from the purest salts available.

3 Large (even gross) erratic contamination is revealed by very variable blank values. Subtraction of the mean blank may then lead to negative sample values or other anomalies and is not acceptable. Instead the contamination source should be traced if possible and the analyses repeated under cleaner conditions. Spurious contamination affecting the zero standard may cause its value to exceed the blanks whose predicted values will then be negative. A computer program will *add* negative blanks to sample values. The operator must check if this is introducing bias. The corrected value for the control sample is a guide.

Control charts

Control procedures become most efficient with control charts which remove the subjective assessment of the operator. There are several types which can be applied to routine analysis, but the Shewart chart (Fig. 9.3) is very straightforward and is widely used. A control chart is a graphical representation of the quality control values under investigation which are plotted sequentially. The Shewhart chart consists of a central line representing the mean value and two lines above and below the central line. These inner and outer control limits are commonly set at 2 and 3 standard deviations from the mean. The choice of these rounded off sigma limits is arbitrary but realistic in that 95% of the values under a normal curve are confined by $\pm 1.96\sigma$ and 99% by $\pm 2.575\sigma$. Data to set control limits must be obtained prior to commencing routine analysis, and preferably accumulated over a period of time. Limits may need to be reviewed from time to time.

A point falling outside the outer limit probably indicates bias affecting the run although it may be just spurious contamination affecting the control solution. A trend of two or three successive points between the inner and outer limits is a clearer indicator of bias. In these cases further evidence is required before the analytical run can be accepted. This may include some repeat analyses

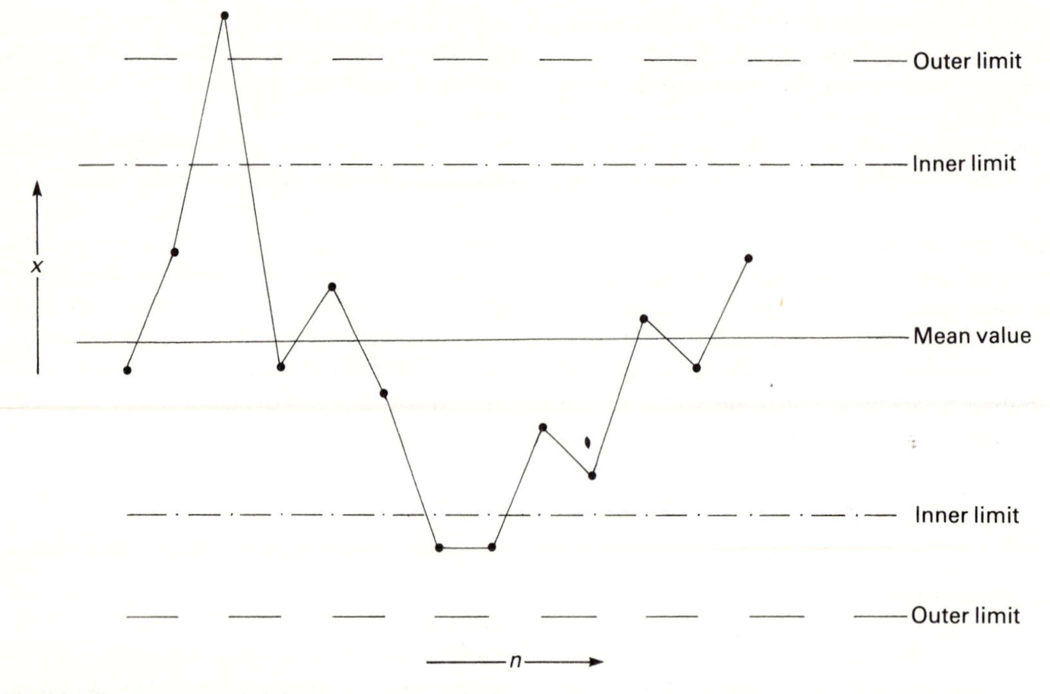

Fig. 9.3. Shewart type control chart.

but calibration solutions and lines and all calculations should first be checked.

The principle of the control chart may be applied to other experimental values if the need arises. These may include blank values, the spread of sample replicates or instrumental readings of standards. However, acceptable limits for predictions of standard values can also be built into the computer program. Many texts discuss control charts but see also Davies and Goldsmith (1972).

Other methods for detection of errors

The use of reference samples and blanks will reveal many errors but others may still escape notice. However, certain general relationships exist between various chemical values of both soils and vegetation so an examination of the analytical data is worthwhile. Useful relationships are given below.

1 Carbon to loss-on-ignition ratios in soils and vegetation (approx. 1:2).

2 Carbon to nitrogen ratios in soils (from about 10:1 in fertile soils to about 50:1 in peats).

3 Cation exchange capacity and loss-on-ignition in soils (high CEC with high LOI).

4 pH and calcium carbonate in soils (high pH with high calcium carbonate content).

5 pH and free carbonate in waters (pH $>$ 7.0 if free carbonate present).

6 Total cation and anion balance in waters and soil extracts (sums of cations and anions roughly equal as meq).

7 Nitrogen to phosphorus ratio in unfertilized vegetation (roughly constant).

8 Iron and manganese in vegetation normally low in calcareous samples.

Computer applications

The introduction of computer technology offers many potential benefits in the laboratory. Adoption has varied in scope depending on the role and size of the laboratory but many modern instruments contain sophisticated microprocessor facilities to control instrument operation and data production (see below). Equipment lacking these facilities can be interfaced to a computer. This section briefly outlines the role of computers in routine analysis and examines those functions which have particular applications. Topics mentioned will include equipment control, data acquisition, reduction and reporting together with the use of networks. Meanwhile an extensive literature exists which should be consulted and includes major reviews such as that of Barker (1983). Analytical Chemistry published a series of articles entitled A/C Interface in Volumes **58** and **59** (1986, 1987). Articles on the application of computers in analytical chemistry have also been featured in *Chemistry in Britain* (Vol. **11**, 1987).

Laboratory computing

Manual methods may remain adequate for occasional or small numbers of samples but otherwise computers offer many potential benefits for analytical laboratories if applied in an integrated way. A general policy for integrating the needs of the laboratory is needed. However, a centralized system attempting to serve all the needs of the laboratory rarely proves to be ideal and periods of breakdown ('down-time') will hinder the overall work of the laboratory. A more flexible system can be devised based on a number of low cost microcomputers each serving one or more instruments of a similar type, e.g. continuous flow lines. If desired, one or more of these could be interfaced to a more central mini-computer in the research centre or to a mainframe system. The latter may be at some distance but network would allow access (see below).

To examine the suitability of a given system it is important to assess the operating speed, available memory, required peripheral items (VDUs and printers), required storage media (floppy or winchester discs) and whether mailing or archiving facilities would be an advantage. These factors will be determined by examining in detail the existing operations in the laboratory. Software will be needed and a decision made to take time to develop this in the laboratory or to purchase a commercial package. Although a proven package has much to commend it, the manufacturer will often limit its scope to essential tasks such as drift correction and calibration. The package may not necessarily meet every requirement for a given analysis. In addition the methods used for statistical stages, such as calibration, should be known to the operator. Finally, trials and evaluation of hardware and software should involve both computing and laboratory staff.

Equipment control

As noted earlier, microprocessors are incorporated into the design of most types of modern analytical equipment. They have many functions including the control of the instrument parameters and display of a menu of operating conditions as well as producing an analytical result. In some cases the instrument is automatically set up for analysis. Data systems such as integrators often have timed event modules to activate relays for control of instrument functions and any attached modules such as an autosampler. Such equipment will not need a separate computing system, but an instrument lacking inbuilt facilities can often be interfaced on-line to a computer (see also below). The latter can monitor various parameters such as gas flows and alert the operator to possible faults. The type of control limit discussed earlier can also be built into a program for standard solutions and reference materials and serve as an 'early warning' system for the analyst.

A room containing equipment with either inbuilt or on-line microprocessors can form the basis of a fully automated laboratory with remote or even robotic control at the sample preparation stage. However, for this discussion it is assumed that small and medium sized laboratories are unlikely to need a fully automated set-up for routine analyses of soils and vegetation.

Data acquisition

Equipment lacking an in-built microprocessor cannot be plugged in directly to a computer as if it were a chart recorder. The electronic output of the instrument may be continuous (analogue form) and compatible with the recorder but the computer handles discrete 'bits' of information (digital form). Hence an analogue to digital converter is needed for interfacing with the computer and software is needed to detect sample peaks. This approach is straightforward for single channel operation although less ideal for multichannel work. Flexibility is increased if the raw data points can be stored and processed later at the convenience of the analyst. However, raw data points occupy large amounts of memory for which floppy discs may not be adequate and a winchester disc unit would then be needed.

Data processing

It is convenient to summarize the three stages here:

1 *Manipulation of raw data points for base line, standards, samples, blanks, etc. and computation of concentrations in the sample solutions.* Software for this is always provided by the manufacturer whenever the microprocessor (in-built or on-line) is designed for this work. However, the operator should know how stages such as drift correction and calibration area are carried out and their limitations (see above).

2 *Conversion of concentration values to content in original sample and including corrections for moisture, etc.* A program for this can easily be written but will already be combined with **1** if a laboratory computing system is purchased.

3 *Statistical processing of elemental values of the original samples.* Basic parametric procedures and some others were outlined earlier and all can be progammed. However, statistical packages including GENSTAT, SAS and others are very widely used. Other packages for multivariate analyses are also available. It is essential for the operator to check that his data meets the assumptions underlying the tests in the package particularly for parametric methods. For example some packages do not have provision for checking departures from normality. This cannot be obtained reliably from the small numbers of samples which most experiments produce. This may require a large sample to be taken for the purpose and a program written for computing coefficients of skewness and kurtosis (see p. 295). Other limitations of the tests were mentioned earlier including, for example constraints for the 't'-test (p. 300).

Common statistical packages include:

MINITAB—used for statistical analysis of small data sets at a relatively simple level. It is mainly restricted to teaching and training.

SAS—universally available package compatible with a wide range of computing systems. This provides a flexible and powerful statistical analysis, capable of handling large sets of data. SAS is excellent at representing data in a graphical form.

GENSTAT—is the most powerful and wide-ranging computer statistical analysis system, which has been developed over many years by the Biometrics department at Rothamsted Experimental Station.

Laboratory management

Computer technology offers many facilities of value for laboratory management and three examples are mentioned here.

Databases

Automation of instrumentation in the laboratory together with the advent of multi-element analysis, e.g. from chromatography and ICP, has led to the generation of large matrices of analytical data. Various options are available to manipulate large data blocks which will order and present the information for statistical analysis. Databases offer a range of management systems to present the data files in an order for processing or output. Relational databases are the most useful because they enable files to be combined easily into complex tables through the high-level database language. ORACLE is a relational database management which is compatible with IBM and DEC systems and others, and which contains a high-level interrogative language known as SQL. Databases are at the heart of Laboratory Information Systems (see below), where the administration and reporting of data can be coordinated. In addition to the power of the program to store and manipulate data, facilities are available to extract data files and link them with statistical and spreadsheet packages such as SAS.

Other relational databases commonly found include INAMS and DBASE2 for microcomputers.

Laboratory information systems (LIMS)

LIMS facilities have evolved and developed over the last few years to the extent that now there are packages available to suit the size and budget of the laboratory. Automation of the analysis and data processing has improved sample throughput. LIMS are particularly useful to cope with the organization and administration associated with rapid throughput. In general the systems offer:

a Efficient registration of work.
b Work scheduling.
c Progress monitoring.
d Coordination of data from various sources.
e Automatic quality control.
f Report generation.
g Facilities for planning and costing.
h Efficient interfacing and networking to other computers.

Other standard computer packages may also operate on the same system either independently of the database or by accessing information from it.

Networks

At the lowest level, local area networks are employed around the laboratory or group of laboratories to allow communication or data collection. Regional/global network communications allow access to national mainframe facilities and larger computing packages such as GENSTAT. Networks also offer the opportunity to transfer chemical analysis files direct to the customers' computer using an electronic mail service.

Appendices

Appendix I
Chemical Composition of Ecological Materials

It is often of value to have access to data from other sources when assessing the results obtained from the analysis of ecological materials. It is for this purpose that the tables in this section have been compiled. Nearly all the data in this appendix have been obtained from samples analysed in the laboratories of the Chemical Section of the Institute of Terrestrial Ecology.

The variability of soil data is much greater than for plants and their classification is far less precise. For this reason the figures given for soils should only be used as a rough guide.

Only extractable nutrient concentrations in soils are given in the tables. In agricultural work these are often determined for the purpose of assessing the fertilizer requirement of a soil in relation to crop response. Many agricultural soils have a long history of fertilizer application and the extractable concentrations may be much higher than those found in soils from a habitat not previously disturbed. For the purpose of this appendix, fertilized soils are excluded from consideration.

Table A1 attempts to classify mineral soils for ecological purposes. Some of the extractants normally used are listed and although they will extract different amounts, these differences are small compared with the variation over a range of soil types. One exception is 2.5% acetic acid, which will extract considerably more calcium from a calcareous soil than the other extractants. It must not be assumed that a low value obtained from particular soil types implies that the nutrient concerned is limiting in its capacity to support the vegetation found on that soil. Table A.2 presents characteristics of a few contrasting soil types and Table A.3 gives some typical examples of soil particle fractionation.

Tables A.4–A.8 present chemical data for a selection of plant species. For each species, photo-

Table A.1. Classification of mg $100\,g^{-1}$ levels of extractable elements in mineral soils

	Low <	Medium	High >	Typical extractants
Potassium	10	10–30	30	M Ammonium acetate (pH 7),
Calcium	10	10–200[1]	200[1]	2.5% (v/v) acetic acid,
Magnesium	5	5–30	30	Morgan's reagent
Manganese	0.1	0.1–2	2	
Ammonium-nitrogen	0.5	0.5–2	2	6% (w/v) potassium chloride
Nitrate-nitrogen	0.2	0.2–1	1	Water
Phosphate-phosphorus	0.2	0.2–2	2	Truog's reagent, 2.5% (v/v) acetic acid. Morgan's reagent, Olsen's reagent (calcareous soils only)

[1] May be considerably higher in a calcareous soil.

Table A.2. Chemical characteristics of some contrasting soil types. Sampling depth 0–15 cm. All results (except pH) given on dry weight basis. Extractions carried out using M ammonium acetate, pH 7.0

		mg 100 g^{-1}					%	
	pH	Na	K	Ca	Mg	P	C	N
Upland limestone soil, Ingleborough	8.0	1	15	310	17	0.2	6	0.3
Lowland chalk soil, Sussex Downs	8.2	1	16	720	40	0.7	9	0.7
Brown-earth (Serpentine), Cornwall	7.5	29	47	92	52	0.4	6	0.4
Iron humus podzol (Moine schists), Sutherland	5.5	7	10	12	8	0.4	5	0.2
Gleyed podzol (Basalt), Ben Harris, Argyll	6.3	8	13	170	130	1.0	12	0.3
Iron podzol (Granite), Kerloch, Kincardineshire	4.8	3	15	14	7	0.8	9	0.3
Upland blanket peat, Moor House, Cumbria	3.5	11	32	70	27	1.0	50	1.1
Upland gley (Silurian slates), Furness, Cumbria	4.4	10	18	22	3	0.3	8	0.2
Lowland podzol (Bunter sst), Delamere	4.9	1	3	6	2	0.3	7	0.3
'Thin' light podzol (Kellaways rock), N. York moors	4.7	2	4	14	4	0.2	6	0.2
Brown forest soil (O.R.S.), Shropshire	5.4	8	19	25	17	0.9	8	0.4
Alluvial warp, Trent Valley	5.9	10	24	70	31	0.7	4	0.5
Peaty-gley (Torridonian) Beinn Eighe, W. Ross	4.7	4	14	42	21	0.8	19	0.7
Shallow podzol (Extrusive lavas), Borrowdale, Cumbria	4.3	10	23	20	11	0.5	10	0.5
Skeletal montane soil (Cairngorms)	6.2	1	2	3	1	0.1	1	0.1
Blown sand regosol, Ainsdale, Lancashire	6.3	8	2	6	6	0.1	2	0.1
Chalk heath, Hampshire	4.7	10	23	84	25	0.3	11	0.7
Peaty-podzol (Calc sst), Northumberland	4.3	4	8	5	2	0.7	8	0.5

synthetic material was collected in late July or early August from sites in Great Britain where the plant was prominent. Nearly all values are a mean of at least three sites but they should be interpreted as only an approximate guide to levels that might be expected for the species. Table A.8 presents data for individual components of one species (*Quercus petraea*) at a single site. Scarcity of data has made it impossible to prepare similar tables for other sample materials, apart from a limited one for animals (Table A.9). However, the concentration ranges for most sample types likely to be examined by ecologists are given for each element in Chapter 5. Some of the concentration ranges to be expected for some organic constituents are given in Chapter 6.

Other data on the composition of ecological materials may be obtained from the following books and papers.

Soil

Scott D.R.M. (1955) Yale School of Forestry. Bulletin 62.

Swaine D.J. (1955) *The Trace Element Content of Soils.* Commonwealth Agricultural Bureau.

Swaine D.J. & Mitchell R.J. (1960) *J. Soil Sci*, **11**, 350.

Plant materials

Fleming G.A. (1963) *J. Sci. Food Agric.*, **14**, 203.

Langille W.M. & McLean K.S. (1976) *Plant Soil*, **45**, 17.

Likens G.E. & Bormann F.H. (1970) Yale School of Forestry. Bulletin 79.

Ovington J.D.O. (1956) *Forestry*, **29**, 22.

Ricklefs R.E. & Matthew K.K. (1982) *Can. J. Bot.*, **60**, 2037.

Rodin L.E. & Bazilevich D.J. (1966) *Production and Mineral Cycling in Terrestrial Vegetation.* Oliver and Boyd, Edinburgh.

Yarie J. (1980). *Ecology*, **61**, 1498.

Fungi

Gronwall O. & Pehrson A. (1984) *Oecologia*, **64**, 230.

Wetlands

Boyd C.E. (1978) in *Freshwater Wetlands* (Ed. by R.E. Good, D.F. Whigham and R.L. Simpson). Academic Press, London.

Invertebrates

Olechowicz E. & Mochnacka-Lawacz H. (1985) *Ekologia Polska*, **33**, 123.

Table A.3. Examples of soil particle classification (I.S.S.S. scale). Sampling depth 0–15 cm. All results given on % dry weight basis

	Brown earth, Lancashire	Loam, Suffolk	Iron podzol, N York Moors	Gleyed soil, Furness	Blown sand, Fife	Estuarine alluvium, Ribble Estuary	Calcareous, Sussex Downs	Peaty-podzol, Cheviot	Blanket bog, NW Pennines	Montane detritus, Cairn-gorms
Organic matter	8	10	12	8	1	4	6	25	95	1
Coarse sand	9	5	14	1	33	4	1	21	—	74
Fine sand	21	28	42	4	55	56	13	22	2	19
Silt	34	38	19	19	9	27	21	18	2	6
Clay	28	18	13	69	1	8	18	14	1	1
$CaCO_3$	—	1	—	—	1	1	41	—	—	—
Cation exchange capacity (me 100 g^{-1})	14	18	13	24	3	7	13	24	150	2
pH	5.6	7.1	4.4	4.9	6.7	7.0	8.1	3.9	3.3	4.8

Table A.4. Chemical characteristics of grasses and other monocotyledons. (Aerial growth, expressed on dry weight basis)

	%							$\mu g\ g^{-1}$				
	Na	K	Ca	Mg	Si	P	N	Fe	Al	Mn	Zn	Cu
Agropyron pungens	0.36	1.4	0.21	0.11	0.3	0.14	1.2	40	200	30	15	3
Agrostis tenuis	0.17	2.2	0.15	0.14	0.3	0.17	1.6	50	80	500	35	8
Ammophila arenaria[1]	0.15	1.4	0.30	0.08		0.14	0.9	70	80	20	10	2
Anthoxanthum odoratum	0.04	1.9	0.33	0.09		0.09	1.6	90	50	220	25	4
Arrhenatherum elatius	0.03	2.7	0.70	0.12	0.3	0.27	2.3	120	130	120	40	8
Brachypodium sylvaticum	0.04	2.2	0.49	0.15	0.3	0.17	2.0	80		90	40	11
Brachypodium pinnatum	0.01	1.9	0.38	0.07	0.3	0.15	1.5	30	100	50	15	6
Carex nigra	0.12	1.6	0.45	0.17	0.4	0.13	1.6	100	70	280	45	12
Dactylis glomerata	0.10	2.6	0.47	0.18	0.3	0.18	1.7	130	110	140	30	7
Deschampsia caespitosa	0.01	2.1	0.22	0.13	0.4	0.18	1.4	60	100	140	30	6
Deschampsia flexuosa	0.01	1.9	0.14	0.10	0.4	0.16	1.4	100	80	500	35	
Endymion nonscriptus[2]	0.07	3.3	0.60	0.19	0.4	0.25	2.8	300		150	30	6
Eriophorum angustifolium	0.08	1.2	0.14	0.11	0.1	0.11	1.4	60		400	80	3
Eriophorum vaginatum	0.03	0.9	0.13	0.12	0.1	0.15	1.7	60	60	180	95	12
Festuca ovina	0.01	1.5	0.27	0.09	0.2	0.17	1.5	130	50	210	65	16
Glyceria maxima[3]	0.04	1.6	0.71	0.21	0.2	0.23	1.8	70	70	130	20	4
Holcus lanatus	0.07	2.1	0.80	0.11		0.13	1.7	250	70	300	28	11
Holcus mollis	0.08	2.0	0.50	0.19		0.18	2.2	240	120	500	25	9
Juncus effusus	0.26	1.3	0.13	0.10	0.2	0.15	1.2	30		260	50	6
Juncus squarrosus	0.18	1.6	0.06	0.15	0.2	0.14	1.3	40	70	300	70	4
Luzula sylvatica	0.04	2.7	0.23	0.17		0.19	1.8	50		600	30	4
Molinia caerulea	0.03	1.3	0.14	0.11	0.3	0.12	1.9	90	170	300	30	6
Nardus stricta	0.01	1.2	0.14	0.07	1.5	0.12	1.4	90	130	300	45	7
Phleum pratense	0.05	1.2	0.23	0.15	0.4	0.20	2.1	220	210	220	28	8
Phragmites australis[3]	0.04	1.6	0.50	0.18		0.20	3.0	70		150	20	4
Poa pratensis	0.05	2.1	0.52	0.15		0.22	1.8	100	55	110	40	8
Schoenus nigricans	0.07	0.9	0.37	0.17	0.4	0.03	0.9	40		250	20	2
Spartina anglica	2.5	1.4	0.33	0.28	0.2	0.27	1.9	510		100	45	7
Zerna erectas	0.01	2.3	0.62	0.14		0.10	1.4	80	90	90	20	11

[1] Inflorescence excluded; [2] Aerial growth, May sampling; [3] Leaves only.

Table A.5. Chemical characteristics of dicotyledons (trees and large shrubs) (leaves only, expressed on dry weight basis)

	%							$\mu g\ g^{-1}$				
	Na	K	Ca	Mg	Si	P	N	Fe	Al	Mn	Zn	Cu
Trees > 4.5 metres												
Acer pseudoplatanus	0.02	2.2	1.8	0.23	0.2	0.20	2.4	80	140	100	45	9
Alnus glutinosa	0.01	1.3	1.0	0.27	0.2	0.17	2.8	140	120	610	65	19
Betula spp.	0.01	1.0	0.69	0.27	0.3	0.19	2.0	110	100	1400	200	6
Corylus avellana	0.03	1.5	1.4	0.33	0.4	0.17	2.3	110	130	1100	35	10
Fagus sylvatica	0.02	1.7	1.1	0.21	0.3	0.20	2.6	100	150	700	40	10
Fraxinus excelsior[1]	0.01	1.7	2.2	0.39	0.2	0.19	2.3	80	80	30	15	11
Ilex aquifolium	0.05	1.8	0.35	0.19		0.19	2.0	50	90	200	140	7
Quercus spp.	0.01	1.6	0.74	0.17	0.3	0.21	2.5	100	150	1100	15	11
Sorbus aucuparia	0.01	1.6	0.92	0.33		0.22	1.9	100	80	100	20	6
Tilia cordata	0.02	1.7	1.0	0.25		0.18	2.2	100	120	300	40	8
Ulmus procera	0.03	1.4	1.6	0.30		0.17	2.1	130		250	50	8
Trees/shrubs < 4.5 metres												
Crataegus monogyna	0.03	1.5	1.5	0.32	0.2	0.20	1.9	130		80	40	15
Hippophae rhamnoides	0.26	1.0	0.38	0.11	0.4	0.17	2.9	170	220	50	10	6
Prunus spinosa	0.04	3.4	1.1	0.29		0.22	2.8	100		90	15	5
Rhododendron ponticum	0.06	1.0	1.4	0.33	0.5	0.15	1.9	180	200		40	5
Salix spp.	0.04	1.6	0.74	0.23	0.2	0.23	2.4	70	100	600	40	8
Sambucus nigra	0.02	1.9	0.70	0.28		0.17	2.2	150		100	30	8

[1] Including rachids.

Table A.6. Chemical characteristics of dicotyledons (low shrubs and herbs) (leaves only, expressed on dry weight basis)

	%							$\mu g\ g^{-1}$				
	Na	K	Ca	Mg	Si	P	N	Fe	Al	Mn	Zn	Cu
Low shrubs												
Calluna vulgaris[1]	0.07	0.72	0.33	0.16	0.4	0.10	1.2	190	200	600	25	12
Empetrum nigrum[1]	0.03	0.65	0.55	0.22	0.3	0.11	1.1	200	140	200	20	6
Erica cinerea	0.12	0.52	0.23	0.14	0.3	0.05	0.7	60	80	130	15	5
Lonicera periclymenum	0.62	2.5	1.1	0.42		0.19	2.2	160		430	35	10
Myrica gale	0.05	0.94	0.31	0.12	0.2	0.10	2.7	100	120	250	45	6
Rubus fruticosus agg.	0.02	1.3	1.2	0.25	0.3	0.19	2.2	250		120	45	12
Ulex europaeus	0.31	1.4	0.24	0.16		0.17	1.9	50	100	90	25	5
Vaccinium myrtillus	0.02	1.1	0.87	0.24	0.3	0.12	1.8	130	100	1710	15	1
Vaccinium vitis-idaea	0.03	0.9	0.75	0.20		0.12	1.3	150	120	1000	25	14

Table A.6. (*continued*)

	%							$\mu g\ g^{-1}$				
	Na	K	Ca	Mg	Si	P	N	Fe	Al	Mn	Zn	Cu
Herbs												
Armeria maritima	1.8	1.9	0.52	0.57		0.20	1.7	550	180	60	30	4
Chamaenerion angustifolium	0.02	1.9	1.3	0.27	0.4	0.38	3.1	160	150	250	55	11
Circaea lutetiana	0.02	3.8	1.8	0.40		0.24	2.5	370	220	260	70	19
Filipendula ulmaria	0.01	1.5	0.85	0.59		0.22	2.6	130		370	100	13
Geranium robertianum	0.04	1.3	0.85	0.35	0.3	0.19	2.0	250		300		
Glechoma hederacea	0.02	5.0	2.3	0.34		0.36	3.0	160	120	90	75	9
Mercurialis perennis	0.02	3.9	2.8	0.45	0.2	0.16	3.0	300	200	60	110	10
Oxalis acetosella	0.04	2.3	0.47	0.37		0.28	1.8	100	150	800	40	5
Potentilla erecta	0.09	1.5	1.1	0.65		0.12	1.9	190		110	60	8
Prunella vulgaris	0.43	3.3	1.2	0.44		0.23	1.8	330	220	200	115	9
Rubus chamaemorus	0.10	1.0	0.20	0.25	0.2	0.25	1.5	300	450	350	25	8
Trifolium pratense	0.05	1.9	1.6	0.4	0.3	0.22	3.1	220		300	65	15
Urtica dioica	0.05	2.7	7.1	0.53		0.44	3.9	140		140	20	8

[1] Current year's growth.

Table A.7. Chemical characteristics of gymnosperms (needles), pteridiophytes (aerial tissue), bryophytes (entire plant). (Expressed on dry weight basis)

	%							$\mu g\ g^{-1}$				
	Na	K	Ca	Mg	Si	P	N	Fe	Al	Mn	Zn	Cu
Gymnosperms												
Juniperus communis[1]	0.01	0.89	0.99	0.14	0.2	0.17	1.4	70	300	300	15	9
Larix decidua	0.10	1.2	0.48	0.17	0.3	0.18	1.3	130		300	20	7
Picea abies[2]	0.02	0.75	0.70	0.14	0.4	0.14	1.45	100	100	250	40	7
Pinus sylvestris[2]	0.01	1.0	0.28	0.11	0.1	0.14	1.1	40	150	300	35	4
Taxus baccata[2]	0.01	1.4	0.65	0.14		0.19	1.9	50		700	65	7
Pteridiophytes												
Dryopteris felix-mas		3.0	1.0	0.32		0.16	1.70	300		250		
Equisetum fluviatile	0.01	2.8	2.2	0.76		0.25	1.5	60		40	30	7
Pteridium aquilinum[3]	0.05	2.3	0.36	0.22	0.5	0.24	2.4	70	160	400	40	8
Bryophytes												
Polytrichum commune	0.05	0.95	0.15	0.08	0.1	0.12	1.15	200	80	80	45	10
Rhacomitrium lanuginosum	0.08	1.20	0.15	0.15	0.7	0.16	1.10	1000	800	150	25	8
Sphagnum spp.	0.04	0.35	0.30	0.10	0.2	0.09	1.05	350	150	120	20	5

[1] First and second year growth; [2] First year needles; [3] Pinnules + rachids to *c.* 1 mm.

Table A.8. Chemical characteristics of parts of oak (*Quercus petraea*) (collected at Girzedale, Cumbria). (All results given on % dry weight basis)

	Na	K	Ca	Mg	P	N
Roots (large)	0.2	0.2	0.4	0.08	0.06	0.5
Roots (small)	0.2	0.4	0.4	0.11	0.10	0.9
Outer bark	0.02	0.08	0.5	0.03	0.17	0.5
Inner bark	0.01	0.3	1.1	0.06	0.03	0.4
Sap wood	0.01	0.14	0.05	0.01	0.02	0.16
Cambium and inner phloem	0.04	0.4	1.3	0.15	0.08	0.9
Outer heartwood	0.01	0.06	0.06	0.01	0.01	0.13
Inner heartwood	0.01	0.06	0.06	0.01	0.01	0.10
Branches	0.02	0.15	0.3	0.03	0.03	0.3
Twigs	0.02	0.2	0.4	0.06	0.05	0.8
Shoots	0.06	1.1	0.5	0.2	0.17	1.7
Leaves	0.03	1.2	0.7	0.2	0.2	2.0
Flowers	0.03	1.3	0.6	0.2	0.2	1.7
Fruit	0.03	1.2	0.5	0.16	0.15	1.5
Litter (branches, twigs)[1]	0.01	0.10	0.4	0.03	0.02	0.3
Litter (leaves)[1]	0.01	0.17	0.7	0.12	0.07	1.1

[1] Samples collected 6 months after falling.

Waters

Taylor E.W. (1958) *The Examination of Waters and Water Supplies*. Churchill, London.

Environmental materials

Bowen H.J.M. (1979) *Environmental Chemistry of the Elements*. Academic Press, London.

A data bank covering the chemical composition of many plant species in Great Britain is maintained by the Chemical Section of the Institute of Terrestrial Ecology at Merlewood Research Station, Grange-over-Sands, Cumbria from whom information can be obtained.

Table A.9. Chemical characteristics of animal materials (expressed on dry weight basis)

	%						μg g^{-1}			
	Na	K	Ca	Mg	P	N	Fe	Mn	Zn	Cu
Vertebrates										
Birds	0.4	0.8	3.1	0.1	2.0	10.5	400	70		
Rabbit	0.2	0.8	2.2	0.2	2.3	2.6	500	23	120	20
Mole	0.2	0.7	2.5	0.2	2.2	2.8	800	20	180	25
Vole	0.4	1.2	3.8	0.2	2.5	2.7	300	18	160	23
Shrew	0.5	1.0	4.3	0.3	2.8	2.7	400	20	110	20
Mouse	0.4	1.0	3.3	0.2	2.3	2.4	600	15	120	30
Invertebrates										
Mollusca	0.2	0.6	1.9	0.2	1.5	6.5				
Arachnida	0.2	0.6	0.4	0.3	1.0	9.0	300	25	150	30
Insecta	0.3	0.7	0.3	0.2	0.9	8.5	200	30	150	50
Myriapoda (Diplopoda)	0.3	0.5	14.0	0.2	1.9	5.8				
Crustacea	0.3	0.4	11.0	0.2	1.7	6.2		40	130	40
Oligochaeta	0.2	0.5	0.3	0.2	1.1	10.5				

Appendix II
Bioenergetics

Ecosystem productivity which is an essential part of ecology depends on the energy availability of the system. Despite this the importance of energy utilization is often underrated or even ignored. The following notes are provided to introduce the subject and references are supplied to more detailed sources of information.

Energy flow in the ecosystem

Although only about 1% of the radiation input per unit area goes into net biomass production (gross production less respiration and other losses) this fraction is of the greatest importance. Ultimately all animal life depends on it.

Organisms that synthesize organic matter from inorganic compounds are the primary producers. They include all photosynthetic organisms and a few bacterial species, which are the chemoautotrophs. Animals and many micro-organisms that feed on the tissue produced are secondary producers. These can be subdivided into other feeding levels and thus an idealized food and energy chain is built up (Fig. A.1). In a complex community the various stages are known as trophic levels.

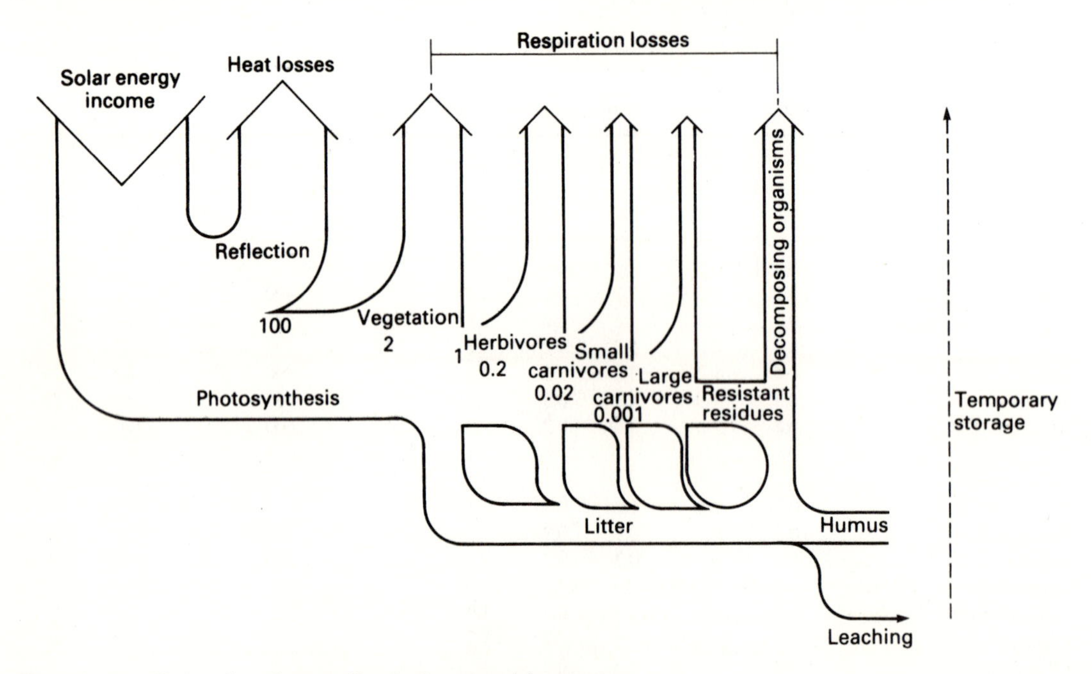

Fig. A.1. Simplified outline of energy flow in the terrestrial ecosystem.

It has gradually become clear that the idealized food chain is much too simple to represent processes known to occur in ecosystems and has been replaced by the concept of a food and energy network. Some of these networks may be extremely complex. Further information is given by Odum (1971) and Phillipson (1966).

Nutrition

The energy flow within communities is now more widely studied. Most attention in the past has been given to the energy requirements and utilization of individual animal species, especially those of economic importance. A major difficulty in this work is to assess the energy losses at various stages of the digestion process. In experimental conditions with large animals it is possible to take into account digestible energy, metabolizable energy or even heat losses during feeding action to arrive at the net energy available for growth and activity. In most ecological studies mainly concerned with smaller organisms this is not practicable and the digestible energy obtained by deducting faecal from gross (input) energy is generally used. Approximate corrections may be applied for respiration and urine losses.

The percentage digestibility of an organic constituent may be calculated as:

$$\frac{\text{Wt constit. in feed} - \text{wt constit. in faeces}}{\text{Wt constit. in feed}} \times 100$$

Sometimes inert constituents are used as indicators of digestion processes. Inorganic indicators such as magnesium and silicon compounds are probably preferable to artificial additives when working with most birds, mammals and microfauna.

Energy values

The range of gross energy values for ecological materials is relatively narrow compared with most of the inorganic constituents. The overall range for most plant materials is about 19–22 kJ g^{-1} (ash-free dry basis) which reflects the dominance of the carbohydrate fraction (approx. 17 kJ g^{-1}) and to some extent the protein (23.5 kJ g^{-1}). Fats release approx. 39.5 kJ g^{-1}, but with a few exceptions plant samples contain only a small amount of these substances. Most animal materials contain a selectively greater proportion of fats which is sufficient to raise the overall range to about 21–23 kJ g^{-1}.

Individual plant components may vary slightly in their energy content but significant differences are only likely where fatty substances are accumulated, as in oily feeds. Animal tissues differ

Table A.10. Typical ranges for energy values of terrestrial bio-materials

	kJ g^{-1}	
	Dry weight	Ash-free dry weight
Algae	16–22	18–23
Fungi	19–22	10–23
Bryophyta	15–17	17–19
Pteridophyta	18–19	20–21
Gymnospermae	22–24	23–25
Angiospermae		
Monocotyledons (grasses)	17–19	19–21
Dicotyledons		
Herbs (aerial growth)	19–20	20–21
Trees (roots, boles, branches)	18–19	19–20
(leaves, flowers)	20–22	21-23
Invertebrata		
Mollusca	15–16	23–24
Arichnidae	19–20	22–24
Insecta	21–22	24–26
Myriapoda (Diplopoda)	12–13	23–25
Crustacea	13–14	22–24
Oligochaeta	18–19	22–24
Vertebrata		
Aves	20–24	22–26
Mammalia		
Herbivores	20–21	24–25
Carnivores	19–21	23–24
Litter (fresh)	17–18	19–20
(partly decomposed)	13–14	18–19
Peat	17–19	18–21

To convert to kcal g^{-1} divide the above values by 4.184.

considerably in their fat content within the one individual but relatively little calorific data are available except for foodstuffs. Different stages in some invertebrate life cycles show variation in energy value. For example, insect pupae may exceed 25 kJ g^{-1}. Seasonal changes in calorific value of plant tissues appear to be slight but are not easy to detect and little data are available. The possibility of inter-year variation should not be overlooked. It has been reported that the calorific value of herbs can change by as much as 10% from one year to the next and this was thought to be related to climatic variation.

Table A.10 gives some typical energy values for a variety of ecological materials. Other data and information can be obtained from the following sources:

Albritton E.C. (Ed.) (1955) *Standard Values in Nutrition and Metabolism.* Saunders.

Cummins K.W. & Wuycheck J.C. (1971) *Mitt. int. Ver. Limnol.* **18**.

Golley F.B. (1961) *Ecology* **42,** 581.

Grodzinski W., Klekowski R.K. & Duncan A. (1974) *Methods for Ecological Bioenergetics.* I.B.P. Handbook No. 24. Blackwell Scientific Publications, Oxford.

Odum E.P. (1971) *Fundamentals of Ecology.* W.B. Saunders, London.

Ovington J.D. & Lawrence D. (1967) *J. Ecol.* **48,** 639.

Philipson J. (1966) *Ecological Energetics.* Edward Arnold, London.

Richman S. & Slobodkin L. B. (1960) *Ecology* **41,** 784.

Appendix III
Laboratory Contamination

Some of the sources of contamination have been referred to in Chapter 1 and elsewhere in the text in conjunction with specific tests, reagents and equipment. Most of the significant contaminants are listed in Table A.11, but for a detailed examination of the subject the reader should refer to Zief and Mitchell (1976).

Table A.11. Substances encountered in or near laboratories which could cause analytical contamination

Source	Contaminating materials	Principal constituents affected	Possible action
Analytical cross-contamination	Chemical reagents (splashing, dusting, dirty glassware)	Esp. inorganic salts	Improve lab training, avoid conflicting tests, improve washing practice
Building operations	Plaster, cement, brick dust, etc.	Ca, Si	Avoid contact
Laboratory cleaning	Bench polishes	Oils, waxes	Restrict use, cover benches with plastic-coated paper sheets ('Benchkote')
	Dust, cleaning materials	Ca, Si, Na, P, Cl	Wash, vacuum—not sweep, cover affected items
Laboratory equipment and sundries	Rust, dusting from galvanized and alloy materials	Fe, Pb, Al, Cu, Zn	Clean, paint, scrap old items
	Glassware (see p. 4)	Na, Ca, Si, Zn, B	Wash well before use (see p. 5)
	Rubber bungs, tubing (with talc)	Zn, Ca, Mg	Wash well in dil. acid
Laboratory fittings	Rust, dusting from pipes	Fe, Pb, Cu, Zn	Wash, paint, or replace with plastic
	Paint dusting and flaking	Pb, Ti, heavy metals	Use emulsion, epoxy, chlorinated rubber paints with synthetic org. pigments
Nearby farms and industry	Fertilizers (lime, basic slag)	Ca, P	Avoid contact
	Dust, smoke, fumes	Variable	
Protective clothing	Dusting from freshly laundered overalls	Ca, starch	Omit finishing treatments
Staff	Tobacco smoke, ash, matches	Mineral elements	Discourage
	Cosmetics (powders, handcream, etc.)	Zn, Ca, Mg, oils	Avoid excessive use
	Perspiration	Na, Cl, organics	Improve ventilation, wash hands, use plastic gloves
Washing materials	Soap, detergents, scouring powders, etc.	Na, Ca, P, Cl	Use alternatives, P-free detergents

Appendix IV
Laboratory Safety

The Health and Safety at Work act of 1974 provided a framework for the introduction of comprehensive controls in Britain. Codes of Practice are now required for operating equipment and highlighting hazardous operations in the laboratory. Other Western countries have similar regulations. The methods given in this book, together with necessary precautions, may form a basis for a local code of practice. The legislation is far reaching and covers all aspects of laboratories, ranging from design and management to the requirements for fire doors and the handling of gas cylinders. For more detailed information the reader is referred to *Hazards in the Chemical Laboratory*, edited by L. Bretherick (The Royal Society of Chemistry, 1986).

The notes given here highlight some of the more likely or serious laboratory hazards. However, most chemicals and laboratory operations can, in certain circumstances, be dangerous. It is not practicable to mention every possible risk and it is assumed that those using, or directing the use of methods given in this book, have a basic knowledge of safe laboratory practice. It is particularly important that inexperienced students or staff are supervised adequately and dangers are pointed out as they arise. It is also essential that appropriate protective clothing and eye shields are used when handling concentrated acids and other hazardous chemicals. Heat resistant gloves and face masks should also be available. For the safe handling of chemicals, it is essential to have a clear system of labelling. Bottles without labels, or with damaged labels, are a hazard. Good housekeeping by cleaning up spillages immediately and regular cleaning will minimize problems. The prevention of accidents is largely a matter of common sense and experience.

Poisons

Scheduled poisons such as cyanides, arsenical compounds, etc. must be kept in a locked cupboard and keys only issued to authorized persons against a signature. Many compounds can be immediately toxic due to ingestion by mouth, absorption by skin or inhalation. Ingestion should be virtually eliminated by proper attention to personal technique. Pipetting by mouth should never be used for hazardous liquids. Absorption through the skin can be minimized by wearing gloves or protective clothing. Adequate fume extraction facilities designed to the latest air-flow specifications ($0.5\ m\,s^{-1}$) should be provided to reduce the hazards due to inhalation.

It is necessary constantly to check that chemical reagents are not carcinogenic. If in doubt, information should be sought from a recognized authority (for example, The International Agency for Research on Cancer, Lyons, France). Apart from reagents, other products about the laboratory may be health hazards. For example, early construction materials based on asbestos are now condemned and legislation requires their removal from laboratory structures.

Fire hazards

Organic solvents constitute the greatest laboratory fire risk, the vapour being more dangerous than the liquid. Of the solvents mentioned in this book, ether and petroleum ether are probably the most hazardous.

Solvents with a flash point below ambient can liberate sufficient vapour to form a flammable mixture with air. It is most important therefore to keep the quantity of solvents present in the laboratory to the absolute minimum.

The fire fighting equipment should be strategically located with relation to the nature of the work in the vicinity. A number of small extinguishers may be more effective than a large extinguisher since rapid action is crucial for laboratory fires (see Table A.13).

Do not use or store halon extinguishers in confined spaces as the vapours and products of pyrolysis are toxic.

The local fire officer will advise on the number and type of fire extinguishers required for a specific laboratory.

Table A.12. Boiling and flash points of common solvents

Solvent	Boiling point (°C)	Flash point (°C)
Acetone	56.5	−16.7
Benzene	80.1	−11.1
Diethyl ether	35.0	−40
Ethyl alcohol	78.3	+12.7
Methyl alcohol	64.7	+15.5
Ethyl acetate	77.2	−4.4
Toluene	110.5	+4.4
Xylene	138.1	+17.2
Petroleum ether	40–60	−45.5

Oxidizing agents

The strong oxidizing agents (perchloric, nitric and chromic acids, hydrogen peroxide and metal peroxides) which are used in the destruction of organic materials need very careful handling. Dangers involved in their use are not always appreciated. Some comments have been made about the use of these reagents in Chapter 3, but the following points should also be noted:

1 Stand reagent bottles on plastic trays which are designed to contain any spillage.

2 Perchloric acid should be stored in glass bottles away from organic and reducing materials. Hydrogen peroxide is better stored in polythene bottles to prevent formation of metal peroxides. The containers should not stand directly on wooden benches, but on a suitable protective tile or dish. 100 volume strength is best stored at just above freezing.

3 Wooden fume cupboards should be avoided when handling perchloric acid or hydrogen peroxide. Suitable materials are: plastics, stainless steel, and glass fibre. PVC is often used for fume cupboards and some models are also provided with a water spray to wash away the perchloric acid. PVC is also a replacement for asbestos inducting. Dust should not be allowed to accumulate in enclosures where oxidizing materials are used.

4 Hydrofluoric acid is particularly corrosive as it may penetrate deep into the skin before the effects are noticed. Tubes of 2% calcium gluconate gel should be available in case of skin contact. Before

Table A.13. Applications of different types of fire extinguisher

	Type of fire			
Extinguisher	Paper wood, textile, fabric	Flammable liquids	Flammable gases	Electrical
Water	✓			
Foam	✓	✓		
CO_2 gas		✓	✓	✓
Powder	✓	✓	✓	✓
Halon	✓	✓	✓	✓

handling, test gloves by filling with water to ensure they form an effective barrier. This acid attacks glass, and so the fume cupboard will need to be lined with a plastic film to prevent damage.

Waste disposal

For significant quantities of toxic substances, contact the Local Authority who will advise on the best method for disposal. Flammable or water immiscible substances should be collected in a safety trap or vessel and may be incinerated. Waste chlorinated hydrocarbons should be kept in a separate vessel. Meanwhile the routine disposal of digest solutions, toxic reagents, etc. down the laboratory sink always requires excess dilution with tap water and sometimes prior neutralization. The end-point of laboratory drainage systems should be known. In rural districts local areas may be sensitive to pollution in water courses and disposal of some substances in this way is restricted by legislation.

Gas cylinders

Cylinders of bottled gas need to be handled with care in order to avoid damage to the valves. Cylinder trolley, stands, straps or chains can all be obtained for moving or securing the cylinders. Oil or grease should never be used for lubricating valves. This particularly applies to oxygen and nitrous oxide. Wherever possible, cylinders should be situated in a lockable external shelter. It is important to store cylinders in the upright position, especially for acetylene. The metal cylinders themselves are heavy and should be firmly anchored to the wall or bench, or held in a floor stand. Once the cylinder has been connected, check for leaks with a leak detection foam.

Additional precautions must be taken when using flammable gases. Hydrogen–air mixtures can cause violent explosions and acetylene too can be very hazardous because it has a wide flash-point when mixed with air. In addition acetylene must not be used with copper fittings and should always be kept upright when in use. Another gas requiring special care is nitrous oxide which replaces air in providing a hot flame for certain atomic absorption methods.

The colour coding for industrial and medical gas cylinders is standardized in most countries. In general red indicates flammability and yellow is used for poisonous gases. Only modern gas cylinders should be used because older cylinders and valves may not meet current safety regulations. Further information on the use and handling of cylinders and valves can be obtained from the manufacturers.

Appendix V
Chemical Constants

A few tables are included in this appendix giving data which the authors have found to be frequently required in connection with the type of analyses covered. For more detailed information on chemical constants, reference should be made to the specialized handbooks. These include:

Handbook of Chemistry and Physics. (Annual) Chemical Rubber Company.

Meites L. (Ed.) (1963) *Handbook of Analytical Chemistry.* McGraw-Hill Book Co, Maidenhead.

Kaye G.W.C. & Laby T.H. (1986) *Tables of Physical and Chemical Constants.* Longman, Harlow.

Table A.14. Selected list of atomic weights

Element	Symbol	At. no.	At. wt	Element	Symbol	At. no.	At. wt
Aluminium	Al	13	26.98	Lithium	Li	3	6.94
Antimony	Sb	51	121.76	Magnesium	Mg	12	24.32
Argon	A	18	39.94	Manganese	Mn	25	54.94
Arsenic	As	33	74.91	Mercury	Hg	80	200.61
Barium	Ba	56	137.36	Molybdenum	Mo	42	95.95
Beryllium	Be	4	9.01	Nickel	Ni	28	58.71
Bismuth	Bi	83	209.00	Nitrogen	N	7	14.008
Boron	B	5	10.82	Oxygen	O	8	16.000
Bromine	Br	35	79.92	Phosphorus	P	15	30.975
Cadmium	Cd	48	112.41	Platinum	Pt	78	195.09
Caesium	Cs	55	132.91	Potassium	K	19	39.100
Calcium	Ca	20	40.08	Selenium	Se	34	78.96
Carbon	C	6	12.01	Silicon	Si	14	28.09
Cerium	Ce	58	140.13	Silver	Ag	47	107.880
Chlorine	Cl	17	35.57	Sodium	Na	11	22.991
Chromium	Cr	24	52.01	Strontium	Sr	38	87.63
Cobalt	Co	27	58.94	Sulphur	S	16	32.066
Copper	Cu	29	63.54	Thallium	Tl	81	204.39
Fluorine	F	9	19.00	Thorium	Th	90	232.05
Germanium	Ge	32	72.60	Tin	Sn	50	118.70
Gold	Au	79	197.00	Titanium	Ti	22	47.90
Helium	He	2	4.003	Tungsten	W	74	183.86
Hydrogen	H	1	1.008	Uranium	U	92	238.07
Iodine	I	53	126.91	Vanadium	V	23	50.95
Iron	Fe	26	55.85	Zinc	Zn	30	65.38
Lanthanum	La	57	138.92	Zirconium	Zr	40	91.22
Lead	Pb	82	207.21				

Table A.15. Chemical conversion factors

A	B	To convert A to B multiply by:	To convert B to A multiply by:
Na	Na_2O	1.3480	0.7419
K	K_2O	1.2046	0.8302
Ca	CaO	1.3992	0.7147
Ca	$CaCO_3$	2.4973	0.4004
CaO	$CaCO_3$	1.7848	0.5603
Mg	MgO	1.6579	0.6032
Fe	Fe_2O_3	1.4297	0.6994
P	P_2O_5	2.2914	0.4364
P	PO_4^{3-}	3.0662	0.3261
N	NO_3^-	4.4266	0.2259
N	NH_4^+	1.2878	0.7765
S	SO_4^{2-}	2.9956	0.3338
Si	SiO_2	2.1392	0.4675

Table A.16. Concentrated acids and ammonia solution

	Specific gravity	Approximate composition by wt (%)	Composition ($g\,l^{-1}$)	Approximate molarity	$ml\,l^{-1}$ for M solution
Sulphuric acid	1.84	98	1800	18	56
Nitric acid	1.42	71	1010	16	63
Hydrochloric acid	1.18	35	413	11	83
o-Phosphoric acid	1.75	88	1540	16	63
Acetic acid	1.05	99	1040	17	59
Ammonia solution	0.88	35	310	16	63

Primary standards should not be prepared from these data.

Table A.17. B.S. sieve conversions

Mesh number	Aperture width (μm)
5	3350
10	1680
20	750
30	500
40	370
50	310
60	250
70	215
100	150
150	105
200	75
250	61
300	53
350	45

Table A.18. Guide to resistance of laboratory materials to chemical reagents

	Temperature (°C)	H_2SO_4 (98%)	HNO_3 (95%)	HCl (conc)	HCl (10%)	HF (40%)	$HClO_4$ (60%)	H_3PO_4 (conc)	Chromic acid (conc)	HOAc	Caustic alkalis	Hypochlorites	Alcohols	Acetone	Ether	Chloroform	Carbon tetrachloride	Trichloroethylene	Phenol	Detergents	Maximum working temperature (°C)
Aluminium	20	X	R	X	X	X	X	X	X	R	X	X	R	R	R	R	R	R	R	R	600
	100	X	R	X	X	X	X	X	X	X	X	X	R	R	R	R	X	R	R	R	
Nickel	20	X	X	X	X	X	X	X	X	R	R	X	R	R	R	R	R	R	R	R	1340
	100	X	X	X	X	X	X	X	X	R	R	X	R	R	R	R	R	R	R	R	
Platinum	20	R	R	R	R	R	R	R	R	R	R	R	R	R	R	R	R	R	R	R	1750
	100	R	R	R	R	R	R	R	R	R	R	R	R	R	R	R	R	R	R	R	
Stainless steel	20	X	R	X	X	X	X	R	X	R	R	X	R	R	R	R	R	R	R	R	1400
	100	X	X	X	X	X	X	X	X	X	R	X	R	R	R	R	R	R	R	R	
Natural rubber	20	X	X	R	R	R	X	R	X	X	R	R	R	R	X	X	X	X	X	R	—
	100	X	X	R	R	—	X	X	X	X	R	—	—	—	X	X	X	X	X	—	
Neoprene	20	X	X	R	R	R	X	R	X	X	R	X	R	X	X	X	X	X	X	R	110
	100	X	X	X	R	—	X	X	X	X	R	X	R	X	X	X	X	X	X	R	
Acrylic sheet, e.g. perspex	20	—	X	R	R	X	—	X	R	X	R	R	X	X	R	X	X	X	X	R	90
	100	—	X	X	X	X	—	X	X	X	X	X	X	X	X	X	X	X	X	X	
Nylon	20	X	X	X	X	X	X	X	X	X	R	X	R	R	R	X	R	R	X	R	250
	100	X	X	X	X	X	X	X	X	X	R	X	—	—	—	X	—	—	X	R	
PTFE	20	R	R	R	R	R	R	R	R	R	R	R	R	R	R	R	R	R	R	R	280
	100	R	R	R	R	R	R	R	R	R	R	R	R	R	R	R	R	R	R	R	
Rigid PVC	20	R	R	R	R	R	R	R	R	X	R	X	R	X	X	X	R	X	R	R	75
	100	X	X	X	X	X	X	X	X	X	X	X	X	X	X	X	X	X	X	X	
Polythylene low density	20	R	X	R	R	R	—	R	R	R	R	R	R	R	R	X	X	R	R	R	80
	100	X	X	X	X	X	X	X	X	X	X	X	X	X	X	X	X	X	X	X	
Polythylene high density	20	R	X	R	R	R	R	R	R	R	R	R	R	R	R	X	R	X	R	R	100
	100	X	X	X	X	X	X	X	X	X	X	X	X	X	X	X	X	X	X	X	
Polycarbonate	20	X	X	R	R	R	—	R	R	X	X	R	R	X	X	X	X	—	X	R	140
	100	X	X	X	X	X	X	X	X	X	X	X	X	X	X	X	X	X	X	X	
Polypropylene	20	R	X	R	R	R	X	R	X	X	R	R	R	R	R	X	X	R	R	R	140
	100	X	X	X	R	—	X	—	X	X	—	X	—	—	—	X	X	R	—	—	
Polystyrene	20	X	X	R	R	—	X	—	X	X	R	R	R	X	X	X	X	X	R	R	70
	100	X	X	X	X	—	X	—	X	X	X	X	X	X	X	X	X	X	X	X	
Glass (borosilicate)	20	R	R	R	R	X	R	R	R	R	—	R	R	R	R	R	R	R	R	R	550
	100	R	R	R	R	X	R	R	—	R	X	R	R	R	R	R	R	R	—	R	
Fused silica	20	R	R	R	R	X	R	R	R	R	R	R	R	R	R	R	R	R	R	R	1200
	100	R	R	R	R	X	R	R	R	R	R	R	R	R	R	R	R	R	R	R	
Glass (soda lime)	20	R	R	R	R	X	R	R	R	R	X	R	R	R	R	R	R	R	R	R	—
	100	R	R	R	R	X	R	R	R	R	X	R	R	R	R	R	R	R	R	R	
Porcelain and stoneware	20	R	R	R	R	X	R	R	R	R	R	R	R	R	R	R	R	R	R	R	—
	100	R	R	R	R	X	R	R	R	R	X	R	R	R	R	R	R	R	R	R	

R = Resistant, X = unsuitable, — = no data.
Based on data mainly abstracted from the Chemical Processing Corrosion Chart published by the *Journal of Chemical Processing*, 33 Bowling Green Lane, London EC1R 0NE.

References

Abbott D.C. & Egan H. (1967) *Analyst (London)*. **97,** 475.

ABCM-SAC (1957) *Analyst (London)*. **82,** 764.

Abdel-Kader M.H.K., Stiles D.A. & Ragab M.T.H. (1984) *Int. J. Environ. Anal. Chem.,* **15,** 281.

Abu-Hilal A.H. & Rilev J.P. (1981) *Anal. Chim. Acta.* **131,** 175.

Acquaye D.K. (1967) *J. Sci. Food Agric.* **18,** 24.

Adeljou S.B., Bond A.M. & Briggs M.H. (1984) *Anal. Chem.* **56,** 2397.

Adlea I. (1966) *X-ray Emission Spectrography in Geology.* Elsevier, London.

Adrian W.J. & Stevens M.L. (1977) *Analyst (London)*. **102,** 446.

Agarwal R.R. (1960) *Soil Fertil.,* **23,** 375.

Agrawal Y.K. & Patke S.K. (1981) *Int. J. Environ. Anal. Chem.* **10,** 175.

Agricultural Development and Advisory Service. (1986) *The Analysis of Agricultural Materials.* HMSO, London.

Aharouson N. & Resnick C. (1972) in *Analytical Chemistry of Phosphorus Compounds* (Ed. by M. Halmann). Vol. 37 in Chemical Analysis: a series of monographs in analytical chemistry and its applications (Ed. by P.J. Elving and J.M. Kolthoff). Wiley-Interscience, New York.

Ahlmwalia B. & Ellis E.E. (1984) *J. Inst. Brew.,* **90,** 254.

Aiken G., Thurman E.M., Malcolm R.L. & Walton H.F. (1979) *Anal. Chem.,* **51,** 1799.

Akerblom M. (1985) *J. Chromatogr.* **319,** 427.

Alexander P. & Block R.J. (1960) *A Laboratory Manual of Analytical Methods of Protein Chemistry.* Pergamon Press.

Alexander R.H. (1969) *Lab. Pract.,* **18,** 63.

Alkemade. C. & Herrman R. (1979) *Fundamentals of Analytical Flame Spectroscopy.* Hilger, Bristol.

Allan J.E. (1959) *Spectrochim. Acta.* **15,** 800.

Allan, J. E. (1961) *Spectrochim. Acta,* **17,** 459.

Allen S. E., Grimshaw H.M., Parkinson J.A. & Quarmby C. (1974) *Chemical Analysis of Ecological Materials,* 1st edn. Blackwell Scientific Publications, Oxford.

Allen S. E. & Grimshaw H. M. (1962) *J. Sci. Food Agric.* **13,** 525.

Allen T. & Khan A.A. (1970) *Chem. Eng.* 108.

Allison L.E. (1960) *Proc. Soil. Sci. Soc. Am.* **24,** 36.

Alvarez R. (1980) *J. Assoc. Off. Anal. Chem.* **63,** 80.

Alvarez R. (1986) *Fresnius' Z. Anal. Chem.* **324,** 376.

Ambrus A., Lantos J., Visi E., Csatlos J. & Sarvari L. (1981) *A.O.A.C.* **64,** 733.

Analytical Methods Committee (1954) *Analyst (London)*. **79,** 393.

Analytical Methods Committee (1965) *Analyst (London)*. **90,** 515.

Analytical Methods Committee (1968) *The Determination of Particle Size.* 1. A Critical Review of Sedimentation Methods. Society for Analytical Chemistry, London.

Analytical Methods Committee (1972) *Analyst (London)*. **97,** 734.

Analytical Methods Committee (1978a) *Analyst (London)*. **103,** 521.

Analytical Methods Committee (1978b) *Analyst (London)*. **103,** 643.

Analytical Methods Committee (1987) *Analyst (London)*. **112,** 199.

Analytical Quality Control Committee (1984) *Analyst (London)* **109,** 431.

Anderman G. & Kemp K.W. (1958) *Anal. Chem.* **30,** 1306.

Andersen J.M. & Pedersen W.B. (1982) *J. Chromatogr.* **259,** 131.

Anderson G. (1960) *J. Sci. Food Agric.* **11,** 497.

Anderson N.H. & Girling J. (1982) *Analyst (London)*. **107,** 826.

Andreae M.O., Asmode J.F., Foster P. & Van't dack L. (1981) *Anal. Chem.* **53,** 1766.

Anton A. (1960) *Anal. Chem.* **32,** 725.

Appel B.R. & Tokiwa Y. (1981) *Atmos. Environ.* **15,** 1087.

Apte S.C. & Howard A.G. (1986) *J. Anal. At. Spectrum.* **1,** 379.

Archiv für Hydrobiologie (1980).

Armour J.A. & Burke J.A. (1970) *J. Assoc. Off. Anal. Chem.* **53,** 761.

Arshad M.A., St. Amaud R.J. & Huang P.M. (1972) *Can. J. Soil Sci.* **52,** 19.

Askinazi D.L. & Guinsbourg K.E. (1957) *Pl. Soil.* **9,** 3.

Aspinall G.O. (1959) *Adv. in Carbohyd. Chem.* **14,** 429.

Atkinson T., Fowler V.R., Garton. G.A. & Lough A.K. (1972) *Analyst (London)* **97,** 562.

Bache B.W. & Williams. E.G. (1971) *J. Soil Sci.* **22,** 289.
Bache B.W. (1984) *Plant Cell and Environ.* **7,** 391.
Bache B.W. (1985) *Soil Use and Management.* **1,** 1.
Bailey B.W., Rankin J.M. & Weinbloom R. (1971). *Int. J. Environ. Anal. Chem.,* **1,** 3.
Bailey P.H., Hughes M. & McDonald A.N.C. (1957) *J. Br. Grassld. Soc.* **12,** 157.
Bailey P.L. (1980) *Analysis with Ion selective Electrodes.* Heyden, London.
Bailey S., Bungan P.J. & Fishwick F.B. (1970) *Chem. Ind.* **22,** 705.
Baker A.J.M. (1981) *J. Plant Nutrition.* **3,** 643.
Baker R.T. (1975) *J. Soil Sci.* **26,** 432.
Ball D.F. (1964) *J. Soil Sci.* **15,** 84.
Ball D.F. & Beaumont P. (1972) *J. Soil. Sci.* **23,** 298.
Ball D.F. & Perkins D.F. (1962) *Nature.* **194,** 1163.
Ball D.F. & Williams W.M. (1968) *J. Soil Sci.* **19,** 379.
Bander H. (1980) *J. Chromatogr.* **189,** 414.
Banerjee G. (1975) *Fresenius' Z. Anal. Chem.* **277,** 207.
Bank M.S. (1966) *Proc. 5th Conf. on X-ray Analytical Methods.* Philips, Eindhoven.
Bard A.J. & Faulkner. (1980) *Electrochemical Methods.* John Wiley, New York.
Barker J.E., Mott R.A. & Thomas W.C. (1955) *Fuel,* **34,** 283.
Barker P.G. (1983) *Computers in Analytical Chemistry.* Pergamon Press, Oxford.
Barnett V. (1974) *Elements of Sampling Theory.* Hodder and Stroughton, London.
Barrow N.J. & Shaw T.C. (1980). *Common Soil. Sci. Plant Anal.* **11,** 347.
Bartlett R. & James B. (1980) *Soil Sci. Soc. Am. J.,* **44,** 721.
Bascomb C.L. (1961) *Chem. Ind.* **45,** 1826.
Bascomb C.L. (1964) *J. Sci. Food Agric.* **15,** 821.
Bascomb C.L. (1968) *J. Soil Sci.* **19,** 251.
Bascomb C.L. & Thanigasalam K. (1978) *J. Soil Sci.* **29,** 382.
Basson W.D. Böhmer R.G. & Stanton D.A. (1969) *Analyst (London).* **94,** 1135.
Basson A.T. & Kempster P.L. (1980) *Water S.A.* **6,** 88.
Bate-Smith E.C. (1948) *Nature.* **161,** 835.
Bates R.G. (1964) *Determination of pH: Theory and Practice.* John Wiley, New York.
Bath I.H. (1960) *J. Sci. Food Agric.* **11,** 560.
Baumann E.W. (1974) *Anal. Chem.* **46,** 1345.
Baur J.R., Baker R.D. & Davis F.S. (1971) *J. Assoc. Off. Anal. Chem.* **4,** 5713.
Bear F.E. (1964) *Chemistry of the Soil* (A.C.S. Monograph no. 160), Reinhold.
Beaton J.D., Burns G.P. & Platov J. (1968) *Determination of Sulphur in Soils and Plant material.* Sulphur Inst. Tech. Bulletin no. 14.
Beaty R.D. (1973) *Anal. Chem.* **45,** 234.
Beckett P.H.T. & Tinker P.B. (1964) *J. Soil Sci.* **15,** 1.
Beckman H. & Bevenue A. (1963) *J. Chromatogr.* **10,** 231.
Belcher R., Close R.A. & West T.S. (1958) *Talanta.* **1,** 238.
Benham B.G. & Mellanby K. (1978) *Weather,* **33,** 351.
Bentley E.M. & Lee G.F. (1967) *Environ. Sci. Technol.,* **1,** 721.
Benville P.E. & Tindle R.C. (1970) *J. Agric. Food Chem.* **18,** 948.
Bergseth H. & Kristiansen J. (1978) *Acta Agric. Scand.* **28,** 404.
Beringer H. (1985) *Plant Soil.* **83,** 21.
Berry J.W. (1966) *Advances in Chromatagraphy* Vol 2, (Ed. by J.G. Giddings and R.A. Keller). Arnold, London.
Bertin E.P. (1978) *Introduction to X-ray Spectrometric Analysis.* Plenum Press, New York.
Bertoni G. & Morard P. (1982) *Commun. Soil. Sci. Plant Anal.* **13,** 539.
Best E.K. & Cranwell E.T. (1985) *Commun. Soil Sci. Plant Anal.* **16,** 1189.
Bevan E.J. & Cross C.F. (1880) *J. Chem. Soc.* **38,** 666.
Bevenue A. (1967) in *Analytical Methods for Pesticides, Plant Growth Regulators and Food Additives,* Vol. 5. (Ed. by G. Zweig and J. Sherma). Academic Press New York.
Bevenue A. & Ogata J.N. (1970) *J. Chromatogr.* **50,** 142.
Bevenue A., Kelley T.W. & Hylin J.W. (1971) *J. Chromatogr.* **54,** 71.
Bevenue A. & Washauer B. (1950) *J. Assoc. Off. Agric. Chem.* **33,** 986.
Bieleski R.L. & Ferguson I.D. (1983) in *Inorganic Plant Nutrition. Part A.* (Ed. by A. Lanchli and R.L.Bieleski). Springer, Berlin.
Bilham E.G. (1932) *Met. Mag. (London)* **67,** 86.
Binder H. (1980) *J. Chromatogr.* **189,** 414.
Birchfield H.P., Johnson D.E., Rhoades J.W. & Wheeler R.J. (1965) *J. Gas Chromatogr.* **3,** 28.
Birchfield H.P., Johnson D.E. & Storrs E.E. (1965) *Guide to the Analysis of Pesticide Residues,* Vols. 1 and 2. U.S. Govt. Printing Office.
Birnie S.E. & Hodges D.J. (1981) *Environ. Technol. Lett.* **2,** 433.
Bjorkman A. (1956) *Svensk Papperstidn.* **59,** 477.
Black C.A. (Ed.) (1965) *Methods of Soil Analysis,* Part 1. American Society of Agronomy.
Black C.A. & Goring C.A.I. (1953) *Agronomy.* **4,** 123.
Bligh E.G. & Dyer W.J. (1959) *Can. J. Biochem. Physiol.* **37,** 911.
Block R.J. & Bolling D. (1951) *The Amino Acid Composition of Proteins and Foods.* Thomas, Springfield, Ill.
Blondeau R. (1986) *Agrochimica.* **30,** 128.
Blount C.W., Leyder D.E., Thomas T.L. & Guill S.M. (1973) *Anal. Chem.* **45,** 1045.
Boar P.L. & Ingram L.K. (1970) *Analyst (London).* **95,** 124.
Bodewig F.G., Valenta P. & Neuerenberg H.W. (1982) *Z. Anal. Chem.* **311,** 187.
Bohn H.L. (1971) *Soil Sci.* **112,** 39.
Bokelman G.H., Ryan W.S. & Oakley E.T. (1983) *J. Agric. Food Chem.* **31,** 897.
Bolton J., Brown G., Pruden G. & Williams C. (1973) *J. Sci. Food Agric.* **24,** 557.
Boltz D.F. & Howell J.A. (1978) *Colorimetric Determination of Non-metals,* Vol. 8. John Wiley, New York.
Bolnja-Santos C., Gonzales-Portal A., & Bermejo-Martinez F. (1984) *Analyst (London)* **109,** 797.
Bond A.M. (1980) *Determination of pH.* Wiley, New York.

Boniface M., Boniface B., Erb F. & Hanquez N. (1975) *Anal. Abstr.* **29,** 6H27.

Bould C., Bradfield E.G. & Clarke G.M. (1960) *J. Sci. Food Agric.* **11,** 229.

Boumans P.W.J.M. (Ed.) (1987) *Inductively Coupled Plasma Emission Spectroscopy.* Part 1, *Methodology, Instrumentation and Performance*; Part 2, *Applications and Fundamentals.* Wiley-Interscience, New York.

Bouyoucos G.J. (1926) *Science.* **64,** 362.

Bouyoucos G.J. (1938) *Soil Sci.* **46,** 107.

Bowden J., Brown G. & Stride T. (1979) *Ecol. Entomology.* **4,** 199.

Bowditch D.C., Edmund C.R., Dunstein P.J. & McGlynn J.A. (1976) *Austr. Water Resource Counc. Tech. Paper* No. 16.

Bowen H.J.M. (1966) *Trace Elements in Biochemistry.* Academic Press, London.

Bowen H.J.M. (1979) *Environmental Chemistry of the Elements.* Academic Press, London.

Bowen G.D. & Nambiar E.K.S. (1984) *Nutrition of Plantation Forests.* Academic Press, London.

Bower C.A., Reitemeier R.F. & Fireman M. (1952) *Soil Sci.* **73,** 251.

Box J.D. (1983) *Water Res.* **17,** 511.

Bradfield E.G. (1957) *Analyst (London).* **82,** 254.

Bradfield E.G. & Bould C. (1963) *J. Sci. Food Agric.* **14,** 729.

Bradshaw A.D. & Chadwick M.J. (1980) *Studies in Ecology* No. 6: *Restoration of the Land.* Blackwell Scientific Publications, Oxford.

Bradstreet R.B. (1965) *The Kjeldahl Method for Organic Nitrogen.* Academic Press, London.

Braman R.S. & Tompkins M.A. (1978) *Anal. Chem.* **50,** 1088.

Brar S.P.S. & Bishnoi S.R. (1987) *Analyst (London).* **112,** 917.

Brauns F.E. (1939) *J. Am. Chem. Soc.* **61,** 2120.

Brauns F.E. (1952) *The Chemistry of Lignin.* Academic Press, London.

Brauns F.E. & Brauns D.A. (1960) *The Chemistry of Lignin*-Supplement Volume. Academic Press, London.

Bray J.T., Bricker O.P. & Troup B.N. (1973) *Science.* **180,** 1362.

Bray R.H. & Kurtz L.T. (1945) *Soil Sci.* **59,** 39.

Bray R.H. & Willhite F.M. (1929) *Ind. Eng. Chem. Anal. Edn.* **1,** 144.

Bremner J.M. (1949a) *J. Agric. Sci., Camb.* **39,** 183.

Bremner J.M. (1949b) *J. Agric. Sci., Camb.* **39,** 280.

Bremner J.M. (1960) *J. Agric. Sci., Camb.* **55,** 1.

Bremner J.M. & Jenkinson D.S. (1960) *J. Soil Sci.* **11,** 403.

Bremner J.M. & Lees H. (1949) *J. Agric. Sci., Camb.* **39,** 247.

Bremner J.M. & Shaw K. (1958) *J. Agric. Sci., Camb.* **51,** 22.

Bremner J.M. & Tabatabai M.A. (1971) in *Instrumental Methods for Analysis of Soils and Plant Tissue.* (Ed. by L.M. Walsh). *Soil Sci. Soc. Amer.*

Brenner M. & Niederwieser A. (1960) *Experientia.* **16,** 378.

British Standards. (1972) *BS 1747*, Part 5.

British Standards Institute. (1980) *BS 5666.* Part 7.

Bridson J.N. (1985) *Soil Biol. Biochem.* **17,** 285.

Brobst K.M. & Scobell H.D. (1982) *Starch-Starke.* **32,** 243.

Brody S.S. & Chaney J.E. (1966) *J. Gas Chromatogr.* **4,** 42.

Brooks R.R., Presley B.J. & Kaplan I.R. (1967) *Talanta.* **14,** 809.

Brooks R.R., Reeves R.D., Morrison R.S. & Malaisse F. (1980) *Bull. Soc. R. Bot. Belg.* **113,** 166.

Brouko I.A., Tursunov A., Rish M.A. & Davison A.D. (1985) *Anal. Abstr.* **47,** 11D51.

Brown A.A. & Morton S.F.N. (1985) *Lab. Sci. Technol.* **1,** 24.

Brown G. & Kanaris-Sotiriou R. (1969) *Analyst (London).* **94,** 782.

Brown I.C. (1943) *Soil Sci.* **56,** 353.

Brown J., Lewis M., Hargrave B.T. & MacKinnon M.D. (1981) *Can. J. Fish Aquat. Sci.* **38,** 205.

Brown W.L., Young M.K. & Seraile L.G. (1957) *J. Lab. Clin. Med.* **49,** 630.

Brownell P.F. (1979) *Adv. Bot. Res.* **7,** 117.

Browning B.L. (1967) *Methods of Wood Chemistry*, Vols. I & II. Wiley-Interscience, New York.

Buckman H.O. & Brady N.C. (1969) *The Nature and Properties of Soils.* Macmillan, New York.

Bunton N.G., Crosby N.T. & Patterson S.J. (1969) *Analyst (London).* **94,** 585.

Burgett C.A., Smith D.H. & Beute H.B. (1977) *J. Chromatogr.* **134,** 57.

Burney C.M. & Sieburth J. McN. (1977) *Mar. Chem.* **5,** 15.

Busnan L.M., Dick R.P. & Tabatabai M.A. (1983) *Soil Sci. Soc. Amer. J.* **47,** 1167.

Butler L.R.P. & Mathews P.M. (1966) *Anal. Chim. Acta.* **36,** 79.

Butters B. & Chenery E.M. (1959) *Analyst (London).* **54,** 507.

Byast T.H. Cotterill E.G. & Hance R.J. (1977) *Technical Report* No. 15. ARC, Weed Research Organisation.

Cameron D.R. (1971) *Can. J. Soil Sci.* **51,** 165.

Campbell J.A. & Bewick M.W.M. (1978) *Spec. Publ. Commonw. Bur. Soils.* No. 7. 36.

Cartwright B., Tiller U.G., Zarcinas B.A. & Spencer L.R. (1983) *Austr. J. Soil Res.* **21,** 321.

Carey E.E., Grunes E.L., Bohman V.R. & Sanchirico C.A. (1986) *Agron. J.* **78,** 933.

Case F.H. (1951) *J. Org. Chem.* **16,** 1541.

Caulcutt R. & Boddy R. (1983) *Statistics for Analytical Chemists.* Chapman and Hall, London.

Caverly D.J. & Denney R.C. (1978) *Analyst (London).* **103,** 368.

Chabra R., Pleysier J.L. & Cremers A. (1976) *Proc. Int. Clay Conference*, Mexico City 1975 (Ed. by S.W. Bailey). Applied Publishing.

Champion K.P., Taylor J.C. & Whittem P.N. (1966) *Anal. Chem.* **38,** 109.

Chang S.C. & Jackson M.L. (1957) *Proc. Soil Sci. Soc. Amer.* **21,** 265.

Chapman H.D., Axley J.H. & Curtis D.S. (1941) *Proc. Soil. Sci. Soc. Amer.* **5,** 191.

Chapman H.D. & Pratt P.F. (1961) *Methods of Analysis for Soils, Plants, and Waters.* University of California.
Chapman M. & Mahon B. (1986) *Plain Figures.* HMSO, London.
Charles M.J. & Simmons M.S. (1986) *Analyst (London).* **111,** 385.
Cheam V. & Chau A.S.Y. (1987) *Analyst (London).* **112,** 993.
Cheeseman R.V. & Nicholson N.J. (1968). Inter-Laboratory Testing of Analytical Methods-I, Calcium and magnesium. *Water Research Association Tech.,* paper 65.
Cheng B.T. (1982). *J. Plant Nutrition.* **5,** 1345.
Cheng Y. & Barak P. (1982) *Adv. in Agron.* **35,** 217.
Cheremisinoff P.N. & Morresi A.C. (1978) *Air Pollution Sampling and Analysis Deskbook.* Ann Arbor Sc. Publ. Michigan.
Cherney J.H. & Robinson D.L. (1983) *Agron. J.* **75,** 145.
Cheshire A.V., Goodman B.A., & Mundic J. (1975) *Report Welsh Soils Discussion Group* **16,** 73.
Chesin L. & Yien C.H. (1950) *Proc. Soil Sci. Soc. Am.* **15,** 149.
Chesters G., Konrad J.G., Schrag B.D. & Everett L. (1971) in *Instrumental Methods for Analysis of Soils and Plant Tissues.* (Ed. by L.M. Walsh). *Soil Sci. Soc. Amer.*
Chian E.S.K. & DeWalle F.B. (1978) *J. Water Pollut. Control Fed.* **50,** 1026.
Chion K.Y. & Manuel O.K. (1984) *Anal. Chem.* **56,** 2721.
Christian G.D. & Feldman F.J. (1970) *Atomic Absorption Spectroscopy: Applications in Agriculture, Biology and Medicine.* Wiley-Interscience, New York.
Christian G.D., Knoblock E.C. & Purdy W.C. (1965) *J. Assoc. Off. Agric. Chem.* **48,** 877.
Christopher D.H. & West T.S. (1966) *Talanta.* **13,** 507.
Cieslak M.E. (1983) *J. Am. Soc. brew. Chem.* **41,** 10.
Citron I., Tai H., Day, R.A. & Underwood, A.L. (1961) *Talanta.* **8,** 798.
Claisse F. & Quintin M. (1967) *Can. Spectrosc.* **12,** 129.
Clarke G.R. (1975) *The Study of Soil in the Field.* Oxford University Press.
Classen A. & Bastings L. (1966) *Analyst (London).* **91,** 725.
Clement A. & Loubinoux B. (1983) *J. Liq. Chromatogr.* **6,** 1705.
Coakley J.E., Campbell J.E. & McFarren E.F. (1964) *J. Agric. Food Chem.* **12,** 262.
Coakley W.A. (1981) *Handbook of Automated Analysis: Continuous Flow Techniques.* Dekker, New York.
Cochrane W.P. (1979) *J. Chromatogr. Sci.* **17,** 124.
Cochran W. & Cox D. (1957) *Experimental Design.* John Wiley, New York.
Codell M. (1959) *Analytical Chemistry of Titanium Metals and Compounds.* Wiley-Interscience, New York.
Cohen H.R. (1943) *Arch. Biochem.* **2,** 1.
Coleman R.F. (1974) *Anal. Chem.* **46,** 989A, 994 A.
Collins G.B., Holmes D.C. & Jackson F.J. (1972) *J. Chromatogr.* **71,** 443.
Collings G.F., Yokoyuma M.T. & Bergen W.G. (1978) *J. Dairy Sci.* **61,** 1156.
Colin H., Krstulovic A.M., Excoffier J.L. & Guiochon, G. (1984–85) *A Guide to the HPLC Literature,* Vols 1–3. John Wiley, New York.
Colovas N.F., Keener H.A. & Davis H.A. (1957) *J. Dairy Sci.* **40,** 173.
Colowick S.P. & Kaplan N.O. (1957) *Methods in Enzymology.* Vol III. Academic Press, London.
Comar C.L. & Zscheille F.P. (1942) *Plant Physiol.* **17,** 198.
Comber N.M. (1961) *An Introduction to the Scientific Study of the Soil* (revised by W.W. Townsend.) Arnold, London.
Commonwealth Bureau of Soils. (1975) *Mechanical Analysis of Soil* (1974-65), Suppl. No. 983.
Commonwealth Bureau of Soils (1978) *Determination of Experimental Control of Soil Redox Potential* (1959–1976).
Commonwealth Bureau of Soils. (1978) *Automated Methods for Chemical Analysis of Soils and Plants.* (1971 to 1977). Report SA 1951.
Conroy L.E., Maier W.J. & Shih, Y.T. (1981) in *Chemistry in Water Re-use,* Vol 1. (Ed. by W.J. Cooper). Ann. Arbor. Sci. Publ., Michigan.
Consden R., Gordon A.H. & Martin A.J.P. (1944) *Biochem. J.* **38,** 224.
Conway E.J. (1962) *Microdiffusion Analysis and Volumetric Error.* Crosby Lockwood, London.
Cooke G.W. (1951) *J. Soil Sci.* **2,** 254.
Cooke J.A., Johnson M.S., Davison A.W. & Bradshaw A.D. (1976) *Environ. Pollut.* **11,** 257.
Cookson G.A. (1987) *Int. Lab.* **17,** 86.
Cooper M.J. & Anders M.W. (1975) *J. Chromatogr. Sci.* **13,** 407.
Cope F. & Trickett E.S. (1965) *Soils Fertil.* **28,** 201.
Corbett J.A. & Godbeer W.C. (1977) *Anal. Chim. Acta.* **91,** 211.
Cosovic B. (1985) in *Chemical Processes in Lakes.* (Ed. by W. Stram) John Wiley, New York.
Cotteril E.G. & Byast T.H. (1984) in *Liquid Chromatography on Environmental Analysis,* (Ed. by J.F. Lawrence). Humana Press, Clifton, New Jersey.
Coulter B.S. (1969) *Soils Fertil.* **32,** 215.
Covington A.K., Whalley, P.D. & Davison, W. (1983) *Analyst (London).* **108,** 1528.
Cowgill V.M. (1968) *Appl. Spectrosc.* **22,** 415.
Cowling H. & Miller E.J. (1941) *Ind. Eng. Chem., Anal. Edn.* **13,** 145.
Cox D.R. (1958) *Planning of Experiments.* John Wiley, New York.
Coyne R.V. & Collins J.A. (1972) *Anal. Chem.* **44,** 1093.
Craggs A., Moody G.J. & Thomas J.D.R. (1979) *Analyst (London).* **104,** 412.
Crane L. & Dewey D.J. (1980) *Tech. Rep. 127,* Water Res, Cent., Medmenham. UK.
Cranwell P.A. (1978) *Geochim Cosmochim. Acta.* **42,** 1523.
Creaser C.S. & Fernandes A.S. (1987) *Anal. Proc. (London).* **24,** 41.
Cresser M.S. & Parsons, J.W. (1979) *Anal. Chim Acta.* **109,** 431.
Croll B.T. (1971) *Chem. Ind.* 789.

Crooke W.M. (1964) *Plant Soil.* **21,** 43.
Curtis K.E. (1969) *Analyst* (*London*). **94,** 1068.

Dagnali R.M., West T.S. & Young P. (1965) *Talanta.* **12,** 583.
Daigle D.J. & Cankerton E.J. (1983) *J. Liq. Chromatogr.* **6,** 105.
Dalal R.C. (1977) *Adv. Agron.* **29,** 83
Daniel R.C. (1980) *Mitt. Geb. Lebensmittelunters Hyd.* **71,** 242.
Darbre A. & Blau K. (1965) *Biochem. biophys. Acta.* **100,** 298.
Darbre A. & Islam A. (1968) *Biochem. J.* **106,** 923.
David D.J. (1962) *At. Abs. Newsletter.* Dec.
David D.J. (1978) *Prog. Anal. At. Spectrosc.* **1,** 225.
Davies O.L. (Ed.) (1956) *Design and Analysis of Industrial Experiments.* Oliver and Boyd, Edinburgh.
Davies O.L. & Goldsmith P.L. (1972) *Standard Methods in Research and Production.* Longmans, London.
Davies R.I. Coulson C.B. & Lewis D.A. (1960) *Scient. Proc. R. Dub. Soc., Ser.* **A1,** 183
Davison W. (1978) *J. Electroanal. Chem.* **87,** 395.
Davison W. (1987) in *Chemical Analysis in Environmental Research.* (Ed. by A.P. Rowland). Inst. Terr. Ecology. UK.
Davison W. & Tipping E. (1984) *Freshw. Biol. Assoc. Ann. Rep 1984..*
Davison W. & Whitfield M. (1977) *J. Electroanal. Chem.* **75,** 763.
Davison W. & Woof, C. (1985) *Anal. Chem.* **57,** 2567.
Dean J.A. (1960) *Flame Photometry.* McGraw-Hill, New York.
Deb B.C. (1950) *J. Soil Sci.* **1,** 212
Decock P.C. (1981) *J. Plant Nutrition* **3,** 513.
De Faubert Maunder M.J., Egan H., Cudley E.W., Hammond J. & Thompson, J. (1964) *Analyst* (*London*). **89,** 168.
De Galen L., Van Dalen H.P.J. & Kornblum G.R. (1985) *Analyst* (*London*). **110,** 323.
De los Angeles B.M. (1982) *J. Chromatogr.* **238,** 175.
Demeter J. & Heyndrickx A. (1979) *Vet. Hum. Toxicol.* **21,** 151.
Dempster J.P., Lakhani K.H. & Coward, P.A. (1986) *Ecol. Entomol.* **11,** 51.
Den Tonkelaar W.A.M. & Bergshoeff G. (1969) *Water Resources.* **3,** 31.
Department of the Environment (1972) *Analysis of Raw, Potable and Waste Waters.* HMSO, London.
Department of the Environment (1983) *United Kingdom Review Group on Acid Rain.* HMSO, London.
Deriaz R.E. (1961) *J. Sci. Food Agric.* **12,** 152.
De Stefano J., Goldberg A.P., Larmann J.P. & Parris N.A. (1980) *Ind. Res/Dev.* **22,** 99.
Dewis J. & Freitas F. (1970) *Physical and Chemical Methods of Soil and Water Analysis.* Soils Bulletin No. 10. FAO.
Dible W.T., Truog E. & Berger K.C. (1954) *Anal. Chem.* **26,** 418.
Dickson W. (1983) in *Ecological Effects of Acid Deposition.* Rep. PM 1636. National Swedish. Environ. Protection Board.
Diehl H. & Smith G.F. (1960) *Proc. Int. Sym. Microchem.* Birmingam, 1958. Pergamon Press, Oxford.
Dinnin J.I. (1960) *Anal. Chem.* **32,** 1475.
Dolling P.J. & Ritchie G.S.P. (1985). *J. Austr. Soil Res.* **23,** 309.
Dolmat M.T., Patrick W.H. & Peterson F.J. (1980) *Soil Sci.* **129,** 229.
Doran J.W. & Mielke L.N. (1984) *Soil Sci. Soc. Amer. Proc.* **48,** 717.
Doree C. (1947) *The Methods of Cellulose Chemistry.* Chapman and Hall, London.
Doshi G.R. (1969) *Curr. Sci.* **38,** 206
Dougan W.K. & Wilson A.L. (1974) *Analyst* (*London*). **99,** 413.
Doutre P.A., Hay G.W., Hood A. & Van Loon G.W. (1978). *Soil Biol. Biochem.* **10,** 475.
Draper N. & Smith H. (1981) *Applied Regression Analysis.* John Wiley, New York.
Dreywood R. (1946) *Ind. Eng. Chem., Anal. Edn.* **18,** 499.
Driscoll C.T. (1984) *Inť J. Environ. Anal. Chem.* **16,** 267.
Dubach P. & Mehta N.C. (1963) *Soil Fertil.* **26,** 293.
Dubois M., Gillies K.A., Hamilton J.K., Rebers, P.A. & Smith, F. (1956) *Anal. Chem.* **28,** 350.
Ducret, L. (1957) *Anal. Chim. Acta.* **17,** 213.
Durst, R.A. (1969) *Ion Selective Electrodes.* NBS Spec. Publ. No. 314, Washington D.C.
Durst, R.A. (1971). *Amer. Sci.* **59,** 353.
Dutton H.J. (1983) *AOAC Monogr.* **10,** 209.
Dyer, B. (1894) *Trans. Chem. Soc.* **65,** 115.
Dzubay, T.G. (Ed.) (1977) *X-Ray Fluorescence Analysis of Environmental Materials.* Ann Arbor Sc. Publ. Michigan.

Eagle D.J., Jones, J.L.O. & Jewell E.J. (1983) in *Soil Analysis, Instrumental Techniques and Related Procedures.* (Ed. by K.A. Smith). Dekker, New York.
Eaton A., Oelker G. & Leong L. (1982) *At. Spectrosc.* **3,** 152.
Edlin V.M., Karamonos R.E. & Halstead E.H. (1983) *Commun. Soil Sci. Plant Anal.* **14,** 1167.
Edmeades, D.C. & Clinton, D.E. (1981) *Commun. Soil Sci. Plant Anal.* **12,** 683.
Edwards C.E. (1974) *Persistent Pesticides in the Environment.* CRC Press, Cleveland.
Edwards P.J. (1965) *S. Afr. J. Agric. Sci.* **8,** 337.
Egger K. (1961) *Z. anal. Chem.* **182,** 161.
Elder J.F., Perry S.K. & Brody P.F. (1975) *Environ. Sci. Technol.* **9,** 1039.
Elliot J.M. (1971) Publication No. 25. Freshwater Biological Association, Ambleside, U.K.
Ellis R. & Olson R.V. (1950) *Anal. Chem.,* **22,** 328.
Erdahl W.L., Stolyhwo A. & Privett O.S. (1973) *J. Amer. Oil. Chem. Soc.* **50,** 513.
Evans C.C. (1970) *Analyst* (*London*). **95,** 919.
Evans C.C. & Grimshaw H.M. (1968) *Talanta.* **15,** 413.

Evans C.M. & Sparks, D.L. (1983) *Commun. Soil. Sci. Plant Anal.* **14,** 827.

Fairing J.D. & Warrington H.B. (1950) *Adv. chem. Ser.* **1,** 260.

Faithfull N.T. (1971) *Lab. Pract.* **20,** 41.

Farrar K. (1975) *ADAS Quarterly Rev.* **19,** 93.

Fassel V.A. & Becker D.A. (1969) *Anal. Chem.* **41,** 1522.

Faust S.D. & Suffet I.H. (1972) in *Water and Water Pollution Handbook,* Vol 3. (Ed by L.L. Ciaccio). Dekker, New York.

Fehringer N.V. & Westfall J.E. (1971) *J. Chromatogr.* **57,** 397.

Feldman C.R. (1974). *Instrumental Analysis of Chemical Pollutants.,* Rept. No. EPA/430/1-74-001. Environmental Protection Agency.

Feldman F.J. (1970) *Anal. Chem.* **42,** 719.

Feldman F.J. & Purdy W.C. (1965) *Anal. Chim. Acta.* **33,** 273.

Fieldes M. & Furkert R.J. (1971) *N.Z.J. Sci.* **14,** 280.

Filby R.H. & Shah K.R. (1974) *Toxicol Environ. Chem. Rev.* **2,** 1.

Finocchiaro J.M. & Benson W.R. (1967) *J. Assoc. Off. Anal. Chem.* **50,** 888.

Fishbein L. (1984) *Int. J. Environ. Anal. Chem.* **17,** 113.

Fishwick M.J. & Wright A.J. (1977) *Phytochemistry.* **16,** 1507.

Flannery R.L. & Markus D.K. (1980) *J. Assoc. Off. Anal. Chem.* **63,** 779.

Flatt W.P. (1957) *J. Dairy Sci.* **40,** 612.

Flowers T.J., Troke P.F. & Yeo A.R. (1977) *Ann. Rev. Plant Physiol.* **28,** 99.

Fogg A.G., Jillings J., Marriott, D.R. & Thorburn Burns D. (1969) *Analyst (London)* **94,** 768.

Folin O. & Denis W. (1912) *J. Biol. Chem.* **12,** 239.

Food & Drug Administration (1985) *Pesticide Analytical Manual.* US Dept of Health, Education and Welfare Washington DC.

Forest J., Tanner R.L., Spandau D. & Newman L. (1982) *Atmos. Environ.* **16,** 1473.

Forster W. & Zeitlin H. (1966) *Anal. Chem.* **38,** 649.

Fox J.G.M. (1963) *J. Inst. Metals.* **91,** 239.

Fox R.H. & Piekielek W.P. (1978) *Soil Sci. Soc. Amer. J.* **42,** 751.

Foy C.D., Chaney R.L. & White M.C. (1978) *Ann. Rev. Plant Phys.* **29,** 511.

Framstad E., Engen S. & Stenseth N.C. (1985) *Oikos.* **44,** 318.

Frank G. & Strubert W. (1973) *Chromatographic.* **6,** 522.

Frankland J.C., Ovington J.D. & Macrae C. (1963) *J. Ecol.* **51,** 97.

Freer–Smith P.H. (1985) *New Phytol.* **99,** 417.

Frei R.W., Lawrence J.F. & Legay D.S. (1973) *Analyst (London).* **98,** 9.

French M.C. & Jeffries D.J. (1971) *Bull. Environ. Contam. Toxicol.* **6,** 460.

Fried M. & Broeshart H. (1967) *The Soil-Plant System in Relation to Inorganic Nutrition.* Academic Press, London.

Fritz J.S., Cyerde D.T. & Becker R.M. (1980) *Anal. Chem.* **52,** 1519.

Frye W.W. & Hutcheson T.B. (1981) *Soil Sci-Soc. Am. J.* **45,** 889.

Fuchs W. & Kohler E. (1957) *Milt. Ver. Grosskesselbets.* **47,** 107.

Fukuzaki N., Suzuki T., Sugai R. & Oshima T. (1979) *Anal. Abstr.* **37,** 4B 131.

Fuller C.W. (1977) *Electrothermal Atomisation for Atomic Absorption Spectrometry.* The Chemical Society, London.

Furman W.B. (1976) *Continuous Flow Analysis: Theory and Practice.* Dekker, New York.

Furman W.B. & Fehringer M.V. (1967) *J. Assoc. Off. Anal. Chem.* **50,** 903.

Gaines T.P. & Gascho G.J. (1985) *J. Sci. Food Agric.* **36,** 157.

Gaines T.P. & Mitchell G.A. (1979) *Commun. Soil Sci. Plant Anal.* **10,** 1099.

Gale M.E., Kaylor W.H. & Longbottom J.E. (1968) *Analyst (London).* **93,** 97.

Galen L. de, Dolen H.P.J. & Kornblum G.R. (1985) *Analyst (London)* **110,** 323.

Galensa R. & Herrmann K. (1980) *J. Chromatogr.* **189,** 217.

Gallagher P.H. & Walsh T. (1943) *Proc. R. Ir. Acad.* **49B,** 1.

Games, D.E. (1979) *Mass Spectrom.* **5,** 285.

Garton G.A. & Duncan W.R.H. (1957) *Biochem. J.* **67,** 340.

Garton G.A., Lough A.K. & Vioque E. (1961) *J. Gen. Microbiol.* **25,** 215.

Gawen D. (1965) *Lab. Pract.* **14,** 1397.

Geary P.J. (1956) *Determination of Moisture in Solids.* B.S.I.R.A. Research Report M24.

Geissman T.A. (1962) *The Chemistry of Flavonoid Compounds.* Oxford, Pergamon Press.

Gestring W.D. & Soltanpur P.N. (1981) *Commun. Soil Sci. Plant Anal.* **12,** 743.

Giddings J.C. & Keller R.A. (Eds.) (1970) *Advances in Chromatography.* **IX.** Dekker, New York.

Giger S. (1969). *Focus* (July) 6.

Gilliam F.C. & Richter D.D. (1985) *Soil Sci. Am. J.* **49,** 1576.

Gillman G.P., Bruce R.C., Davey B.G., Kimble J.M., Searle P.L., & Skjemstid, J.O. (1983) *Commun. Soil Sci Plant Anal.,* 14, 1005.

Giron H.C. (1973) *At. Abs. Newsletter.* **12,** 28.

Giuffrida L. (1964) *J. Assoc. Off. Agric. Chem.* **47,** 293.

Gleit C.E. & Holland W.D. (1962) *Anal. Chem..* **34,** 1454.

Glowa W. (1974) *J. Assoc. Off. Anal. Chem.* **57,** 1228.

Godden R.G. & Thomson D.R. (1980) *Analyst, (London).* **105,** 1137.

Godfrey–Sam–Aggrey W. & Garber M.J. (1979) *Commun. Soil Sci. Plant Anal.* **10,** 1079.

Goerlitz D.F. & Brown E. (1972) *Methods for the Analysis of Organic Substances in Water.* Techniques of Water Resources. Investigations of the US Geological Survey Book 5, Chapter A3. U.S. Department of the Interior.

Goldin, A. (1987) *Commun. Soil Sci. Plant. Anal.* **18,** 1111.
Golterman H.L., Clymo R.S. & Ohnstad M.A.M. (1978) *Methods of Physical and Chemical Analysis of Freshwater.* Blackwell, Oxford.
Goodwin T.W. (1965) *Chemistry and Biochemistry of Plant Pigments.* Academic Press, London.
Gorbach S. (1980) *Pure Appl. Chem.* **52,** 2569.
Gore A.J.P. (1968) *J. Ecol.* **56,** 483.
Gorsuch T.T. (1959) *Analyst (London).* **84,** 135.
Gorsuch T.T. (1962) *Analyst (London).* **87,** 112.
Gorsuch T.T. (1970) *The Destruction of Organic Matter.* Pergamon Press, Oxford.
Goss M.J. & Phillips M. (1936) *J. Assoc. Off. Agric. Chem.* **19,** 341.
Götz W., Sachs A. & Wimmer H. (1980) *Thin Layer Chromatography.* John Wiley, New York.
Goulden P.D., Anthony D.H.J. & Austen K.D. (1981) *Anal. Chem.* **53,** 2027.
Gran G. (1952) *Analyst (London).* **77,** 661.
Grassland Research Institute (1961) *Research Techniques in Use at the Grassland Research Institute,* Hurley, Berks., England. Bulletin 45, Pasture and Field Crops. Commonwealth Agricultural Bureau.
Grava J., Spalding G.E. & Caldwell A.C. (1961) *Agronomy J.* **53,** 219.
Greeley R.H. (1974) *J. Chromatogr.* **88,** 229.
Greenberg A. (1985). See under *Standard Methods for the Examination of Water and Wastewater.*
Greenhill W.L. (1960) *J. Br. Grassld. Soc.* **15,** 48.
Greenland D.J. & Hayes M.H.B. (1981) *The Chemistry of Soil Processes.* John Wiley, New York.
Greve P. & Goewie C.E. (1985) *Int. J. Environ. Anal. Chem.* **20,** 29.
Griffiths E.J., Beeton A., Spence J.M. & Mitchell D.T. (Eds.) (1973) *Environmental Phosphorus Handbook.* John Wiley, New York.
Grigg J.L., Flewitt H.J., Baird G.A., Jordan R.B. & Vo K.B. (1980) *Analyst (London).* **105,** 1.
Grigg J.L. & Morrison J.D. (1982). *Commun. Soil Sci. Plant Anal.* **13,** 351.
Grigg J.P. (1953) *N.Z.J. Sci. Technol.* **34A,** 405.
Grimaldi F.S. & Wells R.C. (1943) *Ind. Eng. Chem. Anal Edn.* **15,** 315.
Grimshaw H.M. & Allen S.E. (1987) *Vegetatio.* **70,** 157.
Grinstead R.R. & Snider S. (1967) *Analyst (London).* **92,** 532.
Grunden N.J. & Asher C.J. (1981) *Commun. Soil. Sci. Plant Anal.* **12,** 1181.
Guha M.M. & Mitchell R.L. (1965) *Plant Soil.* **23,** 323.
Guha M.M. & Mitchell R.L. (1966) *Plant Soil.* **34,** 90.
Guicherit R., Jettes, R. & Lindqvist, F. (1973) *Environ. Pollut.* **3,** 91.
Gunstone F.D. (1977) *Progr. Chem. Fats Other Lipids.* **15,** 75.
Gunther, H. & Schweiger A. (1968) *J. Chromatogr.* **34,** 498.
Gupta P.L. & Rorison I.H. (1974) *J. Appl. Ecol.* **11,** 1197.
Gustafsson K.H. & Thompson R.A. (1981) *J. Agric. Food Chem.* **29,** 729.
Gustavsson I. & Hansson L. (1984) *Int. J. Environ. Anal. Chem.* **17,** 57.
Guyon J.C. & Shults W.D. (1969) *J. Am. Water Wks. Assoc.* **61,** 403.
Gyorgy B., Ardre L., Stehli L. & Pungar E. (1969) *Anal. Chim. Acta.* **46,** 318.

Hach C.C., Brayton S.V. & Kopelove A.B. (1985) *J. Sci. Food Chem.* **33,** 1117.
Hagenmaier H., Kant H. & Krauss P. (1986) *Int. J. Environ. Anal. Chem.* **23,** 331.
Haines S.G., Haines C.W. & White G. (1979) *Forensic Sci.* **25,** 154.
Hall R.C. & Harris D.E. (1979) *J. Chromatogr.* **169,** 245.
Hall R.J. (1968) *Analyst (London).* **93,** 461.
Hall R.J. & Gupta P.L. (1969) *Analyst (London).* **94,** 292.
Halliday C.G. & Leonard M.A. (1987) *Analyst (London)* **112,** 83.
Hamilton E.I., Minski M.J. & Cleary J.J. (1972) *Sci. Total Environ.* **1,** 1.
Hamilton J.G. & Comai K. (1984) *J. Lipid Res.* **25,** 1142.
Hammes J.K. & Berger K.C. (1960) *Proc. Soil Sci. Soc. Am.* **24,** 361.
Handbook of Chemistry and Physics (1982) 63 Edn. (Ed in chief R.C. Weast) CRC Press, Florida.
Hanley K. (1962) *Irish J. Agric. Res.* **1,** 192.
Hanson J.B. (1984) *Adv. Plant Nutrition.* **1,** 149.
Harborne J.B. (1958) *J. Chromatogr.* **1,** 473.
Harborne J.B. (1964) in *Methods of Polyphenol Chemistry.* (Ed. by J.B. Pridham). Pergamon Press, Oxford.
Harborne J.B. (1967) *Comparative Biochemistry of Flavonoids.* Academic Press, London.
Hardin J.M. & Stutte C.A. (1980) *Anal. Biochem.* **102,** 171.
Harper H.J. (1924) *Ind. Eng. Chem., Anal. Edn.* **16,** 180.
Harris A.M. & Baines D.F. (1976) X-ray *Spectrosc.* **5,** 129.
Harrison A.F. (1985) *J. Environ. Manag.* **20,** 163.
Harrison A.F. (1986) *Soil Organic Phosphorus: A World Literature Review.* Commonwealth Agricultural Bureau.
Harrison A.F. & Bocock K.L. (1981) *J. Appl. Ecol.* **8,** 919.
Harrison A.F. & Pearce T. (1979) *Soil Biol. Biochem.* **11,** 405.
Harrison R.M. & Laxen D.P.H. (1977) *Water, Air, Soil Pollut.* **8,** 387.
Hartley R.D. & Buchan H. (1979) *J. Chromatogr.* **180,** 139.
Hartmann C.H (1966) *Bull. Env. Contam. Toxicol.* **1,** 141.
Harvey M.C. & Stearns S.D. (1984) *Anal. Chem.* **56,** 837.
Harwood J.E. (1969) *Water Res.* **3,** 273.
Hasegawa K. (1980) *Methods Enzymol.* **67F,** 138.
Hatch W.R. & Ott W.L. (1968) *Anal. Chem.* **40,** 2085.
Hatcher J.T. & Wilcox L.V. (1950) *Anal. Chem.* **22,** 567.
Hawkins D.M. (1980) *Identification of Outliers.* Chapman and Hall, London.
Haworth C. & Heathcote J.G. (1969) *J. Chromatogr.* **41,** 380.
Hayes M.R. & Metcalfe J. (1962) *Analyst (London).* **87,** 956.
Haynes, R.J. (1980) *Common Soil Sci. Plant Anal.* **11,** 459.
Haynes R.J. (1982) *Plant Soil.* **68,** 289.
Haynes R.J. & Swift R.S. (1985) *Geoderma.* **35,** 145.

Heathcote J.G. & Haworth C. (1969) *J. Chromatogr.* **43,** 84.
Heftmann E. & Hunter I.R. (1979) *J. Chromatogr.* **165,** 283 (*Chromatogr. Rev.* Vol. **23**).
Hegemann D.A. & Keenan J.D. (1983) *Water, Air & Soil Pollut.* **19,** 259.
Heinonen J. & Suschny O. (1974) *J. Radioanal. Chem.* **20,** 499.
Heinrich K.F.T. (1981) *Electron Beam X-ray Microanalysis.* Van Nostrand Reinhold, Wokingham.
Henler P.K. & Wayne R.O. (1985) *Ann. Rev. Plant Physiol.* **36,** 397.
Henriksen A. (1965) *Analyst (London).* **90,** 83.
Henriksen A. (1970) *Analyst (London),* **95,** 601.
Henriksen A., Bergmann P. & Inger, M. (1975) *Valten.* **31,** 339.
Herczeg A.L., Broecker W.S., Anderson R.F., Schiff S. L.'& Schindler D.W. (1985) *Nature.* **315,** 133.
Herglotz H.K. & Birks L.S. (Eds.) (1978) *X-ray Spectrometry.* in *Practical Spectroscopy Series* (Ed. by E.G. Brame). Dekker, New York.
Heron J. (1962) *Limnol. and Oceanogr.* **7,** 316.
Hesse P.R. (1971) *A Textbook of Soil Chemical Analysis.* Murray, London.
Hetherington J.C. & Owens C.A. (1979) *Q. Jl. For.* **73,** 101.
Hiermann A. (1980) *Oesterr. Chem. Z.* **81,** 138.
Hillebrand W.F., Lundell G.E.F., Bright H.A. & Hoffman J.I. (1953) *Applied Inorganic Analysis.* John Wiley, New York.
Hine D.T. & Bursill D.B. (1984) *Water Res.* **18,** 1461.
Hitchman M. L. (1978) *Measurement of Dissolved Oxygen.* John Wiley, New York.
Hodge J.E. & Davis H.A. (1952) *Selected Methods for Determining Reducing Sugars.* USDA Bur. Agric. Ind. Chem. A.I.C. 333.
Hoff D.J. & Medenski H.J. (1958) *Proc. Soil Sci. Soc. Am.* **22,** 129.
Holak W. (1980) *Anal. Chem.* **52,** 2189.
Holford I.C.A. & Cullis B.R. (1985) *Austral. J. Soil Res.* **23,** 417.
Holman G.T. & Elliott G.L. (1983) *Lab. Prac.* **32,** 91.
Holmes D.C., Simmons J.H. & Tatton J.O'G. (1967) *Nature (London)* **216,** 227.
Holtz F. (1983) *Lebensum-Unters. Fors.* **176,** 262.
Hood S.L., Parks R.Q. & Hurwitz C. (1944) *Ind. Eng. Chem., Anal. Edn.* **16,** 202.
Hora F.B. & Webber P.J. (1960) *Analyst (London).* **85,** 567.
Horning E.C., Karmen A. & Sweeley G.C. (1964) *Progress. Chem. Fats and Other Lipids,* Vol. 7. (Ed. by R.T. Holmann). Pergamon Press, Oxford.
Horning F.C. & Van den Heuval W.J.A. (1965) *Advances in Chromatography,* Vol. I. (Ed. by J.C. Giddings & R.A. Keller). Arnold, London.
Horvat, R.D. & Senter S.D. (1980) *J. Agric. Food Chem.* **28,** 1292.
Hoste J. & Gillis J. (1955) *Anal. Chim. Acta.* **12,** 158.
Hosettmann K., Doman B., Schaufelberger D. & Hosettmann M. (1984) *J. Chromatogr.* **283,** 137.
Houghton R.A., Hobbie J.E., Melrose J.M., Moore B., Peterson B.J., Shaver G.R. and Woodwell G.M. (1983) *Ecol. Monogr.* **53,** 235.
Howard A.G. & Arbub-Zavar M.H. (1981) *Analyst (London).* **106,** 213.
Howard P.J.A. (1966) *Oikos.,* 15. 229.
Huff D. (1965) *How to Take a Chance.* Penguin Books, London.
Hughes A.D. (1979) *Sampling of Soils, Soil-less Growing Media, Crop Plants and Miscellaneous Substances for Chemical Analysis.* MAFF/ADAS Bulletin 2082. HMSO, London.
Hunt D.T.E. & Wilson A.L. (1986) *The Chemical Analysis of Water.* The Royal Society of Chemistry, London.
Hunt J. (1982) *Commun. Soil Sci. Plant Anal.* **13,** 49.
Hurlbut J.A. (1978) *At. Abs. Newsletter.* **17,** 121.
Hurst W.J., Snyder K.P & Martin R.A. (1983) *J. Liq Chromatogr.* **6,** 2067.
Hutton J.T. & Norrish K. (1977) *X-Ray Spectrom.* **6,** 12.
Hutton R.C. & Preston B. (1983) *Analyst (London)* **108,** 1409.

Ingemells C.O. (1964) *Talanta.* **11,** 665.
Ingram G. (1948) *Analyst (London).* **73,** 548.
Ingram G. (1956) *Chem. Ind.* **103.**
Innes J. (1987) *For. Comm. Bull.* No. 70. HMSO. London.
Insley M., Boswell R.C. & Gardiner J.B.H. (1981) *Plant Soil.* **61,** 377.
International Union of Pure and Appl. Chem. (1984) *Pure Appl. Chem.* **56,** 645.
Irving D.M. (1983) *J. Environ. Qual.* **12,** 442.
Isaacs R.A. (1980) *J. Assoc. Off. Anal. Chem.* **63,** 788.
Isaacs R.A. & Jones J.B. (1972) *Comm. in Soil Sci. and Plant Anal.* **3,** 261.
Iskandar I.K., Syers J.K., Jacobs L.W., Keeney D.R. & Gilmour J.T. (1972) *Analyst (London).* **97,** 388.
Ives D.J.G. & Janz G.J. (1961) *Reference Electrodes.* Academic Press, London.
Ivie, K.F. (1980) in *Analytical Methods for Pesticides and Plant Growth Regulators.* Vol. XI. (Ed. by G.Zweig and J. Sherma). Academic Press, New York.
Iyengar S.S., Martens D.C. & Miller W.P. (1981) *Soil Sci.* **131,** 95.

Jacks G.V. (1954) *Soil.* Nelson, London.
Jackson A.J., Michael L.M. & Schumacher H.J. (1972) *Anal. Chem.* **44,** 1064.
Jackson M.L. (1958) *Soil Chemical Analysis.* Prentice Hall, London.
Jacobs M.B. (1967) *The Analytical Toxicology of Industrial Inorganic Poisons.* John Wiley, New York.
Jacobs S. (1978) *Crit. Rev. Anal. Chem.* **7,** 297.
Jacobson J.S. & Heller L.I. (1971) *Environ. Letters.* **1,** 43.
Jaklin J. & Kronmayr P. (1985) *Int. J. Environ. Anal. Chem.* **21,** 33.
Jarvis M.C. & Duncan H.J. (1974) *J. Chromatogr.* **92,** 432.
Jeffrey D.W. (1970) *J. Ecol.* **58,** 297.

Jeffrey J.G. & Kerr G.O. (1967) *Analyst (London)*. **92,** 763.

Jenkins R. (1974) *An Introduction to X-ray Spectrometry*. Hayden, London.

Jenkins R. & De Vries J.L. (1967) *Practical X-ray Spectrometry*. Philips,

Jenkins R., Gould R.W. & Gidke D. (1981) *Quantitative X-ray Spectrometry*. Dekker, New York.

Jennings D.H. (1976) *Biol. Rev.* **51,** 453.

Jermyn M.A. (1975) *Anal. Biochem.* **68,** 332.

Jin K. & Taga M. (1980) *Bunseki Kagaku*. **29,** 522. Abstract: Anal. Abstr. (1981) 40, No. 4B110.

Johnson G.W. & Vickers C. (1970) *Analyst (London)*. **95,** 356.

Johnson J.E., Bowles J.A. & Knutson J.A. (1985) *Commun. Soil Sci. Plant. Anal.* **16,** 1029.

Johnson W.C. (Ed.) (1964) *Organic Reagents for Metals*. Hopkins and Williams, London.

Johnston M.B. (1974). *Environ. Pollut. Manag.* **4,** 9.

Jones A.A. (1982) in *Methods of Soil Analysis, Part 2. Chemical and Microbiological Properties*. (Ed. by A.L. Page). Soil Science Society of America, Madison, Wisconsin.

Jones K. & Heathcote J.G. (1966) *J. Chromatogr.* **24,** 106.

Jones P.W. & Leber P. (Eds.) (1979) *Polynuclear Aromatic Hydrocarbons*. Ann Arbor Science, Ann Arbor MI.

Jones R.F., Gale P., Hopkins P., Powell L.N. (1966) *Analyst (London)* **91,** 399.

Josefsson B.O., Upsstrom L. & Ostling G. (1972) *Deep Sea Res.* **19,** 385.

Jungalwala F.B., Turel R.T., Evans J.E. & McClure R.H. (1975) *J. Biochem.* **145,** 517.

Juvvik P. (1965) *Acta. Chem. Scand.* **19,** 645.

Kabata-Pendies A. (1963) *Pam. Pulawski.* **9,** 31.

Kadoum A.M. (1967) *Bull. Environ. Contam. Toxicol.* **2,** 264.

Kahler H.L. (1941) *Ind. Eng. Chem., Anal. Edn.* **13,** 536.

Kalam M.A., Hussam A., Kahliquizzaman M., Kahn A.H., Islam M.M., Zaman M.B. & Husain M. (1978) *J. Radioanal. Chem.* **46,** 285.

Kalembasa S.J. & Jenkinson D.S. (1973) *J. Sci. Food Agric.* **24,** 1085.

Kaneko E. (1982) *Chem. Abstr.* **97,** 60668 m.

Karasek F.W., Onuska F.I., Yang F.J. & Clement R.E. (1982) *Anal. Chem.* **54,** 309A, 314A, 316A, 319A, 322A, 324A.

Karasek F.W., Onuska F.J., Yang F.J. & Clement R.E. (1984) *Anal. Chem.* **56,** 174R.

Karmen A. & Saroff H.A. (1964) *New Biochemical Separations*. (Ed. by A.T. James & L.J. Morris). Van Nostrand, New York.

Katz R., Hagin J. & Kurtz L.T. (1983) *Soil Sci.* **136,** 131.

Keil R. (1967) *Z. Anal. Chem.* **229,** 117.

Kelley O.J., Hardman J.A. & Jennings D.S. (1948) *Proc. Soil Sci. Soc. Am.* **12,** 85.

Kempton S., Slerritt A.M. & Lester J.N. (1982) *Talanta*. **29,** 675.

Kent-Jones D.W. & Amos A.J. (1967) *Modern Cereal Chemistry*. Food Trade Press.

Keogh J.L. & Maples R. (1980) *Commun. Soil Sci. Plant Anal.* **11,** 557.

Kiff P.R. (1973) *Lab. Practice.* **22,** 259.

King D.L. (1971) in *Water and Water Pollution Handbook*, Vol. 22. (Ed. by L.L. Ciaccio). Dekker. New York.

King R.R. (1978) *J. Agric. Food Chem.* **26,** 1460.

Kingsford M., Nielson J.S., Pritchard A.D. & Stevenson C.D. (1977) *Sampling of Surface Waters*. Tech Publ. No. 2., NZ. Natl. Water Soil Conserv. Org.

Kingston D.G.I. (1975) *J. Nat. Prod.* **42,** 237.

Kitson R.E. & Mellon M.G. (1944) *Ind. Eng. Chem. Anal. Edn.* **379.** 16.

Kleeman A.W. (1967) *J. Geol. Soc. Austral.* **14,** 43.

Knapp G. (1985) *Int. J. Environ. Anal. Chem.* **22,** 71.

Kobayashi, Y., Takeno H., Kitagawa T., Kanno S., Tani T., Tsuchiya S. & Fukui S. (1966) *Japan Anal.* **15,** 1245.

Koch W.F., Marinenko G. & Wu Y.C. (1984) *Environ. Int.* **10,** 117.

Kokot M.L. (1976) *Atom. Abs. Newsletter.* **15,** 105.

Kolb B., Auer M. & Popisil P. (1977) *J. Chromatogr. Sci.* **15,** 53.

Kolthoff I.M. & Elving P.J. (Eds.) (1961) *Treatise on Analytical Chemistry*, Part II, Vol. I. Wiley-Interscience, New York.

Komor J. (1940) *Biochem. Z.* 305, 381.

Kononova M.M. (1966). *Soil Organic Matter*. Pergamon Press, Oxford.

Konrad J.G., Pionke H.B. & Chesters G. (1969) *Analyst (London)*. **94,** 490.

Kramer J.R. (1982) in *Water Analysis, Vol. I, Inorganic Species*. (Ed. by R.A. Minear and L.H. Keith). Academic Press, London.

Kretschmer A.E. & Randolph J.W. (1954) *Anal. Chem.* **26,** 1862.

Kruse J.M. & Mellon M.G. (1952) *Sewage Ind. Wastes* **24,** 1254.

Kubiena W.L. (1953) *The Soils of Europe*. London, Allen and Unwin.

Kuksis A. (1977) *Sep. Purif. Methods.* **6,** 353.

Kuksis A. (1978) *Handbook Lipid Res.* **1,** 1.

Kurschner K. & Hoffer A. (1929) *Tech. Chem. Papier-Zellstoff Fabr.* **26,** 125.

Lachance G.R. & Traill R.J. (1966) *Can. Spectrosc.* **11,** 43.

Lachance G.R. (1982) *Introduction to Alpha Coefficients*. Corporation Scientific Claisse, Quebec.

La Fleur P.D. (1973) *Anal. Chem.* **45,** 1534.

Laidlaw R.A. & Smith G.A. (1965) *Holzfershung.* **19,** 129.

Laker M.F. (1980) *J. Chromatogr.* **209,** 41.

Lamb J. (1968) *J. Agric. Res.* **7,** 227.

Lancaster L.A. & Balasubramaniam R. (1974) *J. Sci. Food Agric.* **25,** 381.

Langmuir D. & Jacobson R.L. (1970) *Environ. Sci. Technol.* **10,** 834.

Lanigan G.W. & Jackson R.B. (1965) *J. Chromatogr* **17,** 238.

Larsen S. (1967) *Adv. in Agron.* **19,** 151.

Lauren D.R. (1984) *J. Assoc. Off. Anal. Chem.* **67,** 655.

Law R.M. & Goerlitz D.F. (1970) *J. Assoc. Off. Anal. Chem.* **53,** 1276.

Lawrence J.F. & Turton D. (1978) *Chromatogr. Rev.* **22,** 207.

Lawrence J.F., Panopio L.G. & McLeod H.P.J. (1980) *J. Agric. Food Chem.* **28,** 1019.

Lawrence J.F. (Ed.) (1984) *Liquid Chromatography in Environmental Analysis.* Humana Press, Clifton, New Jersey.

Leathard D.A. & Shurlock B.C. (1970) *Identification Techniques in Gas Chromatography.* John Wiley, New York.

Lee M.L., Young F.J. & Bartle K.D. (1984) *Open Tubular Gas Chromatography: Theory and Practice.* John Wiley, New York.

Lee R.E., Caldwell J.S. & Morgan L. (1972) *Atmos. Environ.* **6,** 593.

Lee R. & Sharp G.S. (1985) *Commun. Soil Sci. Plant Anal.* **16,** 261.

Leenheer J.A. (1981) *Environ. Sci. Technol.* **15,** 578.

Leggett G.E. & Westerman D.T. (1973) *J. Agric. Food Chem.* **21,** 65.

Leigh R.A. & Wyn-Jones R.G. (1984) *New Phytol.* **97,** 1.

Leithe W. (1973) *The Analysis of Organic Pollutants in Water and Waste Waters.* Ann Arbor Sc. Publ., Michigan.

Lenstra J.B. & De Wolf J.N.M. (1965) *Pharm. Week bl. Ned.* **100,** 232.

Lercker G. (1983) *J. Chromatogr.* **279,** 543.

Levi I., Mazur P.B. & Nowicki T.W. (1972) *J. Assoc. Off. Anal. Chem.* **59,** 794.

Lewis D.H. (1980) *New Phytol.* **84,** 209.

Leyden D.E. & Wegsheider W. (1981). *Anal. Chem.* **53,** 1059 A.

Li S. & Smith K.A. (1984) *Commun. Soil Sci. Plant Ecol.* **15,** 1437.

Lichtenhaler H.K. & Wallburn A.R. (1983) *Biochem. Soc. Trans.* **11,** 591.

Lieser K.H., Breitwieser E., Burba P., Roeber M. & Spatz R. (1978) *Mikrochim. Acta.* **I** (3–4), 363.

Likens G.E. & Bormann F.H. (1970) *Bull. Yale Univer. Sch.* No. 79.

Likussar W., Raber H., Huser H. & Grill D. (1976) *Anal. Chim. Acta.* **87,** 247.

Lin C. & Coleman N.T. (1960) *Proc. Soil Sci. Soc. Am.* **24,** 444.

Lindquist F. (1972) *Analyst (London).* **97,** 549.

Lindquist J. (1984) *Swed. J. Agric. Res.* **14,** 171.

Lindsay W.L. & Norvell W.A. (1969) *Agron. Abstr.* **84.**

Lindsay W.L. & Norvell W.A. (1978) *Soil Sci. Soc. Am. J.* **42,** 421.

Lockman R.B. & Molloy M.G. (1977) *Commun. Soil Sci. Plant Anal.* **8,** 437.

Longwell J. & Maniece W.D. (1955) *Analyst (London).* **80,** 167.

Lott P.F., Lott J.W. & Doms D.J. (1978) *J. Chromatogr. Sci.* **16,** 390.

Loveday J. (1972) *Moisture Content and Grinding Conditions for Saturation Extracts.* 1972. Divn. of Soils Tech. Paper No. 12., C.S.I.R.O.

Lowry O.H., Rosebrough N.J., Fair A.L. & Randell R.J. (1951) *J. Biol. Chem.* **193,** 265.

Lubecki A. (1969) *J. Radioanal. Chem.* **3,** 317.

Luciano R. & Bosetto M. (1968) *Ricera Sci.* **38,** 855.

Luke M.A., Froberg J.E., Doose G.M. & Masumoto H.T. (1981) *J. Assoc. Off. Anal. Chem.* **64,** 1187.

Lustinec J., Hadacova V., Kaminek M. & Prochazka K. (1984) *Anal. Abstr.* **46,** 4D 151.

L'vov B.V. (1970) *Atomic Absorption Spectrochemical Analysis.* Hilger, London.

L'vov B.V. (1978) *Spectrochim. Acta.* **33B,** 153.

Lytle F.E., Dye W.B. & Siem H.J. (1962) *Adv. X-ray Anal.* **5,** 433.

Mackereth, F.J.H. (1963) *Water Analysis for Limnologists.* Freshwater Biological Association. Sci. Publ. No. 21.

Mackereth F.J.H. (1964) *J. Sci. Instrum.* **41,** 38.

Mackereth F.J.H., Heron J. & Talling J.F. (1978) *Water Analysis.* Freshwater Biological Association. Sci. Publ. No. 36.

Maclean K.S. & Robertson R.G. (1981) *Commun. Soil. Sci. Plant Anal.* **12,** 39.

Madansky A. (1959) *J. Am. Stat. Assoc.* **54,** 173.

Maher W.A. (1984) *Anal. Lett.* **17,** 979.

Majors R.E. (1980) *J. Chromatogr. Sci.* **18,** 488.

Malcolm R.L. & McCracken R.J. (1968) *Proc. Soil Sci. Soc. Am.* **32,** 834.

Malissa H. & Schoffman E. (1955) *Mikrochim. Acta.* **187.**

Mangold H.K. (1984) *Lipids, CRC Handbook of Chromatography Series.* (Ed. by G. Zweig and J. Sherma). CRC Press, Boca Raton, Florida.

Mantoura R.F.C. & Riley J.P. (1975) *Anal. Chim. Acta.* **76,** 97.

Marcie F.J. (1967) *Environ. Sci. Technol.* **1,** 164.

Margler L.W. & Mah R.A. (1981) *J. Assoc. Off. Anal. Chem.* **11,** 195.

Marks J.Y. & Welcher G.G. (1970) *Anal. Chem.* **42,** 1033.

Marsili R.T., Ostapenkpo H., Simmons R.E. & Green D.E. (1981) *J. Food Sci.* **46,** 52.

Martin A.J.P. & Synge R.L.H. (1941) *Biochem. J.* **35,** 91.

Martin I.R. & Wilson A.L. (1963) *Analyst (London).* **88,** 88.

Martin L. & Sparks J. (1985) *Commun. Soil Sci. Plant Ecol.* **16,** 133.

Mason A.C. (1958) *J. Hort. Sci.* **33,** 128.

Massmann H. (1967) *Z. anal. Chem.* **225,** 203.

Math K.S., Bhatki K.S. & Freiser A. (1969) *Talanta.* **16,** 412.

Matisova E., Krupcik J. & Liska O. (1979) *J. Chromatogr.* **173,** 139.

Matthews A.D. & Riley J.P. (1969) *Anal. Chim. Acta.* **48,** 25.

Maxwell J.A. (1968) *Rock and Mineral Analysis.* New York, Interscience.
Mayland H.F., Westermann D.T. & Florence A.R. (1978) *Commun. Sol. Sci. Plant. Anal.* **9**, 551.
McBee G.G. & Maness N.O. (1983) *J. Chromatogr.* **264,** 474.
McCrae J.C., Smith D. & McReady R.M. (1974) *J. Sci. Food Agric.* **25,** 1465.
McCready R.M., Gaggotz J., Silveira V. & Owens H.S. (1950) *Anal. Chem.* **22,** 1156.
McCully K.A. & McKinley W.P. (1964a). *J. Assoc. Off. Agric. Chem.* **47,** 652.
McDonald G.E., Peck N.H. & Vittum M.T. (1978) *Commun. Soil Sci. Plant Anal.* **9,** 717.
McDonald P., Stirling A.C., Henderson A.R., Dewar W.A., Stark G.H., Davie W.A., MacPherson, H.T., Reid A.M. & Slater J. (1960) *Studies on Ensilage.* Edinburgh School of Agriculture, Tech Bulletin No. 24.
McDonald R.C. & Rempas S.P. (1977) *J. Chromatogr.* **131,** 157.
McEwan F.L. & Stephenson G.R. (1979) *The Use and Significance of Pesticides in the Environment.* John Wiley, New York.
McKeague J.A. & Day J.H. (1966) *Can. J. Soil. Sci.* **46,** 13
McKenzie S.L. (1981) *Biochem Anal.* **27,** 1.
McKinley W.P., Coffin D.E. & McCully K.A. (1964) *J. Assoc. Off. Agric. Chem.* **47,** 863.
McKone C.E., Byast T.H. & Hance R.J. (1972) *Analyst (London).* **97,** 653.
McLachlan K.D. & Crawford M. (1970) *J. Sci. Food. Agric.* **21,** 408.
McLean E.O. (1959) *Proc. Soil Sci. Soc. Amer.* **23,** 289.
McLean E.O. (1975) *Commun. Soil Sci. Plant Anal.* **6,** 207.
McLeod H.A., Wales P.J., Graham R.A., Osadchuk M. & Bluman N. (Eds.) (1969) *Analytical Methods for Pesticide Residues in Foods.* The Department of National Health and Welfare, Canada.
McNair H.M. (1982) *J. Chromatogr. Sci.* **20,** 537.
McRae J.C., Smith D. & McReady R. M. (1974) *J. Sci. Food. Agric.* **25,** 1465.
Mead R. (1971) *J. Ecol.* **59,** 215.
Mead R. & Curnow R.N. (1983) *Statistical Methods in Agricultural and Experimental Biology.* Chapman and Hall, London.
Meakin J.C. & Pratt M. (1973) *Manual of Laboratory Filtration.* W.R. Balston Ltd.
Mehlich A. (1945) *Soil Sci.* **66,** 429.
Mehlich A. (1984) *Commun. Soil Sci. Plant Anal.* **15,** 1513.
Mehta N.C., Legg J.O., Goring C.A.I. & Black C.A. (1954) *Proc. Soil Sci. Soc. Amer.* **18,** 443.
Meints V.W. & Peterson G.A. (1972) *Soil Sci. Soc. Am. Proc.* **36,** 43.
Meites L., Smith H.C. & Kateman G. (1985) *Anal. Chim. Acta.* **164,** 287
Meltzer A. & Steinberg C. (1983) in *Physiological Plant Ecology IV* (ed by O.L. Lange *et al.*). Springer, Berlin.
Melvin J.F. & Simpson B. (1963) *J. Sci. Food Agric.* **14,** 228.
Mendoza C.E., McCully K.A. & Wales P.J. (1968) *Anal. Chem.* **40,** 2225.
Mengel K. & Kirkby E.A. (1980) *Adv. Agron.* **33,** 59.
Metson A.J. (1956) *Methods of Chemical Analysis for Soil Survey Samples.* NZ. Dept. Sci. Ind. Res., Soil Biol. Bull. 12.
Metson A.J., Blakemore, L.C. & Rhoades D.A. (1979) *NZ. J. Sci.* **22,** 205.
Middleton G. & Stuckey R.E. (1953) *Analyst (London).* **78,** 532.
Midgley D. (1975) *Analyst (London).* **100,** 306.
Miller D.S. & Payne P.R. (1959) *Br. J. Nutrition* **12,** 501.
Miller G.W., Pushnik J.C. & Welkie L. (1984) *J. Plant Nutrition.* **7,** 1.
Miller J.C. & Miller J.N. (1984) *Statistics for Analytical Chemistry.* Ellis Horwood, Chichester.
Miller V.L., Lillis D. & Csonka E. (1958) *Anal. Chem.* **30,** 1705.
Milner C. & Hughes R.E. (1968) *Methods for the Measurement of the Primary Production of Grassland.* IBP Handbook No. 6. Blackwell Scientific Publications, Oxford.
Milton R.F. & Waters W.A. (1955) *Methods of Quantitative Analysis.* Arnold, London.
Minear R.A. & Keith L.A. (1982) *Water Analysis.* Vol II, Part 2. Academic Press, London.
Mitchell H.L. & Chandler R.F. (1939) *The Nitrogen Nutrition and Growth of Certain Deciduous Tress of NE United States.* Black Rock Forest Bulletin. No 11.
Mitchell J. (1951) *Anal. Chem.* **23,** 1069.
Mitchell M.C., Barrow M.L. & Shend C.A. (1987) *J. Anal. Atom Spec.* **2,** 261.
Moir K.W. (1971) *Lab. Pract.* **20,** 801.
Molloy M.G. & Lockman R.B. (1979) *Commun. Soil Sci. Plant Anal.* **10,** 545.
Moon F.E. & Abou-Raya A.K. (1952) *J. Sci. Food Agric.* **3,** 407.
Moore S. & Stein W.H. (1951) *J. Biol. Chem.* **192,** 663.
Mor E., Scotto V., Marcenaro G. & Alabino G. (1975) *Anal. Chim. Acta.* **75,** 159.
Morgan M.F. (1937) *Conn. Agric. Exp. Str. Bull.,* 392.
Morris A.G.C. (1983) *Analyst (London).* **108,** 546.
Morris L.J. (1966). *J. Lipid Res.* **7,** 717.
Morrison I.R. & Wilson A.L. (1963) *Analyst (London).* **88,** 100.
Morrison J.L. & George G.M. (1969) *J. Assoc. Off. Anal. Chem.* **52,** 930.
Mortimer, C.H. (1942) *J. Ecol.* **30,** 147.
Mossel D.A.A. (1950) *Recl. Trav. Chim. Pays-Bas Belg.* **69,** 932.
Mott, R.A. (1958). *Fuel.* **37,** 3.
Moye, H.A. (1981). *Analysis of Pesticide Residues.* John Wiley, Chichester.
Moye H.A. & Jones M.J. (1971) *J. Agric. Food Chem.* **19,** 459.
Mudroch, A. & Mudroch O. (1977) *X-ray Spectrom.* **6,** 215.
Mullin, J.B. & Riley J.P. (1955) *Anal. Chim Acta.* **12,** 162.
Mullins C.E. & Hutchinson B.J. (1982) *J. Soil. Sci.* **33,** 547.

Munyinde K., Aramanos R.E. & Bettany J.R. (1982) *Commun. Soil Sci. Plant Anal.* **17,** 735.

Murata T. (1980) *Kagaku to Kogyo* (*Osaka*). 54, 208.

Murphy L.S., Ellis R. & Adrian D.C. (1981) *J. Plant Nutrition* **3,** 593.

Nadkarni R.A. (1982) *Anal. Chim. Acta.* **135,** 363.

Nangniot P. (1967) *Ind. Chim. Belge.* **32,** 1323.

Nardishaw M. & Cornfield, H.A. (1968) *Analyst* (*London*). **93,** 475.

National Research Council Committee on Biological Effects of Atmospheric Pollutants (1971) *Fluorides.* National Academy of Sciences, Washington, D.C.

Neher R. (1967) in *Advances in Chromatography*, Vol. 4. (Ed by J.G. Giddings and R.A. Keller). Arnold, London.

Neis U. (1978) *Wasser Abwasser Forsch.* **11,** 3.

Nelson D.W. & Bremner J.M. (1972) *Agronomy J.* **64,** 196.

Nelson D.W. & Romkens M.J.H. (1974) *J. Environ. Qual.* **1,** 323.

Neu R. (1958) *Nature* **182,** 660.

Neurath H. & Hill R.L. (Eds.) (1977) *Proteins.* Academic Press, London.

Neuray M., Pirard N. & Polinard M-A. (1986) *Ann. Mines Belg.* **5–6,** 684.

Neve J., Hanocq M. & Molle L. (1980) *Int. J. Environ. Anal. Chem.* **8,** 177.

Newman E.J. & Jones P.P. (1966) *Analyst* (*London*). **91,** 406.

Nicholls B.W. (1966) *Br. Med. Bull.* **22,** 137.

Nicholson G. (1984) *Methods of Soil, Plant and Water Analysis.* NZ. Forest Service. FRI. Bulletin 70.

Niesner B., Brueller W. & Bubleter O. (1978) *Chromatographia.* **11,** 400.

Nijkamp H. J. (1961) *Anal. Chim. Acta.* **24,** 529.

Nishi S., Horimoto Y. & Kobayashi R. (1971) in *Identification and Measurement of Environmental Pollutants.* (Ed. by B. Westley). Proceedings of International Symposium, Canada, June 1971. National Research Board of Canada, Ottawa.

Norrish K. & Hutton J.T. (1969) *Geochim. et Cosmochim. Acta.* **33,** 431.

Norrish K. & Hutton J.T. (1977) *X-ray Spectrom.* **6,** 6.

Novazamsky I., Houba V.J.G., Eck R.V. & Vark W.V. (1983) *Commun. Soil Sci. Plant Anal.* **14,** 239.

Nusbaum I. (1953) *Sewage Ind. Wastes.* **25,** 311.

Nutting, P.-G. (1943) *Some Standard Thermal Dehydration Curves of Minerals.* Profess. Paper 197-E. US. Geol. Surv.

Oades J.M. & Townsend W.N. (1967) *J. Soil Sci.* **14,** 134.

Obenaus R. (1963) *Schweizer Ztschr. Hydrol.* **25,** 9.

O'Connor P.W. & Syers J.K. (1975) *J. Environ. Qual.* **4,** 347.

Oelschlager W. (1965) *Landw. Forsch.* **18,** 79–88.

Oelschlager W. & Lantzsch H-J. (1974) *Landw. Forsch.* **27,** 31.

Official Methods of Analysis of the AOAC (1984) (14th edn., Ed by S. Williams). Association of Official Analytical Chemists, Arlington USA.

Okazaki R., Smith H.W. & Moodie C.D. (1964) *Anal. Chem.* **36,** 202.

Olsen S.R., Cole C.V., Watanabe F.S. & Dean L.S. (1954) *Estimation of Available Phosphours in Soils by Extraction with Sodium Bicarbonate.* U.S. Dep. Agric. Cir. No. 939.

Olson O.E. (1969) *J. Assoc. Off. Anal. Chem.* **52,** 627.

Osawa K., Kuzikawa K. & Imaeda K. (1980) *Chem. Abstr.* **93,** 125098q.

Osawa K., Kuzikawa K. & Imaeda K. (1981) *Chem. Abstr.* **95,** 54155.

Osselton M.D. & Snelling R.D. (1986) *J. Chromatogr.* **368,** 265.

Oswiecimska M. & Golcz L. (1969) *Farmacja pol.* **25,** 123.

Othman M.R.B., Hill J.O. & Magee R.J. (1984) *J. Electroanal. Chem. Interfacial Electrochem.* **168,** 219.

Owen J.A., Igoo B., Scandrett F.J. & Stewart C.P. (1954) *Biochem. J.* **58,** 426.

Oxenham D.J., Catchpole V.R. & Dolby G.R. (1983) *Commun. Soil. Sci. Plant Anal.* **14,** 153.

Padfield T. & Gray A. (1971) *Major Element Rock Analysis by X-ray Fluorescence—A simple Fusion Method.* Philips Analytical Equipment Bulletin FS 35. Philips, Eindhoven.

Paech K. & Tracey M.V. (1955) *Modern Methods of Plant Analysis*, Vols. 1 to 5. Springer-Verlag, Berlin.

Page A.L., Miller R. & Keeney D.R. (Eds.) (1982) *Methods of Soil Analysis.* Part 2: *Chemical and Microbiological Properties.* American Society of Agronomy.

Paine R.T. (1964) *Limnol. Oceanog.* **11,** 126.

Palm A.W. & Beckwith R.S. (1956) *Anal. Chem.* **28,** 1637.

Palmes E.D., Gunnison A.F., Dimattio J. & Tomezyk C. (1975) *Am. Ind. Hyg. Assoc. J.* **37,** 70.

Pantony D.A. & Hurley P.W. (1972) *Analyst* (*London*). **97,** 497.

Parker C.A. & Barnes W.J. (1960) *Analyst* (*London*). **85,** 828.

Parker M.M., Humuller F.L. & Mahler D.J. (1967) *Clin. Chem.* **13,** 40.

Parkinson J.A. (1987) in *Chemical Analysis in Environmental Research.* (Ed. by A.P. Rowland). Inst. Terr. Ecology. UK.

Parkinson J.A. & Grimshaw H.M. (1986) *Commun. Soil Sci. Plant Anal.* **17,** 735.

Parisis N.E. & Heyndrickx A. (1986) *Analyst* (*London*) **111,** 281.

Parris N.A. (1976) *Instrumental Liquid Chromatography.* Elsevier, Amsterdam.

Pasztor L., Bode J.D. & Fernando Q. (1960) *Anal. Chem.* **32,** 277.

Pataki G. (1969) *Techniques of Thin Layer Chromatography in Amino Acid and Peptide Chemistry.* Ann Arbor Sc. Publ., Michigan.

Patterson P.L. & Howe, R.L. (1978) *J. Chromatogr. Sci.* **16,** 275.

Patton G.M., Fasulo J.M. & Robins S.J. (1982) *J. Lipid Res.* **23,** 190.

Pearson E.S. & Hartley H.O. (1966) *Biometrika Tables for Statisticians*, Vol 1. Cambridge University Press.

Peech M., Alexander L.T., Dean L.A. & Reeds J.F. (1947) *Methods of Soil Analysis of Soil Fertility*. U.S. Dep. Agric. Circ. 757.

Pepper J.M., Baylis P.E.T. & Alder E. (1959) *Can. J. Chem.* **37,** 1241.

Perry R. & Young R.J. (1977) *Handbook of Air Pollution Analysis*. Chapman and Hall, London.

Peter G. & Tuchscheerer T. (1966) *Z. Anal. Chem.* **220,** 351.

Peterson G.L. (1979) *Anal. Biochem.* **100,** 201.

Pfeifer R.F. & Hill D.W. (1983) in *Advances in Chromatography*, Vol. 21. (Ed. by J.C. Giddings, E. Grushka, J. Cazes & R.P. Brown). Dekker, New York.

Phillipson J. (1966) *Ecological Energetics*. Arnold, London.

Pietta P., Culatroni A. & Zio C. (1983) *J. Chromatogr.* **280,** 172.

Pik A.J. & Hodgson G.W. (1976) *J. Assoc. Off. Anal. Chem.* **57,** 264.

Pimplaskar M.S., Floate M.J.S. & Newbould P. (1982) *J. Sci. Food Agric.* **33,** 957.

Piper C.S. (1950) *Soil and Plant Analysis*. University of Adelaide.

Pleysier J.L. & Juo A.S.R. (1980) *Soil Sci.* **129,** 205.

Pomerantz Y. (1965) *J. Food. Sci.* **30,** 823.

Poole C.F., & Schuette S.A. (1984) *Contemporary Practice of Chromatography*. Elsevier, Amsterdam.

Powers G.W., Martin R.L., Piahl F.J. & Griffin J.M. (1959) *Anal. Chem.* **31,** 1589.

Powers M.C. (1960) *X-ray Fluorescent Spectrometer Conversion Tables*. Philips, Eindhoven.

Powers R.F., Van den Gent L. & Towsend R.F. (1981) *Commun. Soil Sci. Plant Anal.* **12,** 9.

Prasad, M. & Spiers, T.M. (1978) *Soil Sci. Soc. Am. J.*, **42,** 661.

Pratt P.F. & Bair F.L. (1961) *Soil. Sci.* **91,** 357.

Preez C. du., Burgea R. du T. & Laubscher D.J. (1987) *Commun. Soil Sci. Plant Anal.* **18,** 483.

Premi P.R. & Cornfield A.H. (1968) *Spectrovision.* **19,** 15.

Pribil R. (1972) *Analytical Applications of EDTA and other compounds*. Pergamon Press, Oxford.

Price W.J. (1979) *Spectrochemical Analysis by Atomic Absorption*. Heyden, London.

Privett O.S., Dougherty K.A., Erdahl W.L. & Stalyhwo A. (1973) *J. Amer. Oil Chem-Soc.* **50,** 516.

Probert M.E. (1976) *Plant Soil.* **45,** 461.

Proctor J. & Woodell S.R.J. (1975) *Adv. Ecol. Res.* **9,** 255.

Prodan M. (1968) *Forest Biometrics*. Pergamon Press, Oxford.

Pruden G., Kalembasa S.J. & Jenkinson D.S. (1985) *J. Sci. Food Agric.* **36,** 71.

Pryde A. & Gilbert M.T. (1969) *Applications of High Performance Liquid Chromatography*. Chapman and Hall, London.

Pucher G.W., Leavenworth C.S. & Vickery H.B. (1948) *Anal. Chem.* **20,** 850.

Puck J., Hoste J. & Gillis J. (1960) in *Proc. Int. Sym. Microchem.*, Birmingham, 1958. p. 48. Pergamon Press, Oxford.

Pungor E. (1970) in *Proc. Int. Cont. Atomic Absorption Spectroscopy*, Sheffield 1969. (Ed by R.H. Dagnall and G.F. Kirkbright). Butterworth, London.

Putzien J. (1986) *Z. Wasser Abwasser Forsch.* **19,** 228.

Quackanbush F.W. & Miller S.L. (1972) *J. Assoc. Off. Anal. Chem.* **55,** 617.

Quarmby C. (1968) *J. Chromatogr.* **34,** 52.

Quarmby C. & Grimshaw H.M. (1967) *Analyst (London).* **92,** 305.

Quin B.F. & Brooks R.R. (1975) *Anal. Chim. Acta.* **74,** 75.

Rainwater F.H. & Thatcher L.L. (1960) *Methods for Collection and Analysis of Water Samples*. U.S. Geol. Survey Water Supply, Paper 1454.

Ramirez–Munoz J. (1968) *Atomic Absorption Spectroscopy and Analysis by Atomic Absorption Flame Photometry*. American Elsevier, New York.

Randall P.J. & Spencer K. (1980) *Commun. Soil Sci. Plant Anal.* **11,** 257.

Randall P.J. & Sakai H. (1983) in *Sulphur in SE Asia and Pacific Agriculture*. (Ed. by G.J. Blair and A.R. Till). Austr. Dev. Ass. Bureau.

Randerath K. (1966) *Thin Layer Chromatography*. Academic Press, London.

Ranney T.A. & Bartlett R.J. (1972) *Commun. Soil Sci. Plant Anal.* **3,** 183.

Rasberry S.D. & Heinrich K.F. (1974) *Anal. Chem.* **46,** 81.

Rattenberg J.M. (1981) *Amino Acid Analysis*. Ellis Horwood, London.

Raun W.R., Olsen R.A., Sander D.H. & Westerman R.L. (1987) *Commun. Soil Sci. Plant Anal.* **18,** 543.

Raymakers A. (1974) *Pharm. Weekbl.* **109,** 1229.

Reay P.F. (1974) *Anal. Chim. Acta.* **72,** 145.

Reed B. (1973) *J. Sci. Food Agric.* **24,** 139.

Reed S.J.B. (1975) *Electron Microprobe Analysis*. Cambridge University Press.

Reeve R.C. (1957) *Agronomy.* **7,** 404.

Reifer I. & Melville J. (1947) *Proc. 11th Cong. Pure and Applied Chem.* **3,** 233.

Reith J.F., Mossel D.A.A. & Van de Kamer J.H. (1948) *Anal. Chim. Acta.* **2,** 359.

Rennenberg H. (1984) *Ann. Rev. Plant Physiol.* **35,** 121.

Reynolds L.M. (1969) *Bull. Environ. Contam. Toxicol.* **4,** 128.

Rich C.I. (1962) *Soil Sci.* **93,** 87.

Richards G.E. & McLean E.O. (1963) *Soil Sci.* **95,** 308.

Richards L.A. (1954) *Diagnosis and Improvement of Saline and Alkaline Soils*. Handbook No. 60. U.S. Dept. Agriculture.

Ridgman W.J. (1975) *Experimentation in Biology: An Introduction to Design and Analysis*. Blackie, London.

Rietz P. & Scheidegger C. (1980) *Bull. Schweiz. Ges. Klin. Chem.* **21,** 236.

Riley J.P. & Skirrow G. (1965) *Chemical Oceanography*. Academic Press, London.
Riley J.P. & Williams H.P. (1959) *Mikrochim. Acta*. a, 504; b, 525; c, 804; d, 825.
Ripley B.D. & Thompson M. (1987) *Analyst (London)*. **112,** 377.
Ritter G.J. & Barbour J.H. (1932) *Ind. Eng. Chem. anal. Edn.* **2,** 238.
Ritter G.J. & Kurth E.F. (1934) *Ind. Eng. Chem. anal. Edn.* **6,** 1250.
Ritter G.J., Seaborg R.M. & Mitchell R.L. (1932) *Ind. Eng. Chem. Anal. Edn.* **4,** 202.
Robberecht H.J. & Van Grieken R.E. (1980) *Anal. Chem.* **52,** 449.
Roberts J.L. (1971) *Talanta*. **18,** 1070.
Robertson T., Thomas C.J., Caddy B. & Lewis A.J.M. (1984) *Forensic Sci. Int.* **24,** 209.
Robinson J.B.D. (1967) *Plant. Soil.* **27,** 53.
Robinson T. (1979) *Plant Sci. Lett.* **15,** 211.
Rodin L.S. & Bazilevich N.I. (1967) *Production and Mineral Cycling in Terrestrial Vegetation.* Oliver and Boyd, Edinburgh.
Roe J.H. (1955) *J. Biol. Chem.* **212,** 335.
Rorison I.H. & Robinson D. (1984) *Plant Cell & Environ.* **7,** 381.
Rorison I.H. (1985) *J. Ecol.* **73,** 83.
Rose L. (1983) *Trans. Brit Cave Res. Assoc.* **10,** 21.
Ross, J.W. (1969) in *Ion Selective Electrodes.* (Ed. by R.A. Durst) NBS Spec. Publ. No. 314. Washington, D.C.
Rourke D.R., Mueller W.F. & Yang S.H. (1977) *J. Assoc. Off. Anal. Chem.* **60,** 233.
Rowland A.P. (1983) *Commun. Soil Sci. Plant Anal.* **14,** 49.
Rowland A.P. & Grimshaw H.M. (1985) *Commun. Soil Sci. Plant Anal.* **16,** 551.
Rowland A.P. (1987) *Reference Materials for Ecological Analysis.* in *Chemical Analysis in Environmental Research.* Inst. Terr. Ecology. UK.
Royal Society Study Group (1963) *The Nitrogen Cycle of the U.K.* Royal Society, London.
Rueppel M.L., Brightwell B.B., Schaeffer J. & Marvell J.J. (1977) *J. Agric. Food. Chem.* **25,** 517.
Russell E.J. (1961) *Soil Conditions and Plant Growth.* Longmans. London.
Ryan P.J. & Honeyman T.W. (1985) *J. Chromatogr.* **312,** 461.
Ryan T.A., Jones B.L. & Ryan B.F. (1986) Minitab Student Handbook, Duxburg Press, MA.
Ryden J.C., Syers J.K. & Harris R.F. (1972) *Analyst (London)* **97,** 903.

Sabre W. (1980) *J. Assoc. Off. Anal. Chem.* **63,** 763.
Sagar M. & Toelg G. (1982) *Mikrochim. Acta.* **II** (3–4), 231.
Sagar M. (1984) *Mikrochim. Acta.* **II** (5–6), 381.
Salbu B. (1981) *Mikrochim. Acta.* **II,** 351.
Salcedo I.H. & Warncke D.D. (1979) *Soil Sci. Soc. Am. J.* **43,** 135.
Salinas F. & March J.G. (1984) *Environ. Anal. Chem.* **18,** 209.
Salles L.C. & Curtis A.J. (1983) *Mikrochim Acta.* **II** (1–2), 125.
Salonius P.O., McDonald C.C. & Fisher R.A. (1979) *Biomonthly Research Notes.* **35,** 4.
Saltzman B.E. (1952) *Anal. Chem.* **24,** 1016.
Samsoni Z. (1975) *Soils and Fertilizers* **38,** 4534.
Samuelson O. (1963) *Ion Exchange Separations in Analytical Chemistry.* John Wiley, New York.
Sandell E.B. (1959) *Colorimetric Determination of Traces of Metals.* Wiley-Interscience, New York.
Sauerer A. & Troll G. (1984) *Talanta.* **31,** 249.
Saunders W.M.H. & Metson, A.J. (1971) *NZ. J. Agric Res.* **14,** 307.
Saunders W.M.H. & Williams E.G. (1955) *J. Soil Sci.* **6,** 254.
Sawyer D.T. & Roberts J.L. (1974) *Experimental Electrochemistry for Chemists.* John Wiley, New York.
Scharrer K. & Goltschell R. (1935) *Z. Pflernahr Dung Bodenk.* **39,** 178.
Schierup H-H. & Riemann B. (1979) *Arch. Hydrobiol.* **86,** 204.
Schmidt W. & Dietl F. (1983) *Fresnius' Z. Anal. Chem.* **315,** 687.
Schneider T., DeKening H.W. & Brasser L.J. (Eds.) *Air Pollution Reference Measurements: Methods & Systems.* Elsevier, Amsterdam.
Schneider T.F., Bourne S. & Bopari A.S. (1984) *J. Chromatogr. Sci.* **22,** 203.
Schnitzer M. & Khan S.V. (1972) *Humic Substances in the Environment.* Dekker, New York.
Schofield R.K. & Taylor A.W. (1955) *Proc. Soil Sci. Soc. Am.* **19,** 164.
Schollenberger C.J. (1927a) *Soil Sci.* **24,** 65.
Schollenberger C.J. (1927b) *Science.* **65,** 552.
Schollenberger C.J. & Dreibelbis R.H. (1930) *Soil Sci.* **30,** 161.
Schollenberger C.J. & Simons R.H. (1945) *Soil Sci.* **59,** 13.
Schomburg G. (1983) *J. Chromatogr. Sci.* **21,** 97.
Schoniger W. (1955) *Mikrochim Acta.* 123.
Schoniger W. (1956) *Mikrochim. Acta.* 869.
Schramel P. & Xu L-G. (1982) *Anal. Chem.* **54,** 1333.
Schroeder B., Thompson G. & Sulanowska M. (1982) *Int. Lab.* **14,** 24.
Schultze E. (1892) *Phys. Chem.* **16,** 387.
Schurtz J. (1977) *Bioconvers Cellul. Subst. Energy. Chem. Microb. Protein, Symp. Proc.* (1st). 37.
Schwartz S.J. &Van Elbe J.H. (1982) *J. Liq. Chromatogr.* **5,** (Suppl 1), 43.
Schwarzenbach G., Biederman W. & Bangertrer F. (1946) *Helv. Chim. Acta.* **29,** 811.
Searle P.L. (1984) *Anal. (London).* **109,** 549.
Searle P.L. & Sparling G.P. (1987) *Commun. Soil Sci. Plant Anal.* **18,** 725.
Sedberry J.G., Miller B.J. & Said M.B. (1979) *Commun. Soil Sci. Plant Anal.* **10,** 689.
Seikel M.K. (1962) *The Chemistry of Flavonoid Compounds.* (Ed. by T.A. Geisman). Pergamon Press, Oxford.
Sekerka I. & Lechner J.F. (1978) *J. Assoc. Off. Anal. Chem.* **61,** 1493.

Selvendran R.R., Ring S.G. & Du Pont M.S. (1979) *Chem. Ind.* (*London*) **7,** 225.

Shafik M.T., Sullivan H.C. & Enos H.F. (1971) *Int. J. Environ. Anal. Chem.* **1,** 23.

Shapiro L. & Brannock W.W. (1956) *Rapid Analysis of Silicate Rocks.* Geological survey Bulletin No. 1036-C. U.S. Dept. Interior.

Sharkey M.J. (1970) *J. Br. Grassld. Soc.* **25,** 289.

Sharpe R.R. & Parks W.L. (1982) *Agron. J.* **74,** 785.

Shaw K. (1959) *J. Soil Sci.* **10,** 316.

Shaw W.M. (1951) *J. Assoc. Off. Agric. Chem.* **34,** 595.

Shaw W.M. (1952) *J. Assoc. Off. Agric. Chem.* **35,** 397.

Sheldrick B.H. & McKeague J.A. (1975) *Can J. Soil Sci.* **55,** 77.

Shellard E.J. (1968) *Quantitative Paper and Thin Layer Chromatography.* Proc. of Symposium. Academic Press, London.

Shendrikar A.D. (1974) *Sci. Total Environ.* **3,** 155.

Shendrikar A.D. & West P.W. (1975) *Anal. Chim. Acta.* **74,** 189.

Sheppard, S.C. & Bates T.E. (1982) *Commun. Soil Sci. Plant Anal.* **13,** 1095.

Sherma J. (1986) *Anal. Methods Pestic. Plant Growth Regul.* **14,** 1.

Sherman G.D. & Harmer P.M. (1943) *Proc. Soil Sci. Soc. Am.* **7,** 398.

Shrift A. (1964) *Nature* (*London*). **27,** 552.

Shimoishi Y. & Toei K. (1978) *Anal. Chim. Acta.* **100,** 65.

Sichere M-C., Cesbrom F. & Zuppi G-M. (1978) *Anal. Chim. Acta.* **98,** 299.

Siegal S. (1956) *Non-parametric Statistics.* McGraw Hill, New York.

Simon B.M. & Jones J.G. (1983) *Hydrobiologia.* **101,** 189.

Sinclair A.G. (1973) *N.Z.J. Agric. Res.* **16,** 287.

Singh K.D., Goulding K.W.T. & Sinclair A.H. (1983) *Commun. Soil. Sci. Plant Anal.* **14,** 1015.

Sirois J.C. (1962) *Analyst* (*London*) **87,** 900.

Sissons D.J., Telling G.M. & Usher C.D. (1968) *J. Chromatogr.* **33,** 435.

Sisterson D.L. & Wurfel B.E. (1984) *Int. J. Environ. Anal. Chem.* **18,** 143.

Skeggs L.T. (1957) *Am. J. Clin. Pathol.* **28,** 311.

Slavin M. (1979) *Atomic Absorption Spectroscopy.* John Wiley, New York.

Sloneker J.H. (1971) *Analyt. Biochem.* **43,** 539.

Small H., Stevens T.S. & Baumann W.D. (1975) *Anal. Chem.* **47,** 1801.

Small R.A., Lowry T.W. & Ejzak E.W. (1986) *Int. Lab.* **16,** 56.

Smee B.W., Hall G.E.M. & Koop D.J. (1978) *J. Geochem. Explor.* **10,** 245.

Smith A.E. & Walker A. (1977) *Pestic. Sci.* **8,** 449.

Smith A.E. & Milward L.J. (1983) *J. Chromatogr.* **265,** 378.

Smith B.F.L. & Bain D.C. (1982) *Commun. Soil Sci. Plant Anal.* **13,** 185.

Smith D., Paulsen G.M. & Raguse C.A. (1964) *Plant Physiol.* **39,** 960.

Smith D.D. & Bremner R.F. (1984) *Anal. Chem.* **56,** 2702.

Smith E.D., Switzer G.L. & Nelson L.E. (1970) *Forest Sci.* **16,** 483.

Smith I. (1969) *Chromatographic and Electrophoretic Techniques,* Vol. 1. Heinemann, London.

Smith J.D. (1971) *Anal. Chim. Acta.* **57,** 371.

Smith K.A. (Ed.) (1983) *Soil Analysis Instrumental and Related Procedures.* Dekker, New York.

Smith K.A. & Scott A. (1983) in *Soil Analysis* (Ed. by K.A. Smith). Dekker New York.

Smith R. & James G.V. (1981) *The Sampling of Bulk Materials.* Royal Soc. of Chemistry, London.

Smith R.M. (1983) *J. Pharm. Biomed. Anal.* **1,** 143.

Smith V.R. (1979) *Commun Soil Sci. Plant Anal.* **10,** 1067.

Smith W.O. & Stallman R.W. (1954) *Measurement of Permeabilities in Ground Water Investigations.* Amer. Soc Test Mat. Spec. Tech. Publ. 143.

Smyth W.F. & Healy J.A. (1984) *Sci. Total Environ.* **37,** 71.

Snedecor G.W. & Cochran W.G. (1980) *Statistical Methods,* University Press, Iowa State.

Snyder D. & Reinert R. (1971) *Bull. Environ. Contam. Toxicol.* **6,** 385.

Snyder J.D. & Trofymow J.A. (1984) *Commun. Soil Sci. Plant Anal.* **15,** 587.

Snyder L., Levine J., Stoy R. & Canetta A. (1976) *Anal. Chem.* **48,** 942 A.

Society for Analytical Chemistry (1959) Analytical Methods Committee. *Analyst* (*London*). **84,** 214.

Society for Analytical Chemistry (1960) Analytical Methods Committee. *Analyst* (*London*). **85,** 643.

Society for Analytical Chemistry (1963) Analytical Methods Committee. *Analyst* (*London*) **88.**

Society for Analytical Chemistry. (1968) Analytical Methods Committee. *Analyst* (*London*). **93,** 414.

Society for Analytical Chemistry (1982) Analytical Methods Committee. *Analyst* (*London*) **107.**

Soldin S.J. & Hill J.G. (1978) *Clin Chem.* **24,** 747.

Soltanpur P.N., Khan A. & Lindsay W.L. (1976) *Commun. Soil Sci. Plant Anal.* **7,** 797.

Somogyi J. (1952) *J. Biol. Chem.* **195,** 19.

Sonneveld C. & Van Dijk D.A. (1982) *Commun. Soil Sci. Plant Anal.* **13,** 487.

Spackman D.H., Stein W.H. & Moore S. (1958) *Anal. Chem.* **30,** 1190.

Sparks C.J. & Ogle J.C. (1973) *Proc. 1st. Annual NSF Contaminants Contf.* Oak Ridge, 1973.

Springer V. (1943) *Bodenk. Pflernahr.* **32,** 129.

Stahl. E. (Ed.) (1970) *Thin layer Chromatography, A Laboratory Handbook.* Springer-Verlag, Berlin.

Standard Methods for the Examination of Water and Wastewater (1985) Ed. by A. Greenberg). American Public Health Association.

Standing Committee of Analysts (1978) *Determination of pH Value of Sludge, Soil, Mud & Sediment.* HMSO, London.

Standing Committee of Analysts (1980) *The Instrumental Determination of Total Oxygen Carbon, Total Oxygen Demand and Related Determinants.* HMSO, London.

Standing Committee of Analysts (1981) *Silica in Water and Effluents.* HMSO, London.
Standing Committee of Analysts (1982) *Analysis of Surfactants in Waters, Wastewaters and Sludges.* HMSO, London.
Standing Committee of Analysts (1982) *Hardness in Waters by EDTA Titrimetry.* HMSO, London.
Standing Committee of Analysts (1987) *The Determination of Trace Metals in Marine and other waters by Stripping Voltammetry or AAS.* HMSO, London.
Standing Committee of Analysts (1988a) *The Determination of Carbonates, Thiocarbonates, Related Compounds and Ureas in Waters.* HMSO, London.
Standing Committee of Analysts (1988b) *The Determination of Six Specific Polynuclear Aromatic Hydrocarbons in Waters.* HMSO, London.
Stanton R.E. (1966) *Rapid Methods of Trace Analysis for Geochemical Applications.* Arnold, London.
Stark N. & Spitzer C. (1985) *Can. J. For. Res.* **15,** 783.
Steel R.D. & Torrie J.H. (1960) *Principles and Procedures of Statistics.* McGraw-Hill, New York.
Steele P.J. (1978) North Div. Rep. ND-R-240(R), UKAEA.
Steikel J.E. & Flannery R. (1971) in *Instrumental Methods for Analysis of Soils and Plant Tissues.* (Ed. by L.M. Walsh). Soil Sci. Soc. Am.
Steinberg S., Venkatesan M.I. & Kaplan I.R. (1984) *J. Chromatogr.* **298,** 427.
Steinhardt G.C. & Mengel D.B. (1981) *Commun. Soil. Sci. Plant Anal.* **12,** 71.
Stern A.L. (Ed.) (1968) *Air Pollution.* Academic Press, London.
Sterrett S.B., Smith C.B., Mascianica M.P. & Demchek K.T. (1987) *Commun. Soil Sci. Plant Anal.* **18,** 287.
Stevens L.J. & Hodgeson J.A. (1973) *Anal. Chem.* **45,** 443A.
Stevenson F.J. (1982) in *Methods of Soil Analysis.* Part 2. (Ed. by A.L. Page). American Society of Agronomy, Madison, Wisconsin.
Stevenson I.L. (1956). *Plant Soil.* **8,** 170.
Steward J.H. & Oades J.M. (1972) *J. Soil. Sci.* **23,** 38.
Steyn W.J.A. (1959) *Ph.D. Thesis,* Rhodes University, South Africa.
Strickland J.D.H. & Parsons T.R. (1960) *A Manual of Seawater Analysis.* Bull No. 125. Fish Res. Board, Canada.
Stickland J.D.H. (1968) *A Practical Handbook of Sea-water Analysis.* Bull No. 167 Fish Res. Board, Canada.
Stockwell P.B. (1980) *Talanta.* **27,** 835.
Stupar J. & Ajlec R. (1982) *Analyst (London).* **107,** 144.
Sturgeon R. & Chakrabarti C.L. (1978) *Prog. Anal At. Specrosc.* **1,** 5.
Sturgeon R. (1977) *Carbohydrate Chem.* **9,** 208.
Sturgeon R. (1978) *Carbohydrate. Chem.* **10,** 208.
Sturgeon R. (1979) *Carbohydrate Chem.* **11,** 236.
Subramanian K.S. & Meranger J.S. (1979) *Int. J. Environ. Anal. Chem.* **7,** 25.
Sugino K. (1976) *Chem. Abstr.* **85,** 83074 f.
Sullivan T.J., Seip H.M. & Muniz I.P. (1986) *Int. J. Environ. Anal. Chem.* **26,** 61.
Swain D.J. & Mitchell R.L. (1960) *J. Soil Sci.* **11,** 347.
Swain T. (Ed.) (1963) *Chemical Plant Taxonomy.* Academic Press London.
Swain T. & Goldstein J.L. (1963) in *Methods in Polyphenol Chemistry.* (Ed. by J.B. Pridham) Pergramon Press, Oxford.
Swaminathan K., Sud K.C. & Verma B.C. (1981) *Commun. Soil Sci. Plant Anal.* **12,** 373.
Swift R.S. (1985) in *Humic Substances in Soil Sediment and Water,* (Ed. by G.R. Aiken, D.M. McKnight, R.L. Wershaw, P. McCarthy). John Wiley, New York.
Swift R.W. (1956) *Penn. State Univ. Coll. Agric Bull.* 616.
Sykes G. (1958) *Disinfection and Sterilization.* Spon, London.

Tabatabai M.A. & Bremner J.M. (1970) *Proc. Soil Sci. Soc. Am.* **34,** 608.
Tabatabai M.A. & Chae Y.M. (1982) *Agron. J.* **74,** 404.
Tabatabai M.A. & Dick W.A. (1983) in *Ion Chromatographic Analysis of Environmental Pollutants,* Vol. 2. (Ed. by J.D. Mulik and E. Sawacki). Ann Arbor Sc. Publ. Michigan.
Tamm O. (1922) *Mddr. St. Skogsfor Anst.* **19,** 385.
Tarapchak S.J. (1983) *J. Environ. Qual.* **12,** 105.
Tatton J.O.G. & Wagstaffe P.T. (1969) *J. Chromatogr.* **44,** 284.
Taylor D.L. & Anderman G. (1973) *Appl. Spectr.* **27,** 352.
Tempel A.S. (1982) *J. Chem. Ecol.* **8,** 1289.
Tertian R. (1972) *Spectrochim Acta.* **27B,** 159.
Thomas M.D. & Amtower R.E. (1969) *J. Air Pollut. Control Assoc.* **19,** 439.
Thomas T.A. (1977) *J. Sci. Food Agric.* **25,** 381.
Thompson A. & Raven A.M. (1955) *J. Sci. Food Agric.* **6,** 768.
Thompson M. & Walsh J.N. (1983) *A Handbook of Inductively Coupled Plasma Spectrometry.* Blackie, Glasgow and London.
Thompson M.J., Patterson G.W., Dutky S.R. Svoboda J.A. & Kaplanis J.N. (1980) *Lipids.* **15,** 719.
Thompson T. & Bankston D.E. (1970) *Appl. Spectrosc.* **24,** 210.
Thornburg W.W. (1963) in *Analytical Methods for Pesticides, Plant Growth Regulators and Food Additives.* (Ed. by G. Zweig and J. Sherma). Academic Press, New York.
Thornton J.S. & Anderson C.A. (1968) *J. Agric. Food Chem.* **17,** 895.
Thurman E.M. (1984) *U.S. Geological Survey Water Supply,* Paper No. 2262.
Tinsley J. (1950) *Trans. 4th Int. Cong. Soil Sci.* **1,** 161.
Tobia S.K. & Milad N.E. (1956) *J. Sci. Food Agric.* **7,** 314.
Tomimori T., Jin H., Miyaichi Y., Toyafuku S. & Namba T. (1985) *Anal. Abstr.* **47,** 12E24.
Tonamura K., Maeda K., Futai F., Nakagami T. & Yamada M. (1968) *Nature (London).* **217,** 644.
Tong H.Y. & Karasek F.W. (1984) *Anal. Chem.* **56,** 2129.
Touchstone J.C. & Rogers D. (1980) *Thin Layer Chromatography.* John Wiley, New York.

Townsend W.N. (1973) *An Introduction to the Scientific Study of the Soil.* Arnold, London.

Trabalka J.R. (Ed.) (1985) *Atmospheric Carbon Dioxide and the Global Carbon Cycle.* U.S. Dept. Energy.

Tranchant J. (Ed.) (1969) *Practical Manual of Gas Chromatography.* Elsevier, Amsterdam.

Troedsson T. & Tamm C.O. (1969) *Studia for. suec.* No. 74.

Truesdale V.W. & Smith C.J. (1975) *Analyst (London)* **100,** 203.

Truesdale V.W. & Smith C.J. (1975) *Analyst (London)* **100,** 797.

Truog E. (1930) *J. Am. Soc. Agron.* **22,** 874.

Tucker B.M. (1985) *Austral. J. Soil Res.* **23,** 195.

Turner D.R. & Whitfield M. (1974) in *Marine Electrochemistry.* (Ed. by M. Whitfield and D. Jagner). John Wiley, New York.

Turnock W.J., Gerber G.H. & Bickis M. (1979) *Can. Entomol.* **3,** 113.

Tyree S.Y. (1981) *Atmos. Environ.* **5,** 57.

Uriano G.A. & Gravett C.C. (1977) *The Role of Reference Materials and Reference Methods in Chemical Analysis.* CRC Critical Review in Analytical Chemistry. 6, 4, 361.

Ure A.M. & Bacon J.R. (1978) *Analyst (London).* **103,** 807.

Urquart C. (1966) *Nature (London).* **211,** 550.

Urquart C. & Gore A.J.P. (1973) *Soil Biol. Biochem.* 5.

Usepa (1971) *Methods for Organic Pesticides in Water and Wastewater.* National Research Center, Cincinnati.

Uthe J.F. (1971) *Identification and Measurement of Environmental Pollutants* (Ed. by B. Westley). Proc. of Int. Symp. Ottawa. June 1971. National Research Board of Canada.

Van Damme J.C. & Galoux M. (1980) *J. Chromatogr.* **190,** 401.

Van den Berg C.M.G. (1985) *Anal. Chem.* **57,** 1532.

Van den Driessche R. (1974) *Bot. Rev.* **40,** 347.

Van der Borght B.M. & Grieken R. (1977) *Anal. Chem.* **49,** 311.

Van der Mark W. & Das H.A. (1975) *J. Radioanal. Chem.* **23,** 7.

Van der Veen N.G., Keukens H.J. & Ver G. (1985) *Anal. Chem. Acta.* **171,** 285.

Van Grieken R., Bresseleers K., Smits I., Van der Borght B.M. & Van der Stappen M. (1976) *Adv. X-ray Anal.* **19,** 435.

Van Grieken R. (1982) *Anal. Chem. Acta.* **142,** 3.

Van Grieken R. & La Brecque J.J. (1985) in *Trace Analysis,* Vol. 4. (Ed. by J.F. Lawrence). Academic Press, London.

Van Horne K.C. (1985) *Sorbent Extraction Technology.* Analytichem International, Harbor City, California.

Van Loon J.C. (1985). *Selected Methods of Trace Metal Analysis.* John Wiley, New York.

Van Middelem C.H. (1971) *Pesticides in the Environment.* (Ed. by R. White-Stevens) Vol. I. part II. Dekker, Maidenhead.

Van Nierkerk P.J. & Du Plessis L.M. (1976) *S. Afr. Food Rev.* **3,** 167.

Van Slyke D.D. (1929) *J. Biol. Chem.* **83,** 425.

Van Slyke D.D. & Folch J. (1940) *J. Biol. Chem.* **136,** 509.

Van Soest P.J. & Wine R.H. (1967) *J. Anim. Sci.* **26,** 940.

Van Soest P.J. (1973) *J. Assoc. Off. Anal. Chem.* **56,** 781.

Van Zyl J.D. (1978) *Wood Sci. Technol.* **12,** 251.

Vaz Henin J.G. & Karaseva G.I. (1959) *Pochvovedenie.* **8,** 87.

Veith J.A. (1977) *Soil Sci. Soc. Am. J.* **41,** 865.

Veno S. & Ishizaki M. (1979) *Anal. Abstr.* **37,** 6D20.

Verbyla D. (1986) *Can. J. For. Res.* **16,** 1255.

Verma B.C., Swaminathan K. & Sud K.C. (1977) *Talanta.* **24,** 49.

Vestal M.L. (1984) *Science.* 226, 275.

Violanda A.T. & Cooke W.D. (1964) *Anal. Chem.* **36,** 2287.

Voelter W. & Zech K. (1975) *J. Chromatogr.* **112,** 643.

Vogel A.I. (1978) *Quantitative Inorganic Analysis.* Longmans Green, London.

Von Loesecke H.W. (1957) *Anal. Chem.* **29,** 647.

Vos L., Komy Z., Reggers G., Roekens E. & Van Grieken R. (1986) *Anal. Chem. Acta.* **184,** 271.

Wagner H., Tittel G. & Bladt S. (1983) *Dtsch. Apoth-Ztg.* **123,** 515.

Wainwright M. (1984) *Adv. Agron.* **37,** 349.

Wainwright S.J. (1980) *Adv. Bot. Res.* **8,** 221.

Waksman S.A. & Stevens K.R. (1930) *Soil Sci.* **30,** 97.

Waliszewski S.M. & Szymczynski G.A. (1982) *J. Assoc. Off. Anal. Chem.* **60,** 1277.

Walker A. (1976) *Pestic. Sci.* **7,** 41.

Walker P.H. & Hutka J. (1971) *Use of the Coulter Counter for Particle Size Analysis of Soils.* Divn. Soils Tech. Paper No. 1. CSIRO.

Walkley A. & Black C.A. (1934) *Soil Sci.* **37,** 29.

Wall L.L., Gehrke C.W. & Suzuki J. (1980) *J. Assoc. Off. Anal. Chem.* **63,** 847.

Walsh A. (1955) *Spectrochim Acta.* **7,** 108.

Walsh L.M. & Beaton J.D. (Eds.) (1973) *Soil Testing and Plant Analysis.* Soil Sci. Soc. Amer., Madison.

Ward R.S. & Pelter A. (1974). *J. Chromatogr. Sci.* **12,** 570.

Watling H.R. & Wardale I.M. (1977) in *Proc. Spectroscopic Soc.* Pretoria SA. (Ed. by L.R.P. Butler). Pergamon Press, Oxford.

Watson C.A. (1968) *Ammonium Pyrollidine Dithiocarbamate,* Monograph 74.

Watson, M.E. (1981) *Commun. Soil Sci. Plant Anal.* **12,** 601.

Weast R.C. (Ed.) (1987) *CRC Handbook of Chemistry & Physics.* CRC Press Boca Raton.

Webber M.D. (1972) *Can. J. Soil Sci.* **52,** 282.

Webster R. (1977) *Quantitative and Numerical Methods in Soil Classification and Survey.* Clarendon Press, Oxford.

Webster P.T.B. (1980) *Determination of Metals in Sewage. Water Pollut. Control.* **79,** 511.

Wehrmann J. (1963) *Landw. Forsch.* **16,** 130.

Weinstein B. (1966) in *Methods of Biochemical Analysis,* Vol. 14. (Ed. by D. Glick). Interscience, New York.

Weller A. (1894) *Z. Analyt. Chem.* **23,** 410.

Westcott C.C. (1978) *pH Measurement.* Academic Press, New York.

Westley B. (Ed.) (1971) *Identification and Measurement of Environmental Pollutants.* Int, Symp. Proc., Ottawa. June 1971. *National Research Board of Canada.*

West P.W. (1971) *Identification and Measurement of Environmental Pollutants* (Ed. by B. Westley). Proc. of Int. Symp. Ottawa. June 1971. National Research Board of Canada.

West T.S. & Nürnberg H.W. (Eds.) (1988) *The Determination of Trace Metals in Natural Waters.* Blackwell Scientific Publications, Oxford.

Westöö G. (1967) *Acta Chem. Scand.* **21,** 1790.

Wetzel R.G. & Likens G.E. (1979) *Limnological Analysis.* Holt Saunders, Eastbourne.

White C.E., McFarlane H.C.E., Fogt T. & Fuchs R. (1967) *Anal. Chem.* **39,** 367.

White C.E. & Weissler A. (1970) *Anal. Chem.* **42,** 57.

White R.E. (1980) in *Soils and Agric.* (Ed. by P.B. Tinker). *Crit. Rep. Appl. Chem,* 2. Blackwell Scientific Publications, Oxford.

White-Stevens R. (Ed.) (1971) *Pesticides in the Environment,* Vol. I, Part 2. Dekker, New York.

Wickbold R. (1971) *Tenside Detergents* **61.**

Whitehead D.C. (1981) *J. Sci. Food. Agric.* **32,** 359.

Whitfield M. & Jagner D. (Eds.) (1981) *Marine Electrochemistry.* John Wiley, New York.

Wilkins C. (1983) in *Soil Analysis Instrumental and Related Procedures.* (Ed. by K.A. Smith). Dekker, New York.

Willard H.H., Merritt L.L. & Dean J.A. (1965) *Instrumental Methods of Analysis.* Van Nostrand-Reinhold, London.

Willets C.O. (1951) *Anal. Chem.* **23,** 1058.

Williams C. (1976) *J. Sci. Food Agric.* **27,** 561.

Williams C.H. & Steinbergs A.(1964) *Plant Soil.* **27,** 50.

Williams D.E. & Vlamis J. (1961) *Anal. Chem.* **33,** 967.

Williams D.H. (Ed.) (1981) *Mass Spectrometry: Principles and Applications.* McGraw-Hill, London.

Williams E.G. & Stewart A.B. (1941) *J. Soc. Chem. Ind. (London)* **60,** 291.

Williams E.G. (1967) *Anal. Edutologia Agrobiol.* **26,** 1.

Williams J.D., Syers J., Walker T.W. & Rex R.W. (1970) *Soil Sci.* **110,** 13.

Williams P.C. (1973) *J. Sci. Food Agric.* **24,** 343.

Williams R. (1928) *J. Agric. Sci.* **18,** 439.

Williams W.J. (1979) *Handbook of Anion Determination.* Butterworth, London.

Willis R.B. & Gentry C.E. (1987) *Commun. Soil Sci. Plant Anal.* **18,** 625.

Wilson A.D. & Sergeant G.A. (1963) *Analyst (London).* **88,** 109.

Wilson B.O. (1984) *Commun. Soil Sci. Plant Anal.* **15,** 12.

Wilson D.L. (1979) *At. abs. Newsletter.* **18,** 13.

Wilson M.F., Haikala E. & Kirialo P. (1975) *Anal. Chim. Acta.* **74,** 395.

Winkler L.W. (1888) *Ber. Dt. Chem. Ges.* **21,** 2843.

Wise L.E. & Jahn E.C. (1952) *Wood Chemistry,* Vols 1 and 2. Van Norstrand-Reinhold, London.

Wise L.E., Peterson F.C. & Harlow W.M. (1939) *Ind. Eng. Chem., Anal. Edn.* **11,** 18.

Wolf W.R. (Ed.) (1985) *Biological Reference Materials.* John Wiley, New York.

Wolfschoon-Pumbo A., Klostermeyer H. & Weiss G. (1982) *Milchwissenschaft* **37,** 80.

Woodford F. (1964) in *Fatty Acids.* (Ed. by K.S. Markey) Interscience, New York.

Woodruff C.M. (1948) *Soil Sci.* **66,** 53.

Woodwell G.M. (Ed.) (1984) *The Role of Terrestrial Vegetation in the Global Carbon Cycle.* Scope 23, John Wiley, New York.

Woolhouse H.W. (1983) in *Physiological Plant Ecology* III. (Ed. by O.L. Lange *et al.*) Springer, Berlin.

Woolson E.A. & Kearney P.C. (1969) *J. Assoc. Off. Anal. Chem.* **52,** 1202.

Wulf L.W. & Nagel C.W. (1976) *J. Chromatogr.* **110,** 401.

Yainans H.L. (1973) *Statistics for Chemists.* Merrill, Ohio.

Yamada H. & Hattori T. (1987) *J. Chromatogr.* **411,** 401.

Yamamoto K., Kameda Y. & Hikasa Y. (1983) *Int J. Environ. Anal. Chem.* **16,** 1.

Yofe H.Y. & Finkelstein R. (1958) *Anal. Chim. Acta.* **19,** 166.

Youmans H.L. (1973) *Statistics for Chemistry.* Merrill, Ohio.

Young D. & Bache B.W. (1985) *J. Soil Sci.* **36,** 261.

Yu T.R. (1985) *Ion Selective Electrode Rev.* **7,** 165.

Yuan T.L. & Fiskell J.G.A. (1959) *J. Agric. Food Chem.* **7,** 115.

Zeichmann W. (1980). *Huminstoffe: Probleme, Methoden, Ergebnis.* Verlag Chemie, Weinheim.

Zeichmann W. & Weichelt T. (1977) *Z. Pflanzenernaehr Bodenkol.* **140,** 645.

Zief M. & Mitchell J.W. (1976) *Contamination Control in Trace Element Analysis, A series of monographs on analytical chemistry and applications.* John Wiley, New York.

Zill L.P. (1956) *Anal. Chem.* **28,** 1577.

Zweig G. & Sherma J. (1967–1986) *Analytical Methods for Pesticides Plant Growth Regulators and Food Additives,* Vols 1–15. Academic Press, New York.

Zweig G. & Sherma J. (Eds-in-Chief) *CRC Handbook of Chromatography.* Volumes include *General Principles* (1973) *Polymers* (1982), *Carbohydrates* (1982), *Phenols and Organic Acids* (1982), *Amino Acids* and *Amines* (1983), *Terpenoids* (1984), *Lipids* (1984), *Pesticides and Related Organics* (1984), *Steroids* (1985), *Inorganics* (1986) and *Peptides* (1986). CRC Press, Boca Raton, Florida.

Zweig G. & Whitaker J.R. (1967)(1971) *Paper Chromatography and Electrophoresis.* in *Analytical Methods for Pesticides etc.,* Vols 1 and 2. Academic Press, New York.

Subject Index

Abbreviations 6
Abnormal values 295
Accuracy, limitations to 286–7
Acetic acid extractants 34
Acid oxidation techniques 57–60
  for carbon analysis 92–5
  mixed acid digestion 59
  semi-micro Kjeldahl digestion 58
  sulphuric–peroxide digestion 58, 59–60
Acid solutions
  constants for 340
Acidity of water 69–72
Adiabatic bomb calorimeter 285–6
Adsorbents 222–3, 273
Aerosols 239
Air-drying 11, 50
Alkalinity of water 69–72
Aluminium 81–4
  ashing and digestion of vegetation 60
  by atomic absorption 83–4
  by colorimetry 82–3
  by continuous flow 82
  fusion of soils for 29
  soil extraction procedure for 37
Amino-acids 179–84
  extraction 180
  fractionation of 181–2
  by high performance liquid chromatography 183
Ammonia solution
  constants for 340
Ammonium acetate extractants
  pH 7 33–4
  pH 9 34
Ammonium-nitrogen
  by colorimetry 124, 127
  by continuous flow 125–6
  by distillation 126–7
  soil extraction procedure for 41, 126
Ammonium pyrrolidine dithiocarbamate (APDC) 73
Animal materials 49
  chemical characteristics of 331
Anionic surfactants 49
Anions in water
  soluble constituents 75–6
  total constituents 76–8
Anodic stripping voltammetry 267
Anthrone 79
Antimony 202–3
  by atomic absorption 202–3
  by colorimetry 203
Arithmetic mean 295
Arsenic 203–5
  dry-ashing 203
  preparation by molybdenum blue method 204–5
Ascorbic acid reduction 139
Ash
  determination in vegetation 54–5
  silica-free 54–5
Atmospheric deposition 237–9
Atmospheric pollutants
  aerosols 239
  dust 239
  gases 238–9
  rainwater 237
Atomic absorption 252–5
  method for aluminium 83–4
  method for cadmium 205–6
  method for calcium 89–91
  method for chromium 207
  method for copper 106–7
  method for iron 109–11
  method for lead 209
  method for magnesium 112–14
  method for manganese 115–16
  method for mercury 211
  method for nickel 214
  method for zinc 157–9
Atomic spectrometry 249–57
  automated 291
Atomic weights 339
Automated instruments
  continuous flow analysers 245–6
  discrete flow (batch) analysers 290
Automation 289–91
  of calculating procedures 290
  of diluting and dispensing procedures 290
  see also computer applications
Available nutrients 30

Bar chart 293
Barium 216

Baseline drift 247, 311–2, 320
Beer–Lambert Law 241
Beryllium 216
BF_4–methylene blue colorimetric method 85–6
Bioenergetics 332–4
Bismuth 158
Blank values 1, 313
Blocking 309–10
Bomb calorimetry 61, 284–9
Boron 84–6
  by colorimetry 84–6
  fusion of soils for 29
  ignition of vegetation for 60, 61
  soil extraction procedure for 37
Bouyoucos hydrometer 23
Bryophytes
  chemical data for 330
Buffer solutions 17
Bulk density 20
Bulk reduction
  of soil samples 11–12
  of vegetation samples 49–50
Burners
  for atomic absorption 253
  for flame emission systems 250

Cadmium 205–6
  by atomic absorption 205–6
Calcareous soils
  extraction procedures for 32
Calcium 86–91
  ashing and digestion of vegetation for 60
  by atomic absorption 89–91
  fusion and digestion method for 29
  by ion-selective electrode 88–9
  soil extraction method for 38
  by titration 87–8
Calculation procedures 313–14
Calibration procedures 310–13
Calorific value(s)
  determination of 333–4
  of terrestrial biomaterials 333
Carbamate colorimetric method 104–6
Carbamate herbicides 218, 227–8
Carbohydrates 172–6
  anthrone determination 79
  by chromatography 175–6
  soluble 164–6
Carbon 91–7
  by acid oxidation and titration 92–5
  by combustion and gravimetry 92, 96
  and organic matter from loss-on-ignition 95–6
  rapid titration method 96–7
  in water 76–7
Carbonate-carbon
  gravimetric method for 96
  rapid titration method for 96–7
Carbonate soils 28
Carbon dioxide in waters 72
Carbon–nitrogen relationship 91
Carcinogenic reagents 132, 215, 336
Carotenoids 197–8
Catechol violet 82–3
Cation exchange capacity 34–5
Cations in water 72–4
Cellulose 166–7
  alpha- 168
  hemi- 169
  holo- 167–8
Chemical data
  of animal materials 331
  of plant materials 328–31
  of soils 325–7
Chemical oxidation demand (COD) 78–9
'Chi' square 307
Chlorinated hydrocarbons 215, 217
Chlorine and chloride 97–100
  by colorimetry 77, 98–9
  by continuous flow 99–100
  fusion of soils for 29
  ignition of vegetation for 60
  by ion-selective electrode 100
  soil extraction method for 38
Chlorophyll 79–80, 195–7
  total 196–7
Chromatography 271–84
  automated 291
  fractionation of carotenoids by 197
  fractionation of sterols by 198
  gas- 176, 183–4, 189, 274–8
  herbicides 226
  high performance liquid- 175–6, 183, 189, 194, 278–80
  ion- 280–4
  of amino-acids 183
  of carbohydrates 175–6
  of fatty acid methyl esters 188–9
  of flavonoids 192–4
  of organic pesticides 223–9
  of organo-mercurial compounds 212
  of phenoxylalkanoic acids and esters 226
  paper- 175, 182, 192–4
  thin-layer 175, 182–3, 194, 272–4
Chromium 206–7
  by atomic absorption 207
  by colorimetry 206–7
Clay
  determination of 23–4
Cleaning agents 5
Clean-up stage for organic pesticides 221–3
Cobalt 101–4
  ashing and digestion of vegetation for 60
  by atomic absorption 102
  by colorimetry 101–2
  electrothermal atomization 102
  fusion of soils for 29
  by polarography 102–4
  soil extraction method for 38
Cold vapour technique for mercury 211
Colorimetry 241–4
  automated 290–1
Coloured waters 67–8
Comminution 12
Comparison tests 299
Computer applications 312–13, 318–21
Concentrated acids 340
Conductivity 71–2
Confidence limits 296

Constants 339–42
Contamination 1
  of soil samples 9, 25
  of vegetation samples 49, 52–3
  laboratory 335
Continuous flow technique 244–9
  fault-finding 246–8
  for ammonium-nitrogen 125–6
  for chloride 99–100
  for iron 109
  for nitrate and nitrite-nitrogen 130
  for organic-nitrogen 125–6
  for phosphate-phosphorus 137–9
  for silicon 147–8
Control charts 317–18
Control procedures 316–18
Conversion factors
  B. S. sieves 342
  chemical 340
Copper 104–7
  ashing and digestion of vegetation for 60
  by atomic absorption 106–7
  by colorimetry 104–6
  fusion and digestion of soils for 29
  soil extraction procedure for 38
Correlation 306
Covariance
  analysis of 301–2
Creatine and creatinine 184–5
Crude fat 163–4
Crude fibre 171–2
Crude protein 164
Curcumin 60, 85
Curvilinear regression 302
Cyanide
  by colorimetry 231–2

Databases 320–1
Data acquisition 320
Data processing 320
Deb's extraction 39
Deionized water 3
Density 20–1
Detection limit 316
Detectors
  for gas chromatographs 225–6, 277
  for high performance liquid chromatographs 280
  for ion chromatographs 284
  in spectrophotometry 242–3
Detergents 5
Deviation
  mean 296
  standard 296
Diaminobenzidine colorimetric method 215
Dicotyledons
  chemical characteristics of 329–30
Digestion methods
  hydrofluoric–perchloric for soils 27–8
  perchloric–nitric–sulphuric for soils 27–8
  perchloric–nitric–sulphuric for vegetation 59
  semi-micro Kjeldahl 58, 120–1
  sulphuric–hydrogen peroxide for vegetation 59–60
Diluting equipment 290
Dimethylglyoxime colorimetric method 213–14
Diphenylcarbazide colorimetric method 206–7
Discrete data 295
Dispensing equipment 290
Disaccharides 172
Dissolved gases in water
  carbon dioxide 72
  oxygen 72
Dissolved solids 68–9
Distillation 2
  for ammonium-nitrogen 126–7
  for nitrate-nitrogen 127, 128–9
  for total organic nitrogen 122–4
Distilled water 3
Dithizone colorimetric methods 73, 156–7, 208–9, 210–11
Diurnal variation 48
Drift correction 312
Dry-ashing 56–7, 61
Dry-deposition 63, 238–9
Drying
  samples for bomb calorimetry 287
  samples for organic analysis 161
  soils 11
  vegetation 50

EDTA titration
  for calcium 87–8
  for magnesium 111–12
Electroanalytical techniques 265–71
Electrothermal atomization 256
  method for aluminium 84
  method for chromium 207
  method for cobalt 102
  method for copper 107
  method for lead 209
Emission spectrometry 256–7
Energy 332–4
Energy dispersive XRF 260–1
Errors
  random 315–18
  systematic 314–15
Exchangeable hydrogen 36
Exchangeable nutrients 30
Experimental design 308–10
Explosion hazards 338
Expression of results 5–6
Extractants
  acetic acid 34
  ammonium acetate (pH 7) 33–4
  ammonium acetate (pH 9) 34
  choice of 31–2
  for aluminium 37
  for ammonium-nitrogen 41
  for boron 37
  for calcium 38
  for copper, zinc and cobalt 37
  for iron 38–40
  for magnesium 40
  for manganese 40
  for molybdenum 40
  for nitrate-nitrogen 42
  for phosphorus 42–4, 139–40

Extractants (*continued*)
for potassium 44
for sodium 44
for sulphur 45
Extraction methods
for carbohydrates 172–3
for chlorophyll 196
for organic pesticides 219–21
for polyphenolic compounds 191–2
for protein 179
for soils 30–45

Factorial experiments 310
Fat *see* Crude fat
Fatty acids 79, 186, 187–9
fractionation of 188–9
free 188
total 188
Fibres *see* Crude fibre
Filter papers 4
Filters, optical 242
Fire hazards 336
Flame emission 249–52
method for potassium 142–3
method for sodium 148–9
Flameless atomization *see* Electrothermal atomization
Flavonoids and related compounds 189–94
Fluorimetry 244
Fluorine 232–4
by ion-selective electrode 233–4
Folin–Denis reagent 190
Formaldoxime colorimetric method 114–15
Freeze storage 10, 49
Frequency distributions 294
Fungicides 218
Fusion
potassium pyrosulphate 27
sodium carbonate 25–6
sodium hydroxide 26–7

Gas cylinders 338
Gas-liquid chromatography 274–8
of amino-acids 183–4
of carbohydrates 176
of fatty acid methyl esters 189
of herbicides 226
of organo-mercurial compounds 212
of pesticides 223–6
Gases 238
dissolved 72
Germanium 158
Glassware 4
cleaning for 5
Graphs 293–4, 311–12
Grasses
chemical characteristics of 328
Graphite furnace *see* Electrothermal atomization
Gravimetric method
for carbonate-carbon 96
for silica 145
Greiss–Ilosvay reaction 132
Grinders
for soils 12–14
for vegetation 51–2
Gymnosperms
chemical characteristics of 330

Heavy metals 201–15
Herbicides 217, 227–8
Herbs
chemical characteristics of 329–30
High performance liquid chromatography 278–80
of amino-acids 183
of carbohydrates 175–6
of fatty acids 189
of flavonoids 194
of herbicides and pesticides 226–7
Histograms 293
Hollow cathode lamps 253
Homogenization 161–2
Humus 80
fractionation of 199–200
Hydrogen peroxide colorimetric method 154

Indophenol colorimetric method 124–5, 127
Insecticides
chlorinated hydrocarbon 215, 217
organo-phosphorus 215, 217
Instruments
atomic absorption 252–3
bomb calorimetric 284–6
continuous flow 245–6
gas chromatographic 278
high pressure liquid chromatographic 279–80
ion-chromatographic 283–4
ion-selective electrode 269–70
thin-layer chromatographic 274
visible and ultra-violet spectrophotometric 242
voltammetric 267
X-ray fluorescence spectrophotometric 257–61
Interferences 251
Inter-year variation 48
Invertebrates
chemical characteristics of 331
Iodine 158
Ion-selective electrodes 269–71
method for calcium 88–9
method for chloride 100
method for fluoride 233–4
method for nitrate-nitrogen 130–2
Ion chromatography 280–4
Ion-exchange 74
Iron 107–11
ashing and digestion of vegetation for 60
by atomic absorption 109–11
by colorimetry 108–9
by continuous flow 109
fusion and digestion of soils for 29
soil extraction procedures for 38–40
in water 77

Kjeldahl digestion 58, 120–1
Kruskal–Wallis test 307–8
Kurtosis coefficient 295

Laboratory Information Systems 319, 321
Lanthanum chloride
  use as a releasing agent 90, 113
Latin squares 310
Lead 207–9
  by atomic absorption 209
  by colorimetry 208–9
Least squares 303–4
Lignin 169–71
Linear regression 302, 305–6
Lipids 185–9
  fractionation of 187
  total 186–7
Lithium 158
Litter 7, 9, 46
Loss-on-ignition 15–16, 95
Low temperature precipitation 223

Magnesium 111–14
  ashing and digestion of vegetation for 61
  by atomic absorption 112–14
  fusion and digestion of soils for 29
  soil extraction procedure for 40
  by titration 111–12
  in water 77
Manganese 114–16
  ashing and digestion of vegetation for 61
  by atomic absorption 115–16
  by colorimetry 114–15
  fusion and digestion of soils for 29
  soil extraction procedures for 40
Mann–Whitney test 307
Matrix effects 262–3
Maximum likelihood 304
Mean deviation 296
Median 295
Membrane filters 66
Mercury 209–12
  by atomic absorption (cold vapour technique) 211
  by colorimetry 210–11
  organic *see* organo-mercurial compounds
Micro bomb calorimeter 286
Millipore filters 66
Mineral nutrients in the ecosystem 325–6
Mode 295
Moisture
  in soils 14–15
  in vegetation 53–4, 55
Molybdenum 116–18
  by colorimetry 116–17
  by polarography 118
  soil extraction procedure for 40–1
Molybdenum-blue, colorimetric methods 135–9, 145–7, 204–5
Monocotyledons
  chemical characteristics of 328
Monosaccharides 172
Morgan's reagent 31
Multiple regression 305–6

Nebulizer 250
Nernst equation 18, 269
Network communications 321
Nickel 213–14
  by atomic absorption 214
  by colorimetry 213–14
  by polarography 214
Nitrate-nitrogen 127–32
  by colorimetry 127, 129–30
  by continuous flow 130
  by distillation 127, 128–9
  by ion-selective electrode 130–2
  soil extraction method for 41–2
Nitrite-nitrogen 132–3
  by colorimetry 132
  by continuous flow 132 (130)
  soil extraction method for 41–2
Nitrogen 118–34
  α-amino-acid 180–1
  ammonium- 77, 126–7
  availability 118–19
  by colorimetry 124–5
  by continuous flow 125–6
  by distillation 122–4, 126–7
  digestion procedures for 120–1
  ecosystem cycling 119
  inorganic 126
  mineralizable 133–4
  nitrate- 77, 127–32
  nitrite- 77, 132–3
  organic- 119–26
  to carbon ratio 91
  soil extraction procedure for 41–2
Nitrogenous (organic) compounds 176–85
  amino-acids 179–84
  creatine and creatinine 184–5
  protein 179
  uric acid 184
Nitroso-R-salt colorimetric method 101–2
Non-parametric tests 306–8
Normal disfribution 294
Nutrition 333

Oak (*Quercus petraea*)
  chemical characteristics of 47, 326, 331
Odour of waters 67–8
Olsen's extraction 44
Organic acids 194–5
Organic compounds in the ecosystem 160–200
Organic matter—carbon relationship 95
Organic nitrogen *see* Nitrogen
Organic phosphorus 77
  by colorimetry 140–1
  extraction procedure for 139–40
Organo-mercurial compounds 211–12
  by gas chromatography 212
  by thin-layer chromatography 212
Organophosphorus insecticides 215, 217
  TLC separations of 223

Outlying values 296, 298
Oven drying 50
Oxalic acid extraction 39–40
Oxidizing agents 337
Oxygen in water 72

Paper chromatography
  of amino-acids 182
  of carbohydrates 175
  of flavonoids 192–4
Particle density 21
Particle fractionation 21–4
Particle size, reduction of soil 12–14
Peat 9, 46
Peak height 311
Percolation 19–20
Pesticides 215–29
pH
  of soils 16
  of waters 69
Phenoldisulphonic colorimetric method 129–30
Phenyl mercury 212
Phospholipids 189
Phosphorus 134–42
  by colorimetry 135–7, 140–1
  by continuous flow 137–9
  digestion of vegetation for 61
  ecosystem cycling 134–5
  fusion and digestion of soils for 29
  organic- 77, 139–41
  phosphate 135–9
  soil extraction procedures for 42–4
  total 141–2
Physical data
  of soils 14–24
Phytic acid 198–9
Plant litter *see* Litter
Plastic ware 4
Poisons 336
Polarography
  method for cobalt 102–4
  method for molybdenum 118
  method for nickel 213
Pollutants
  atmospheric 201, 237–9
  heavy metal 201–15
  laboratory chemical 335
  organic pesticide 215–29
  in water 229–31
Polychlorinated biphenyl compounds 228–9
Polycyclic aromatic hydrocarbons 236–7
Polyphenolic compounds 191–4
Polysaccharides 172
Porosity 21
Potassium 142–4
  ashing and digestion of vegetation for 61
  by atomic absorption 143–4
  by flame emission 142–3
  fusion and digestion of soils for 30
  soil extraction procedure for 41, 44, 126
Precision, limitations to 286–7
Preservation
  of samples for organic analysis 161
  of waters 65–7
Probability 294
Protein 179
Protein factor 164
Proximate constituents 162–72
  ash 162
  α-cellulose 168–9
  cellulose 162, 166–7
  crude fat 163–4
  crude fibre 171–2
  crude protein 164
  hemi-cellulose 167
  holo-cellulose 167–8
  lignin 162, 169–71
  soluble carbohydrates 164–6
Pteridiophytes
  chemical characteristics of 330
Pure water 2–3
Pyknometer 21
Pyridine–pyrazolone colorimetric method 74, 231–4
Pyrocatechol violet colorimetric method 82–3

Quality control 316

Random errors 304–5, 315
Randomization 308–9
Range 295
Reagents 2
  blank 1
Records, sampling 8–9
Redox potential 17–19
Reducing sugars 173–4
Reference electrodes 270
Reference samples 1–2, 316–17
Regression 302–6
Replication 309
  of samples 8
Reproducibility 264
Resistance (chemical) of laboratory materials 341
Reverse osmosis water 3
R_f values 272, 274
Rhodamine B colorimetric method 203

Safety
  flammable solvents 336–7
  gas cylinders 338
  oxidizing agents 337–8
  poisons 336
  waste disposal 338
Sampling
  of atmospheric materials 237–9
  of soils 7–14
  of vegetation 46–8
  of waters 62–5
  statistical considerations of 309
Sand
  determination of fine and coarse 24
Saturation capacity 21
Scatter diagram 293
Seasonal variation 48
Selective ion electrodes *see* Ion-selective electrodes

Selenium 214–15
by colorimetry 215
Sensing electrodes 269–70
Sensitivity 264, 316
Shewart control chart 317–8
Shrubs
chemical characteristics of 329–30
SI units 5–6
Sieve conversions 342
Sieving
of soil 12
Significance 298–302
Silica-free ash 54–5
Silicon 144–8
by colorimetry 145–7
by continuous flow 147–8
fusion of soils for 30
fusion of vegetation for 61
by gravimetry 77, 145
Silt
determination of 23–4
Skewness coefficient 295
Sodium 148–50
ashing and digestion of vegetation for 61
by atomic absorption 149–50
by flame emission 148–9
fusion of soils for 30
soil extraction procedure for 44
Sodium chloride extraction for ammonium-nitrogen 41, 126
Soil
augers 9
chemical composition of 24–5
classification triangle 22
corers 9–10
dissolution 24–30
extraction procedures 7–14, 30–45, 220
particle classification of 21–2, 327
water solutions 64–5
Soluble carbohydrates 173
Soluble tannins 189–91
Spatial design 309–10
Spearman's coefficient 308
Standard deviation 296
Standard error 296, 299
Starch 174–5
Sterols 198
Stoke's law 22
Storage of samples 10
for organic analysis 161
for pesticide analysis 218–19
soils 10
vegetation 49
waters 65–7
Strontium 158, 216
Substituted urea herbicides 227–8
Sulphide 234–5
iodimetric method for 234
Sulphonated bathophenanthroline 108–9
Sulphur 150–1
ashing of vegetation for 61
fusion of soils for 30
by ion chromatography 280–4
soil extraction procedure for 44–5, 151
sulphate 151–3
sulphide and sulphite 78
total 78
by turbidity 78, 150
by X-ray fluorescence spectrometry 151
Surfactants 235–6
Suspended solids 68
Systematic errors 314–15
Systems analysis 320–1

Tables 293–4
Tannins 189–91
Tellurium 216
Thallium 216
Thermochemical standard 287
Thermometer calibration 286
Thin-layer chromatography
of amino-acids 182–3
of carbohydrates 175
as a clean-up technique 223
of flavonoids 194
of organo-mercurial compounds 212
Thiocyanate colorimetric method 98–9, 116–17
Time of sampling
soil 10
vegetation 47–8
Tin 216
Titanium 153–5
by colorimetry 154
fusion and digestion of soils for 30
fusion and digestion of vegetation for 61
by X-ray fluorescence spectrometry 155
Total dissolved solids (TDS) 68–9
Total exchangeable bases (TEB) 35–6
Total organic matter (TOM) 69
Total suspended solids (TSS) 68
Transport
of animal materials 49
of soil samples 10, 49
of vegetation samples 49
Treatment blank 1
Trees
chemical characteristics of 329
Triazine herbicides 226
Trisaccharides 172
Truog's extraction 43
Turbidimetry 244
method for sulphate-sulphur 151–2
Turbid waters 67–8

Uric acid 184

Vanadium 216
Variance 296, 301
Vertebrates
chemical characteristics of 331
Volatilization losses 56
Voltammetric methods 265

Water equivalent 287–8
Water pollution 62, 229–31

Water purity standards 230
Wavelength dispersive XRF 258–60
Weighing 3–4
Wet deposition 237–8
Wilcoxon test 307
Woody tissues 47

X-ray fluorescence spectrometry 7, 25, 257–65
  method for sulphur 152–3
  method for titanium 155

Zinc 155–9
  ashing and digestion of vegetation for 61
  by atomic absorption 157–9
  by colorimetry 155
  by dithizone method 156–7
  fusion and digestion of soils for 30
  soil extraction procedure for 45
Zirconium 158